Storytelling to Accelerate Climate Solutions

Storytelling to Accelerate Climate Solutions

Emily Coren • Hua Wang

Editors

Storytelling to Accelerate Climate Solutions

 Springer

Editors
Emily Coren
Department of Psychiatry and Behavioral
Sciences
Stanford University
Stanford, CA, USA

Hua Wang
Department of Communication
University at Buffalo, The State University
of New York
Buffalo, NY, USA

ISBN 978-3-031-54789-8   ISBN 978-3-031-54790-4   (eBook)
https://doi.org/10.1007/978-3-031-54790-4

This Springer imprint is published by the registered company Springer Nature Switzerland AG
The registered company address is: Gewerbestrasse 11, 6330 Cham, Switzerland

Paper in this product is recyclable.

"The story of our future has not yet been written. To have the future we want, we need to start writing it. This book—Storytelling to Accelerate Climate Solutions—will provide the inspiration."
—Edward maibach, *Distinguished Professor and Director of Center for Climate Change Communication, George Mason University*

"Telling the stories of the future we want is how we will bring it about. In Storytelling to Accelerate Climate Solutions, *we discover storytelling that inspires hope and determination."*
—Tom Bowman, *author of What If Solving the Climate Crisis Is Simple? and co-editor of Empowering Climate Action in the United States*

"Storytelling to Accelerate Climate Solutions *is an important collection of insights and contributions from diverse perspectives across scholars and practitioners to effectively provide spaces of empowerment for climate action. Their commitments to storytelling advance mindful and compassionate understandings of present-day dimensions of climate change through stories of people who are struggling to re-make their realities in the face of worldly change. Each of these chapters opens up vulnerabilities as the storyteller reveals what they care about and value; each chapter also builds reliance and trust through creativity and authenticity."*
—Max Boykoff, *Professor and Chair, Department of Environmental Studies, and Cooperative Institute for Research in Environmental Sciences (CIRES) Fellow, University of Colorado Boulder*

"It is an all-hands-on-deck to stop the climate crisis, and that means we need the full creative potential of the human race. Coren and Wang do a masterful job curating the powerful tactics to use culture, storytelling, humor, fiction, and much more to activate people and decision-makers now. If you need inspiration that we can solve this with the tools we have now, this book is for you."*
—Sean Kosofsky, *Executive Director of Climate Advocacy Lab*

"This is a really comprehensive and helpful book on a crucial topic—how to express a global emergency in stories, across all media. The answers here go deep and will serve to spark ideas in people from all kinds of backgrounds."
—Kim Stanley Robinson, author of The Ministry for the Future

"An excellent volume providing a much-needed overview of methods for climate communication and storytelling which work—stories which model actions and solutions, rather than portraying doom and gloom. May it lead to much more positive action among storytellers and communicators."
—Kris De Meyer, Director of Climate Action Unit, University College London

"Despite all our new, shiny communication technologies, storytelling—what humans have been doing since time immemorial—remains one of the most powerful methods ever developed to communicate critical ideas, emotions, experiences, and meaning. Storytelling to Accelerate Climate Solutions helps us see the power and promise of effective storytelling in the global movement to ensure a safe and stable climate."
—Anthony Leiserowitz, Director of Yale Program on Climate Change Communication, Yale University

Foreword: Storytelling to Realize Our Better Nature

In 1973, when I was a 7th grader in a boarding school in Mussoorie—a picturesque town located in the mid-Himalayan range in India's State of Uttarakhand, a class excursion took us to an adjoining oak forest. As we sat around a campfire, our teacher narrated the inspiring story of the few brave women of Mandal village, who one week previously, some 175 miles east of our location, put their bodies on the line, quite literally, to protect their forest. When lumberjacks working for a commercial contractor, wielding menacing chainsaws, arrived in the forest to fell trees, a group of women spontaneously put their arms around and hugged them tightly. Their message to the lumberjacks: You will need to cut us down before you cut the tree. The lumberjacks backed down. Stirred by the story, I remember each of us— the 7th graders—ran from one tree to another hugging them. Some pretended to be lumberjacks; most of us preferred to model the brave and unyielding steadfastness of these peasant women from Mandal.

Unknown to me at that time, the story of the brave Mandal women along with photographs of their bodies interposed between the trees and the axes of lumberjacks, spread like wildfire in neighboring communities and towns. The local, regional, national, and international media were reporting on the *Chipko* ("hugging") grassroots movement, birthing the novel language and new meme of "tree hugger" and "tree hugging."

In the fall of 1985, when I enrolled in the doctoral program in communication at the University of Southern California's Annenberg School, I vividly remember sharing the *Chipko* story in a diffusion of innovations class taught by Professor Everett M. Rogers. The class discussion was animated and deeply inspiring. I remember arguing that the *Chipko* movement was turbocharged as it was spearheaded by two charismatic local activists and influencers, Chandi Prasad Bhatt and Sundarlal Bahuguna, both deeply inspired by Mahatma Gandhi's notion of peaceful, non-violent non-cooperation with "evil." In class, I emphasized that both Bhatt and Bahuguna were well networked with journalists, both wielded a prolific pen and wrote with ease in both Hindi and English, and thus the *Chipko* movement not just mobilized India's rural citizenry but also India's urban elites. About this time, I also encountered the writings of British novelist and essayist, G.K. Chesterton. One of his statements (paraphrased) spoke to me loudly: "Fairy tales are more than true: not because they tell us that dragons exist, but because they can tell us that dragons can be beaten." Clearly, I was beginning to make the connections between the narrative potential vested in stories to vanquish social "monsters" (i.e., the entertainment-education strategy), and their intrinsic value in accelerating the diffusion of social innovations—the topic of this volume (Singhal & Rogers, 1999; Singhal et al., 2000).

Why in writing this book's Foreword do I share with you my encounters with *Chipko* from decades ago? Simply, to illustrate the power of stories—their power to transport us to places we have never been, their power to generate new language and memes, to create new realities and possibilities, to enable and inspire new actions—both individual and collective, and thereby prevent, mitigate, and even overcome social "monsters." In hindsight, it is abundantly clear that the narrative appeal of *Chipko* was epic, mythical, and universal. The notion of "cut me down before you cut down a tree" brought a new humanized morality to take on abstract environmental and social monsters; that to protect the tree, or a Mother Earth was a deeply sacred act. I was inspired by how the feminist movement popularized *Chipko*, pointing out that poor rural women walk long distances to collect fuel and fodder, and thus are the frontline victims of forest destruction (Singhal & Lubjuhn, 2010). I was awe-struck by how the "tree hugging" meme climbed rapidly on the public and policy agendas of the regional, national, and global publics. For instance, as a direct consequence of *Chipko*, Indian Prime Minister Indira Gandhi legislated a 15-year ban on felling of green trees growing over 1000 meters above sea level in the Himalayan forests (Singhal & Duedahl, 2022). This decree was then extended to the tree-covered forests in other regions of India.

Such is the power of stories to launch movements. This book—full of stories—delivers that same thrill of potentiality to accelerate climate solutions. Each chapter tells the story of a life-giving idea, a strategy, an approach, a debate, a dilemma, and why our collective destinies are joined, no matter who we are and where we live. The collection represents a stark reminder that the best ideas and innovations for us to realize our better nature reside in the kernel of a story. I invite you to be inspired to realize our better nature—much like the 7th graders in Mussoorie half a century ago.

Arvind Singhal is the Samuel Shirley and Edna Holt Marston Endowed Professor of Communication and Director of the Social Justice Initiative at The University of Texas at El Paso, USA, and Visiting Professor 2 at the School of Business and Social Sciences at Inland Norway University of Applied Sciences, Norway. He also has courtesy appointments as the William J. Clinton Distinguished Fellow at Clinton School of Public Service, University of Arkansas-Little Rock, USA, and Chancellor's Honorary Professor at Amity University, India.

Arvind Singhal
The University of Texas, El Paso, USA
Inland Norway University of Applied Sciences, Norway

References

Singhal, A., & Duedahl, E. (2022). Building movements to protect biodiversity: The diffusible, the stewardable, and the possible. In C. Lindberg and E. Hagen (Eds.), *Our better nature*: *Hopeful excursions in saving biodiversity*. Vermont Alliance for Half-Earth.

Singhal, A., & Lubjuhn, S. (2010). Chipko environmental movement media. In J. D. Downing (Ed.), *Encyclopedia of social movement media*. Sage Publications.

Singhal, A., & Rogers, E. M. (1999). *Entertainment-education: A communication strategy for social change*. Routledge.

Singhal, A., Pant, S., & Rogers, E. M. (2000). Environmental activism through 'Yeh Kahan Aa Gaye Hum,' an entertainment-education radio soap opera in India. In M. Oepen and W. Hamacher (Eds.), *Communicating the environment: Environmental communication for sustainable development* (pp. 176–183). Peter Lang.

Acknowledgements

We are grateful to all organizations and individuals who helped us to make this book possible. First, we thank the chapter authors for their important work and the great efforts they made to share their work in this volume. Yishin Wu and Faye Habermehl provided invaluable editorial assistance. Moreover, we are indebted to the Department of Psychiatry and Behavioral Sciences at Stanford University, the Department of Communication at University at Buffalo, The State University of New York, the Natural Resources Defense Council (NRDC), Population Media Center, and PCI Media for their generous financial support that enabled the Open Access to this book. We thank Zachary Romano and the legal counsel at various institutions who worked diligently to help our contributing authors and affiliated organizations complete their Open Access agreements. This book would not have been possible without the love and support from our families, friends, and mentors. We would like to express deep gratitude to our parents (Ann and Bob Coren for Emily, Naihong Wang and Xiuyun Zhang for Helen). A very special thank you to Jessica Bean, Summer Marsh, and Melissae Stuart. We also appreciate the encouragement from Edward Maibach, Frank Niepold, Max Boykoff, Debra Safer, Arvind Singhal, Tom Bowman, Krista Myers, Jennie Dusheck, Amandeep Gill-Lang, Suruchi Sood, Amy Henderson Riley, Roel Lutkenhaus, Mark McCaffrey, Kris De Meyer, Abraham Mwaura, and Abby Ruskey.

Contents

Storytelling as a Catalyst for Climate Change Communication and Empowerment

Hua Wang and Emily Coren

Storytelling is an essential part of human communication. We are all storytellers, and we all love stories! Since time immemorial, we have relied on storytelling for information sharing, sense-making, and social transformation (Fisher, 1987; Gottschall, 2012; Polkinghorne, 1988). Today, stories are everywhere and more prominent in our daily lives than ever before. They flood our news feeds in almost every waking moment and dominate our screens with endless options at the click of a finger. They travel in time and space as we share with each other, face to face and through media. In addition, they affect the way we think and how we feel (Green et al., 2019). Storytelling is at the core of our very being. The art and science of storytelling in recent decades have taught us that, when told well, stories are powerful. They can not only change people's minds and actions but also fight social ills and save millions of lives (e.g., Green et al., 2002; Kreuter et al., 2007; Riley et al., 2022; Singhal & Rogers, 1999; Singhal et al., 2013). This might just be the secret to the survival of the human species and all life on Earth.

At this very moment, we are all writing the story of our time on this planet. Climate change is one of the most urgent and also one of the most complex global crises. Progress has been made in recent years, but these efforts are often disconnected. Our goal for putting together this book, *Storytelling to Accelerate Climate Solutions,* is three-fold:

1. We hope to help build a "Community of Practice for Climate Storytelling" and see this book as our first step in action by inviting practitioners and scholars from different fields to share their knowledge, experience, and insight about how sto-

H. Wang (✉)
University at Buffalo, The State University of New York, Buffalo, NY, USA
e-mail: hwang23@buffalo.edu

E. Coren
Stanford University, Palo Alto, CA, USA
e-mail: emilycoren@gmail.com

ries can be purposefully designed and effectively told to engage, enable, and empower various populations in climate communication.

2. We want to showcase a wide range of climate storytelling strategies and exemplary applications in terms of professional practices (e.g., education, literature, journalism, and popular media), narrative genres (e.g., drama, comedy, and fiction), media platforms (e.g., television, radio, and mobile), and communication modalities (e.g., text, visual, audio, and multisensory).

3. We plan to synthesize best practices, lessons learned, and what is needed now to advocate for increased funding, improved messaging, coordinated efforts, and necessary policy change that focus on human agency, effective solutions, positive actions, and sustainable initiatives moving forward.

In this introductory chapter, we use storytelling as an umbrella term for any type, form, genre, and practice of narrative communication. After briefly reviewing the current scientific understanding of climate change, recent public opinions about the issue, and the international community's efforts, we point out three critical and interrelated gaps in climate change communication and articulate how narrative strategies, especially entertainment-education for social and behavior change, can effectively inspire and mobilize individuals and communities worldwide to engage in climate action and empowerment.

Current Scientific Knowledge about Climate Change

Widespread and rapid changes in the climate have been occurring since the Industrial Revolution, and syntheses of climate science such as the Intergovernmental Panel on Climate Change (IPCC) assessment reports have shown definitively supporting evidence from paleoclimate archives, observational data, as well as peer-reviewed academic publications (Myers et al., 2021). With improved tools, technologies, and knowledge, scientists have reached a consensus that climate change is real; the evidence is unequivocal that human activities, such as population growth, rapid urbanization, and unsustainable consumption, caused the increased greenhouse gas emissions (IPCC, 2023; Myers et al., 2021). Human influence is the principal driver of global warming and has warmed the climate at an unprecedented rate over centuries (IPCC, 2023; Myers et al., 2021). Higher global temperatures, significant loss of ice volume, rising sea levels, and extreme events such as heat waves, floods, droughts, and tropical cyclones are adversely affecting all life on Earth, across the ecosystems and human civilization in every inhabited region of the world, especially the most vulnerable among us (Friel et al., 2008; IPCC, 2022, 2023).

Without taking immediate and transformative actions, the frequency and intensity of these climate extremes will continue to increase and lead to catastrophic ecological, evolutionary, socioeconomic, and health consequences with biodiversity loss, land degradation, exacerbated inequities and injustices, and the largest human health crisis in history (Frumkin et al., 2008; Maibach et al., 2021; U.S. Global

Change Research Program, 2016; World Health Organization, 2021). There are many ways that we can respond: *climate mitigation* by reducing greenhouse gas emissions and human exposure to climate hazards and *climate adaptation* by conserving biodiversity and developing regional and community resilience (Hawken, 2017). Improving the upstream effects such as reducing carbon emissions is more efficient than treating the downstream effects such as the compounding health impacts from extreme weather disasters, air pollution, water-borne pathogens, vector-borne, and zoonotic diseases (Luber & Lemery, 2015). Both the urgency and complexity of climate change demand effective, swift, scalable, and coordinated actions across all aspects and all levels of society.

Recent Public Opinions about Climate Change

In the United States, communication scientists have identified distinct groups that they called "global warming's six Americas" (Leiserowitz et al., 2009): (1) The *alarmed* groups are most worried and motivated, strongly believe in climate change, and support climate action. (2) The *concerned* groups are worried and agree that global warming is a threat but may still be distant to take action. (3) The *cautious* groups are aware of climate change but uncertain about its causes to make up their minds. (4) The *disengaged* groups are largely unaware of global warming and disconnected from the reporting on this issue. (5) The *doubtful* groups question the claim of global warming and see it as a low risk. (6) The *dismissive* groups deny that global warming is happening, human-caused, or a threat, and oppose most climate change policies. Over time, more Americans are moving toward the alarmed and concerned groups (Leiserowitz et al., 2023). Their survey data from December 2022 also showed that one in ten Americans reported feeling distressed about global warming, and the climate distressed are more likely to take actions such as signing a petition or volunteering at an organization on global warming (Ballew et al., 2023).

Reports from recent international surveys have consistently shown that a majority of the public across the globe is concerned about climate change, and many people would like to contribute to this cause (Andi, 2020; Bell et al., 2021; Flynn et al., 2021; Leiserowitz et al., 2021). According to the UNDP, in People's Climate Vote (the largest public opinion survey on climate change to date), 64% of their 1.2 million respondents across 50 countries said that climate change is an emergency. Of those, 59% said that the world should do everything necessary and urgently in response (Flynn et al., 2021). Similarly, 69% of 80,155 respondents in a Reuters Institute survey in 40 news media markets said that they consider climate change to be an extremely or very serious problem (Andi, 2020). Furthermore, 76,328 Facebook users in 31 countries and territories worldwide who participated in a survey by the Yale Program on Climate Change Communication also showed that most of them agreed that climate change is happening (79–94%) and will harm future generations (53–89%; Leiserowitz et al., 2021). A Pew Research Center survey found 72% of their 16,254 respondents across 17 advanced economies worried about how climate change will personally harm them, and 80% are willing to change how they live and work, including a median of 34% willing to make "a lot of changes" to their daily life, to help curb global warming (Bell et al., 2021). In other words, more people around the globe are willing and ready to act now.

Intergovernmental Organizing for Climate Action

Thirty years ago, the United Nations Framework Convention on Climate Change (UNFCCC) was signed as an international environmental treaty, having been recti-fied by 197 countries since it took effect in 1994. The UNFCCC (n.d.) has set the foundation for subsequent milestones such as the Kyoto Protocol in 1997 (setting global targets for greenhouse gas emission reduction) and the Paris Agreement in 2015 (limiting global warming well below 2 °C, preferably 1.5 °C to sustain a func-tioning biosphere). Most recently, the COP26 in 2021 was convened in Glasgow to bring countries together and accelerate climate action toward the goals of the Paris Agreement. The resulting Glasgow Climate Pact shows some progress; however, many scientists and activists criticized its lack of stronger commitments and are calling for a more inclusive and just movement (e.g., Hawken, 2021; Masood & Tollefson, 2021). An ongoing effort in that direction is the Action for Climate Empowerment (ACE) framework, which prioritizes and advocates for equitable community-centered engagement and active empowerment at the local level through education, training, public awareness, public participation, public access to infor-mation, and international cooperation on these issues (Bowman et al., 2021; Bowman & Morrison, 2021; Cintron-Rodriguez et al., 2021). These existing inter-governmental frameworks provide high-level guidance for funding, coordinating, and monitoring progress toward climate mitigation and adaptation goals. However, many citizens worldwide are not confident in the international community's efforts to reduce global warming. In fact, 52% of the Pew Research Center survey said they were not too confident or not confident at all (Bell et al., 2021). What is needed now is to mobilize all members of society in collective action, especially at the local and regional levels, through actionable climate mitigation and adaptation pathways (IPCC, 2022).

Critical Gaps in Climate Change Communication

Climate change communication is a growing field of scholarship and praxis cen-tered around how we communicate about climate change, such as the public under-standing of the issue, news coverage, and message framing, as well as risk perceptions and media effects (Chadwick, 2017; Yale Program on Climate Change Communication, n.d.). There are at least three critical and interrelated gaps in climate change communication that warrant immediate attention and action: funding, messaging, and coordination.

First, funding support is severely lacking, especially for the science communicators to conduct applied work with what we already know that works. There is no incentive for media producers to invest in research and development to create accurate content while keeping up with the rapid production cycles. Practitioners such as visual illustrators in the field are either not paid or not paid adequately for their work on climate change. Moreover, there is an urgent need for programmatic investment in full-time jobs dedicated to climate change communication and systematically developed networks of these professionals to create a truly meaningful and sustainable impact over time.

Second, messaging is significantly limited in terms of overall quantity, quality, and efficacy. We need more messages about climate change, messages that go beyond merely raising awareness and sharing scientific findings. Current messaging about climate change is predominantly about fear and anxiety, in both the language and imagery used in scientific reports, news stories, and entertainment programs. This can be difficult to translate into responsive behaviors. We must shift the messages in the stories we tell about climate change from issue-oriented to action-oriented and focus on positive framing and actionable solutions to foster human agency and facilitate real change (De Meyer et al., 2021). More specifically, we need "simple clear messages, repeated often, by a variety of trusted and caring messengers" that make the behaviors we are promoting "easy, fun, and popular" (Maibach et al., 2023, p. 54).

Third, coordination is almost nonexistent among existing teams and initiatives across various disciplines. Due to financial and capacity constraints, even climate communication professionals are mostly working in silos. This is a major challenge for knowledge sharing. Many must reinvent the wheel and experience difficulty when disseminating their products or interventions. In addition, the messages from different groups and fields are scattered, and there are no clear pathways to connect the dots, leverage existing resources, and amplify positive impact. The most vulnerable groups to climate change are often left out of important discussions and initiatives (IPCC, 2022). We need to coordinate our efforts in engaging communities, curating content, and linking resources across national, regional, and local levels as well as the full range of media platforms to ensure the messages are clear, consistent, and compelling.

Climate change mitigation and adaptation are a planetary race with physical, cognitive, emotional, geopolitical, and sociocultural barriers. Using the exemplary approaches and applications in this book, we argue that storytelling is an effective catalyst for climate change communication and empowerment. We advocate for it to be integrated into climate change funding, planning, and monitoring to help avert this global crisis.

Entertainment-Education for Social and Behavior Change

In the span of over 50 years, the idea of intentionally combining entertainment with education for health promotion and behavior change, known as entertainment-education or edutainment, has evolved into a field of research and practice around the globe, making a significant impact at the individual, interpersonal, community, and societal levels (Frank & Falzone, 2021; Riley et al., 2022; Singhal et al., 2013; Storey & Sood, 2013; Wang & Singhal, 2021a). In its early years, from the 1970s to the early 2000s, practitioners mainly worked with government agencies, nongovernmental organizations, and creative professionals in developing countries to create prosocial radio and television serial dramas (Singhal et al., 2004; Singhal & Rogers, 1999). Given that context, entertainment-education was defined as "the process of purposely designing and implementing a media message to both entertain and educate in order to increase audience knowledge about an educational issue, create favorable attitudes, and change overt behavior" (Singhal & Rogers, 1999, p. 9; also see Singhal et al., 2004, p. 5). The focus was to find the "sweet spot" that helped balance the entertaining and the educational elements in the story so the audience members would relate to the plots, fall in love with the characters, not feel they were being preached at, and could see new possibilities to enhance their lives (Wang & Singhal, 2021a).

With the growing accessibility and popularity of digital technologies and interactive media, storytelling on web-based platforms with immersive environments rose rapidly. In 2009, a reformulated definition was proposed to emphasize that entertainment-education is "a theory-based communication strategy for purposefully embedding educational and social issues in the creation, production, processing, and dissemination process of an entertainment program, in order to achieve desired individual, community, institutional, and societal changes among the intended media user populations" (Wang & Singhal, 2009, pp. 272–273). And most recently, in keeping with the global trend and through the lens of translational research, an updated statement defined entertainment-education as "a social and behavioral change communication (SBCC) strategy that leverages the power of storytelling in entertainment and wisdom from theories in different disciplines—with deliberate intention and collaborative efforts throughout the process of content production, program implementation, monitoring, and evaluation—to address critical issues in the real world and create enabling conditions for desirable and sustainable change across micro-, meso-, and macro-levels" (Wang & Singhal, 2021a, p. 227).

Decades of work in entertainment-education have proven it to be a cost-effective way to foster positive change in communities around the world. For example, a flagship institution Population Media Center (n.d.) estimated the cost of only $2.54 USD for each person to begin discussing family planning with important people in their social circles as the direct result of one of their entertainment-education programs. In addition, the embedded educational messages are typically grounded in formative research with a deep understanding of the unmet needs of the intended audience and clear social objectives for behavior change, which is then modeled through character development and story arcs (Singhal et al., 2013). Mexican writer–producer–director Miguel Sabido pioneered the entertainment-education production methodology (commonly referred to as the Sabido Methodology), which includes a framework with competing moral values related to the educational theme; a set of positive, negative, and transitional characters as role models; a narrative structure that confronts the status quo and progresses through stages of suffering, doubting, and overcoming obstacles to achieve the ultimate triumph; and the use of epilogues to guide and facilitate public discourse, and promotion of resources to enable desired actions (Sabido, 2021; Singhal et al., 2013).

Novel and effective solutions can be introduced and modeled through powerful characters and storylines that boost self and collective efficacy and facilitate behavior change. A famous example is how *Soul City* used neighbors' banging pots and pans as a bystander intervention for domestic violence in their entertainment-education television drama, which led to people emulating the same behavior, naming their community as Soul City, and the featured hotline ringing off the hook (Singhal et al., 2013). A critical point here is that when behavior modeling is coupled with supporting infrastructure, it can effectively facilitate change at scale. For example, Sabido's first entertainment-education telenovela alone "shifted the shameful norms around illiteracy and inspired half a million viewers to enroll in Mexico's national adult education program" (Wang & Singhal, 2021b, p. 822). Three of his early entertainment-education telenovelas were able to slow down Mexico's population growth rate from 3.7 to 2.4 in 5 years; without those interventions, Mexico would have had 50 million more people today (Sabido, 2021; Wang & Singhal, 2021b).

Taken together, entertainment-education is a narrative communication strategy that can promote and accelerate climate action and empowerment. The design, implementation, and impact of different entertainment-education initiatives may vary considerably according to individual project goals, available resources, collaborative processes, and issue- and audience-based contextualization and adaptation. Some of them are illustrated in the chapters of this book (Bish, 2024; Brown, 2024; Garg et al., 2024; Falzone et al., 2024; Hinerfeld et al., 2024; Sood et al., 2024). For illustrative purposes, in this chapter, we describe one example of entertainment-education used in the northern state of Uttar Pradesh in India that modeled actionable solutions and inspired individual and collective action to reduce air pollution.

Example of Entertainment-Education for Climate Change

I am the air
I help you breathe and help you sing
I am a life-giver like your mother

I love greenery and thrive on grasses and trees

But my child, when you do not listen to me
I am starving and in pain

I am trapped in pollution
Suffocated by smoke
Lost in the loud noise
I have become dark, very dark...

I am the air
Like your own mother
A mother who bears so much pain
For her children...

These are the lyrics from *Main Hawa Hoon* (*I Am the Air*), the theme song of a 2017 radio drama series *Ek Zindagi Aisi Bhi* (*A Life We Aspire For*) broadcast in India (The Change Designers, 2017a). These words painted air as a suffering mother, full fleshed, with color and sound, in time and space. It delineated a clear picture of air—an essential element of life on this planet but often invisible to its inhabitants. This personification of air, coupled with a somber melodic tone, helped evoke empathy among listeners to care deeply about the deteriorating quality of air in their communities (The Change Designers, 2017b).

Over the course of six 30-min audio episodes (The Change Designers, 2017a), stories of deteriorating air quality and its impact were told through a constellation of compelling characters that not only echoed people's real-life experiences but also demonstrated the possibility of alternative realities. Sudhir, 28, has recently moved to live with his mother in the city of Varanasi in Uttar Pradesh, the heart of India's northern belt. This is also where Chutki lives. Chutki is 26, educated, independent, and witty, the young woman he has been seeing for 4 years. Through twists and turns as well as community gossip, Sudhir and Chutki manage to get married in court and celebrate their union by planting a native tree (peepal) that produces oxygen 24 h a day. Instead of having a lavish wedding that can cause a lot of pollution and waste, Sudhir and Chutki emerge as a new normal in their community. During one of the episodes, the couple are stuck in heavy traffic at a four-way intersection, as it happens so often on the streets in India, Sudhir hops out of the auto-rickshaw and spontaneously starts a "social intervention" by directing the crowd and clearing the traffic. Chutki, on the other hand, in one scene, demonstrates waste segregation and composting while sharing with the women about the danger of burning waste and cooking on a wooden stove. Produced and broadcast 3 years before the COVID-19 pandemic, publicly wearing masks was modeled as a novel solution to prevent people from inhaling polluted air and protect them from diseases such as asthma and lung cancer (Cohen & Pope III, 1995; Guarnieri & Blames, 2014). Also woven into the plots were a veteran taxi driver's decision to adopt the new technology and switch to an e-rickshaw after his asthma diagnosis, and his daughter Rupa's strong attitude against a village head because of his selfish and socially irresponsible motive to build a factory even though she is in love with his son (The Change Designers, 2017b).

Ek Zindagi Aisi Bhi (*A Life We Aspire For*) reached approximately 20 million people in just 6 weeks. Their program evaluation showed that the 194 participants in their listener groups shared a total of 162 behavior change references; and categorically speaking, adopted 24 eco-friendly behaviors, 15 actions to control emissions, 10 waste management solutions, 10 steps to discuss these issues with others, 9 ways to increase personal and social accountability, and 3 specific new things to reduce their health risk due to environmental hazards—some of these changes may be a result of listeners emulating the attitudes and behaviors modeled by the characters in the show that they aspired to become while others may be creative adaptations or completely new discoveries (The Change Designers, 2017b).

To paint a picture of the show's impact, most listeners adopted the use of cloth bags for everyday shopping in the market as an alternative to plastic bags, which are non-biodegradable and a major issue of littering, landfills, and burning waste that cause pollution. All participants in the listener groups agreed on the benefits of new technologies such as e-rickshaws and solar panels to reduce harmful emissions. One woman went ahead and filled out the form to buy an e-rickshaw! In another group discussion, a listener shared that she installed a compost bin and her family enjoyed sewing herbs and vegetables in her terrace garden, all thanks to her new habit of composting waste. Another listener told the group that she chose to gift a plant at a neighborhood birthday celebration and said to everybody that "whenever there is a moment of happiness around us, we should take that opportunity to plant a tree and grow a new life". Positive changes also took place beyond individual actions. Inspired by the characters in the show, one group was motivated to hold their local leaders accountable for improving environmental conditions. A rally with community members to demand real change grew to be a series of marches that led to a signed petition to the municipal authorities because of the intervention (The Change Designers, 2017b).

Moreover, good stories are generative—they plant seeds to nurture more creative solutions and inspiring stories. A 60-year-old woman said she loved her listener group. They felt they learned so many new things and wanted to share with others. Such enthusiasm turned into a community event where they actually reenacted and performed the story of *Ek Zindagi Aisi Bhi* (*A Life We Aspire For*) on the stage as part of the International Women's Day celebration in their district. There they were holding the microphones and narrating their character's stories in front of more than 1000 women in the audience. The community leader expressed that this was the first time that air pollution was ever discussed publicly, and even more unusual and rare is that these efforts were led by women, whose voice was never valued or heard in the community. As the audience rose to applaud the performers on the stage, they recorded videos on their phones and whistled as they marveled at the drama. In those moments, the invisible air became not just visible through the airwaves but bold and colorful visuals co-created and co-owned by the community, while the silenced members became creatively and powerfully vocal (The Change Designers, 2017b).

This is one example of how inspiring and powerful stories, when designed and shared with deliberate prosocial objectives and coordinated collaborative efforts, can change not just people's minds but also their spirits and behaviors. We encourage you to learn more about entertainment-education as well as other narrative communication strategies in this book that may help the climate change endeavors for you and your communities.

Organization of This Book

As shown in Table 1, we have organized this book to begin with a global overview of entertainment-education and how it has been used to address climate change based on 87 programs from 2000 to 2020 with different approaches and practices in the Global South and the Global North (Sood et al., 2024). Following this overview, we have four chapters demonstrating how organizations that champion entertainment-education implement it in the Global South: PCI Media uses its "My Community Methodology" to create authentic and locally driven radio dramas to support climate action (Brown, 2024). Population Media Center uses the Sabido/PMC Methodology to promote family planning and girls' education in their radio dramas for a healthier and more sustainable population living on this planet (Bish, 2024). BBC Media Action engages young people in storytelling from television to social media to promote public discussions and government accountability for climate change (Garg et al., 2024). Peripheral Vision International developed children's television series with inspirational models for climate education and agency (Falzone et al., 2024). Then, we transition to explore how entertainment-education and social impact entertainment strategies have been used and can potentially be improved in the Global North, represented by the United States. A comedy-drama prototype for climate communication tailored toward young adult Americans is proposed with detailed character development and story arcs based on theory and research (Coren, 2024). A newly established Rewrite the Future program at the Natural Resources Defense Council (NRDC) has been helping Hollywood better incorporate climate change themes into their entertainment programming to promote effective solutions (Hinderfeld et al., 2024). Despite the seriousness of the climate crisis, comedians are having a ball firing their secret weapons to wake up their audiences and get them involved in climate action (Gurney & N'Diaye, 2024).

In addition to entertainment-education, we have included many other creative storytelling strategies for climate change communication in the second half of this book. Climate fiction (or cli-fi) is a literary genre that uses fictional characters and storylines to inspire green actions (Baden & Brown, 2024). Visual illustrations created by professionals have a long history in science communication and should be increasingly supported to produce quality imagery to effectively convey messages about climate change (Monoyios et al., 2024). Music, with powerful melodies and lyrics, can evoke deep human emotions to care about planet Earth, as shown in The ClimateMusic Project (Dixon et al., 2024). The evolution of the food we consume can also tell compelling stories about the impact of global warming (Eiseman & Hoffman, 2024). Practices in journalism can benefit from cross-disciplinary teamwork, audience engagement, and artificial intelligence-driven technologies to

Table 1 Summary of narrative strategies, communication platforms, and geographic locations of climate storytelling programs featured in this book

Chapter	Author(s)	Narrative strategy	Communication platform	Geographic location
02	Sood et al.	Entertainment-education/social impact entertainment	Popular media (television, radio, film, game; drama, comedy, etc.)	Bangladesh, Cambodia, Caribbean, D.R.C., India, Malawi, Mozambique, Nigeria, Peru, Rwanda, Tanzania, UK, US, Vietnam, Zambia
03	Brown	Entertainment-education/my community methodology	Radio dramas	Côte d'Ivoire, Guinea, Liberia, Sierra Leone
04	Bish	Entertainment-education/PMC methodology	Radio dramas	D.R.C., Rwanda, Uganda
05	Garg et al.	Entertainment-education/transmedia storytelling	Television drama, social media forum, mobile app	Indonesia
06	Falzone et al.	Entertainment-education	Children's television, STEM education	Kenya, Nigeria, Tanzania, South Africa, Uganda
07	Coren	Entertainment-education/Social impact entertainment	Comedy drama	U.S.
08	Hinerfeld et al.	Entertainment-education/social impact entertainment	Documentary, film, television series	U.S.
09	Gurney & N'Diaye	Entertainment-education/social impact entertainment	Comedy	U.S.
10	Baden & Brown	Fictional writing/literature	Novels, graphic novels	U.K.
11	Monoyios et al.	Visual illustration	Still images	U.S.
12	Dixon et al.	Music composition and performances	Art-science collaborations, live and streamed performances	U.S.
13	Eiseman & Hoffman	Food storytelling	Books, website, digital art exhibit	U.S.
14	Whitwell	News reporting	BBC news	U.K.
15	Cosentino et al.	Positive deviance and collective storytelling	Face-to-face communication,	U.S., U.K.
16	Wolf-Jacobs et al.	Geospatial visualization	Maps, mapping tools	U.S.

(continued)

Table 1 (continued)

Chapter	Author(s)	Narrative strategy	Communication platform	Geographic location
17	Spiegel & Wang	Interactive storytelling in immersive environments	Game, immersion theatre, AR, VR, MR/XR	Canada, U.S., U.K.
18	Osnes et al.	Performance and interspecies friendship	Face-to-face communication, participant observation, journals	U.S.
19	Bean	Instructional storylines	Personal storytelling in K-12 education	U.S.

generate hyperlocal news reporting that promotes actionable solutions (Whitwell, 2024). Understanding community-based resilience and learning from positively deviant initiatives can help us boost collective efficacy and reduce mental health challenges related to climate change (Cosentino et al., 2024). Maps and geospatial software applications can be powerful tools for climate storytelling and empowerment (Wolf-Jacobs et al., 2024). Interactive storytelling in immersive environments can create conditions for self-directed deep learning and large-scale public engagement in climate science communication (Spiegel & Wang, 2024). Inviting youth to observe and embody birds through puppetry, costumes, and movement in outdoor settings can better integrate the themes of equity and inclusion into environmental preservation (Osnes et al., 2024). In addition, instructional storylines developed by the Understanding Global Change Project use coherent sequences of lessons to help students connect their lived experiences with scientific concepts of systematic change (Bean, 2024).

We invite you to explore different narrative communication approaches and applications for climate change in this book. These projects provide insight into how we can expand climate change communication through many media types, adapted regionally and demographically to support a wide range of climate mitigation and adaptation outcomes. Through partnerships with existing media and storytelling infrastructure, improving the visualization of strategies already underway, and incorporating insights from entertainment-education and other narrative communication strategies, we can effectively accelerate climate solutions.

References

Andi, S. (2020). *How people access news about climate change*. In Digital news report. Reuters Institute and University of Oxford. https://www.digitalnewsreport.org/survey/2020/how-people-access-news-about-climate-change/

Baden, D., & Brown, J. (2024). Climate fiction to inspire green actions: Tales from two authors. In E. Coren & H. Wang (Eds.), *Storytelling to accelerate climate solutions*. Springer Nature.

Ballew, M., Myers, T., Uppalapati, S. S., Rosenthal, S., Kotcher, J., Campbell, E., Goddard, E., Maibach, E., & Leiserowitz, A. (2023, August 3). *Is distress about climate change associated with climate action?* https://climatecommunication.yale.edu/publications/distress-about-climate-change-and-climate-action/

Bean, J. (2024). Instructional strategies for climate education: Storytelling about our place in the earth system. In E. Coren & H. Wang (Eds.), *Storytelling to accelerate climate solutions*. Springer Nature.

Bell, J., Poushter, J., Fagan, M., & Huang, C. (2021). *In response to climate change, citizens in advanced economies are willing to alter how they live and work: Many doubt success of international efforts to reduce global warming.* Pew Research Center. https://www.pewresearch.org/global/2021/09/14/in-response-to-climate-change-citizens-in-advanced-economies-are-willing-to-alter-how-they-live-and-work/

Bish, J. (2024). Positively life-changing stories today – Intergenerational benefits tomorrow. In E. Coren & H. Wang (Eds.), *Storytelling to accelerate climate solutions*. Springer Nature.

Bowman, T., & Morrison, D. (Eds.) (2021). *Empowering climate action in the United States.* Changemakers.

Bowman, T. E., Cintron-Rodriguez, I., Crim, H., Damon, T., Dandridge, C., Kretser, J., Morrison, D., Niepold, F., Poppleton, K., Spitzer, W., & Weiland, L. (2021). Building capacity, momentum and a culture of climate action in the United States. *Environmental Research Letters, 16*(4), 041003. https://doi.org/10.1088/1748-9326/abe961

Brown, N. (2024). The power of locally-driven narratives to support and sustain climate action. In E. Coren & H. Wang (Eds.), *Storytelling to accelerate climate solutions*. Springer Nature.

Chadwick, A. E. (2017). Climate change communication. *Oxford Research Encyclopedia.* https://doi.org/10.1093/acrefore/9780190228613.013.22

Cintron-Rodriguez, I. M., Crim, H. A., Morrison, D. L., Niepold, F., Kretser, J., Spitzer, W., & Bowman, T. (2021). Equitable and empowering participatory policy design strategies to accelerate just climate action. *Journal of Science Policy & Governance, 18*(02). https://doi.org/10.38126/JSPG180203

Cohen, A. J., & Pope, C. A., III. (1995). Lung cancer and air pollution. *Environmental Health Perspectives, 103*(Suppl 8), 219–224. https://doi.org/10.1289/ehp.95103s8219

Coren, E. (2024). Rhythm and Glue: An entertainment-education prototype for climate communication. In E. Coren & H. Wang (Eds.), *Storytelling to accelerate climate solutions*. Springer Nature.

Cosentino, M., Gal-Oz, R., & Safer, D. L. (2024). Community-based resilience: The influence of collective efficacy and positive deviance on climate change-related mental health. In E. Coren & H. Wang (Eds.), *Storytelling to accelerate climate solutions*. Springer Nature.

De Meyer, K., Coren, E., McCaffrey, M., & Slean, C. (2021). Transforming the stories we tell about climate change: From 'issue' to 'action'. *Environmental Research Letters, 16*, 015002. https://doi.org/10.1088/1748-9326/abcd5a

Dixon, C. E., Goldman, L. S., Crawford, S., & Lease, P. C. (2024). Music as a vehicle for climate change communication: The ClimateMusic Project. In E. Coren & H. Wang (Eds.), *Storytelling to accelerate climate solutions*. Springer Nature.

Eiseman, D., & Hoffman, M. (2024). Telling the story of climate change through food. In E. Coren & H. Wang (Eds.), *Storytelling to accelerate climate solutions*. Springer Nature.

Falzone, P., Kiano, J., & Lukomska, G. (2024). Let's Go! Let's Know! N*Gen as an EE tool for climate education and agency. In E. Coren & H. Wang (Eds.), *Storytelling to accelerate climate solutions*. Springer Nature.

Fisher, W. R. (1987). *Human communication as narration: Toward a philosophy of reason, value, and action.* The University of South Carolina Press.

Flynn, C., Yamasumi, E., Fisher, S., Snow, D., Grant, Z., Kirby, M., Browning, P., Rommerskirchen, M., & Russell, I. (2021). *People's climate vote: Results.* UNDP and University of Oxford. https://www.undp.org/publications/peoples-climate-vote

Frank, L. B., & Falzone, P. (Eds.). (2021). *Entertainment-education behind the scenes: Case studies for theory and practice.* Palgrave Macmillan.

Friel, S., Marmot, M., McMichael, A. J., Kjellstrom, T., & Vågerö, D. (2008). Global health equity and climate stabilisation: A common agenda. *The Lancet, 372*(9650), 1677–1683. https://doi.org/10.1016/S0140-6736(08)61692-X

Frumkin, H., Hess, J., Luber, G., Malilay, J., & McGeehin, M. (2008). Climate change: The public health response. *American Journal of Public Health, 98*(3), 435–445. https://doi.org/10.2105/AJPH.2007.119362

Garg, A., Godfrey, A., & Eko, R. (2024). *Kembali Ke Hutan* (Return to the Forest): Using storytelling for youth engagement and climate action in Indonesia. In E. Coren & H. Wang (Eds.), *Storytelling to accelerate climate solutions*. Springer Nature.

Gottschall, J. (2012). *The storytelling animal: How stories make us human*. Houghton Mifflin Harcourt.

Green, M. C., Strange, J. J., & Brock, T. C. (Eds.). (2002). *Narrative impact: Social and cognitive foundations*. Psychology Press.

Green, M. C., Bilandzic, H., Fitzgerald, K., & Paravati, E. (2019). Narrative effectives. In M. B. Oliver, A. A. Raney, & J. Bryant (Eds.), *Media effects: Advances in theory and research* (pp. 130–145). Routledge.

Guarnier, M., & Balmes, J. R. (2014). Outdoor air pollution and asthma. *The Lancet, 383*(9928), 1581–1592.

Gurney, C., & N'Diaye, M. (2024). LOLs: Secret weapon against CFCs and CO$_2$? In E. Coren & H. Wang (Eds.), *Storytelling to accelerate climate solutions*. Springer Nature.

Hawken, P. (Ed.). (2017). *Drawdown: The most comprehensive plan ever proposed to reverse global warming*. Penguin Books.

Hawken, P. (2021). *Regeneration: Ending the climate crisis in one generation*. Penguin Books.

Hinderfeld, D., Slean, C., & Jacobs, K. (2024). Rewrite the future: Helping Hollywood accelerate climate solutions through storytelling. In E. Coren & H. Wang (Eds.), *Storytelling to accelerate climate solutions*. Springer Nature.

IPCC. (2022). *Climate change 2022: Impacts, adaption, and vulnerability*. https://www.ipcc.ch/report/ar6/wg2/downloads/report/IPCC_AR6_WGII_FinalDraft_FullReport.pdf

IPCC. (2023). *AR6 synthesis report: Climate change 2023*. https://www.ipcc.ch/ar6-syr/

Kreuter, M. W., Green, M. C., Cappella, J. N., Slater, M. D., Wise, M. E., Storey, D., Clark, E. M., O'Keefe, D. J., Erwin, D. O., Holmes, K., Hinyard, L. J., Houston, T., & Woolley, S. (2007). Narrative communication in cancer prevention and control: A framework to guide research and application. *Annals of Behavioral Medicine, 33*(3), 221–235.

Leiserowitz, A. Maibach, E., & Roser-Renouf, C. (2009). *Global warming's six Americas 2009: An audience segmentation analysis*. Yale University and George Mason University. Yale Program on Climate Change Communication. https://climatecommunication.yale.edu/publications/global-warmings-six-americas-2009/

Leiserowitz, A., Carman, J., Buttermore, N., Wang, X., Rosenthal, S., Marlon, J., & Mulcahy, K. (2021). *International public opinion on climate change*. Yale Program on Climate Change Communication and Facebook Data for Good. https://climatecommunication.yale.edu/publications/international-public-opinion-on-climate-change/

Leiserowitz, A., Maibach, E., Rosenthal, S., Kotcher, J., Lee, S., Verner, M., Ballew, M., Carman, J., Myers, T., Goldberg, M., Badullovich, N., & Marlon, J. (2023). Climate Change in the American Mind: Beliefs & Attitudes, Spring 2023. Yale University and George Mason University. New Haven, CT: Yale Program on Climate Change Communication.

Luber, G., & Lemery, J. (Eds.). (2015). *Global climate change and human health: From science to practice*. Jossey-Bass, a Wiley brand; APHA Press, an imprint of American Public Health Association.

Maibach, E., Miller, J., Armstrong, F., El Omrani, O., Zhang, Y., Philpott, N., Atkinson, S., Rudoph, L., Karliner, J., Wang, J., Pétrin-Desrosiers, C., Stauffer, A., & Jensen, G. K. (2021). Health professionals, the Paris agreement, and the fierce urgency of now. *The Journal of Climate Change and Health, 1*, 100002. https://doi.org/10.1016/j.joclim.2020.100002

Maibach, E., Uppalapati, S. S., Orr, M., & Thaker, J. (2023). Harnessing the power of communication and behavior science to enhance society's response to climate change. *Annual Review of Earth and Planetary Sciences, 51*, 53–77.

Masood, E., & Tollefson, J. (2021, November 14). 'COP26 hasn't solved the problem': Scientists react to UB climate deal. *Nature*. https://www.nature.com/articles/d41586-021-03431-4

Monoyios, K., Carlson, K., Litwak, T., Marien, T., & Martin, F. (2024). Visual storytelling as a catalyst for climate science communication. In E. Coren & H. Wang (Eds.), *Storytelling to accelerate climate solutions*. Springer Nature.

Myers, K. F., Doran, P. T., Cook, J., Kotcher, J. E., & Myers, T. A. (2021). Consensus revisited: Quantifying scientific agreement on climate change and climate expertise among Earth scientists 10 years later. *Environmental Research Letters, 16*(10), 104030. https://doi.org/10.1088/1748-9326/ac2774

Osnes, B., with Hackett, C, McDermott, M. T., & Safran, R. (2024). *Bird's Eye View*: Engaging youth in storying a survivability future through performance and interspieces friendship. In E. Coren & H. Wang (Eds.), *Storytelling to accelerate climate solutions*. Springer Nature.

Polkinghome, D. E. (1988). *Narrative knowing and the human sciences*. State University of New York Press.

Population Media Center. (n.d.). *How do we know if we succeeded?* https://www.populationmedia.org/our-approach/evaluation/

Riley, A. H., Sood, S., & Wang, H. (2022). Entertainment-education (effects). In J. van Weert, E. Ho, C. Bylund, N. Bol, M. D. Kruzel, & I. Basnyat (Eds.), *The International encyclopedia of health communication*. Wiley.

Sabido, M. (2021). Miguel Sabido's entertainment-education. In L. B. Frank & P. Falzone (Eds.), Entertainment-education behind the scenes: Case studies for theory and practice (pp.15–22). Palgrave Macmillan.

Singhal, A., & Rogers, E. M. (1999). *Entertainment-education: A communication strategy for social change*. Erlbaum.

Singhal, A., Cody, M. J., Rogers, E. M., & Sabido, M. (2004). *Entertainment-education and social change: History, research, and practice*. Erlbaum.

Singhal, A., Wang, H., & Rogers, E. M. (2013). The rising tide of entertainment-education in communication campaigns. In R. Rice & C. Atkin (Eds.), *Public communication campaigns* (pp. 321–333). Sage.

Sood, S., Riley, A., & Birkenstock, L. (2024). Entertainment-education and climate change: Program examples, evidence, and best practices from around the World. In E. Coren & H. Wang (Eds.), *Storytelling to accelerate climate solutions*. Springer Nature.

Spiegel, S., & Wang, H. (2024). Exploring climate science in the metaverse: Interactive storytelling in immersive environments for deep learning and public engagement. In E. Coren & H. Wang (Eds.), *Storytelling to accelerate climate solutions*. Springer Nature.

Storey, D., & Sood, S. (2013). Increasing equity, affirming the power of narrative and expanding dialogue: The evolution of entertainment education over two decades. *Critical Arts, 27*, 9–35.

The Change Designers. (2017a). *Ek Zindagi Aisi Bhi – Air Quality* [full episodes and sound tracks]. Posted by PCI Media, https://soundcloud.com/mediaimpact/sets/ek-zindagi-aisi-bhi-air

The Change Designers. (2017b, July). *Improving air quality through air saves: Ek Zindagi Aisi Bhi*. A program evaluation report on the entertainment-education radio drama to improve air quality in Uttar Pradesh, India.

U.S. Global Change Research Program (2009-). (2016). *The impacts of climate change on human health in the United States: A scientific assessment*.

UNFCCC. (n.d.) United Nations Climate Change. https://unfccc.int/

Wang, H., & Singhal, A. (2009). Entertainment-education through digital games. In Ritterfeld, U., Cody, M. J., & Vorderer, P. (Eds.) Serious games: Mechanisms and effects (pp. 271–292). Routledge.

Wang, H., & Singhal, A. (2021a). Mind the gap! Confronting the challenges of translational communication research in entertainment-education. In L. B. Frank & P. Falzone (Eds.), *Entertainment-education behind the scenes: Case studies for theory and practice* (pp. 223–242). Palgrave Macmillan.

Wang, H., & Singhal, A. (2021b). Theorizing entertainment-education: A complementary perspective to the development of entertainment theory. In P. Vorderer & C. Klimmt (Eds.), *The Oxford handbook of entertainment theory* (pp. 819–838). Oxford University Press.

Whitwell, J. (2024). Three ways to introduce more stories of climate action into climate change news reporting. In E. Coren & H. Wang (Eds.), *Storytelling to accelerate climate solutions.* Springer Nature.

Wolf-Jacobs, A., Glock-Grueneich, N., & Uchtmann, N. (2024). Mapping out our future: Using geospatial tools and visual aids to achieve climate empowerment in the U.S. In E. Coren & H. Wang (Eds.), *Storytelling to accelerate climate solutions.* Springer Nature.

World Health Organization. (2021). *COP26 special report on climate change and health: The health argument for climate action.* World Health Organization. https://apps.who.int/iris/handle/10665/346168

Yale Program on Climate Change Communication. (n.d.). *What is climate change communication?* https://climatecommunication.yale.edu/about/what-is-climate-change-communication/

Hua Wang is a communication scientist who is passionate about using innovative strategies for health promotion, behavior change, and social justice. She specializes in the design, implementation, and evaluation of initiatives that leverage powerful storytelling, emerging technologies, and communication networks to facilitate positive change, particularly in the field of entertainment-education. She holds a Ph.D. from the University of Southern California's Annenberg School for Communication & Journalism and is currently a Professor of Communication at the University at Buffalo, The State University of New York. Her interdisciplinary research has been funded by federal agencies, private foundations, and nonprofit organizations; appeared in high-impact journals; and received prestigious awards from the American Public Health Association and the International Communication Association.

Emily Coren is a science communicator and an affiliate in the Department of Psychiatry and Behavioral Sciences at Stanford University where she has been working to adapt entertainment-education strategies for health promotion and social change to create more effective climate communication. She has a B.S. in Ecology and Evolutionary Biology and is a certified professional Science Illustrator. She has worked in science communication for almost 20 years, contributing to collections at the Smithsonian Institution's Museum of Natural History, consulting on a World Health Organization clean air campaign, and developing educational content for children's films. In recent years, her work has led to new methods in developing frameworks at a national level, connecting community-led experiences to federal, local, and nonprofit sector programs for climate change communication. She is a member of the National Association of Science Writers and the Society of Environmental Journalists.

Entertainment-Education and Climate Change: Program Examples, Evidence, and Best Practices from around the World

Suruchi Sood, Amy Henderson Riley, and Lyena Birkenstock

> With all due respect, Mr. Vice President, the cost of doing nothing could be even higher. Our climate is fragile.

The Day After Tomorrow was a fictional dystopian American blockbuster movie released in 2004 that warned of the dangers of climate change. Combining cinematic elements, such as the dramatic dialogue spoken by the main character Jack Hall, alongside award-winning visual effects, the film takes viewers through a journey where they are both entertained and prompted to consider their own actions and the future of our planet (Leiserowitz, 2004). Stories like these have been used for millennia to impart wisdom and encourage actions among people through oral traditions, written texts, and traditional and new media (Riley et al., 2017a, b). Entertainment-education (EE) is the term used to describe the specific communication strategy whereby educational messages are purposely integrated into entertainment platforms to influence knowledge, attitudes, behaviors, and social norms (Singhal & Rogers, 1999, 2002, 2004). EE has been applied in countries around the world on a host of programmatic topics, including health, education, gender, water, and sanitation, and—most recently—climate change (Sood et al., 2017).

Climate change is a broad term to describe shifts in climate patterns over a period of time. These shifts may be natural. However, since the 1800s, human activities primarily due to burning fossil fuels such as coal, oil, and gas have been the main

S. Sood (✉)
Johns Hopkins University, Baltimore, MD, USA
e-mail: ssood3@jh.edu

A. H. Riley
Thomas Jefferson University, Philadelphia, PA, USA

Population Media Center, Burlington, VT, USA
e-mail: amy.riley@jefferson.edu; ahendersonriley@populationmedia.org

L. Birkenstock
Thomas Jefferson University, Philadelphia, PA, USA
e-mail: lyena.birkenstock@students.jefferson.edu

© The Author(s) 2024
E. Coren, H. Wang (eds.), *Storytelling to Accelerate Climate Solutions*,
https://doi.org/10.1007/978-3-031-54790-4_2

driver of climate change. Global warming, or a rise in temperatures, is one aspect of climate change, as are changing weather patterns with more extreme weather events, rising sea waters, increasing water scarcity, and shifts in wind patterns, among others. Climate change is at the forefront of the global agenda and is reflected in documents such as the United Nations Framework Convention on Climate Change (UNFCCC; United Nations, 2020) and the Sustainable Development Goals (United Nations, n.d.). As a planetary issue, addressing climate change requires collaboration and coordination across all levels of the social–ecological model, from an individual change in knowledge, attitudes, and behaviors, to pressure on governments to enact climate-friendly policies. Despite political debate on the topic, there is widespread agreement among the scientific community that climate change is an urgent issue that transcends geopolitical boundaries. Communicating climate science messages in ways that people can relate to and feel spurred to act in their everyday lives requires creative, tailored, and evidence-based solutions. Social and behavioral change communication, which is the strategic use of communication to promote changes in individual knowledge, attitudes, norms, beliefs, and behaviors, interpersonal communication, as well as community-level action, is one approach.

In this chapter, we give a brief history and overview of EE as a theory-driven, evidence-based, social and behavioral change communication approach for climate change. We situate EE under a larger umbrella of storytelling or narrative approaches (we use the terms narrative and storytelling interchangeably). We conducted a formal search of the peer-reviewed literature on EE and climate change over the past 20 years. We also informally contacted colleagues working in this space to catalog evidence-based EE programs addressing climate change from around the world. We present our programmatic findings and conclude with implications for EE theory, practice, and future research and evaluation on this pressing global issue.

Overview of Entertainment-Education

While storytelling for change has existed throughout human history (Fisher, 1985), a specific strategy called entertainment-education emerged in the mid-twentieth century. Miguel Sabido, a telenovela producer for the Mexican media company *Televisa*, and other colleagues initiated its formal study by observing real-world outcomes of popular mass media programs. For example, viewers of the Peruvian telenovela *Simplemente María (Simply Maria)*, one of the most popular telenovelas ever broadcast in Latin America, purchased sewing machines en masse after watching their beloved protagonist do the same (Singhal et al., 1995). Sabido originally called this approach "entertainment with proven social benefit" (Sabido, 2004, p. 61), which was later adapted into the term "entertainment-education."

Put simply, EE embeds didactic information into entertainment formats. Although there are differing views on what is considered "educational" and who determines the content of EE campaigns (Dutta, 2006), traditional EE programs include a cast of relatable positive, negative, and transitional characters who evolve throughout storylines that can last months, and even years (Ryerson, 1994). EE has evolved over time from long-running soap opera-type programs focused on individual change and dealing mostly with sexual and reproductive health to a variety of issues and media representing changing global priorities, funders, and technology and is now considered an established approach to research and scholarship (Storey & Sood, 2013). This century has seen a rapid rise in the popularity of storytelling on web-based platforms. EE theory and research have adapted to the possibilities of immersive environments created by digital technologies. An updated definition of EE by Wang and Singhal (2021) states, "EE is a social and behavioral change communication (SBCC) strategy that leverages the power of storytelling in entertainment and wisdom from theories in different disciplines—with deliberate intention and collaborative efforts throughout the process of content production, program implementation, monitoring, and evaluation—to address critical issues in the real world and create enabling conditions for desirable and sustainable change across micro-, meso-, and macro-levels" (Wang & Singhal, 2021, p. 227).

Since its formal inception, EE has covered a wide range of educational topics. A review we conducted of the literature from 2005 to 2016 found that HIV/AIDS and sexual and reproductive health were the most common topics covered in EE (Sood et al., 2017). The field of EE has evolved over time, from soap opera-type programs focused on individual change based on the tenets of social cognitive theory and dealing mostly with sexual and reproductive health, to a variety of issues representing changing global priorities and funders (Storey & Sood, 2013).

EE is one type of narrative approach, among a suite of narrative approaches, for climate change communication (Fig. 1). Closely related narrative approaches are covered in other chapters in this text and elsewhere. For this chapter, we focus on EE for climate communication, also referred to as climate change communication (Yale Program on Climate Change Communication, 2020), the formal sub-discipline

Fig. 1 Umbrella of climate change communication approaches

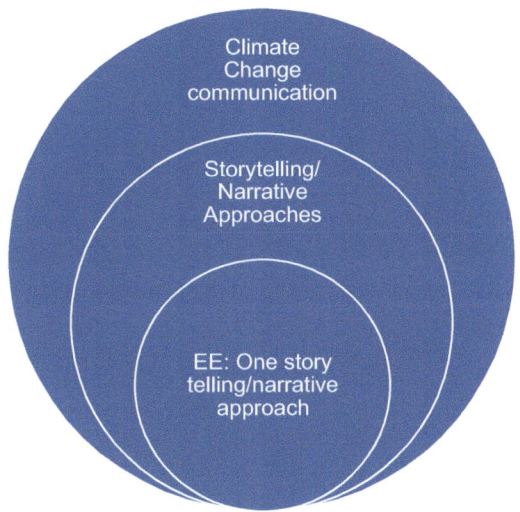

of environmental and science communication (Jickling & Wals, 2008). We acknowledge an intersection of several disciplines at each level of our figure, which includes environmental education, climate education, public health education, and social and behavioral change communication. Our aim here is not to exhaustively list the various intersections of fields and viewpoints that undergird this work but to highlight the EE strategy as a multi-disciplinary approach to communicating climate change.

Theoretical Roots of Entertainment-Education

Since its inception, EE has been designed, implemented, and evaluated as a communication strategy using a rich theoretical foundation with theories spanning several social science disciplines. Full reviews of EE theory exist elsewhere but in short: the theories that have been used to understand the mechanisms of how and why EE engenders change encompass each level of the social ecological model (Fig. 2). Individual-level theories are often used in EE to understand individual knowledge, attitudes, and behaviors that campaign designers hope will be impacted by an EE program. For example, *the theory of planned behavior* says that individual behavior is shaped by a combination of attitudes and intention to change (Ajzen, 1991) and, thus, has often been used to explain the effects of EE programs on those individual-level indicators.

Previous research has shown that the most commonly used theory to understand the mechanisms of how and why EE engenders change is Albert Bandura's *social cognitive theory* (Bandura, 1977), which states that people learn behavior from their environment and through watching others, i.e., observational learning. Bandura's work was instrumental in developing the EE approach and characters that role

Fig. 2 Social ecological model

model behavior for audiences (Bandura, 2004). Recent research has focused on the influence of EE via theories of social norms (Riley et al., 2020), the unwritten rules that guide human behavior (WHO, 2010). Through theories, such as *diffusion of innovations* (Rogers, 2003) and *bounded normative influence* (Kincaid, 2004), EE characters role model and promote new norms for audiences and demonstrate the benefits and sanctions, or rewards and punishments, of following or not following social norms. For example, the Rwandan EE program *Umurage Urukwiye (Rwanda's Brighter Future)* role modeled a community where planting trees to protect the environment became normal. This 312-episode EE radio drama aired from July 2007 to August 2009, focused on the impacts of environmental degradation and promoted wildlife habitat preservation. The drama mirrored an existing government-driven reforestation program by incorporating characters planting tree seedlings to save their farmlands. An independent evaluation of *Umurage Urukwiye* found that 11% of those buying tree seedlings during the broadcast period had been motivated to do so by listening to the show (Barker et al., 2013).

Other theories from communication seek to understand the underlying mechanisms supporting stories and storytelling in EE. These include the *extended-elaboration likelihood model* (Slater & Rouner, 2002), *narrative transportation* (Green & Brock, 2000), and *narrative engagement* (Busselle & Bilandzic, 2009), as well as theories that describe audience members' perceived relationship with characters in EE, including *parasocial interaction* (Papa et al., 2000), *audience involvement* (Sood, 2002), and *identification* (Cohen, 2001). Moyer-Gusé's (2008) *entertainment overcoming resistance model* takes several of these concepts into account. For example, *Punta Fuego (Fire Point)* in Belize demonstrated the benefits of sustainable fishing to mitigate the local impact of climate change on personal livelihoods and included a radio drama serial, call-in shows, theatre productions, and community mobilization, in an effort to educate local fishing communities on the environment, overfishing, and their own rights and responsibilities. Quantitative evaluation results showed significant improvements in knowledge regarding fishery regulations and the benefits of marine protection areas. Fishermen explained their positive response to the show because it was about them and spoke to them about issues that they considered important (Cheung et al., 2018).

A rich body of evidence through different methodological approaches including randomized control trials, mixed methods, and cost-effectiveness analysis, has demonstrated that EE is effective at a relatively low cost. For example, the Indian EE television program *Jasoos Vijay (Detective Vijay)* costs just $2.49 per person who changes their behavior (started using condoms) as a result of exposure (Sood & Nambiar, 2006). A literature review of 126 published EE research programs found that 18.3% of studies reported high levels of effective behavioral and social changes, and 61.1% reported intermediate effectiveness of these outcomes (Sood et al., 2017). While previous EE research has tended to focus on health topics related to sexual and reproductive health, HIV/AIDS, and family planning, there is a burgeoning interest in applying EE to climate issues as evidenced by recent events including a 2018 EE conference at Stanford University and a panel at the 2021 Sundance Film

Festival, sponsored by the National Resources Defense Council. Cross traditions, EE is standing on firm theoretical ground to understand the effects of three different types of present-day EE.

Types of Present-Day Entertainment-Education

EE began as a long-running communication strategy. Sabido utilized the telenovela format, popular across Latin America, for his early EE programs on *Televisa* (Singhal et al., 1993). These programs included a cast of positive characters, negative characters, and transitional characters who evolve throughout storylines that can last months and even years (Ryerson, 1994). The Indian television show *Hum Log* (*We People*) was one of the earliest replications of Sabido's methodology, and utilizing this long format ran for 154 episodes in 1984. As the EE strategy spread around the world, the long-running format was copied and adapted to local context and emerging media. This first type of EE is typically programs with several dozen episodes or more of a serial drama or soap opera on television or radio specifically designed from inception with behavioral and/or social change objectives. Funders of these efforts often include international aid organizations. An example using the Sabido methodology from the climate literature comes from India. *Yeh Kahan Aa Gate Hum* (*Where Have We Arrived?*) was a 52-episode radio serial in 1998 that covered issues across environmental domains including air, water, deforestation, and other topics. A popular actress provided 30- to 60-s epilogues about environmental behaviors covered in each episode (Singhal et al., 2000a, b).

A second branch of EE veers away from the original EE format and is called social impact entertainment by some media professionals (UCLA, 2019). While this is an evolving space, these efforts involve strategically incorporating educational messages into existing media and entertainment. Producers and media professionals may create successful EE with or without any knowledge of the global history and research supporting EE. This branch of EE includes a wide range of genres addressing climate change. Documentary and nonfiction film examples include *An Inconvenient Truth*, *When the Levees Broke*, and *Before the Flood*. A comedy example is the live show *Ain't Your Mama's Heat Wave*, which was made into a feature film. Other genres include television dramas, reality TV, podcasts, and emerging media. These efforts tend to be funded by private funders and nonprofits with a climate, or otherwise aligned, mission (Riley & Borum Chattoo, 2019). Research to understand these efforts is often done post hoc (Dudo et al., 2017) and is researched using similar theories and social science methods as the traditional EE space, such as research conducted on the documentary series *Years of Living Dangerously* (Bieniek-Tobasco et al., 2019, 2020).

A third type of EE is popular media and entertainment, which was created purely for entertainment purposes. In these examples, climate change is at the heart of the narrative, not necessarily for prosocial purposes but instead to tell an entertaining or informative story. Climate change examples in this category include late-night comedy shows, broadcast and cable news and newspapers (Farnsworth & Lichter, 2012), entertainment and prime-time TV (McComas et al., 2001), and shows on the Internet/YouTube (Shapiro & Park, 2015). As these are not purposely planned efforts with strategic objectives, funders do not include international development organizations or private foundations with climate missions. This is a growing group, and research is emerging on the unanticipated media effects of popular entertainment programs.

Methods

To catalog and describe EE and climate change efforts, we gathered information through three methods: (1) a formal review of the peer-reviewed literature on EE and climate change; (2) a keyword search on the Communication Initiative (CI), a network dedicated to creating a shared knowledge base of communication and media development for social and behavioral changes among local, national, and international communities (Feek, 2019); and (3) informal outreach via emails and conversations with colleagues working in the field of entertainment-education. Our formal literature review and keyword search on the CI network provided a retrospective look at research published between 2000 and the present, while our communication with colleagues presented both retrospective and prospective looks into existing examples, as well as perspectives on future directions and the role of EE in directing climate change understanding and behavior.

Formal Literature Review

We searched three academic databases to review the literature: PubMed, ScienceDirect, and Scopus. To operationalize the field of EE, we used three terms: entertainment education, entertainment-education, and edutainment. To operationalize climate change, we used four terms: climate, climate change, global warming, and environment. We used the *and* identifier to connect EE and climate change terms. This resulted in 12 separate searches per database, for a total of 36 searches. Searches were limited to English only and between the years 2000 and 2020 or the present. Of the 1509 total search results returned from the three databases, 1487 were removed because they were duplicates and/or not relevant to our topic of interest based on title, leaving 22 articles remaining.

We also performed a search in Google Scholar using the same combinations of operational search terms and limitations (English only articles between 2000 and 2020). For the Google Scholar searches, we searched for relevant articles within the top 300 results, as proposed by Haddaway et al. (2015). We prioritized peer-reviewed, published journal articles in the literature review that evaluated a program related to climate change and EE. We performed the same identification and title screening steps outlined above, which resulted in 134 shortlisted articles.

Using RefWorks, we then screened the abstracts of the 156 articles, identifying and removing 8 more duplicates, 10 results that were not original or peer-reviewed research articles, 14 articles that were too broad and unspecific to climate change and EE, 58 articles that were not focused on climate change and EE, and 9 articles that were unavailable from institutional libraries. This left 57 articles to include in our review.

Keyword Search

We next ran a keyword search on the Communication Network (CI) network. The CI network allows people and organizations to share articles, opinions, and media/communication programs for discussion and review (Feek, 2019). This provided another source for potential programs related to EE and climate change. Although these articles do not undergo peer review, many provide links to peer-reviewed articles regarding programs of interest. We used the same operational search terms to conduct a brief search of the CI literature, including the first 100 articles for each of the 12 separate searches. We found 15 programs identified through the CI network for inclusion.

Informal Outreach

Finally, we contacted 83 colleagues informally via email for any examples of work on climate change and EE. The co-authors drafted a list of colleagues from our respective networks, and our emails were also forwarded on to new connections and organizations. We received 70 responses, of which 58 offered information in the form of project or program details, published journal articles (31 additional articles), or referrals to other colleagues working on media projects associated with climate change. We conducted 9 online conversations with leaders of flagship EE organizations[1] regarding current or past productions of EE that involved a climate change component.

---

[1] We spoke to colleagues at the following organizations: BBC Media Action; Conserv Congo; Hollywood, Health & Society; Johns Hopkins University Bloomberg School of Public Health; Minga Peru; PCI Media; Population Media Center; and Wise Entertainment.

Results

Compiling programs from the literature review, keyword search, and colleague conversations, we identified 87 different programs involving some component of climate change and some component of entertainment-education. In terms of locations of entertainment programs addressing climate change, the United States was most widely represented, with 37 efforts. Other countries that reported multiple projects include Nigeria and India. This is not surprising given the size and scope of the entertainment industry in both countries. Radio ($n = 22$) and TV ($n = 27$) emerged as the most commonly reported medium for climate change programs included in our search. A relatively smaller number of programs indicated a multi-channel focus, leveraging transmedia options for addressing climate change. Results around the timing of the programs varied. A handful of popular entertainment examples were from the late 1990s. For example, *Yeh Kahan Aa Gate Hum (Where Have We Arrived?)* covered themes of air/water/noise pollution, deforestation, solid waste disposal, and organic farming (Singhal et al., 2000a, b). Anecdotally, there appears to be a growing interest in climate change and EE. In our sample, 22 programs began in the decade spanning 2000–2010 and 51 from 2010–2020.

In the course of analyzing the programs, it became clear that all branches of EE required intentional storytelling and close collaboration among different partners. All the examples of traditional EE messaging through long-running narratives that are built around a goal of promoting changes in knowledge, attitudes, behavior, and/or social norms surrounding climate change came from the Global South. The majority of the examples of existing television or media incorporating an EE narrative related to climate change to reach broad audiences that already may watch the program came from the United States. As a whole, it appears that while climate change communication using EE remains relatively uncommon around the world, there are unique media industries and related reasons driving this work in specific geographical areas.

Climate Change and EE in the Global North

Climate change communication in the Global North, specifically in the USA and UK, can be categorized into three discrete strategies. These include the purposeful inclusion of climate change storylines within popular entertainment, late-night political comedy, and unplanned climate change messaging within popular entertainment. Table 1 summarizes these findings.

Table 1 Entertainment-education and climate change programs

Program name	Implementation country/region	Medium	Year	Citation (where applicable)
Climate change and EE programs in the global North				
1. A Plastic Ocean#	U.K.	Film	2017	–
2. Ain't Your Mama's Heat Wave#*	U.S.	Comedy Show	2019	Borum Chattoo (2020)
3. An Inconvenient Truth#*	U.S.	Film	2006	Jacobson (2011) and Griep and Reimer (2016)
4. Blue Planet#	U.K.	Television	2001	–
5. Catan: Oil Springs and Catan: Global Warming+*	U.S.	Board Game	2017	Chappin et al. (2017)
6. Chasing Coral#	U.S.	Television	2015–2019	–
7. Chasing Ice#*	U.S.	Film	2012	Parant et al. (2016)
8. Cowspiracy	U.S.	Film	2014	–
9. Dirty Little Secrets*	U.S.	Collaborative Reporting Project	2015	Borum Chattoo and Green-Barber (2018)
10. Earth (Official Music Video) by Lil Dicky	U.S.	Song	2019	–
11. EAT Forum	Sweden	Podcast & Global Platform	2018–2020	–
12. Episode from 30 Rock*	U.S.	Television	2007	Moyer-Gusé et al. (2019)
13. Episode from Black-ish	U.S.	Television	2016	–
14. Episode from Doc McStuffins+	U.S.	Television	2014	–
15. Episode from Last Week Tonight with John Oliver*	U.S.	Television	2014	Brewer and McKnight (2015)
16. Episode from Madam Secretary	U.S.	Television	2019	–
17. Episode from Modern Family	U.S.	Television	2014	–
18. Episodes from The Daily Show, the Colbert Report*	U.S.	Television	2011	Feldman et al. (2011) and Feldman (2013)
19. Fly Away Home#*	U.S.	Film	1996	Chen and Lin (2014)
20. Food, Inc.	U.S.	Film	2008	–
21. Hot Dish*	U.S.	Facebook Application	2009	Robelia et al. (2011)

(continued)

Table 1 (continued)

Program name	Implementation country/region	Medium	Year	Citation (where applicable)
22. How I Fell in Love with a Fish: A Surprising Parable of Foie Gras	U.S.	TedTalk by Dan Barber	2008	–
23. Inuk	Greenland	Film	2012	–
24. IPCC Special Report on Global Warming of 1.5C*	U.S.	YouTube	2018	Bounegru et al. (2020)
25. Keep Cool+*	Germany	Board Game	2004	Meya and Eisenack 2017
26. School Strike from Climate: Save the World by Changing the Rules	Sweden	TEDxStockholm TedTalk by Greta Thunberg	2018	–
27. Stand Up for Climate Change	U.S.	Video Competition	2016–2018	–
28. State of Fear#*	U.S.	Book	2004	Farnsworth and Lichter (2012)
29. The Day After Tomorrow*	U.S.	Film	2004	Leiserowitz (2004)
30. The Devil We Know#	U.S.	Film	2018	–
31. The Third Plate	U.S.	Book	2014	–
32. Winds of Change+*	Europe	Board Game	2005	Winds of Change (2005)
33. Years and Years	U.S.	Television	2019	–
34. Years of Living Dangerously*	U.S.	Television	2014–2016	Bieniek-Tobasco et al. (2019)
Climate change and EE programs in the global South				
1. Amrai Pari (Together We Can Do It)*	Bangladesh	Television	2010–2017	Whitehead (2017)
2. Bennde Mutale Theater Group*	South Africa	Theater	2009–2020	Bennde Mutale Theater Group (2011)
3. Bienvenida Salud (Welcome Health)	Amazon region of Peru	Radio	1998–2020	–
4. Bravos do Zambeze (Zambezi Braves)*	Mozambique	Radio	2009–2011	Bravos do Zambeze (2011)
5. Callaloo	St. Lucia / Caribbean	Radio	2011–2013	–
6. Chuyen Que Minh (Homeland Story)*	Vietnam	Radio	2004	Heong et al. (2008)
7. ClimateConscious Programme*	South Africa, Namibia, Tanzania, Kenya	Photostories, short films	2009–2011	ClimateConscious Programme (2010)

(continued)

Table 1 (continued)

Program name	Implementation country/region	Medium	Year	Citation (where applicable)
8. ConservCongo+	Democratic Republic of Congo	Board Games, Puzzles	2014–2020	–
9. Don't Wait for Rain	Cambodia	Television	2017–2019	–
10. Ek Zindagi Aisi Bhi (A Life We Aspire For)*	India	Radio	2017	Sachdev (2018)
11. Forest Blessings	Mano River Union (Guinea, Sierra Leone, Liberia, Cote d'Ivoire)	Radio	2020	–
12. Hanh Trinh Xanh (Green Journey)	Vietnam	Radio	2008–2010	–
13. Jol-Danga (Water and Earth)*	Bangladesh	Television	2012–2020	Jayedi (2014)
14. Kaun Kitney Paani Mein (How Much Water is He in?)	India	Film	2015	–
15. Kidi Ya Chanza (When the Drumbeat Changes, You Must Change Your Dance Steps)*	Nigeria	Radio	2007–2009	ARDA (2008)
16. Koko Et Les Lunettes Magiques (Koko and the Magic Glasses)*	Cote d-Ivoire	Television	2014	Essé et al. (2017)
17. Lifeline Programming: Together Nepal, Linking Hand to Keep Living*	Myanmar, Nepal	Radio	2015	BBC Media Action (2016)
18. My Gorilla My Community	Nigeria	Radio	2014–2019	–
19. N*Gen Africa+	Kenya, Namibia, Nigeria, South Africa, Uganda, Zambia, Zimbabwe	Television	2020	–
20. Nyakati Zinabadilika (The Times/Winds are Changing)*	Tanzania	Radio	2013–2015	Whitehead (2017)
21. Pambazuko (The Dawn)*	Democratic Republic of Congo	Radio	2018–2020	Barker et al. (2018)
22. Protect Our Planet (POP) Movement	Africa, Latin America	Varied	2016–2020	–

(continued)

Table 1 (continued)

Program name	Implementation country/region	Medium	Year	Citation (where applicable)
23. Punta Fuego (Fire Point)*	Belize	Radio	2013–2019	Southey and Lewis (2016)
24. Shamba Chef*	Kenya	Television & Radio	2017–2020	Evans et al. (2020)
25. Shamba Shape-Up*	Kenya	Television	2012–2020	Areal et al. (2020)
26. Story Story: Voices from the Market*	Nigeria	Radio	2003–2017	BBC Media Action (2017)
27. Talking About the Weather and Doing Something*	Zambia	Radio	2013–2017	DRP (2017)
28. Thank you for the Rain#	Kenya (main), Norway, France	Film	2017	–
29. The Green Campaign, Zika Superheroes+	Latin America	Television	2017	–
30. The Rough Season*	English-speaking Caribbean	Radio	2010–2011	The Rough Season (2011)
31. The World Has Malaria#*	Tanzania	Film	2011	The World has Malaria (2011)
32. Umurage Urukwiye "Rwanda's Brighter Future"*	Rwanda	Radio	2007–2009	Barker et al. (2013)
33. Wanji+	Malawi, Madagascar, Uganda, Tanzania, Ghana, Nepal	Game	2017–2020	–
34. Watasay Ston (Immovable Rock)*	Sierra Leone	Radio	2015–2020	*USAID/WA BiCC (2018)
35. Yeh Kahan Aa Gaye Hum (Where have we arrived?)*	India	Radio	1998	Singhal et al. (2000a, b)
36. Zachilengedwe Tsogolo Lathu (Our Environment, Our Future)*	Zambia, Malawi	Radio, Mobile Phone SMS	2012	Knoblich and Siemering (2012)
Global climate change and EE programs				
1. 8 Billion Angels#	Global	Film	2019	–
2. Comics Uniting Nations*	Global	Comic Books	2015	PCI Media Impact (2016)
3. Drowning World	U.K., India, Haiti, Pakistan, Australia, Thailand, Nigeria, Germany, Philippines	Art	2007	–

Note. *#documentary; + children's programming; *reference listed in References*

Purposeful Climate Change Storylines Within Popular Entertainment

In the domestic US market, Hollywood Health and Society (HH&S) is an industry leader that has worked for many years with prime-time drama and comedy shows to incorporate messages into popular entertainment. Of specific note is HH&S' *Climate Change Initiative*, which provides expert opinions, top scientists, and researchers in education, government, and health care, as well as environmentalists and activists, as tools for the television and film industry. Film screenings are often followed by Q&As with filmmakers and materials like tip sheets on environmental health issues. The *Climate Change Initiative* has devoted special editions of their HH&S newsletter featuring climate stories, reports, and studies on global warming (HH&S, 2020). Some specific EE and climate change examples from HH&S include:

- Cameo appearances on NBC's *30 Rock* (2006–2013) by Al Gore, who has championed for changes in climate change policy for decades (see research by Moyer-Gusé et al., 2019)
- A 2014 episode of ABC's *Modern Family* (Season 5, Episode 12, "Under Pressure"), where the character Asher, played by actor Jesse Eisenberg, is portrayed as a climate crusader, who provides cues to action to address climate change, specifically associated with potable water
- Episodes of *Madam Secretary*, a popular drama on CBS, including Season 1, Episode 17 (2015), where one of the storylines involved the prevention of an ecological disaster in the Amazon, and Season 5, Episode 16, "The New Normal" (2019), which focused on the global impact of climate change, as illustrated by changing weather patterns and extreme weather events

Late-Night Political Comedy

The second strategy of climate change communication we found in the Global North included late-night political comedy television, which engages in satirizing doubts about global warming. Research conducted by Brewer and McKnight (2015) found that satirical television news (specifically *The Daily Show*, hosted by Jon Stewart) shaped public opinion about climate change, regardless of the political and ideological leanings of viewers. In May 2019, based on a U.N. report that predicted that lasting climate change is coming to the planet as soon as 2040 (United Nations, n.d.-a, n.d.-b), *Last Week Tonight* host John Oliver teamed up with Bill Nye the Science Guy (whose media effects scholars credit with making science enjoyable for children) Bryant et al., 2012). Oliver and Nye encouraged viewers to take unprecedented steps to halt climate change.

Climate Change Messaging Within Popular Entertainment

The third set of entertainment programs in the Global North that incorporate climate change themes is somewhat serendipitous. One example of this accidental inclusion of climate change is Tom Perrotta's novel *The Leftovers*. According to the author, this novel was not meant to be an allegory for climate change, but when the book was converted to an HBO TV show, some critics connected it to climate change, a view the author now endorses.

Journalists and media bloggers commenting on popular TV have noted that streaming shows like *Black Mirror* have included specific episodes that deal with climate change as part of interesting narratives. For example, the first and last episodes of Season 5. In Episode I, "Nosedive," the widespread availability of clean energy is juxtaposed with inequality in a totalitarian society. And the last episode, "Hated in the Nation," imagines a world where bees have gone extinct bringing the U.K. to the brink of an ecological and food crisis.

John Mitchell, an editor and writer with the Washington, D.C.-based *Climate Reality Project,* in a blog from Spring 2019, specifically sought to examine major TV shows addressing climate change and isolated just three: *Game of Thrones*, a National Geographic docu-series called *Life Below Zero,* and the Norwegian political thriller *Occupied* (Climate Reality Project, 2019). *Game of Thrones* creator George R.R. Martin claimed the climate crisis was a great parallel to the threat of "winter is coming" in the fictional world of Westeros. *Life Below Zero* was a popular National Geographic docu-series that followed the life of Alaskans who found their lives impacted by the melting permafrost and warmer temperatures. *Occupied* was a critically acclaimed Norwegian political thriller, available on Netflix, that explores a future world where environmentally friendly politicians come to power in Norway, but their progressive actions around climate change result in Norward being held hostage by Russian and E.U. oil interests.

As a whole, we found a need for more planned climate change communication in the United States. However, our review shows that this observation is not unique to the United States. A review of over 100,000 distinct programs from 40 channels in the U.K. conducted by the British Academy of Film and Television Awards (BAFTA) in 2019 and revealed that between 2017 and 2018, the term climate change was rarely mentioned. Summarizing their conclusions from this report, BAFTA chair Pippa Harris said, "Reducing our impact is a given, but our real opportunity lies in the programs we make, and in our ability to use powerful human stories to connect audiences with the world around them" (BAFTA, 2019, p. 8).

While considering the relative dearth of climate change communication in popular media in the North, we highlight one notable exception: the BBC and HBO show *Years and Years,* a 6-part series, that imagines a future in which changes in climate are occurring. The drama imagines a world where the North Pole has entirely melted, bananas no longer exist, and 80 days of relentless rainfall hits. Moreover, we see an upward trend in the inclusion of climate change communication in popular television: for example, Apple TV's dramatic adaptation of Nathaniel Rich's nonfiction book *Losing Earth*, which narrates the attempt to stop climate change in the 1980s (Rich, 2018).

Climate Change and EE in the Global South

The trend with entertainment-education for climate change in the Global South is somewhat different. These efforts are led by organizations with long histories of working in low- and middle-income countries on climate change communication, such as PCI Media (Cheung et al., 2018), Population Media Center (Barker et al., 2013), BBC Media Action (Whitehead, 2017), and the Johns Hopkins Center for Communication Programs (Johns Hopkins Center for Communication Programs, 2020). The main impetus driving their decision to implement these global climate change-focused EE projects appeared to be related to donor interest in increasing audience participation in climate adaptation and mitigation efforts on the ground, combined with the salience of the issues, as evidenced by reliance on robust formative research. Some 25 examples we elicited are summarized in the second half of Table 1.

A majority of these shows tended to situate the narrative within the lived experiences of their audiences and focus on climate adaptation within broader international development strategies. In these shows, notably *Bienvenida Salud* (*Welcome Health*) in the Peruvian Amazon and *Punta Fuego* (*Fire Point*) in Belize (Cheung et al., 2018), climate issues are real and happening now, which makes it easier for programmers to write a consistent and coherent human story, than if climate change is treated as an abstract phenomenon. At the same time, *Ek Zindagi Aisi Bhi* (*A Life We Aspire For*), a radio program in Uttar Pradesh, India, covered the topic of air pollution, a major environmental issue as a main theme in its narrative, while also including climate mitigation strategies such as the use of e-rickshaws and solar panels. In addition, they inspired villagers to demand policy changes (Sachdev, 2018; The Change Designers, 2017). The *Shamba Chef* program in Kenya was designed with safer cooking methods and environmental issues at the forefront (Evans et al., 2020).

CrossCutting Findings

Across both the Global North and South, we found some interesting trends among the programs. These are summarized below under findings related to documentary storytelling, children's programming, and future EE efforts for climate change.

Documentaries

Across both the Global North and South, we identified 13 documentaries dealing with climate change. While documentaries are not traditionally considered to be entertainment-education, given that a documentary is a movie or a television or radio program that provides a factual record or report (Borum Chattoo, 2020) and they consist of sub-genres such as docu-dramas and docu-comedies, these were included in our review and are marked with a hashtag (#) in Table 1. Some examples of documentaries include:

- The docu-comedy show and film *Ain't Your Mama's Heat Wave*, a collaboration between the Center for Media & Social Impact at American University and Hip Hop Caucus, a racial justice group that works on climate change awareness and mobilization for marginalized communities in the United States
- *Chasing Ice*, Jeff Orlowski's 2012 documentary on Arctic glaciers and global warming
- *An Inconvenient Truth* and its sequel *Truth to Power* (2017), both of which are about former United States Vice President Albert A. Gore Jr.'s continuing mission to battle climate change
- *8 Billion Angels*, a feature film about overpopulation and sustainability

A large proportion of documentaries dealing with climate change were from Western Europe and the United States. Two notable exceptions include a documentary from Kenya *Thank You for the Rain* (2017), which followed the journey of a Kenyan farmer as he transitions into becoming a climate change activist; and one from Tanzania, *The World Has Malaria*, a 20-min documentary-drama to explain causes of climate change and present adaptation options (Communication Initiative, 2011).

Children's Programming

Despite younger audiences' interest in environmentalism and climate narratives, evidence of climate change in children's programming is limited. We indicate our findings in Table 1 with the + symbol. One example from our search is *Doc McStuffins*, a children's TV show that made a passing reference to climate change in an episode titled "The Big Storm." A review of 25 children's programs, however, found that only four had an episode with climate change content (Poirier, 2019). At the same time, there were several compelling global projects reported engaging children though cartoons, comics, books, and board games. Wu and Lee (2015) review climate change games for children's entertainment including a variety of emerging formats including online, computer, and mobile games.

Depiction of anxiety associated with climate change or eco-anxiety was a phenomenon we found increasingly described. The American Psychological Association warns that children can be overwhelmed by climate change implications (Whitmore-Williams et al., 2017). In a 2015 review of the literature on young people and climate change, Corner et al. report that climate change is a major cause of concern and, in some cases, is "associated with feelings of anxiety, stress and despair" (p. 525). For example, a controversial season 4 episode of ABC's family sitcom, *Black-ish* "Please Baby Please" includes climate change as a core issue confronting the world. Set during a thunderstorm the two preteen characters in the show confess to their dad that they're scared by the storm, but not because they fear thunder but because the extreme weather reminds them of how terrified they are of climate change, while being resigned to cleaning up the messes of older generations. Another example comes from the HBO show *Big Little Lies,* where a child is visibly distressed and hides in a closet due to her concern that the planet is doomed. Millennials and Gen Z have forced conversations about eco-anxiety into the mainstream. According to the Pew Research Center (Funk & Hefferon, 2019), this is the demographic most likely to understand the link between human activity and climate change. We found little evidence in the literature of EE focused on children and youth, with a few exceptions, most notably a national high school assembly program that was based on EE principles and presented at 779 schools across the United States (Flora et al., 2014).

Future Implementation Plans

Throughout our informal conversations, colleagues mentioned both domestic and international EE programs that were in the planning phases. While many of these details are "under wraps" as of this writing, suffice it to say, that we learned about more than a few programs in proposal, pre-production, and/or production phases. One of these is the program proposed by Coren and Safer (2020) that outlines a plan for a long-running EE program for climate change in the United States. As of this writing, there has yet to be such a long-running EE program or a transmedia effort on this topic in the Global North.

Discussion

Implications for Theory

EE was initially conceptualized by Sabido and other pioneers as a long-running narrative strategy for behavioral and social changes. While some practitioners, mostly in the Global South, continue to apply a long-running EE approach over many episodes and interwoven stories, we found growing examples of EE formats that include documentary storytelling, social impact storytelling, and short-form storytelling using emerging and new media. A re-examination of theory is needed to understand if and how these newer EE adaptations engender change. EE today is much more complex than the model depicted in early programs, and climate change is a topic that requires more than simple awareness-raising campaigns. People around the world know that the climate is changing and can see it through their own eyes, through extreme weather events, such as hurricanes, floods, and fires, and the impact of external events on their families, communities, livelihoods, and homes. The *protection motivation theory* proposes that individuals rely on two factors to make decisions and change their behavior. These include threat appraisal and coping appraisal (Prentice-Dunn & Rogers, 1986). Threat appraisal assesses perceived severity and perceived susceptibility to a situation, for example, climate change, while coping appraisal is how one responds to the situation through both perceived response efficacy and self-efficacy, which are expectations that carrying out specific recommended actions will remove the threat combined with the confidence to execute the recommended courses of action successfully. While focusing on storylines depicting fear about an uncertain future (threat efficacy), EE narratives could rely on coping appraisal to promote action.

Newer models, such as the social–ecological model, provide an updated understanding of how people act, and how to spur people to act, by considering the interactive influences of families, communities, and society as a whole. Research indicates that EE is most effective when it is designed to address multiple levels of the social–ecological model, for example, communication initiatives that include complementary activities such as call-in shows to incite dialogue and discussion, community events to promote behaviors, and support for policies and initiatives to increase the uptake of services and supplies raised in EE stories. Recent research by Gesser-Edelsburg et al. (2021) based on studying childhood injuries among the Bedouin in Israel has proposed a hybrid model incorporating positive deviance and community-based participatory research with entertainment-education to engender social change.

EE has a long history of changing community social norms and applying pressure on governments to take action around complex and difficult topics, from child marriage to gender-based violence (Bouman et al., 2016; Usdin et al., 2005). While, in this review, we found the evidence based on EE and climate change to be relatively small, EE is theoretically poised to shift norms on climate change as well. Most of the scholarship on EE and social norms on other topics has been from long-running programs (Riley et al., 2020). For example, we previously found social norms change on maternal and child health topics as a result of exposure to *Kyunki … Jeena Issi Ka Naam Hai* (Because… That's What Life Is), a long-running EE TV program in India, but that show ran for over 500 episodes (Riley et al., 2017a, b). Riley et al. (2021) report on four different EE case studies that have successfully promoted social change across several topics. As new and shorter EE programs seek to address social norms and climate change, theory indicates that a more direct connection is required. As one example, social sanctions are the punishments for not following social norms, such as social exclusion that may be experienced if one does not have their daughter married by a certain age (Bouman et al., 2016). EE programs working in climate change must thoughtfully consider how to demonstrate social sanctions in stories to encourage, and not discourage, the adoption of new and critical mitigation and adaptation norms.

Theories related to storytelling are additionally useful for understanding this topic. Audiences need to identify and relate with characters in EE stories. As theory indicates EE is effective when characters are designed with the audience in mind, climate characters should not be annoying or preachy but rather relatable and admirable. Our review found several examples of EE programs with characters designed with formative research, such as the characters in the Mozambican EE program *Bravos de Zambeze* (*Zambezi Braves*), on natural disasters and climate change. We recommended a continued focus on connecting the theoretical underpinnings of characters and storytelling in future EE efforts for climate change.

Implications for Practice

Our review of past and current EE efforts to address climate change yielded some critical implications for practice. A central tenet for EE is that the entertainment aspect has to be front and center. Audiences will get turned off by programming that appears preachy. Another key aspect that came to the fore in the Global North was the relatively minor impact climate change has had on individual everyday lives. This lack of immediate and tangible impacts leads audiences to the conclusion that climate change is a long-term issue and not a matter of urgent concern. Experimental research by Chu and Yang (2020) reports that when communicating distant and abstract risks of climate change, highlighting their long-term disastrous impacts may be more effective in motivating action. In contrast, when communicating impending and concrete risks of climate change, stressing coping appraisal might have stronger potential. It is important therefore to start focusing on the tangible impacts of climate change on people's everyday lives.

In line with this implication is the need for climate change messaging to highlight the rewards associated with taking action and be relevant and actionable in both the short and long term. Climate change messages should be framed in a way that speaks to the values, interests, and experiences of youth. Almost all EE tackling climate change falls within the realm of *cli-fi* (Baden & Brown, 2024), which translates into narratives of a dystopian future arising from failure to address climate change. These narratives, while compelling, may not be entirely relevant for several reasons. First, they display a post-disaster world that human beings are helpless to change and cannot do anything about. Second, this type of narrative does not engender self-efficacy by making audiences feel confident in their ability to take positive actions that help address a problem. Finally, these narrativea do not provide audiences with concrete *cues to action*, that is small measurable steps that they can take to address this broad and overwhelming issue. It is important to reduce eco-anxiety by focusing on science-based tangible, actionable, and concrete *cues to action*. This is where the Sabido methodology of transitional characters who change their behaviors during the course of an EE narrative is critical. For example: to address climate change specifically what can individuals do, how these actions impact their livelihood, and what are the different actions that people can take.

Another programmatic implication is that the language used to address climate change needs to be consistent and positive. In particular, communication with children needs to have a positive tone. A lack of positive messages in environmental education about climate change may prevent many young people from engaging on the issue. According to the Sapir-Whorf Hypothesis, language impacts people's world views and individuals perceive reality relative to the language they speak (Kay & Kempton, 1984). The term climate change is politically charged, and the complexity of climate change jargon and information can be a barrier to learning and engagement. This is why we used four terms—climate, climate change, global warming, and environment—for our formal literature review. A 2014 Yale Program on Climate Change Communication and the George Mason University Center for Climate Change Communication survey found that *global warming* drew out more emotional engagement and support for action than *climate change* (Leiserowitz et al., 2014). While scientists prefer the term climate change, activists talk in terms of a climate crisis. The need for communication with children to have a positive tone is well-established in environmental education. A lack of positive messages about climate change in the media may prevent many young people from engaging on the issue. EE implementers have used *markers* as potential triggers for conversation. Markers are "distinctive and identifiable message elements" that serve as potential triggers for conversation (Singhal & Rogers, 2002, p. 131). The proactive inclusion of markers in EE programs is challenging. It needs to meet four criteria: markers have to be unique, attractive, attuned to the intended audiences, and fit organically into an EE narrative (Bouman, 2021).

A final programmatic implication is about linking climate change with other development issues. EE has been used successfully to address health issues, specifically around sexual and reproductive, as well as maternal and child health. Several EE programs from the Global North rely on this technique by emphasizing climate change as one of the many issues facing the world. For example, the *Black-ish* episode mentioned above added climate change among other cross-generational concerns such as the rise in white supremacy and the Black Lives Matter movement. In the Global South, narrative programs such as *Ek Zindagi Aisi Bhi* (*A Life We Aspire For*) have incorporated environmental concerns within a broader EE narrative (Coren & Wang, 2024). Instead of tackling climate change head-on, there may be a way for EE to explore the link between population growth and climate change. Such thematic intersectionality will allow for exploring climate change within the context of population growth, food systems, agricultural issues, migration, and the overall health and well-being of society.

Implications for Research and Evaluation

In a 2014 journal article titled, "Entertainment-education: storytelling for the greater, greener good," the authors highlighted the role of EE in sustainability and climate change goals and called for more research in this area (Reinermann et al., 2014). A descriptive review of our data suggests a possible uptick in EE and climate change programs in the years since this call, and our findings lead us to recommend several new areas of focus for future research. Notably, as EE as a strategy continues to proliferate across the globe, it is important to consider who will pay for research and evaluation. EE projects are traditionally funded by a variety of organizations including private foundations, the United Nations (U.N)., government agencies, and NGOs with interests across development issues. EE programs focused on climate change may bring new funders to the table, but research and evaluation to determine if and how initiatives work will remain crucial for the future of the strategy.

Formative Evaluation

Our findings in this chapter highlight the pivotal role of formative research for EE programs. Formative work determines audience needs and format choices. With a topic as complex as climate change, one solution will not fit all, and planned research helps to tailor efforts for the intended audience. At the same time, an understanding is needed that media projects in the Global North are developed differently than in the Global South. Traditionally, in the Global South, EE programs are planned after projects are funded over the course of several months and years with research and field work guiding and shaping the programs. This is not possible in Hollywood, for instance, where shows are piloted and only funded once a series or project is ordered by a network or studio. The tip sheets, Q&A's, and reports produced by Hollywood, Health & Society appear to be a nice bridge to connect storytellers and experts working in a Western model. More work is needed, however, to help shape climate change projects regardless of geography to be as impactful as possible.

Process Evaluation Process evaluation, or monitoring, answers the question, *Is the EE program being implemented as planned?* Monitoring happens while a program is in progress to determine fidelity, reach, and recruitment. In other words: Is the program airing when it is supposed to air? Are people tuning in? Do they like the program? How might the program course correct, if needed, to stay on track and increase listenership/viewership? We found few examples of process evaluation in this review but think this neglected area of research remains necessary. Without process evaluation, an EE program could air an entire season of a show only to realize at the end that adjusting midcourse could have greatly improved the outcomes. To urgently address climate change, programs cannot waste time and resources and, instead, should set aside funds for process evaluation to be conducted along the way.

Impact Evaluation This review found a small mix of diverse EE programs delivered on television, radio, film, and social media; designed using fiction and nonfiction approaches; applied drama and comedy; and produced for children and adults alike. Across these programs, we call attention to the importance of impact, or outcome, evaluation to determine whether these initiatives met their climate change goals and objectives. Paramount here is listener/viewership. Without an audience, researchers cannot attribute change to the program. We found several examples in our search of EE programs that struggled to keep audiences or lacked large enough sample sizes to make claims about change over time. Future efforts will likely include more transmedia approaches (Wang & Singhal, 2016) and, thus, will require researchers to think about attribution differently. Previous EE work has used markers (outlined earlier) to make claims about attribution. For example, pots and pans were used in episodes of the South African program *Soul City* to alert neighbors of domestic violence. Evaluation questions used audiences' answers to questions about the pots and pans as proxies for whether they had seen or heard about the program. In short, the markers bolstered the researchers' ability to make claims that it was, in fact, the EE program that shifted the audience's knowledge or behavior and not another, or competing, domestic violence prevention program. As EE programs shift formats and move online, researchers should work together with storytellers to plan markers in advance to make attribution claims in evaluation. These may include GIFs, sound bits, graphics, and other digital media (Lutkenhaus et al., 2019). Finally, we encourage the measurement of indicators across recent theorizing including the social–ecological model, such as knowledge, attitudes, behaviors, and social norms, using methods supported by the peer-reviewed literature (Riley et al., 2021).

Conclusion

This review indicated that while there is a current need for more attention, entertainment-education may be a promising vehicle for climate change communication. Climate change transcends geopolitical boundaries and impacts everyone around the globe. EE can place the issue of climate change in an entertaining and informal context while reaching many people simultaneously. Applying best practices from the entertainment industry, in combination with a growing theoretical understanding of how to communicate climate change with audiences, can foster compelling communications that help audiences understand what climate change is, encourage them to share their questions and concerns, and leave them feeling empowered and hopeful about the future. It is our perspective that there is a place for all types of storytelling or narrative approaches to climate change, and all are needed at this critical juncture for our planet. Such a future requires a multidisciplinary set of change-makers with different skills, including academics, scriptwriters, producers, directors, actors, and climate activists, working together to create meaningful and relevant EE programs for climate change based on theory and best practices.

References

*African Radio Drama Association (ARDA). (2008). In Kidi Ya Chanza (When the drumbeat changes you must change your dance-steps). *The Communication Network Initiative.* https://www.comminit.com/global/content/kidi-ya-chanza-when-drumbeat-changes-you-must-change-your-dance-steps

Ajzen, I. (1991). The theory of planned behavior. *Organizational Behavior and Human Decision Processes, 50*, 179–211.

*Areal, F. J., Clarkson, G., Garforth, C., Barahona, C., Dove, M., & Dorward, P. (2020). Does TV edutainment lead to farmers changing their agricultural practices aiming at increasing productivity? *Journal of Rural Studies, 76*, 213–229.

Baden, D., & Brown, J. (2024). Climate fiction to inspire green actions: Tales from two authors. In E. Coren & H. Wang (Eds.), *Storytelling to accelerate climate solutions*. Springer Nature.

BAFTA. (2019). *Subtitles to save the world: Understanding how the broadcasting community is covering climate*. A report by albert, BAFTA's TV industry-backed project on environmental sustainability. https://wearealbert.org/planet-placement/wp-content/uploads/sites/6/2019/05/Subtitles-to-Save-the-World-Report-FINAL.pdf

Bandura, A. (1977). *Social learning theory*. Prentice Hall.

Bandura, A. (2004). Social cognitive theory for personal and social change by enabling media. In A. Singhal, M. J. Cody, E. M. Rogers, & M. Sabido (Eds.), *Entertainment-education and social change* (pp. 75–96). Lawrence Erlbaum Associates, Inc.

*Barker, K., Connolly, S., & Angelone, C. (2013). Creating a brighter future in Rwanda through entertainment education. *Critical Arts, 27*(1), 75–90.

*Barker, K., Jah, J., & Connolly, S. (2018). A radio drama for apes? An entertainment-education approach to supporting ape conservation through an integrated human behaviour, health, and environment serial drama. *The Journal of Development Communication, 29*(1), 16–24

*BBC Media Action. (2016, October). *What role does lifeline preparedness play in enabling effective communication in a crisis? Case studies from Myanmar and Nepal*. Evaluation of Lifeline Preparedness. https://www.bbc.co.uk/mediaaction/publications-and-resources/research/briefings/asia/myanmar/evaluation-of-lifeline-preparedness/

*BBC Media Action. (2017, October 24). *The story of Story Story: 13 years of drama making a difference in Nigeria*. The Communication Initiative Network. https://www.comminit.com/global/content/story-story-story-13-years-drama-making-difference-nigeria-0

*Bennde Mutale Theatre Group. (2011). The Communication Initiative Network. https://www.comminit.com/global/content/bennde-mutale-theatre-group

*Bieniek-Tobasco, A., McCormick, S., Rimal, R. R., Harrington, C. B., Shafer, M., & Shaik, H. (2019). Communicating climate through documentary film: Imagery, emotion, and efficacy. *Climatic Change, 154*, 1–18.

Bieniek-Tobasco, A., Rimal, R. N., McCormick, S., & Harrington, C. B. (2020). The power of being transported: Efficacy beliefs, risk perceptions, and political affiliation in the context of climate change. *Science Communication, 42*(6), 776–802. https://doi.org/10.1177/1075547020951794

*Borum Chatoo, C. (2020). Comedy for racial justice in the climate crisis: Leveraging creativity and building community power in *Ain't Your Mama's Heat Wave*. Comedy Think Tank

*Borum Chatoo, C., & Green-Barber, L. (2018). An investigative journalist and a stand-up comic walk into a bar: The role of comedy in public engagement with environmental journalism. *Journalism, 22*(1), 196–214. https://doi.org/10.1177/1464884918763526

Borum Chattoo, C. (2020). *Story movements: How documentaries empower people and inspire social change*. Oxford University Press. https://doi.org/10.1093/oso/9780190943417.001.0001

Bouman, M. (2021). A strange kind of marriage: The challenging journey of entertainment-education collaboration. In L. Frank & P. Falzone (Eds.), *Entertainment education behind the scenes* (pp. 61–83). Palgrave Macmillan.

Bouman, M., Lubjohn, S., & Hollemans, H. (2016). *Entertainment-education and child marriage: A scoping study for Girls Not Brides*. Center for Media & Health.

*Bounegru, L., De Pryck, K., Venturini, T., & Mauri, M. (2020). "We only have 12 years": YouTube and the IPCC report on global warming of 1.5°C. *First Monday*.

*Bravos do Zambeze. (2011). The Communication Initiative Network. https://www.comminit.com/global/content/bravos-do-zambeze

*Brewer, P. R., & McKnight, J. (2015). Climate as comedy: The effects of satirical television news on climate change perceptions. *Science Communication, 37*(5), 635–657. https://doi.org/10.1177/1075547015597911

Bryant, J., Thompson, S., & Finklea, B. W. (2012). *Fundamentals of media effects*. Waveland Press.

Busselle, R., & Bilandzic, H. (2009). Measuring narrative engagement. *Media Psychology, 12*(4), 321–347. https://doi.org/10.1080/15213260903287259

*Chappin, E. J., Bijvoet, X., & Oei, A. (2017). Teaching sustainability to a broad audience through an entertainment game – The effect of Catan: Oil Springs. *Journal of Cleaner Production, 156*, 556–568. https://doi.org/10.1016/j.jclepro.2017.04.069

*Chen, T., & Lin, J. S. (2014). Entertainment-education of altruistic behaviors: An empirical study of the effects of the narrative persuasion of a nature conservation film. *Chinese Journal of Communication, 7*(4), 373–388. https://doi.org/10.1080/17544750.2014.946430

Cheung, L., Riley, A. H., Brown, A., & Schmid, C. (2018). Not enough fish in the sea: Community-based entertainment-education to promote sustainable fishing in Belize. *The Journal of Development Communication, 29*(1), 52–60.

Chu, H., & Yang, J. Z. (2020). Risk or efficacy? How psychological distance influences climate change engagement. *Risk Analysis, 40*(4), 758–770. https://doi.org/10.1111/risa.13446

Climate Reality Project. (2019). *We tried to write a blog about TV shows tackling climate change. Here's what happened.* https://climaterealityproject.org/blog/we-tried-write-blog-about-tv-shows

*ClimateConscious Programme. (2010). The Communication Initiative Network. https://www.comminit.com/global/content/climateconscious-programme

Cohen, J. (2001). Defining identification: A theoretical look at the identification of audiences with media characters. *Mass Communication & Society, 4*(3), 245–264.

Communication Initiative (2011). The world has malaria. Available at: https://www.comminit.com/communicating_children/content/world-has-malaria.

Coren, E., & Safer, D. (2020). Solutions stories: An innovative strategy for managing negative physical and mental health impacts from extreme weather events. In W. L. Filho, G. J. Nagy, M. Borga, D. C. Muñoz, & A. Magnuszewski (Eds.), *Climate change, hazards and adaption options: Handling the impacts of a changing climate* (pp. 441–462). Springer.

Coren, E., & Wang, H. (2024). Storytelling as a catalyst for climate communication and empowerment. In E. Coren & H. Wang (Eds.), *Storytelling to accelerate climate solutions.* Springer Nature.

*Developing Radio Partners (DRP). (2017). *Climate change and community radio project.* The Communication Initiative Network. https://www.comminit.com/global/content/climate-change-and-community-radio-project

Dudo, A., Copple, J., & Atkinson, L. (2017). Entertainment film and TV portrayals of climate change and their societal impacts. *Oxford Research Encyclopedia of Climate Science.*

Dutta, M. J. (2006). Theoretical approaches to entertainment education campaigns: A subaltern critique. *Health Communication, 20*(3), 221–231. https://doi.org/10.1207/s15327027hc2003_2

*Essé, C., Koffi, V. A., Kouamé, A., Dongo, K., Yapi, R. B., Moro, H. M., Kouakou, C. A., Palmeirim, M. S., Bonfoh, B., N'Goran, E. K., Utzinger, J., & Raso, G. (2017). "Koko et les lunettes magiques": An educational entertainment tool to prevent parasitic worms and diarrheal diseases in Côte d'Ivoire. *PLoS Neglected Tropical Diseases, 11*(9), e0005839. https://doi.org/10.1371/journal.pntd.0005839

*Evans, W. D., Young, B. N., Johnson, M. A., Jagoe, K. A., Charron, D., Rossanese, M., Lloyd Morgan, K., Gichinga, P., & Ipe, J. (2020). The *Shamba Chef* educational entertainment program to promote modern cookstoves in Kenya: Outcomes and dose-response analysis. *International Journal of Environmental Research and Public Health, 17*(1), 162. https://doi.org/10.3390/ijerph17010162

*Farnsworth, S. J., & Lichter, S. R. (2012). Scientific assessments of climate change information in news and entertainment media. *Science Communication, 34*(4), 435–459.

Feek, W. (2019). *Overview: The communication initiative.* The Communication Initiative Network. https://www.comminit.com/global/content/overview-communication-initiative

**Feldman, L. (2013). Cloudy with a chance of heat balls: The portrayal of global warming on The Daily Show and The Colbert Report. *International Journal of Communication, 7*, 430–451.

*Feldman, L., Leiserowitz, A., & Maibach, E. (2011, May). *The impact of the Daily Show and the Colbert Report on public attentiveness to science and the environment* [Paper presentation]. International Communication Association.

Fisher, W. R. (1985). The narrative paradigm: In the beginning. *Journal of Communication, 35*(4), 74.

Flora, J. A., Saphir, M., Lappé, M., Roser-Renouf, C., Maibach, E. W., & Leiserowitz, A. A. (2014). Evaluation of a national high school entertainment education program: The alliance for climate education. *Climatic Change, 127*(3), 419–434.

Funk, C., & Hefferon, M. (2019). *US public views on climate and energy*. Pew Research Center. https://www.pewresearch.org/science/2019/11/25/u-s-public-views-on-climate-and-energy/

Gesser-Edelsburg, A., Alamour, Y., Cohen, R., Abed Elhadi Shahbari, N., Hijazi, R., Orr, D., et al. (2021). Creating safe spaces to prevent unintentional childhood injuries among the Bedouins in southern Israel: A hybrid model comprising positive deviance, community-based participatory research, and entertainment-education. *PLoS One, 16*(9), e0257696.

Green, M. C., & Brock, T. C. (2000). The role of transportation in the persuasiveness of public narratives. *Journal of Personality and Social Psychology, 79*(5), 701–721.

*Griep, M.A., & Reimer, K. (2016). An Inconvenient Truth – Is it still effective at familiarizing students with global warming? *Journal of Chemical Education, 93*, 1886–1893.

Haddaway, N. R., Collins, A. M., Coughlin, D., & Kirk, S. (2015). The role of Google Scholar in evidence reviews and its applicability to grey literature searching. *PloS One, 10*(9), e0138237. https://doi.org/10.1371/journal.pone.0138237

*Heong, K. L., Escalada, M. M., Huan, N. H., Ky Ba, V. H., Quynh, P. V., Thiet, L. V., & Chien, H. V. (2008). Entertainment-education and rice pest management: A radio soap opera in Vietnam. *Crop Protection, 27*, 1392–1397.

Hollywood, Health, & Society. (2020). *Climate change*. https://hollywoodhealthandsociety.org/content/climate-change

*Jacobson, G. D. (2011). The Al Gore effect: An inconvenient truth and voluntary carbon offsets. *Journal of Environmental Economics and Management, 61*, 67–78.

*Jayedi, R. (2014). *Jol-Danga* [Water and Earth]. The Communication Initiative Network. https://www.comminit.com/global/content/jol-danga-water-and-earth.

Jickling, B., & Wals, A. E. J. (2008). Globalization and environmental education: Looking beyond sustainable development. *Journal of Curriculum Studies, 40*(1), 1–21.

John Hopkins Center for Communication Programs. (2020). *About CCP*. https://ccp.jhu.edu/about-ccp/.

Kay, P., & Kempton, W. (1984). What is the Sapir-Whorf hypothesis? *American Anthropologist, 86*(1), 65–79.

Kincaid, D. L. (2004). From innovation to social norm: Bounded normative influence. *Journal of Health Communication: International Perspectives, 9*(S1), 37–57.

*Knoblich, T., & Siemering, B. (2012). *Combining local radio and mobile phones to promote climate stewardship*. The Communication Initiative Network. https://www.comminit.com/global/content/combining-local-radio-and-mobile-phones-promote-climate-stewardship.

*Leiserowitz, A. A. (2004). Day after tomorrow: Study of climate change risk perception. *Environment: Science and Policy for Sustainable Development, 46*(9), 22–39.

Leiserowitz, A., Feinberg, G., Rosenthal, S., Smith, N., Anderson, A., & Roser-Renouf, C. (2014). *What's in a name? Global warming versus climate change*. https://climatecommunication.yale.edu/wp-content/uploads/2014/05/Global-Warming_Climate-Change_Report_May_2014.pdf

Lutkenhaus, R., Jansz, J., & Bouman, M. P. A. (2019). Toward spreadable entertainment-education: Leveraging social influence in online networks. *Health Promotion International, 35*, 1–10. https://doi.org/10.1093/heapro/daz104

McComas, K. A., Shanahan, J., & Butler, J. S. (2001). Environmental content in prime-time network TV's non-news entertainment and fictional programs. *Society and Natural Resources, 14*, 533–542.

*Meya, J. N., & Eisenack, K. (2017). *Effectiveness of gaming for communicating and teaching climate change*. THESys discussion paper no. 2017-3 (pp. 1–31). Humboldt Universität zu Berlin.

Moyer-Gusé, E. (2008). Toward a theory of entertainment persuasion: Explaining the persuasive effects of entertainment-education messages. *Communication Theory, 18*(3), 407–425.

*Moyer-Gusé, E., Tchernev, J. M., & Walther-Martin, W. (2019). The persuasiveness of a humorous environmental narrative combined with an explicit persuasive appeal. *Science Communication, 41*(4), 422–441.

Papa, M. J., Singhal, A., Law, S., Pant, S., Sood, S., Rogers, E. M., & Shefner-Rogers, C. L. (2000). Entertainment-education and social change: An analysis of parasocial interaction, social learning, collective efficacy, and paradoxical communication. *Journal of Communication, 50*(4), 31–55.

*Parant, A., Pascual, A., Jugel, M., Kerroume, M., Felonneau, M., & Guéguan, N. (2016). Raising students awareness to climate change: An illustration with binding communication. *Environment and Behavior, 49*(3), 339–353.

*PCI Media Impact – Headquarters. (2016). *Comics uniting nations*. The Communication Initiative Network. https://www.comminit.com/global/content/comics-uniting-nations

Poirier, S. (2019). Children's educational TV falls short on climate change. *Yale Climate Connections.* https://yaleclimateconnections.org/2019/01/kids-educational-tv-falls-short-on-climate-change/

Prentice-Dunn, S., & Rogers, R. W. (1986). Protection motivation theory and preventive health: Beyond the health belief model. *Health Education Research, 1*(3), 153–161.

Reinermann, J.-L., Lubjuhn, S., Bouman, M., & Singhal, A. (2014). Entertainment-education: Storytelling for the greater, greener good. *International Journal of Sustainable Development, 17*(2), 176–191.

Rich, N. (2018). Losing Earth: The decade we almost stopped climate change. *The New York Times.* https://www.nytimes.com/interactive/2018/08/01/magazine/climate-change-losing-earth.html.

Riley, A. H., & Borum Chattoo, C. (2019). Developing multimedia social impact entertainment programming for Hispanics in the United States. *The Journal of Development Communication, 30*(2), 16–29.

Riley, A. H., Sood, S., & Robichaud, M. (2017a). Participatory methods for entertainment-education: Analysis of best practices. *Journal of Creative Communications, 12*(1), 1–15.

Riley, A. H., Sood, S., Mazumdar, P. D., Chowdary, N., Malhotra, A., & Sahba, N. (2017b). Encoded exposure and social norms in entertainment-education. *Journal of Health Communication: International Perspectives, 22*(1), 66–74.

Riley, A. H., Sood, S., & Sani, M. (2020). Narrative persuasion and social norms in entertainment-education: Results from a radio drama in Mozambique. *Health Communication, 35*(8), 1023–1032.

Riley, A. H., Rodrigues, F., & Sood, S. (2021). Social norms theory and measurement in entertainment-education: Insights from case studies in four countries. In L. Frank & P. Falzone (Eds.), *Entertainment education behind the scenes* (pp. 175–194). Palgrave Macmillan.

Robelia, B., Greenhow, C., & Burton, L. (2011). Environmental learning in online social networks: adopting environmentally responsible behaviors. *Environmental Education Research, 17*(4), 553–575.

Rogers, E. M. (2003). *Diffusion of innovations* (5th ed.). Simon & Schuster.

Ryerson, W. N. (1994). Population communications international: Its role in family planning soap operas. *Population and Environment, 15*(4), 255–264.

Sabido, M. (2004). The origins of entertainment-education. In A. Singhal, M. J. Cody, E. M. Rogers, & M. Sabido (Eds.), *Entertainment-education and social change* (pp. 61–74). Lawrence Erlbaum Associates, Inc.

*Sachdev, A. (2018). *Improving air quality through air waves: Ek Zindagi Aisi Bhi, Entertainment-education radio drama to improve air quality in India*. International Communication Association.

Shapiro, M. A., & Park, H. W. (2015). More than entertainment: YouTube and public responses to the science of global warming and climate change. *Social Science Information, 54*(1), 115–145.

Singhal, A., & Rogers, E. M. (1999). *Entertainment-education: A communication strategy for social change*. Lawrence Erlbaum Associates.

Singhal, A., & Rogers, E. M. (2002). A theoretical agenda for entertainment-education. *Communication Theory, 12*(2), 117–135.

Singhal, A., & Rogers, E. M. (2004). The status of entertainment-education worldwide. In A. Singhal, M. J. Cody, E. M. Rogers, & M. Sabido (Eds.), *Entertainment-education and Social Change* (pp. 3–20). Lawrence Erlbaum Associates.

Singhal, A., Rogers, E. M., & Brown, W. J. (1993). Harnessing the potential of entertainment-education telenovelas. *International Communication Gazette, 51*, 1–18.

Singhal, A., Obregon, R., & Rogers, E. M. (1995). Reconstructing the story of *Simplemente Maria*, the most popular telenovela in Latin America of all time. *International Communication Gazette, 54*(1), 1–15.

*Singhal, A., Pant, S., & Rogers, E. M. (2000a). Environmental activism through an entertainment-education radio soap opera in India. In M. Oepen & W. Hamacher (Eds.), *Communicating the environment: Environmental communication for sustainable development* (pp. 176–183) Peter Lang.

Singhal, A., Pant, S., & Rogers, E. M. (2000b). Environmental activism through 'Yeh Kahan Aa Gaye Hum,' an entertainment-education radio soap opera in India. In M. Oepen & W. Hamacher (Eds.), *Communicating the environment: Environmental considerations for sustainable development* (pp. 176–183). Peter Lang.

Slater, M. D., & Rouner, D. (2002). Entertainment-education and the elaboration likelihood: Understanding the process of narrative persuasion. *Communication Theory, 12*(2), 173–191.

Sood, S. (2002). Audience involvement and entertainment-education. *Communication Theory, 12*(2), 153–172.

Sood, S., & Nambiar, D. (2006). Comparative cost-effectiveness of the components of a behavior change communication campaign on HIV/AIDS in North India. *Journal of Health Communication, 11*(Supp 2), 143–162.

Sood, S., Riley, A. H., & Alarcon, K. (2017). Entertainment-education and health and risk messaging. In R. Parrott (Ed.), *Oxford research encyclopedia of communication*. Oxford University Press.

*Southey, S., & Lewis, R. (2016). *Punta Fuego: Entertainment-education radio drama promotes sustainable fisheries in Belize: Abstract*. The Communication Initiative Network. https://www.comminit.com/global/content/ punta-fuego-entertainment-education-radio-drama-promotes-sustainable-fisheries-belize-ab.

Storey, D., & Sood, S. (2013). Increasing equity, affirming the power of narratives and expanding dialogue. The evolution of entertainment education over two decades. *Critical Arts: South North Cultural and Media Studies, 27*(1), 9–35.

The Change Designers. (2017, July). *Improving air quality through air saves: Ek Zindagi Aisi Bhi*. A program evaluation report on the entertainment-education radio drama to improve air quality in Uttar Pradesh, India.

*The Rough Season. (2011). The Communication Initiative Network. https://www.comminit.com/ global/content/rough-season

*The World has Malaria (2011). The Communication Initiative Network. https://www.comminit. com/global/content/world-has-malaria

UCLA. (2019). *The state of SIE: Mapping the landscape of social impact entertainment*. Skoll Center for Social Impact Entertainment. https://sfo2.digitaloceanspaces.com/skoll-sie-wp/2019/03/the-state-of-sie-report.pdf

United Nations. (2020). United Nations climate change. https://unfccc.int/

United Nations. (n.d.-a). *14. Take urgent action to combat climate change and its impacts**. Department of Economic and Social Affairs. Sustainable Development. https://sdgs.un.org/ goals/goal13

United Nations. (n.d.-b). *Climate change*. https://www.un.org/en/sections/issues-depth/ climate-change/

*USAID/West Africa Biodiversity and Climate Change (WA BiCC). (2018). *Report of radio drama design workshop (for Sierra Leone coastal landscape complex)*, 2nd Labone Link, North Labone, Accra – Ghana, 42p.

Usdin, S., Scheepers, E., Goldstein, S., & Japhet, G. (2005). Achieving social change on gender-based violence: A report on the impact evaluation of Soul City's fourth series. *Social Science & Medicine, 61*(11), 2434–2445.

Wang, H., & Singhal, A. (2016). East Los High: Transmedia edutainment to promote the sexual and reproductive health of young Latina/o Americans. *American Journal of Public Health, 106*(6), 1002–1010.

Wang, H., & Singhal, A. (2021). Mind the gap! Confronting the challenges of translational communication research in entertainment-education. In *Entertainment-education behind the scenes* (pp. 223–242). Palgrave Macmillan.

*Whitehead, S. (2017). *Building resilience: How research has been used to develop and evaluate a media and communication approach*. BBC Media Action. https://dataportal.bbcmediaaction.org/site/assets/uploads/2016/07/Building-Resilience-research-report.pdf

Whitmore-Williams, S. C., Manning, C., Krygsman, K., & Speiser, M. (2017). *Mental health and our changing climate: Impacts, implications, and guidance*. https://www.apa.org/news/press/releases/2017/03/mental-health-climate.pdf

*Winds of Change (Game). (2005). The Communication Initiative Network. https://www.comminit.com/global/content/winds-change-game

World Health Organization. (2010). *Violence prevention: The evidence*. http://www.who.int/violence_injury_prevention/violence/4th_milestones_meeting/evidence_briefings_all.pdf

Wu, J. S., & Lee, J. J. (2015). Climate change games as tools for education and engagement. *Nature Climate Change, 5*, 413–418.

Yale Program on Climate Change Communication. (2020). *What is climate change communication?* https://climatecommunication.yale.edu/about/what-is-climate-change-communication/

Suruchi Sood is the Director for Communication Sciences, Johns Hopkins Center for Communication Programs and Senior Scientist, Department of Health Behavior, and Society at the Bloomberg School of Public Health, Johns Hopkins University, USA. Suruchi earned her Ph.D. in Intercultural Communication from the University of New Mexico. Her research career has spanned areas inside and outside academics for 30 years, specifically around research processes associated with the design, implementation, and evaluation of health communication efforts to promote behavioral and social changes. Her global portfolio includes managing capacity building and research projects in over 30 low- and middle-income countries covering five continents. Dr. Sood's research has been widely published with over 100 journal articles, technical reports, and conference papers. At the American Public Health Association (APHA) Annual Conference in 2019, Suruchi was awarded the Public Health Education and Health Promotion (PHEHP) Section's Everett M. Rogers award for outstanding achievement in the field.

Amy Henderson Riley is the Director of Research, Evaluation, and Impact at Population Media Center and a Lecturer in the College of Population Health at Thomas Jefferson University in Philadelphia, Pennsylvania, USA. Throughout her life, Amy has been interested in storytelling, having attended a public school for the performing arts from 4–12th grade, studying theater as an undergraduate, working in film and television in her 20 s in New York City, and now as a mixed methods researcher who focuses on entertainment-education. Amy holds a B.F.A. in theater from The New School and an M.A. in health education from Teachers College, Columbia University. She completed her Doctor of Public Health degree with Suruchi at the Drexel University Dornsife School of Public Health and a postdoctoral fellowship at the American University School of Communication with Caty Borum Chattoo. Amy delights in demonstrating the powerful impact of storytelling on individuals, families, and communities around the world on a variety of pressing global topics.

Lyena Birkenstock is an M.D./M.P.H. dual degree candidate at Sidney Kimmel Medical College and the College of Population Health at Thomas Jefferson University in Philadelphia, Pennsylvania, USA. Lyena was Amy's graduate assistant during the drafting of this chapter and was integral to its publication. Throughout her life, Lyena has demonstrated her passion for education and healthcare. She studied Human Developmental and Regenerative Biology at Harvard University, where she was also a peer tutor, EMT, and CPR instructor. Lyena's work in public health focuses on improving healthcare experiences for refugees and immigrants, as well as optimizing education for her fellow medical students in cultural humility and caring for these vulnerable populations. Through working on this chapter, Lyena has been impressed by the universal value of storytelling across the globe and how this method of communication can contribute to crucial public health topics affecting us now and in the future.

The Power of Locally Driven Narratives to Support and Sustain Climate Action

Neemesha (Meesha) Brown

Thunder rolls through the air and heavy rains fall from the sky, flooding the earthen streets of a small town in Sierra Leone. A mother cries out for her small child that has been washed away in the swift, rushing waters now running through the streets. A fisherwoman's boats are swept out to sea and a local farmer reacts violently to losing his crops in the flood. The local chief struggles with how best to respond to the loss of life, property, and normalcy his villagers are experiencing. While they had heard about climate change before, they thought it was a foreign idea that described what naturally occurs. No one expected the rainy season to be this severe and the villagers had failed to prepare for more dangerous storms along their beloved coast. Now, in the midst of the tragedy, they wonder what the future holds and if they can rebuild together.

The scene above, from *Watasay Ston* (*Unmovable Rock*), is a pivotal moment in the journey of characters in an entertainment-education (EE) drama produced by PCI Media for coastal communities in West Africa. This drama invited audiences to reconsider the story they believed about climate change and its effect on their lives. Is climate change inevitable and natural—something that is destined by God? Or is climate change something that can be understood and mitigated? The answer to this essential question will make all the difference.

The narratives humans believe about their relationship to the natural world shape the behavior that will, ultimately avert or fail to solve the worst effects of the climate crisis on human civilization as we know it. Because familiar and internalized narratives are such critical drivers of behavior, entertainment-education can be a powerful vehicle by which people can interrogate longhand beliefs, and ways of living and also gain exposure to new ways of relating to the natural world and what stewardship looks like in the face of a growing threat to life on our planet.

N. Brown (✉)
PCI Media, New York, NY, USA
e-mail: mbrown@pcimedia.org

E. Coren, H. Wang (eds.), *Storytelling to Accelerate Climate Solutions*,
https://doi.org/10.1007/978-3-031-54790-4_3

PCI Media and Entertainment-Education

For over three decades, PCI Media has been a leader in impact communications—producing entertainment-education shows for radio and television to inspire audiences to act for a more healthy, just, and sustainable world. PCI Media's productions are designed to engage and entertain audiences while sharing critical knowledge and providing character-based role models for new behaviors and social norms in order to address problems in the real world. PCI Media's approach has evolved over the years along with changes the larger field of EE has experienced. Initially, an approach that centered the perspectives and expertise of those traditionally considered creators to an established field of scholarship and praxis in social and behavioral change communication (Wang & Singhal, 2021).

While entertainment-education can be a useful tool in reframing the climate challenge for wide audiences and reshaping behavior, we cannot overlook what the fields of conservation and development have learned about the value of local knowledge, local narratives, and local ownership to support the adoption and sustain effective resource management (Bixler, 2013).

At the heart of PCI Media's EE methodologies is authentic participation. Our local partners and counterparts are, in a very real sense, responsible for the program. In a typical program, a host of local organizations from several sectors gain new skills to plan, implement, and evaluate various C4D, social marketing, and entertainment-education communications program elements.

Typically, our programs comprise:

- EE serial drama, which hooks audiences through compelling dramatic conflict, leveraging parasocial modeling influences and providing role modeling for desirable and undesirable behaviors.
- Call-in shows, which contextualize issues deeply within local cultures, provide access to experts and information about access to local resources, and deepen the conversation across a community, encouraging interpersonal communication (which has been shown to be a direct precursor to behavior change).
- Social mobilization and social marketing campaigns, which distill messaging, move audiences, and catalyze change on the ground, often involving civil society sectors.

Since 2010, PCI Media has been using our signature "My Community" methodology to produce entertainment-education programming (Fig. 1). This approach centers on local and locally driven narratives while aiming to deepen the communications capacity of local partners.

PCI Media leverages a five-phase approach:

1. Coalition Building and Formative Research
2. Training and Program Design
3. Mentoring and Production
4. Broadcast and Community Mobilization
5. Monitoring and Evaluation

Coalition Building & Formative Research
We identify and partner with groups to drive formative research in its cultural context and create a coalition.

Training & Program Design
We host in-country workshops with coalition partners to analyze formative research, provide training on EE methodology and initiate campaign development.

Mentoring & Production
We mentor coalition partners. Together, we produce scripts and media materials using available radio, television and digital platforms.

Broadcast & Community Mobilization
With our partners, we broadcast and implement all components of the program, including interactive radio and TV call-in shows and community action campaigns.

Monitoring & Evaluation
As communicators, we focus on understanding the audience's needs as they evolve and change throughout the program, measuring impact and adapting our interventions accordingly.

Fig. 1 PCI Media's "My Community" methodology

Phase 1: Coalition Building and Formative Research

This phase is critical for the identification and engagement of local partners that will support the EE initiative as well as developing an evidence-based foundation for effective program design and distribution, harnessing input and buy-in from a wide spectrum of stakeholders and articulating the program's quantitative and qualitative objectives.

Phase 2: Training and Program Design

The design phase yields consensus on messaging, knowledge, attitude and behavior change objectives, target audiences, and campaign approaches. We bring together coalition partners and other relevant stakeholders to develop a research-based communications strategy. Through a participatory workshop, we analyze formative research data, provide training in communications for social change and entertainment-education processes, refine the communications strategy and planning documents, and develop tools for effective community engagement and mobilization.

Phase 3: Mentoring and Production

PCI Media closely mentors coalition partners, supporting the creative development and production of media materials, incorporating careful technical review for successful messaging, conceptual pretesting, and constant reviews to ensure the strategy remains focused on the knowledge, attitude, and behavior change objectives.

Phase 4: Broadcast and Community Mobilization

PCI Media provides continued support during the broadcast and implementation of impact communications and EE programs. This support typically includes training, designing, and producing interactive radio and television talk shows as well as on-the-ground mobilization campaigns that involve people in the desired behaviors. The community engagement, dialogue, and social learning structures are key to PCI Media's approach. From 2012 to 2020, PCI Media produced a collection of locally driven entertainment-education dramas using the "My Community" methodology in West Africa to address various conservation issues, from forest management to coastal resilience to species conservation.

Phase 5: Monitoring and Evaluation

Monitoring and Evaluation is an ongoing process for each program. We use a variety of methods to monitor and evaluate our programs, including measuring changes in social norms and interpersonal communication, and audience behaviors.

This chapter will explore the value of entertainment-education applications for climate action in West Africa that center narratives originating from the local context. The discussion will examine how centering narratives informed by local points of view can create the conditions for climate solutions to be optimally contextualized in the socio-cultural context of intended audiences. The chapter will illustrate these concepts through two examples, the Sustainable and Thriving Environments for West Africa Regional Development (STEWARD) Program administered by the US Forest Service—International Programs Department West Africa Biodiversity and Climate Change (WA BiCC) Program funded and administered by USAID. Using these examples, the chapter will demonstrate how PCI Media's My Community methodology allows local actors to authentically engage in a participatory approach to the narrative development process as part of an entertainment-education program.

Sustainable and Thriving Environments for West Africa Regional Development (STEWARD) Program

PCI Media served as a Communications partner to the US Forest Service—International Programs Department for the implementation of the Sustainable and Thriving Environments for West Africa Regional Development (STEWARD) Program. PCI Media produced radio dramas in four languages; Krio (the primary language of Sierra Leone), French, Sousou (a local Guinean language), and Liberian Simple English. The name of the drama, *Gbengbeh Soyama* in Krio and *Yete Kane* in Sousou translates to "*Putting Pepper into One's Own Eyes.*" The story is set in a forest community and follows the saga of characters who must overcome the challenges of unsustainable forest practices and unhygienic living conditions to improve their health and future livelihoods. Each 52-episode series was complemented by radio call-in shows and discussions, led by radio show hosts throughout the region. Listeners were encouraged to call in to ask questions and share their opinions.

In 2015, PCI Media led Policy Dialogues in West Africa through our STEWARD program—in partnership with US Forest Service and USAID—to create an important space for discussion and consensus among groups from different regions such as Côte d'Ivoire and Liberia. These dialogues contributed to framing positive changes in ways that facilitated new policies and practices adopted by the Mano River Union (MRU), a group of four West African countries that share similar cultures, transborder forests, and economic and security interests. Natural Resource Management is now a priority for MRU.

In addition, these dialogues also offered insights into locally relevant ways of thinking about conservation issues and solutions. These insights were integrated into the entertainment-education work PCI Media produced in the priority zones (Fig. 2), such as Otamba Kilime National Park, Tambaka Chiefdom, Sierra Leone, which is transboundary to the Medina-Oula Forest in Kindia, Guinea, and Mount Nimba Forest, Guinea which is transboundary to the East Nimba Nature Reserve, Liberia.

Ultimately, the space for discussion, dissension, and consensus building, followed by seeing the valuable perspectives of local voices direct the narrative intervention, created the conditions for ownership of new narratives for climate action by local leaders and community members essential for changes in behaviors and practices. This local ownership was invaluable as PCI Media began its next entertainment-education program in West Africa.

After the successful completion of the STEWARD, in 2016, PCI Media joined the West Africa Biodiversity and Climate Change (WA BiCC) consortium as the communication partner. The aims of this regional, multi-year initiative were to increase coastal resilience, combat wildlife trafficking, and improve sustainable management of forests in 15 countries over the course of 5 years.

The EE applications for WA BICC included a set of short first-person narrative videos and two radio dramas, *Watasay Ston* (Unmovable Rock), which focused on coastal resilience, and *Forest Blessings,* which focused on forest management.

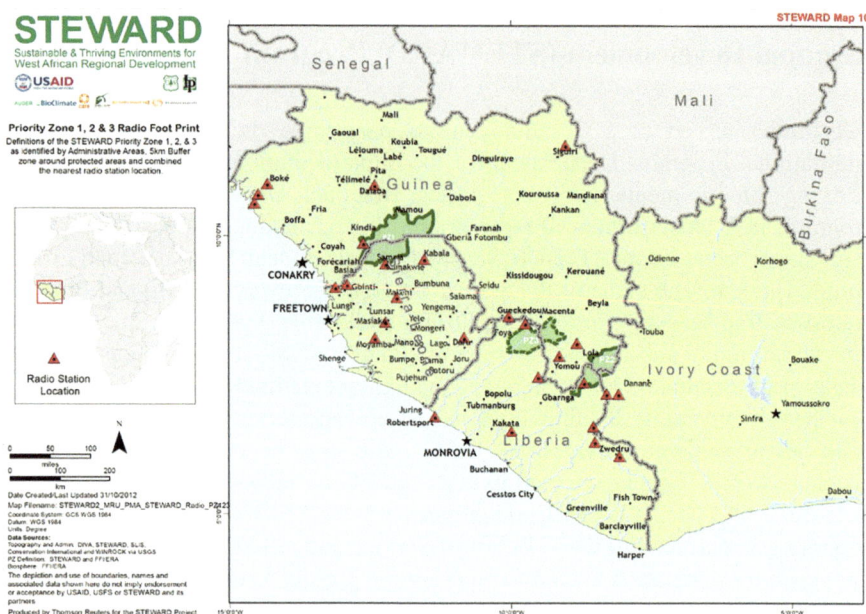

Fig. 2 STEWARD program map showing radio station location for all priority zones

Watasay Ston was a 24-episode radio drama broadcast on four radio stations, reaching 24 coastal communities. The drama was aired in Sierra Leone. It was produced in English and three local languages (i.e., Krio, Mende, and Temne) to expand its reach to a diverse audience in Sierra Leone. The series was created to increase the knowledge of people living in Sierra Leone's coastal communities on sustainable natural resource management practices in order to build their resilience to climate change.

The journey to get *Watasay Ston* on the air began with a radio drama design workshop in October 2018 (Fig. 3). Meeting in rural Port Loko, Sierra Leone, representatives from Sierra Leone's coastal communities, national and local government, and media entities came together to develop the blueprint for the *Watasay Ston* radio drama and call-in show series. PCI Media facilitated dialogue among technical experts, government officials, community representatives, and media professionals, leading the group to design a character set and narrative arc that spoke to the culture, concerns, and aspirations of Sierra Leone. The result was a drama that follows the seemingly hardheaded main character, Marie Dembad, a fish seller who used mangrove wood unsustainably to preserve her catch, and other community members who grapple with threats to their livelihoods as a result of dwindling mangrove forests and the impending impacts of climate change.

Seventeen of 24 intended communities listened to the radio drama. They participated in the call-in shows, where listeners were able to discuss the dramatic characters and conservation dilemmas they recognized in their own lives. These call-in

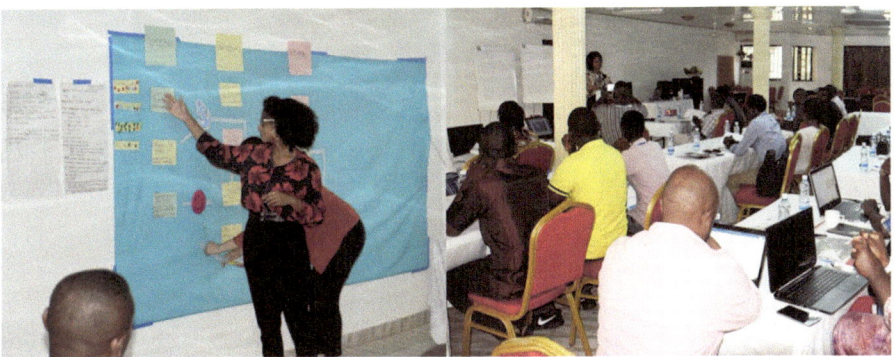

Fig. 3 *Watasay Ston* design workshop in Port Loko, Sierra Leone facilitated by Neemesha Brown

shows are critical to activating opportunities for social learning among the audience. The radio drama and call-in shows highlighted the issues and provided tangible solutions. Listeners reported that the drama influenced them to enforce by-laws surrounding conservation, do selective cutting of mangroves and plant mangroves—all keys to building coastal resilience to climate change. They also said the radio program has generated discussions around the negative effects of cutting mangroves near communities, the danger of flooding, and the construction of embankments to protect the communities from coastal erosion. A total of 287 call-ins (78% males and 22% females) were made within the first 3 months of airing on the four radio stations. These callers included fishermen, farmers, fish sellers, community leaders, and students.

In addition to the increase in knowledge among coastal community members, it was noted that the production and broadcast of the radio drama series created employment for Sierra Leoneans, including radio drama writers, actors, editors, radio station staff, and community animators. It also benefited those involved in different ways.

> Because of the program, I have become so popular. So many people call me to talk about *Watasay Ston*. The program has added value to our programming. In light of this, we volunteer to play Season 1 again. *Watasay Ston* is my baby.—Manager of AYV Radio, Michael Samuels

The radio drama, *Forest Blessings,* followed a similar development journey and aired on 9 radio stations across Liberia, Sierra Leone, Guinea, and Côte d'Ivoire.

In addition to the improvements in knowledge and adoption of new behaviors by listeners, PCI Media was also able to deepen the capacity of our radio station partners to use entertainment-education methodologies. We built strong partnerships with African Young Voices (AYV) radio and Sierra Leone Broadcasting Corporation in Sierra Leone and Liberia Broadcasting Systems (LBS) in particular because we worked with them on many projects. We also worked with many other radio stations in the four MRU Member States. Many of our broadcast partners had not used our typical format (15-min radio drama and 45-min talk show) to feature important guests and engage listeners. Broadcasters found value in adding new formats to their toolbox.

Radio dramas are a part of West Africa's rich culture; hence, it is no surprise that radio drama crafted as entertainment-education has quickly emerged as popular in the region. But the deeper successes we experienced can be attributed to the intentional and authentic effort to engage a range of local stakeholders (government, donors, communities, drama producers, and radio stations) and to facilitate the development process so that local actors took ownership of this series, advocate for its continuation, continue to learn from the process to improve the messages and increase listenership. This is the true value of the *My Community* methodology—it produces a powerful narrative intervention that creates short-term change, while seeding and surfacing the necessary sense of local value and agency that is needed to sustain change once a program ends.

References

Bixler, P. R. (2013). The political ecology of local environmental narratives: Power, knowledge, and mountain caribou conservation. *Journal of Political Ecology, 20*(1), 273–285.
Wang, H., & Singhal, A. (2021). Mind the gap! Confronting the challenges of translational communication research in entertainment-education. In L. B. Frank & P. Falzone (Eds.), *Entertainment-education behind the scenes: Case studies for theory and practice* (pp. 223–242). Palgrave Macmillan.

Neemesha (Meesha) Brown has more than 20 years of experience in education and communication spaces. She has worked with a range of partners spanning government, NGO, and private sector agencies to improve efforts to educate children and adults and to produce award-winning media. Meesha's work has touched and changed the lives of millions of people. Her expertise in learning design, education, and community engagement has helped strengthen natural resource management in West Africa and raise awareness and change behaviors around climate-resilient and adaptation practices. Meesha is also an award-winning producer of #ISurvivedEbola, a global public health campaign that empowered over ten million people to take measures to protect themselves and others from contracting the deadly Ebola virus.

Positively Life-Changing Stories Today, Intergenerational Climate Benefits Tomorrow

Joseph J. Bish

"Every man must have a son!" exclaims the domineering mother of Simba, as she attempts to pressure her grown son into enlarging his family of two. "You do not have an heir—you need to have a boy very soon. You only have girls at the moment!"

"But Mother, my girls are my children. They can become my heirs."

"No, no!" says Mama Simba. "It is our tradition that every man must have a son."

"Momma—things have changed, people have evolved."

"No, culture is culture," says Mama Simba. "I am not going to argue with you anymore. I have told you what must be done."

So begins the tale of Simba, a key character in *Sotakai* (*Footsteps*), a 156-episode radio serial drama that aired in Uganda from October 2019 through January 2022. Over the course of many punctuated, melodramatic highs and lows woven throughout this story, Simba and his wife wrestle with decisions related to the uptake of modern family planning, the patriarchal urge for sons, and what their ideal family size truly is. Along the way, a diverse slate of tertiary characters tries to influence these decisions one way or another, employing unsolicited advice, emotional manipulation, or benevolent comradery—depending on whether they are cast as protagonists or antagonists to Simba and his family.

Created by Population Media Center, a global sustainability nonprofit and broadcast production partner based in the United States, *Sotakai* was designed to offer Ugandan listeners new information, alternative perspectives, and behavioral possibilities on a range of key human development issues. In addition to the family

J. J. Bish (✉)
Population Media Center, Candor, NY, USA
e-mail: joebish@populationmedia.org

E. Coren, H. Wang (eds.), *Storytelling to Accelerate Climate Solutions*,
https://doi.org/10.1007/978-3-031-54790-4_4

planning storyline, this radio drama also offered listeners facts and resources about nutrition, youth sexual reproductive health, and gender-based violence.

In terms of the climate connection central to this book, the family planning storyline is fundamental. The universal uptake of family planning along with improving girls' education around the world are key components of humanity's long-term efforts to achieve a sustainable living scenario with our planet and its climate. As the meticulous research of Project Drawdown has shown, the combination of improving girls' education and increased global use of family planning constitutes the greatest single emission reductions strategy currently available to humanity: with the possibility of avoiding 85.4 gigatons of carbon dioxide between 2020 and 2050 (Project Drawdown, 2022).

Antecedent to these quantifiable climate benefits, however, family planning and girls' education are stand-alone "must-haves" for realizing the full human rights of women and girls around the world. Family planning, one of the greatest public health achievements in human history, allows individuals and couples to anticipate and attain their desired number of children and the spacing and timing of their births. It protects the health of the mother and the outcome of each pregnancy. Family planning also allows parents the time and space to determine if they are able to care for more children, increasing the likelihood that any further children they have will be able to access health care, education, and other services.

In this light, the Population Media Center's work around family planning:

- Is in line with the Sustainable Development Goals, especially SDG targets 3.7 and 5.6.
- Advances The Beijing Declaration and Platform for Action, specifically its call to meet the unmet needs in good quality family planning services and in contraception, and to increase knowledge and use of family planning and contraceptive methods—as well as increasing awareness among men of their responsibility in family planning and contraceptive methods and their use.
- Advances the Family Planning 2030's vision for change, including "Voluntary modern contraceptive use by everyone who wants it, achieved through individuals' informed choice and agency."
- Is in line with the International Union for the Conservation of Nature (IUCN) motion titled "Importance for the conservation of nature of removing barriers to rights-based voluntary family planning."

About Population Media Center

The Population Media Center is a mission-driven, 501 (c)–(3) charitable organization with headquarters in the United States. Creating life-changing, popular entertainment for a more sustainable world, the Population Media Center's stories are rooted in a methodology originally designed by Miguel Sabido in Mexico during the 1970s. Sabido was a Vice President at Televisa, a powerhouse Mexican TV corporation, and one of the world's largest producers of Spanish-language TV content.

Sabido oversaw the creation of a series of telenovelas, or serial dramas, which were widely recognized as increasing social acceptance of family planning in Mexico (Singhal et al., 2013). Sabido's methods were informed by Stanford psychologist Alfred Bandura, a globally recognized expert on the power of human role models in influencing individual behavior (Bandura, 2004).

William "Bill" Ryerson, the Population Media Center's founder and President, had become aware of Sabido's methods while working at the Population Institute in the 1970s. His colleague at the Population Institute, David Poindexter, was the Director of the Communication Center. Together, they promoted the use of radio and television entertainment programs informed by Sabido's methods. In 1985, Poindexter founded Population Communications International in New York, and Ryerson soon joined the organization as executive vice president until 1998 when he established Population Media Center, with Poindexter serving as honorary chair of the board. Since then, Population Media Center's radio, television, and Internet stories have aired in more than 50 countries around the world and are estimated to have reached more than 500 million people.

Global Fertility, Global Population, and Future Propensity for Greenhouse Gas Emissions

But how could family planning and girls' education—uncontroversial, widely recognized social goods—also work so powerfully to avoid such a vast amount of future emissions? And how can a melodramatic, episodic radio show impact listeners' behaviors? In this chapter, readers will learn how unrestricted access to family planning and universal girls' education is instrumental in realizing the full human rights of women and girls around the world. Additionally, by destigmatizing voluntary family planning, correcting misinformation about the safety and efficacy of modern contraception, and strengthening social acceptance and support of girls' education, the long arc of human population size is influenced toward a smaller, more sustainably scaled civilization in the future. This relatively smaller human presence on the planet scales down the basic human need for and propensity to emit greenhouse gas emissions in the future.

Family planning and girls' education have well documented and widely understood influences on fertility or the average number of births per woman per lifetime. In terms of storytelling for the climate, the Population Media Center seeks to help global fertility or the average number of births per woman, per lifetime, decreases by at least 14% from its current level of 2.4 down to 2.1 or lower. A global fertility rate of 2.1 would, eventually, cause population growth to stop. Today's fertility always affects tomorrow's population dynamics.

Expert United Nations demographers currently project that the global population will increase from around eight billion today to 9.7 billion people by 2050—a 20% increase. By 2100, the projections calculate an increase of 10.8 billion—a 35% increase (United Nations, n.d.). These projections are built on expert assumptions about childbearing and human longevity trends. The most influential variable in these models is the total fertility rate. The good news is that by making immediate and transformative improvements in people's lives today, history shows us that we can optimize the chances of population growth ending sooner rather than later. Indeed, Project Drawdown (2022) notes that "Slowing the momentum of human population growth in a way that upholds human rights is an important factor in slowing carbon dioxide-equivalent emissions."

To be philosophically clear, the Population Media Center sees the greenhouse gas emissions emergency as nested within a larger and even more substantial problem: humanity's overexploitation of the Earth. Also known as ecological overshoot, this is a quantifiable and measurable situation in which the Earth cannot regenerate the resources used by the world's population each year or sufficiently absorb our wastes—including greenhouse gas emissions. The unfortunate reality is that humanity is well into ecological overshoot and has been for decades (Global Footprint Network, 2022). We can see the symptoms of this overshoot everywhere we look, from climate change to species extinctions and ocean acidification to poorly performing environmental health indicators. It all ties back to this collective overexploitation of Earth by humans.

The enormous size of humanity and its ongoing net growth—adding one million more people every 5 days in net growth (births minus deaths)—is one of many factors contributing to planetary overexploitation and is a serious obstacle to global sustainability. Population size and growth are not the only systemic strains and stresses on the Earth's ability to provide for us while retaining its health and vibrancy, but they are strongly influential.

That is why they deserve the thoughtful attention of the most forward-looking people on the planet. At the Population Media Center, it is central to our mission. We aspire to be progressive, reverent toward nature and people, and creative. We do not try to "limit," "control," or otherwise manage the population. We operate on a completely different level: We weaken the many identifiable regressive forces that are known to keep fertility elevated. To make progress toward our goals, we focus primarily on destigmatizing voluntary family planning, correcting misinformation about the safety and efficacy of modern contraception, dismantling patriarchal opposition to contraception, and educating parents about the many health (ScienceDaily, 2016) and economic benefits of smaller families.

A Creative Solution for a Sustainable Planet: Population Media Center's Advanced Storytelling for Climate-Related Outcomes

Humans have always told stories as a way to understand, share, and shift beliefs and actions. Since 1998, the Population Media Center's award-winning, locally produced radio and video series—featuring relatable characters, familiar communities, and very real choices—have inspired planet-positive behavior change across 30+ languages, 50+ countries, and 500+ million lives. Combining behavior change theories, media industry insight, and character-driven, culturally relevant storylines, our transformative approach empowers by entertaining and inspiring.

The story of Simba was created, as all Population Media Center's stories are, in partnership with in-country creative teams. It is critical, both operationally and philosophically, to recruit top local talent to manage the overall project, carry out the production of the dramas, and serve as scriptwriters and actors. We believe that the resultant local ownership of projects is vital to creating effective programming and contributing to lasting change.

While our dramas honor and advance the hallmarks of great episodic storytelling, such as captivating characters, cliffhangers, multiple/interwoven storylines, and unexpected plot twists, we go beyond mere entertainment and leverage key components of the fiction to promote and catalyze behavior change in audiences. Our expertise is in entertainment-education (EE), a sub-discipline of social and behavior change communication (Wang & Singhal, 2021a).

As EE practitioners, we create socially relevant dramas whose fictional settings accurately reflect the existing world of the audience. We then develop three types of characters in the story for each issue being addressed: positive, negative, and transitional, all from the audience's point of view.

1. **Positive characters** are highly aspirational, to the point of being "larger than life." Their behaviors, persona, statements, decisions, and general presence in the story are portrayed as almost ideal on the issue(s) we are addressing.
2. **Negative characters** are recognizable but untoward; though perhaps charming, gregarious, or attention-grabbing. Their behavior is ultimately revealed as dysfunctional relative to the issues we are addressing.
3. **Transitional characters** most closely represent the average people in the society. Like positive and negative character types, they speak and dress in familiar ways, live in familiar houses, and eat familiar foods. But they are uniquely designed to resonate strongly with the audience to such an extent that the audience forms emotional bonds with them, often to the point of love and affection, which in turn allows these fictional characters to be presented as the primary behavioral role models for the audience.

The positive and negative characters represent the extremes of behavior on one or more particular issues as practiced in the audience's own community, while also serving as positive and negative role models for the transitional characters. The methodology portrays these transitional characters moving through a sequence of experiences:

1. They encounter a situation that requires or forces a decision on the part of the transitional character (e.g., "Life cannot continue as before").
2. Both the positive and negative characters—purposefully designed to exhibit polarized and opposing values—make attempts to influence the transitional character's decision such that it coincides with their own outlook and worldview.

3. The transitional character makes a decision, wholly and completely of their own accord.
4. The transitional character experiences a reward or punishment based on a good decision (aligned with the positive character and values being promoted) or a bad decision (aligned with the negative character and the values being discouraged).

This dramatic mini-cycle is presented to the audiences repeatedly as the larger drama unfolds—yet with increasingly "higher stakes" for the transitional character (i.e., more dramatic situations and decisions that have increasingly profound and consequential impacts on the fate of the transitional character).

Eventually, the dramatic tension built by these decisions is resolved with a climactic occurrence in the story that results in the transitional character being changed. Importantly, having developed resiliency and self-efficacy through repeated decision-making and experiencing the consequences, the transitional character is able to maintain the new behavior—even in the face of temptation to revert—and becomes an advocate for the newly adopted behavior.

Meanwhile, the negative character is ultimately punished, specifically due to their negative actions and attempted influence on the transitional character, cueing the audience that the negative character's values and behaviors on the issue are indeed negative.

The positive character is ultimately rewarded for their positive actions and influence on the transitional character, cueing to the audience that the positive character's values and behaviors on the issue are positive.

As audience members listen or watch the transitional character experience this sequence of events, spread over successive episodes, their emotional ties and identification with the transitional character sparks an emotive, psychological *desire* to adopt similar values and actions in their real life. Simultaneously, as they follow the transitional character successfully navigating the challenge of repeated decision-making and the resulting consequences, the audience gains a vicarious experience of self-efficacy in implementing change.

Our dramas always employ multiple interweaving plot lines, a characteristic that multiplies the power of our core behavior change convention in several ways. First, it allows for a staggered presentation of experiential sequences for multiple transitional characters in the drama: meaning that at any given time, at least one plot line portrays dramatic changes of fortune to a major character. This complexity of plotting plays into audience retention and loyalty, as the audience must remain attentive to each episode of the program to determine what the resolution of the current "cliffhanger." Likewise, in the interim between episodes, cliffhangers encourage the audience to self-reflection about what might happen in the story and what they hope will happen. This stirs interpersonal communication among the audience and other community members: talking about the issues, as a matter of course, as they discuss the story. Finally, multiple storylines also enhance the pure entertainment value of the serial, helping to deliver large, dedicated audiences and distinguishing our product from many other strategies.

Beyond the individual episodes of the drama, the Population Media Center also regularly employs epilogues, designed to reiterate, and enhances the audience's exposure to the story's content. Unlike a public service announcement, epilogues are generally short, nonfiction spots broadcast at the end of a show episode. Generally, around 30 s in duration, epilogues convey helpful real-world information to the audience, such as the locations of family planning clinics.

Progression of Simba and the Impact of Stories Alike

As the *Sotakai* radio drama unfolds, Simba continues to wrestle with the influences of his mother, a purposefully designed negative character. In addition to her worries about Simba having a male heir, Mama Simba also exaggerates the potential negative health side effects of contraception. She says uptake of family planning methods will breed promiscuity and that Simba's wife will give birth to deformed children if she decides to have more.

While these concerns may seem oversimplified or even tawdry, they reflect many myths, misconceptions, and social norms prevalent in Uganda and in large swaths of the developing world. For example, a study conducted in three African countries found that 62% of female respondents in Kenya thought contraceptives could result in deformed babies. Over 55% of women surveyed in that country suggested contraceptives could cause cancer (Gueye et al., 2015). A study in Pakistan found that women are often dissuaded from using contraception because of the belief that God's will determine fertility or that family-planning decisions should be made solely by husbands (Agha, 2010).

At the Population Media Center, we consider these and similar informational obstacles as regressive—forces that affect an unfair informational and social ecosystem that works against men, women, and families as they deliberate whether family planning is a choice that is right for them. When looking at global maps of contraceptive prevalence and fertility, it is easy to see where these unfair and unjust systems are most at work: sub-Saharan Africa and parts of Asia. In many countries in these geographies, contraceptive prevalence is below 30%, while the global average is around 57% and many high-income countries are above 70% (United Nations, n.d.). Meanwhile, fertility rates in low-income countries are often 80–100% or more above the global average of 2.4.

Simba, and the hundreds of thousands of Ugandan listeners to *Sotakai*, also get the perspectives of several positive characters as the story unfolds. He hears from friends who say parents should not pressure their children for more grandchildren but should support their children in making informed personal decisions. He is told by another sympathetic character that family planning side effects are rare, understood, and manageable.

After several tests of his mettle and other travails, including a brief discontinuation of family planning (and a resultant and ill-fated pregnancy that threatens the life of Simba's wife), the couple opts to keep their family as is and no longer worry about having a son. Once this decision is made, good things happen for Simba and he and his family live the proverbial "happily ever after" life.

While the end line program evaluation of *Sotakai* (*Footsteps*) is still in progress at the time of this writing, we are confident in our chances of positively influenced listeners in Uganda based on our previous projects in the region on family planning.

For example, in Rwanda, which borders Uganda to the south, we produced *Umurage Urukwiye* (*Rwanda's Brighter Future*), a 312-episode radio show, which aired from October 2012 through October 2014. End line research found that the likelihood of respondents saying they "currently use something to delay or avoid pregnancy" was 1.6 times greater compared to baseline. Moreover:

- Listeners were 1.6 times more likely than nonlisteners to say they talked to their spouse or partner "once or twice" or "more often" about family planning in the last 3 months.
- Listeners were 1.5 times more likely than nonlisteners to want three or fewer children.
- Listeners were 2.1 times more likely than nonlisteners to know of a place to get a female condom.

The Population Media Center also produced *Pambazuko* (*New Dawn*) in the Democratic Republic of the Congo. This 156-episode show originally aired from February 2016 to August 2017 and was rebroadcast from 2018 to 2019.

- Listeners were 2.4 times more likely than nonlisteners to say "Yes" when asked "In general, do you approve of family planning?"
- Listeners were 1.8 times more likely than nonlisteners to say that their ideal family size is three children or fewer.
- Listeners were 2 times more likely than nonlisteners to say "Yes" when asked "Do you think you will use a contraceptive method to delay or avoid pregnancy in the next six months?"
- Listeners were 3.2 times more likely than nonlisteners to say that girls should be encouraged to pursue their education to a high level.
- Listeners were 2.1 times more likely than nonlisteners to say they strongly agree with the statement "Investing in a girl's education benefits the entire family."
- Listeners were 1.9 times more likely than nonlisteners to say that they strongly agree with the statement "Girls should have equal opportunity in education as boys."

Improved Lives Today, Intergenerational Results Tomorrow

Although there is a lack of consensus over the precise weight and significance of the various interactions among educational attainment, contraceptive use, and fertility outcomes, it is clear that these complementary interventions are important [to future population dynamics] and emissions reductions.—Project Drawdown

There is urgent work to be done around the many failings of human attitudes, knowledge, and behaviors related to protecting Earth's resources—their use and misuse, distribution, and mal-distribution. Today, far too many outcomes of human enterprise related to resource extraction and use result in unjust, lopsided outcomes among human beings. Just as many are ecologically uninformed, careless, or simply contemptuous. Decades of profligate greenhouse gas emissions by rich nations, even as the impending and likely dire impacts were understood all along (McGreal, 2021), attest to an unspeakably selfish engagement with the natural world.

A person in the United States consumes 5 times as many resources as a person in Uganda (Global Footprint Network, 2022). CO_2 emissions per capita in Uganda are equivalent to 0.13 tons per person—compared to 15.5 tons per person in the United States. If the whole world adopted the material lifestyle of an average American citizen, the planet could only sustainably support 1.5 billion people. Yet, our global population is going to hit eight billion sometime early in the year 2023.

In this light, it is critical to understand that long-term, intergenerational work to slow down and stop population growth—at the global scale—cannot and should not be expected to pay immediate and substantial demographic dividends. We may improve individual lives immediately, but demography does not turn on a dime. Working on population, including the Population Media Center's battle with unfair and unjust information ecosystems around family planning, is certainly not a priority action for near-term mitigation of the greenhouse gas emissions emergency. The emissions crisis must be forcefully addressed and largely solved far before the 30 or 40 years of hard work it will likely take to see population growth come to an end. Moreover, it must be remembered that at least 2.5 billion of the nearly eight billion people on Earth desperately need more material resources to escape poverty and achieve minimally acceptable living standards (IFC, 2009). Until relatively clean, renewable energy is deployed at scale, it is almost certain that poverty reduction will require an increase in local greenhouse gas emissions (Wilkinson, 2017). Yet, we are already in an ecological overshoot and a full-blown climate emergency.

Hence, circling back to the beginning, we see population is one component factoring into humanity's overall overexploitation of the planet. We have reached millions of people with liberating information, alternative perspectives, and newfound confidence to enact meaningful change in their own lives around family planning. Indeed, the history of EE provides reasons for great hope. For example, the telenovela programs produced by Miguel Sabido in Mexico between 1977 and 1986 that spoke about family planning are widely credited with helping the national population growth rate decline by 34% during that time (Wang & Singhal, 2021b). And, as the UN Environment Programme pointed out in their 2019 report (Global Assessment Report on Biodiversity and Ecosystem Services of the Intergovernmental Science-Policy Platform on Biodiversity and Ecosystem Services), slowing expansion of human habitation and human farming is the most critical need for stopping the loss of biodiversity ("the web if life") that makes the planet habitable. Further, deforestation is also a major climate change factor.

Looking forward — beyond having helped so many people achieve near-term victories in their personal lives — we have undoubtedly, in our patient and persevering way, bent the long arc of human population size toward a smaller, more ecologically right-sized number of human beings in the future. Ours is long-term work to help give the people and creatures of the late twenty-first century and early twenty-second century the best possible chance to overcome the numerous ecological, climate, and social challenges they will undoubtedly face.

> *Honoring the dignity of women and children through family planning is not about governments forcing the birth rate down (or up, through natalist policies). Nor is it about those in rich countries, where emissions are highest, telling people elsewhere to stop having children. When family planning focuses on health-care provision and meeting women's expressed needs, empowerment, equality, and well-being are the result; the benefits to the planet are side effects.*
>
> —Project Drawdown

References

Agha, S. (2010). Intentions to use contraceptives in Pakistan: Implications for behavior change campaigns. *BMC Public Health, 10*, 450. https://doi.org/10.1186/1471-2458-10-450

Bandura, A. (2004). Social cognitive theory for personal and social change by enabling media. In A. Singhal, M. J. Cody, E. M. Rogers, & M. Sabido (Eds.), *Entertainment-education and social change: History, research, and practice* (pp. 97–118). Lawrence Erlbaum.

Global Footprint Network. (2022). *Ecological footprint.* https://www.footprintnetwork.org/our-work/ecological-footprint/

Gueye, A., Speizer, I. S., Corroon, M., & Okigbo, C. C. (2015, December). Belief in family planning myths at the individual and community levels and modern contraceptive use in Urban Africa. *International Perspectives on Sexual and Reproductive Health, 41*(4), 191–199. https://doi.org/10.1363/4119115

IFC. (2009). *This is the story of 2.5 billion people living in poverty.* https://www.ifc.org/wps/wcm/connect/ff22faeb-2384-4510-9b7d-ffb7603cf093/AR2009_Introduction.pdf?MOD=AJPERES&CVID=iYSJPMU

McGreal, C. (2021, June 30). Big oil and gas kept a dirty secret for decades. Now they may pay the price. *The Guardian.* https://www.theguardian.com/environment/2021/jun/30/climate-crimes-oil-and-gas-environment

Project Drawdown. (2022). *Solutions: Health and education.* https://drawdown.org/solutions/health-and-education.

ScienceDaily. (2016, January 27). *Life expectancy three years longer for children born into smaller families in developing world.* https://www.sciencedaily.com/releases/2016/01/160127054417.htm

Singhal, A., Wang, H., & Rogers, E. M. (2013). The rising tide of entertainment–education in communication campaigns. In *Public communication campaigns* (pp. 320–333). Sage. https://doi.org/10.4135/9781544308449.n22

United Nations. (n.d.). *World population prospects—Population division.* United Nations. Retrieved January 19, 2022, from https://population.un.org/wpp/Graphs/Probabilistic/POP/TOT/900

Wang, H., & Singhal, A. (2021a). Mind the gap! confronting the challenges of translational communication research in entertainment-education. In L. B. Frank & P. Falzone (Eds.), *Entertainment-education behind the scenes: Cases studies for theory and practice* (pp. 223–242). Palgrave Macmillan.

Wang, H., & Singhal, A. (2021b). Theorizing entertainment-education: A complementary perspective to the development of entertainment theory. In P. Vorderer & C. Klimmt (Eds.), *The Oxford handbook of entertainment theory* (pp. 819–838). Oxford University Press.
Wilkinson. (2017). Fighting poverty might make it harder to fight climate change. *Science*. https://www.science.org/content/article/fighting-poverty-might-make-it-harder-fight-climate-change

Joseph J. Bish has served as an expert on population, family planning, and other global sustainability issues related to Population Media Center's mission for over a decade. From an early age, Joseph gravitated toward an eco-centric ethic. Growing up in rural central New York, he spent much time playing in the woods, far removed from the bustle of cities and towns. These formative experiences impressed on him both the awesome beauty of nature and the intrinsic value of Earth's life system. He achieved a Master's degree of Environmental Advocacy and Organizing from Antioch New England University in 2005. Soon thereafter, he met Bill Ryerson, who employed him at the Population Media Center. From day 1, his remit has been to ensure the organization's communication initiatives included effective advocacy on the importance of deploying progressive solutions to the monumental challenges to global sustainability that are represented by human population size and growth.

Kembali Ke Hutan (Return to the Forest): Using Storytelling for Youth Engagement and Climate Action in Indonesia

Ankur Garg, Anna Godfrey, and Rosiana Eko

Set in a peri-urban area, the TV drama show starts with a loud shriek waking Bodo, our lead character, from a slumber to find flooding in his room and the local area. The slumber is representative of a general apathy toward the environment and sustainable development at multiple levels. Its vivid realization was evident in a Cambridge University and YouGov study (2020), which showed that Indonesia is home to the largest percentage of climate change deniers in the world, who do not believe that human activities could cause natural disasters. The study further showed that over eight percent of Indonesian respondents assumed that human-driven global warming is a swindle and an element of conspiracy theory.

BBC Media Action's *Kembali Ke Hutan* (Return to the Forest) project aims to engage Indonesian millennials on the sustainable development challenges the country faces, help them make informed choices, and create platforms to have their voices heard. To do so, we have co-produced an award-winning TV drama *#CeritaKita* (Our Story) (Bandung Film Festival, 2021), with a companion social media discussion series Ngobrolin #CeritaKita (Chatter—Our Story), created a social media brand *AksiKita Indonesia* (Our Action), and partnered with media and civil society organizations for community engagement and capacity building.

With over 270 million people, Indonesia is one of the most populous countries in the world (Worldometers, n.d.). Young people in urban areas make up a significant proportion of this and by 2030, they are projected to make up 46% of the total population (Badan Pusat Statistik, 2013). As such, engaging meaningfully with young people on some of the most critical issues facing the country—including those related to sustainable development and green growth—is vital, as they will be at the forefront of addressing these and leading on innovative solutions in the years to come. However, this is not an easy feat. Complex reports on deforestation trends,

A. Garg (✉) · A. Godfrey · R. Eko
BBC Media Action, London, UK
e-mail: Ankur.Garg@in.bbcmediaaction.org; Anna.Godfrey@bbc.co.uk; Rosiana.Eko@id.bbcmediaaction.org

© The Author(s) 2024
E. Coren, H. Wang (eds.), *Storytelling to Accelerate Climate Solutions*,
https://doi.org/10.1007/978-3-031-54790-4_5

intermittent, event-driven media coverage, and presenting information in a way that isn't accessible or relevant are unlikely to capture youth interest (Lenhardt, 2020). This, in turn, is a considerable barrier to driving engagement and inspiring action. This is where the role of creative media and communication becomes important in providing young Indonesians with engaging stories, role models, and safe environments for unbiased debate and discussion, which can increase knowledge, build people's confidence and motivation to act, and support informed decision-making.

BBC Media Action (n.d.), the BBC's international charity, believes in the power of media and communication for good. Since 1999, and drawing on the BBC's much longer history of using storytelling to hold a mirror up to society, we have created dramas to increase understanding, challenge attitudes, and help people to take action. Our and others' evidence demonstrates that storytelling can be an important and incredibly powerful force for positive social impact. Stories not only inform, educate, and entertain but also encourage empathy and promote social cohesion (Gowland, 2021), challenge harmful gender norms (Shannon, 2018), start conversations on invisible and hard-to-address topics like fecal sludge management (Newton-Lewis et al., 2021) and inspire new intentions (Godfrey, 2017).

Using BBC Media Action's *Kembali Ke Hutan* (Return to the Forest) project in Indonesia as a case study, this chapter examines how trusted media and creative communication, informed by theory, practice, and rigorous audience research, can be used to engage urban youth (aged 18–30) and:

- Provide stimuli to influence knowledge, attitudes, discussion, and participation on climate, environment, and green growth issues.
- Offer platforms for urban youth to participate in the existing governance systems on sustainable development and green growth.
- Enhance media's ability to act as an accountability mechanism between young people and their decision-makers.

In doing so, this chapter contributes to the evidence and learning on the transformative potential of civic participation and governance interventions, using narrative-based media and public engagement strategies to address climate-related and sustainable development issues in Indonesia. We begin this chapter with an overview of the current context in Indonesia, then summarize BBC Media Action's research approach and findings to inform its communication strategy, describe the project's media and communication outputs, and conclude with the planned evaluation for 2022.

Climate Change and Indonesia

Indonesia has some of the world's highest proportion of forests, natural resources, and fauna biodiversity (Mongabay, 2011). However, it is also one of the largest producers of greenhouse gases (GHGs)—primarily coming from energy, transport, and deforestation (Dunne, 2019). Whilst the country has already shown reasonable progress in reducing deforestation at the national level in recent years, with a declining rate of forest loss between 2015 and 2018, there are key provinces (which contain most of Indonesia's primary forests) that show an increase in forest loss—such as East Kalimantan (43% loss), Maluku (40%), and West Papua (36%; Wijaya et al., 2019). The situation in Indonesia is further exacerbated by the fact that, due to its geography and existing vulnerabilities, the country is also particularly prone to devastating climate impacts, such as floods and droughts. Forest fires have become particularly common in recent years and have created a significant impact on Indonesia's emissions profile across the years, as well as piqued media interest in the issue. There are also increasing instances of erratic rainfall patterns, and urban flooding in recent years (Chamorro et al., 2017).

The Government of Indonesia has made notable commitments to address these issues, including introducing a greenhouse gas emission reduction target of 29% (business as usual) to 41% (with international assistance) by 2030 as part of its Nationally Determined Contribution (NDC) as per UN Framework Convention on Climate Change (2021). At the UN Climate Change Conference (COP26) in 2021, Indonesia further reiterated its commitment to foster low carbon growth and neutralize carbon emissions from deforestation over the next few years. Nevertheless, its ability to successfully fill these commitments depends on large-scale public engagement and enhanced governance and accountability for the sustainable use of its natural resources.

Urban youth, in particular, can play a critical role in accelerating climate action. First, they make up a significant proportion of the Indonesian population—UNFPA (2015) predicts that 46% will be under the age of 30 by 2030 (the current median age is 28.5 years) and 68% will be in urban areas. Second, Indonesian youth have played an important role in political and governance issues over time—from the change to a republican state to the demand for accountability around key livelihood or political decisions (Adioetomo, 2015).

Around the world, young people are leading engagement on climate action. However, climate change doesn't currently fit within the list of priorities for the majority of Indonesian youth (Devai et al., 2019). Communities do not engage with or act on issues that they perceive as irrelevant.

Formative Research on Indonesian Urban Youth and Climate Engagement

BBC Media Action conducted formative audience research to understand urban youth, their online behaviors, and what motivates them to engage with environmental issues. This involved 12 focus groups with young people in two locations—Jakarta and Medan—followed later by in-depth key informant interviews with active members of community youth organizations and a qualitative audience segmentation informed by artificial intelligence (AI) to develop deeper insights (Fig. 1).

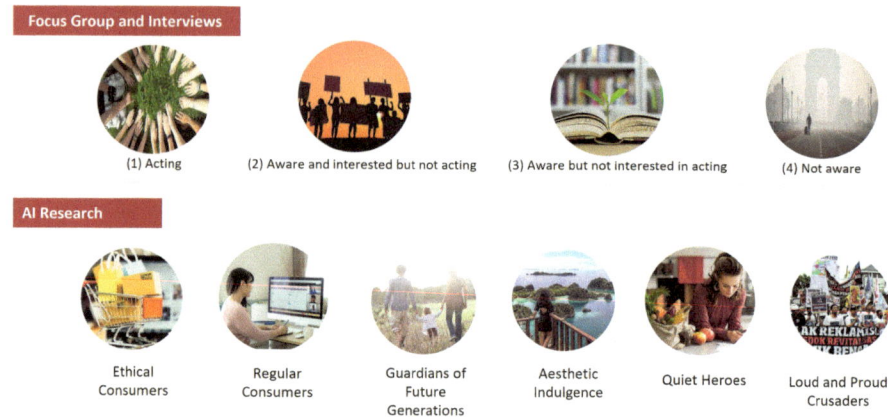

Fig. 1 Audience segments based on the formative research and AI study

This formative research found that Indonesia's young population feel disconnected from natural resources and do not know how they can get involved in accelerating the transition to low-carbon and climate-resilient development. Therefore, issues related to climate change and sustainable development need to be communicated to young people in a way that resonates with them and involves them, and through multiple formats and platforms that a generation of Indonesian youth can meaningfully engage with.

Insights from Focus Groups and Key Informant Interviews

Indonesia's urban youth is a very large, heterogeneous group with varying levels of engagement with issues related to climate, forests, and sustainable development. Our formative research identified four audience segments: (1) *acting*, (2) *aware and interested but not acting*, (3) *aware but not interested in acting*, and (4) *not aware*. Most young, urban Indonesians fall under the categories of (2) aware and interested but not acting and (3) aware but not interested in acting. Those who are aware and interested but not acting were largely found to be 18–21-year-olds who perceive themselves as lacking resources, access to channels and platforms, and confidence to participate in governance and accountability processes. Those who were aware but not interested in acting were mostly 26–30-year-olds and of higher socioeconomic groups in Jakarta who reported being more concerned about issues such as employment and their families. Across both groups, there's also the fear that acting on environmental issues is seen as being and/or doing "too much."

> "Let's say if we have to combat forest fire, but it related to powerful company, such as palm oil. Then, who are we? We are powerless." – Jakarta, 18 – 21 years old, Focus Group

> "If you want to be an agent of change, it starts with you and what you're made of." – Jakarta, 18–21 years old, Key Informant Interview

Insights from Artificial Intelligence-Driven Audience Segmentation

The formative research revealed that young urban Indonesians are digital natives. Digital media is the most easily accessible source of information and entertainment. This general audience segmentation informed a further AI-driven digital audience segmentation based on drivers, interests, and aspirations for their engagement with climate and environmental issues (Eko, 2020). We used 118 sources of information, across 16 sourcing areas such as local community fora, social media, sustainable living blogs and fora, lifestyle bloggers and influencers, to create audience segments. Through machine learning, the rich and diverse sources were explored and clustered to create patterns and connections between the publicly available data. In contrast, each unique pattern was clubbed into a segment. These segments ranged from those who would take climate action to protect the world for their children, to others who would take climate action to protect the biodiversity and beauty of the country.

The first two segments, Ethical Consumers, and Regular Consumers were the largest. While both groups were aware of environmental issues, it was not their main priority. Personal contribution becomes one of the primary focuses of these two segments. They tended to hesitate to become more actively involved, assumed environmental issues were not interesting, and had higher priorities in life such as study, work, or leisure. The segmentation informed the communication strategy and character development.

Six audience segments were found:

1. *Ethical Consumers*, their environmental focus stems from lifestyle-related initiatives especially those that can affect their purchase habits. They have the desire to be seen as socially conscious that can make them vulnerable when they get low acknowledgment from others.
2. *Regular Consumers*, most of them have simple and rigid routines (i.e., school and work), and there is a tendency that they do not want to take risks—they want everything to be simple, hassle free, and less confrontative; entertainment is the main topic for them to share on social media.
3. *Guardian of Future Generations*, their biggest aspiration is to be helpful to many people, especially to their kids or students. Their actions tend to be community based with simple initiatives.
4. *Aesthetic Indulgence*, their passion for traveling introduces them to other interests. They try to focus on doing what they can to protect the environment and to live a healthy life, with hopes of inspiring others to do the same.
5. *Quiet Heroes*, they try to focus on doing what they can to protect the environment and to live a healthy life, with hopes of inspiring others to do the same. They start doing small things for environmental sustainability.
6. *Loud and Proud Crusaders*, they use their activism in the environment to define their personal identity. They are often being labeled as too much or extreme in doing their environmental campaign.

As a follow-up to the AI research, we conducted further qualitative research in two cities with digital users across these segments of urban youth. This helped us to better understand our key audience segments' online behavior and habits. We learned that for digital users, television is a "companion" to other activities and often a second screen.

"Despite my busy activities, I still care about how my choices of affect the environment." – Ethical consumer, Jakarta

"All I can do is put my comment on social media although I don't think my critics would be heard." – Regular consumer, Medan

Overall, the findings from the formative research revealed that:

- Urban youth in Indonesia are largely willing to think about and engage with political and social issues through online platforms.
- They are keen to form a social identity to define their place within Indonesian society.
- They want to influence their social networks.
- Television and social media are the best ways to engage the target audience—and content needs not to be too "serious" (entertainment is key and young people also want to "learn").
- Drama is particularly popular while Instagram is the biggest social media platform among our target audience.
- On the issue of Indonesia's forests, many urban youth are informed and desire actionable solutions, specifically those that they can do themselves.
- Despite their knowledge of Indonesia's forests, many also feel detached from the impacts of deforestation as they do not live in or near forest areas.
- Climate change and forestation don't fit within their priorities and has a "distant" issue.
- Among environmental issues, plastic and pollution are the two issues that interest the audience the most because it has a direct impact on their surroundings.
- Many also feel that they could not directly contribute and respond to curbing forest loss.
- They also lack resources, access to trusted platforms, and agency to participate in governance and accountability processes.

Thus, the major challenge with urban youth was how many perceive forests as needing more personal relevance and urgency for action despite their knowledge and desire for actionable solutions. Our baseline quantitative research validated these formative research insights in 2020: Our survey with more than 2000 respondents across Indonesia found that familiarity with environmental issues is fairly polarized, with 39% believing they are somewhat or very knowledgeable as a whole, but nearly as many admitting to lacking knowledge about the environment (Devai et al., 2020). Issues affecting forests are known to more than 7 in 10 people surveyed, with nearly one-fifth saying they know a great deal about them. Despite the apparent appeal of the idea that development and the environment can co-exist, when faced with a difficult choice, nearly half of the survey respondents would place economic progress ahead of forest preservation.

Communication Strategy

Informed by these findings, BBC Media Action developed a communication strategy where the major focus of the content was to increase the desirability of being interested in, and acting on, climate and sustainable development issues, as well as the perceived capacity to act and make a difference among young people. This would be achieved through digital and TV content which would build young people's understanding on these issues and help them to visualize their role in strengthening climate governance, link climate change to their lived experiences, and to discuss and debate issues that resonate with them.

Based on our formative research, we knew that our challenge in engaging urban youth lay in how despite their knowledge and desire for actionable solutions, they perceived forests as lacking impact on themselves as they do not physically see forests in their everyday lives. They also see themselves as unable to significantly affect forests. Therefore, our Theory of Change centered on using identity as the entry point to engage Indonesian millennials; as we know identity formation is an important part of urban youth's lives and integral to how political participation is understood and carried out. Identity, for our audience, encompasses who they are individually, compared to their peers, and within Indonesian society. Our formative research showed that urban youth conceptualize forests as the "Lungs of the World," part of Indonesia's national identity. Forests' place in Indonesia's national identity allowed for a connection between identity and forestry governance by establishing national identity as a reflection of personal and social identity. Our qualitative research showed that Indonesians care about how the world sees them. For example, they felt concerned about Indonesia's image during the 2015 forest fires when haze from the fires spread internationally. Through this avenue, we decided to prompt urban youth to reflect on the current state of Indonesia's green growth and forests and how they choose to respond to it within the context of their own identity as individuals, community members, and responsible citizens.

Considering this, we conceptualized our lead character, Bodo, as a simple, good-hearted man who is kind and innocent but not stupid. He will do anything for his friends and probably forgives those who wrong him too easily. At the start of the drama, Bodo is quite lazy—he likes sleeping and doesn't stick with any job for long. His dream is to be well off and run his own business, but he has no idea what that business would be and does not have the drive or ambition to achieve it. He has a strong attachment to his neighbors and his community and would like to see it improve. But has no real concern for the environment overall. Bodo represents one of our large audience segments, Regular Consumers. Over time, on witnessing the air pollution, rubbish, and floods in the community, Bodo begins to spark concern about these issues. His love interest, and our key protagonist, Tuji, supports him on this journey. Through a roller coaster of emotions, dissent, and discussions, he decides to contest and eventually wins the local election based on a sustainable development mandate. *#CeritaKita* is much about establishing a role model in the transition of Bodo's own civic identity as it is about strengthening governance and championing climate action. Bodo takes the audience on his transition—sparking discussion and motivation—and sows the seed of change.

On the other hand, our protagonist, Tuji is a clever, smart, and mentally strong young woman. She is a mix of modernity and tradition. On the modern side, Tuji is very interested in technology, gadgets, and social and environmental issues and strongly believes in gender equity. But at the same time, she has a moralistic approach to life, worries about her reputation, and wants to please those around her. She personifies the Ethical Consumer and Quiet Heroes segments from our research. Over the series, Tuji strengthens her ambition to become a vlogger who reviews all the latest gadgets and is fascinated by apps that could predict drought or floods. We aim to build this vlog into a transmedia element in the next season whereby Tuji achieves her dream of solving the world's problems using technology.

Some of our other segments, such as Aesthetic Indulgence and Loud and Proud Crusaders, are more prominent users of social media. Hence, we decided to reach them through complementing formats we created on AksiKita Indonesia, using the same primary insight of identity formation. For example, to reach Aesthetic Indulgence segment, and to respond to the COVID-19 travel restrictions our audience faced, we created an Instagram series called *Kangan Jalan Jalan* (Missing Travel). Through a flash format travel show (5 mins each), we covered environmental champions in various parts of the country, as a role model, while also bringing people closer to Indonesia's natural environment and forests at a time when our audience weren't able to travel and see it for themselves.

Through *Ngobrolin #CeritaKita*, we wanted to give young Indonesians the ability and knowledge to join in a national conversation—giving them the confidence and option to share their own opinions with decision-makers on these issues. The focus was on providing engaging, interactive, and participatory platforms—giving young Indonesians the opportunities to be part of the experience (shaping their civic identities) and that itself is empowering.

Our other challenge was tackling an issue as broad as green growth in a finite number of episodes, where most sub-themes such as deforestation, sustainable agriculture, and sustainable energy, are perceivably quite "distant" from the audience (Devai et al., 2020). Our baseline research identified a few positive behaviors or factors that can catalyze other positive behaviors—and used them as gateway behaviors (Schwandt et al., 2015). This approach drew on the notion that successful engagement on one theme creates motivation to engage or self-efficacy to make a change on other, more distant themes. We focused on the environmental issues that young people are most interested in such as plastic pollution, water conservation, and fast fashion. While these themes were interspersed throughout the series, they featured more heavily in the earlier episodes to strengthen the relevance of the content with our target audience, and build engagement. Table 1 shows the primary themes covered through each of the *#CeritaKita* episodes, and brief profiles of guests featured on related *Ngobrolin #CeritaKita* episodes.

Table 1 *#CeritaKita* primary themes and *Ngobrolin #CeritaKita* guests

#CeritaKita episode	*#CeritaKita* primary theme	*Ngobrolin #CeritaKita* guests
1 + 2	Pollution and waste	Mr. Kevin Julio (actor playing Bodo)
3 + 4	Recycling and food waste	Ms. Agni Pratista (social influencer and former Miss Indonesia who speaks out on and practices a zero-waste lifestyle) Mr. Aang Hudaya (community-level champion on sustainability and role model) founder Bogor Ecobrick community
5 + 6	Food biodiversity and fast fashion	Mr. Dian Sidik (actor playing one of our key characters, Salah in the show, also outspoken on nutrition and healthy diet) Ms. Vania Febriyantie (local champion and role model) founder of 1000 gardens community
7 + 8	Clean water and rain water harvesting	Ms. Cut Syifa (influencer and actor) Mr. Sugiantara (local champion and role model) youth conservation initiative Bali, educates people on water conservation in areas short of clean water Ms. Elisabeth Tarigen (government) Jakarta water resources department
9 + 10	Climate change and drought	Ms. Angela Gilsha (influencer known for her minimalist lifestyle and gardening) Mr. Andhyta Firselly Utami (expert) environmental economist Mr. Adi Ripaldi (government) head of climate analysis of meteorology climatology and geophysical agency
11 + 12	Rice and electronic waste	Mr. Marcel Chandrawinata (influencer and actor) Mr. Rafa Jafar (local champion and role model) founder EwasteRJ, nonprofit organization providing drop boxes for e-waste collection Mr. Kuntoro Boga Andri (government) head of public relations and information, Ministry of Agriculture
13 + 14	Biodiversity loss and food	Mr. Haico Van Der Veken (influencer) Ms. Sharlini Eriza Putri (local champion/role model/expert): Microbiology practitioner
15 + 16	Forest fire and extreme weather	Mr. Marshel Widianto (influencer) Ms. Sumarni Laman (local champion/role model): Youth Action coordinator with the movement to protect Kalimantan Forest
17 and 18	Water pollution, flooding and youth involvement.	Mr. Reybong (social influencer and actor) Mr. Sandiaga Uno (government) minister of tourism and creative economy Ms. Rina Suryani (government): Jakarta environmental services
19 + 20	Future cities, renewable energy and youth involvement	Mr. Haico Van Der Veken (influencer and actor) Ms. Natasha Wilona (influencer and actor) Ms. Dyah roro Esti (political): Member of the house of representatives

Multiplatform Storytelling: The Power of Cross-Platform Narrative Engagement

BBC Media Action developed a social media brand called *AksiKita Indonesia* (Our Action), a TV drama show called *#CeritaKita* (Our Story), a social media discussion program *Ngobrolin #CeritaKita* (Chatter—Our Story) broadcast on Instagram, and a range of offline capacity strengthening and outreach activities for media practitioners and civil society organizations (CSOs) to use media and communication more effectively. These platforms work in parallel through various complementary formats to develop an immersive experience for the audience. For example, at the end of each drama episode, one of the characters promotes the social media page and the hashtag *#CeritaKita*. After the broadcast of each drama episode, the live discussion program commences online. The social media content includes some "behind the scenes" discussions to extend the narrative and the outreach work with local organization extending storytelling into local discussion groups. These transmedia elements—storytelling coordinated across different media (Jenkins, 2006)— are exciting. However, COVID-19 had a significant logistical and operational impact on the production. Restrictions meant all aspects of production needed to be redesigned and adapted to the evolving pandemic. Guidelines and new working methods were introduced without local COVID-19 production guidance. As a result, the project's ambitions for "transmedia" storytelling will be developed further in the project's next phase.

Episodic Television Series #CeritaKita (Our Story)

BBC Media Action partnered with a leading national private broadcaster with Indonesia's second most popular media channel, Surya Citra TV (SCTV) (Kurniawan, 2020), to co-produce a 20-episode TV drama that brings climate change, green growth, and environment issues closer to the urban youth. Our qualitative research revealed that urban youth wanted an informative TV program with elements of humor, realism, drama, and emotional appeal. Therefore, a TV drama/ soap opera format was chosen because it can be simultaneously entertaining, tastefully dramatic, and informative. Indeed, within the Education-Entertainment and narrative engagement literature, the power of humor (Moyer-Guse et al., 2011), perceived realism (Miller-Day & Hecht, 2013), and emotional involvement are well documented. This drama is unique because it is set on a backdrop of real-life environmental and governance storylines—something that has never happened before on Indonesian TV. The problems addressed in the show are the same ones the target audience faces, and—drawing on research around the importance of character identification—the characters have many traits the audience can "identify" with (Busselle & Bilandzic, 2009; Moyer-Gusé, 2008). That's why the program was titled *#CeritaKita* (Our Story).

#CeritaKita is an entertaining drama revolving around the lives of five young friends who live in the slum area of an Indonesian city (Surya Citra Televisi (SCTV), 2021). The program focuses on their relationships and the problems they and their community face. While environmental and climate issues are central in the show, the storyline also covers other issues that matter to young people, based on our formative research, to deepen audience engagement. It follows characters' journeys as they progress from young carefree adults to become engaged and powerful voices representing their area and acting to improve the lives of all their community. In doing so, *#CeritaKita* role-models civic participation through key characters and aims to strengthen efficacy and governance for environmental issues faced in everyday life, from flooding and extreme heat to plastic pollution and fast fashion, while exploring solutions and mitigating actions our audiences could take. The storylines and characters are modeled on the intended youth audience and informed by the audience segments identified through research. The drama is informed by practice and draws on theory including social cognitive learning theory (Bandura, 1986) and Sabido's entertainment-education methodology as well as more recent research on narrative engagement, character identification, and the power of narratives compared to non-narrative formats (Green, 2021; Murphy et al., 2015). For example, early on in the show, our lead character Bodo is apathetic toward governance civic participation and climate-related issues. Over time, he "transitions" into someone who is not only interested but acts and speaks on these issues. He is a key role model and throughout the series shows our audience the changes they could also make in their lives to address climate, green growth, and sustainable development challenges in their own communities. *#CeritaKita* is a story of the joys, successes, and struggles that Indonesia faces daily (Fig. 2). It is a journey of young people together, carving a social identity for themselves, based on climate action and civic participation. It is a journey from personal to political.

#CeritaKita represents an important change, not just for young people across Indonesia, but also for locally produced social and behavior change communication (SBCC) drama in Indonesia. Our findings from quantitative baseline research (Devai et al., 2020) strengthen the previous findings that entertainment is consistently the most sought content on TV, although keeping up to date with current news features highly. However, people don't just look for TV shows that entertain. Other factors that interest people in new programming are TV shows that promote learning and deliver good values. While Indonesia accounts for one of the highest media penetrations among developing countries, mass media platforms have previously seldom been used for partnerships to create social impact using storytelling. This was recognized with a special award at the Bandung Film Festival, one of the most prestigious entertainment media awards in the country—for "social and behaviour change communication." The show was aired over 20 episodes, on weekend mornings to optimize the reach toward the target audience, based on audience viewership data. Television Audience Measurement estimates show over 2.5 million viewers watched each episode of #CeritaKita. The planned evaluation to estimate cumulative audience reach and impact is still underway at the time of writing.

Fig. 2 Television series #CeritaKita *(Our Story)*

Social Media Discussion Ngobrolin #CeritaKita *(Chatter—Our Story)*

Based on BBC Media Action's baseline research (2020), we found that social media is the most accessible source of information, while TV usually works as a "companion." People may turn on the TV while cooking, reading, studying, or sleeping. As

Fig. 3 Social media discussion *Ngobrolin #CeritaKita (Chatter—our story)*

a result, on the one hand, our TV show used catchy audio and popular actors to attract audiences. Nevertheless, we built on this "second screen phenomenon" (Mukherjee et al., 2014) to introduce a closely linked social media discussion series on topics covered in the TV drama episodes. *Ngobrolin #CeritaKita* (Chatter—Our Story) combines behind the scenes elements from the TV drama, real stories of people leading change in their communities, and interactions with decision-makers, show cast, and young influencers. This show is presented by Shafira Umm, a famous youth influencer, and Hifzdi Khoir, an Indonesian comedian, and is broadcast on Instagram directly after the drama. Every episode picks one thematic issue raised in the drama just preceding the broadcast, such as fast fashion, or food waste, provides audiences' key facts, and opens a wider conversation with the issues and stakeholders involved. For example, the episode on food waste was followed by a live discussion circled around a recent report that Indonesia is the second largest producer of food waste in the world, and how that could be better managed (Green, 2016). Audiences were given the opportunity to put their questions about the issues raised in the show to experts, real-life inspiring role models, and government representatives (a list is available in Table 1). *Ngobrolin #CeritaKita* provides an important public platform for young people to voice their thoughts, ideas, actions, and opinions on climate-related issues, also to national government ministers, and provincial governors. Filmed in a social media style, it is a varied mix of content based on the typical young Indonesians' Instagram feed to ensure it resonates and engages with the target audience (Fig. 3).

Social Media Brand AksiKita Indonesia *(Our Action)* for Climate Action Online

Our qualitative research revealed that urban youth notice and seek shareable and useful digital content relevant to their everyday lives as well as current events and trends. We found urban youth trust content with audio and/or visuals, geotags, and live footage—which calls for digital content with verifiable sources and features. Therefore, BBC Media Action created the brand *AksiKita Indonesia* (Our Action) for social media platforms such as Facebook, Instagram, YouTube, and Twitter, reaching millions across the country with compelling information and insights that inform young people about environmental and climate issues, and how to take action in their own lives to address them (Fig. 4). Digital content created for these online platforms has sought to increase knowledge, engagement, empathy, and interest of urban youth in issues, as well as a sense of shared commitment to promoting sustainable development.

Fig. 4 Social media brand *AksiKita Indonesia* (*our action*)

Some of our digital content was built directly from the TV drama show, and through the week, fostered conversations around key challenges our characters faced on screen, and enabled our already engaged community of young followers to start discussions around key topics covered in each episode. We used a variety of our popular Instagram formats such as *#InterAksi* video series, which provides examples of other people (including young people) who act on green growth issues including short interviews with various role models (social influencers, actors/actresses, and innovative government officials/ministers), and *#SuperAksi* video series which amplifies the stories of "positive deviants" (Singhal, 2010) from within the community of how people overcome their barriers to act on climate-related issues. One of the #SuperAksi features discusses the motivations behind and barriers to making a career in e-waste with Rafa Jafar, a young entrepreneur. We created a section titled Gossip or Facts, which would allow people to corroborate key misinformation related to the environment.

Role of Social Media Influencers

Our formative research showed that young people are more likely to act on advice and information when it comes from trusted and/or authentic sources. Social media influencers have played a critical role in strengthening engagement with the *AksiKita* brand, as well as through the discussion program *Ngobrolin #CeritaKita*. Based on our research, we work with social media influencers who are interested in environmental topics or have worked substantively on these issues. Kevin Julio, the Indonesian actor who plays Bodo (the lead role) in #CeritaKita, and adds to the conversations on AksiKita Indonesia, has been working on environmental protection for a long time. This has also been the case with other social media influencers featured such as Nicolas Saputra and Nadine Chandrawinata—both have credible environmental and social impact credentials. The planned evaluation will explore whether exposure to our content has led to an increased propensity among the audiences to follow celebrities or social media influencers interested in the environment and/or wider social issues.

Community Engagement and Capacity Building

In addition to the digital and TV content, BBC Media Action is also supporting CSOs working with youth on environmental issues, and media practitioners, to produce socially and behaviorally impactful communications with their audience. Storytelling approaches are at the heart of these sessions. For example, BBC Media Action is conducting media training and mentoring activities for program makers on innovative media formats and ways to design storylines and characters based on audience research, and tools and techniques to integrate the social impact into entertainment programs. In partnership with local CSOs and media organizations, BBC Media Action has also been organizing community outreach events to take *#CeritaKita* to non-TV viewing audiences. These were designed as roadshows but were implemented completely online due to COVID-19 movement restrictions. Moderated discussions were held around some clips from the show, and local triggers and barriers to climate action were shared with key decision-makers and influencers.

Summative Research for Program Evaluation

Road to Evidence

BBC Media Action is committed to understanding the impact of its work. We have a long history of using mixed-method research led by local research teams and forming research partnerships to promote evaluation independence and learning. For this project, a multipronged evaluation is underway. This includes the following components:

- A process evaluation: to document how an intervention is implemented and what was actually delivered, compared with that intended to be delivered.
- An experimental study: Working in partnership with the Colombia University, this study randomly assigns approximately 800 regular viewers of SCTV who have not previously watched any episodes of *#CeritaKita* or engaged with *Ngobrolin #CeritaKita* online to one of two groups: a control or a treatment group. Participants in the treatment group will watch a distillation of storylines from the drama series along with supporting content from the discussion program. The control group will watch a similar TV program and discussion program on a different subject. The experiment will explore the impact of the programming on young people's understanding, discussion, and engagement with climate change as well as support toward government policy on deforestation and media's future role in covering climate change. It will also look at narrative engagement and whether the programming increases people's willingness to follow these issues on social media. This work will be published in a peer-reviewed journal.
- A quantitative, nationally representative survey ($n = 5000$) of Indonesians (age 16 years or over) to assess audience reach and engagement, as well as to explore the associations between exposure and climate action outcomes (knowledge, attitudes, discussion, and participation in green growth and sustainable development).
- Qualitative research with six focus groups of mixed composition of young urban men and women aged 18–30 years old in two cities, who have been exposed to *#CeritaKita* and/or *Ngobrolin #CeritaKita* and/or *#AksiKitaIndonesia*. This qualitative research aims to understand how and why audiences engage with the content and to better understand how it can support climate action. In addition, 16 in-depth interviews with experts from civil society, academia, and the media familiar with the content will also be interviewed. These interviews aim to

identify experts' perceptions of the discussion-based outputs' diversity, accessibility, and transparency as platforms that aim to support discussion surrounding green growth and deforestation.

- Social media analytics will be used to assess the overall social media performance in terms of reach, engagement, and interaction across Facebook, Instagram, Twitter, YouTube, and TikTok.

Conclusion

"Saving our planet is now a communications challenge. We know what to do, we just need the will."—Sir David Attenborough Broadcaster, natural historian and author, on Instagram.

Lesson Learned: The Communication Challenge

Climate Action is primarily a behavioral challenge. However, crucially it is also a communication challenge. This is as true in Indonesia as it is elsewhere in the world. *Kembali Ke Hutan* is a creative response to this challenge. It uses public interest media to reach large audiences to increase understanding as well as motivation to engage in climate action. Despite high media access and consumption in Indonesia, and the need to engage Indonesians in climate solutions, the use of socially and behaviorally informed media and communication to support the climate development agenda is largely unexploited. BBC Media Action's creative partnership with SCTV, the second largest broadcaster in the country, demonstrates the potential role of media and communication to initiate innovation and change. *Kembali Ke Hutan* is an example of a too rarely funded media-based initiative that takes a people-focused, long-term, and large-scale approach as a pillar in the response to climate change and deforestation. It highlights how a well-designed and adequately resourced media intervention—delivered by outlets that audiences trust—can make complex information clear, host discussions that can spark imagination and innovation, and inspire positive action. It makes a distant and critical issue relatable, relevant, and engaging.

Next Steps: Generating Evidence and Leveraging Transmedia Storytelling

While we await the results from the evaluation, which will truly determine the success of this work, initial audience feedback, reach data, and early experimental results show promise in the approach. The evaluation will inform lessons learned and the project's next phase, including how we can continue to leverage storytelling and our social media brand *AksiKita Indonesia* (Our Action). Inspired by some of the commercial as well as development successes of transmedia storytelling such as *Dr. Who, Star Wars,* and *East Los High,* we are looking to continue to broaden the spectrum of our audience, meeting them where they are and providing them with an enriching experience, to deepen our impact. We hope this work inspires donors, policymakers, and climate experts to enhance the impact of their climate change plans and strategies by leveraging the power of media and communication.

Acknowledgments This report was written by Ankur Garg, Anna Godfrey, and Rosiane Eko. The *Kembali Ke Hutan* project was funded by the Norwegian Development Cooperation Agency (Norad). The authors thank our funders, SCTV, Screenplay Productions, and the Government of Indonesia. Thanks also to BBC Media Action's project, research, and production teams particularly Helena Rea, Elizabeth Burgess (for support on an early draft), Intan Permata Sari; Jimmi Silitonga, Benedek Paskuj, Suzanne Devai and Clemency Fraser. We would also like to thank the research participants for sharing their time over the course of the project, and importantly our audiences, who are at the heart of everything we do.

References

Adioetomo, P. S. M. (2015). *The monograph series no. 2: Youth in Indonesia.* UNFPA. https://indonesia.unfpa.org/en/publications/monograph-series-no-2-youth-indonesia

Badan Pusat Statistik. (2013). *Proyeksi Penduduk Indonesia 2010–2035.* https://www.bps.go.id/publication/2013/10/07/053d25bed2e4d62aab3346ec/proyeksi-penduduk-indonesia-2010-2035.html.

Bandura, A. (1986). *Social foundations of thought and action: A social cognitive theory.* Prentice-Hall.

BBC Media Action. (n.d.) *Meet the BBC's international charity.* https://www.bbc.co.uk/mediaaction/

Busselle, R., & Bilandzic, H. (2009). Measuring narrative engagement. *Media Psychology, 12*(4), 321–347. https://doi.org/10.1080/15213260903287259

Chammoro, A., Minnemeyer, S., & Sargent, S. (2017, February). *Exploring Indonesia's long and complicated history of forest fires.* World Resources Institute. https://www.wri.org/insights/exploring-indonesias-long-and-complicated-history-forest-fires.

Daftar Pemenang Festival Film Bandung 2021. (2021, October). *Festival film bandung.* https://www.festivalfilmbandung.com/2021/10/daftar-pemenang-festival-film-bandung-2021.html

Devai, S., & Eko, R. (2019). *Indonesia formative research.* Internal Research Report., BBC Media Action - Please contact authors for further information.

Devai, S., & Eko, R. (2020). *Quantitative baseline research.* Internal Research Report - Please contact authors for further information.

Dunne, D. (2019). *The carbon brief profile: Indonesia.* CarbonBrief Clear on Climate. https://www.carbonbrief.org/the-carbon-brief-profile-indonesia

Eko, R. (2020). *Audience segmentation through AI research.* BBC Media Action - Please contact authors for further information.

Godfrey, A. (2017). Can a health drama and discussion show affect the drivers of behavior change? *BBC Media Action.* http://downloads.bbc.co.uk/mediaaction/pdf/research/health-drama-behavior-change.pdf

Gowland, S., Colquhoun, A., Nyoi, M. Y. H., & Thawng, V. S. (2021). Using audience research to understand and refine a radio drama in Myanmar tackling social cohesion. In L. B. Frank

& P. Falzone (Eds.), *Entertainment-education behind the scenes: Case studies for theory and practice* (pp. 157–174). Palgrave Macmillan.

Green, A. (2016). Food loss and waste. *The Economist Impact.* https://impact.economist.com/perspectives/sustainability/food-sustainability-index-2016/infographic/food-loss-and-waste

Green, M. C. (2021). Transportation into narrative worlds. In L. B. Frank & P. Falzone (Eds.), *Entertainment-education behind the scenes: Case studies for theory and practice* (pp. 87–101). Palgrave Macmillan.

Jenkins, H. (2006). *Convergence culture: Where old & new media collide.* University Press.

Kurniawan, H. (2020, December). *Data Nielsen: MNC Masih Kuasai Prime Time.* Investor.id. https://investor.id/it-and-telecommunication/229852/data-nielsen-mnc-masih-kuasai-prime-time

Lenhardt, A. (2020). *Barriers to preventing deforestation and degradation of Indonesia's tropical rainforests and peatlands.* K4D Helpdesk Report. Institute of Development Studies.

Miller-Day M, Hecht ML. (2013). Narrative means to preventative ends: a narrative engagement framework for designing prevention interventions. *Health Communication, 28(7), 657–670.* https://doi.org/10.1080/10410236.2012.762861. Epub 2013 Aug 27. PMID: 23980613; PMCID: PMC3795942

Mongabay. (2011). *Indonesia forest information and data.* Mongabay.com. https://rainforests.mongabay.com/deforestation/2000/Indonesia.htm

Moyer-Gusé, E. (2008). Toward a theory of entertainment persuasion: Explaining the persuasive effects of entertainment-education messages. *Communication Theory, 18*(3), 407–425. https://doi.org/10.1111/j.1468-2885.2008.00328.x

Moyer-Gusé, E., Mahood, C., & Brookes, S. (2011). Entertainment-education in the context of humor: Effects on safer sex intentions and risk perceptions. *Health Communication, 26*(8), 765–774. https://doi.org/10.1080/10410236.2011.566832

Mukherjee, Partha Wong, Jian-Syuan Jansen, Bernard J. (2014). *Patterns of Social Media Conversations Using Second Screens. Presented at 2014 ASE BIGDATA/SOCIALCOM/CYBERSECURITY Conference, Stanford University, May 27–31, 2014.* https://faculty.ist.psu.edu/jjansen/academic/jansen_patterns_second_screens.pdf#:~:text=The%20phenomenon%20of%20simultaneously%20engaging%20with%20more%20than,be%20a%20conversation%20with%20others%20regarding%20TV%20programing

Murphy, S. T., Frank, L. B., Chatterjee, J. S., Moran, M. B., Zhao, N., Amezola De Herrera, P., & Baezconde-Garbanati, L. A. (2015). Comparing the relative efficacy of narrative vs nonnarrative health messages in reducing health disparities using a randomized trial. *American Journal of Public Health, 105*(10), 2117–2123. https://doi.org/10.2105/ajph.2014.302332

Newton-Lewis, T., Das Roy, R., Gambhir, V., Godfrey, A., Sanyal, R., Sethi, I., Mitra, R., Pasricha, R., Mehrotra, P., Iyer, S., & Mamidi, P. (2021). Starting conversations to tackle sanitation in India through TV drama. *The Journal of Development Communication, 32*(2), 45–58. http://jdc.journals.unisel.edu.my/ojs/index.php/jdc/article/view/208

Schwandt, H. M., Skinner, J., Takruri, A., & Storey, D. (2015). The integrated gateway model: A catalytic approach to behavior change. *International Journal of Gynecology & Obstetrics, 130*(S3). https://doi.org/10.1016/j.ijgo.2015.05.003

Shannon, M. (2018). Can a television drama help young people identify and challenge harmful and deeply entrenched gender norms? *BBC Media Action.* http://downloads.bbc.co.uk/mediaaction/pdf/research/adhafull-rct-executive-summary.pdf

Singhal, A. (2010). Turning diffusion of innovations paradigm on its head: The positive deviance approach to social change. In A. Vishwanath & G. Barnett (Eds.), *Advances in the study of the diffusion of innovations: Theory, methods, and application* (pp. 192–205). Peter Lang Publishers.

Surya Citra Televisi (SCTV). (2021, June 9). *Cerita Kita Official Teaser [Video].* YouTube. https://www.youtube.com/watch?v=dFuHjtwgDCQ

United Nations Framework Convention on Climate Change (UNFCCC). (2021). *Updated NDC Indonesia 2021.* https://www4.unfccc.int/sites/ndcstaging/PublishedDocuments/Indonesia%20First/S.275%20-%20Indonesia%20Updated%20NDC%20-%20Corrected%20Version.pdf

Wijaya, A., Samadhi, N., & Juliane, R. (2019). *Indonesia is reducing deforestation, but problem areas remain*. World Resource Institute Blog. https://www.wri.org/blog/2019/07/indonesiareducing-deforestation-problem-areas-remain

Worldometers. (n.d.). Indonesia population. https://www.worldometers.info/world-population/indonesia-population/

YouGov Cambridge. (2020). *Where climate change deniers live?* YouGov Cambridge Globalism Project. https://docs.cdn.yougov.com/rhokagcmxq/Globalism2020GuardianClimateand LifestyleafterCOVID.pdf

Ankur Garg is a Country Director at BBC Media Action, and has led orgsanition's strategy, programmes, and partnerships across India and Indonesia. Over the last decade at Media Action, Ankur has led some of the seminal works creating measurable impact in lives of hundreds of millions of people on development challenges as wide as resilience, water and sanitation, information disorder, and climate action. Ankur is an energy and climate policy expert by training and is driven by the role of behavioral science in accelerating climate action across the world.

Anna Godfrey is the Head of Evidence at BBC Media Action. Working with BBC Media Action's global research team, she has delivered research to inform, guide and evaluate digital and broadcast media interventions including storytelling on topics as diverse as maternal and child health/mhealth, sanitation, climate change, social cohesion, civic participation, media freedom and information disorder, and gender norms. Anna's formerly worked in central government on behavioural insights and strategic communication in conflict and post-conflict societies. Her training is in behavioural economics, mathematics and applied statistics.

Rosiana Eko is the Research Manager for BBC Media Action Indonesia. As a research manager, Rosi leads qualitative and quantitative research as well as digital and ethnographic research to inform the design of the project, specifically the target audience, honing the project's theory of change and supporting the development of content. In the past 5 years with BBC Media Action, she conducted research focusing on climate change and information disorder.

Let's Go! Let's Know! N*Gen as an EE Tool for Climate Education and Agency

Paul Falzone, Joy Kiano, and Gosia Lukomska

The theme song will stick in your head for days. That was intentional.

How do we hear? How do we see?
What makes up the air we breathe?
How does a plane fly over the sea?
Why do the planets float so weightlessly?

Let's Go, Let's Know! It's Technology!
Let's Go, Let's Grow! Science You and Me!
Let's Go, Let's Know! We're Learning Every Day!
*Let's Go, Let's Grow! N*Gen Leads The Way!*

So begins another episode of *N*Gen* (pronounced "engine") or *Next Generation Television*, Sub-Saharan Africa's first science TV show for kids and a nominee for the 2023 Peabody Award.

First developed with teachers and children's media experts in the early months of the COVID-19 pandemic and filmed across Kenya, Nigeria, Tanzania, South Africa, Uganda and Zambia, the show's initial goals were to elevate girls and women, increase trust in science and scientists, present COVID-19 prevention in an engaging and memorable way, and to help give people the critical thinking tools to fight misinformation. As the program found an audience and the COVID-19 crisis shifted increasingly from pandemic to endemic status, so too did the program's focus. *N*Gen* pivoted to engage more meaningfully with the climate crisis, exploring a range of topics like ocean conservation, ecosystem change, zoology, misinformation/myth-busting, zoonotic disease, and human–wildlife interactions. It also shifted its frame from one rooted in teacher-centered learning toward one that portrayed scientists and children as protagonists in a more narrative and field-based format.

P. Falzone (✉) · J. Kiano · G. Lukomska
Peripheral Vision International, Brooklyn, NY, USA
e-mail: director@pvinternational.org; joykiano@gmail.com; gosia@pvinternational.org

© The Author(s) 2024
E. Coren, H. Wang (eds.), *Storytelling to Accelerate Climate Solutions*,
https://doi.org/10.1007/978-3-031-54790-4_6

This pivot to climate issues is a response to practical considerations. Sub-Saharan Africa is one of the most vulnerable regions in the world to the increasing impacts of the climate crisis, with rising temperatures, drought, changing rainfall patterns, and other manifestations of climate variability set to collide with traditional liveli-hoods largely dependent on family farming (Chikava, 2021; Fleshman, 2012). With a median age of 19 years old, Sub-Saharan Africa is also home to the largest youth population in the world. This population represents not only the future of this region but of the broader world as climate crisis-induced population shifts to increase the footprint and influence of the African diaspora. Though largely overlooked by other advocacy and entertainment-education climate change campaigns, how the young population of Sub-Saharan Africa understands science, nature, and their relation-ship to these fields can have incredible impacts moving forward.

This chapter will explore the practicalities of producing and distributing this science-based entertainment-education television program aimed at youth. Beyond this case study, this chapter details aspects of the broader media landscape in Sub-Saharan Africa, existing science education efforts, and considerations related to the strategic use of entertainment-education to cultivate consciousness and action related to the climate crisis. The authors maintain that the era of climate crisis-based entertainment-education interventions is only just beginning and that this project will prove to be an early attempt to use the powerful method of entertainment-education at scale and across contexts. Furthermore, engaging with diverse, margin-alized, and/or subaltern audiences can help form a bottom-up movement to push for more effective political policies and scientific innovations.

Using Entertainment-Education to Promote Science

Astronaut Mae Jemison, the first African-American woman to go into space, cited the racial and ethnic diversity of the cast of the television show Star Trek and, in particular, African-American actress Nichelle Nichols' portrayal of Lieutenant Uhura as an early inspiration for her ambition to pursue a career as an astronaut (CNN, 2005). Having turned fiction into reality and become a media figure herself, Jemison then became a real-life inspiration to others. For instance, the first African-American woman to earn a doctorate in electrical engineering from New York University, Ruthie Lyle-Cannon, cited Jemison's example as her inspiration to pur-sue electrical engineering when she was 15 years old (Duke University, 2013).

In her study of entertainment's role in shaping the consciousness of future scien-tists, O'Keeffe (2010) interviewed dozens of professional scientists and science stu-dents. She found that they often became interested in the sciences at an early age and cited specific examples from the media as inspirations for their interest. These mediated figures can provide powerful impressions on populations that might lack access to examples of real-life scientists.

This is particularly true for under-represented populations, who look to media, whether fiction or nonfiction, to seek out stereotype-defying, aspirational characters with whom they can identify. The sense of wonder and possibility these figures provide can become a basis for forming professional aspirations and behavioral intentions of their own. Greenberg (1988) frames this as the "drench hypothesis" wherein "striking, new images can make a difference…[and] cause substantial changes in beliefs, perceptions, or expectations about a group or role" (pp. 100–101). He contrasts the power of this "drench" of powerful examples with the slow drip of media effects in Gerbner's Cultivation Theory (1998), which posits that media's impact is the cumulative result of exposure to large amounts of media over a long period of time.

While entertainment has the power to inspire, whether by drip or by drench, it often does so unintentionally or in an undisciplined manner. In contrast, entertainment-education (EE) combines the power of entertainment's inspirational potential with education's intention. Though definitions abound, Wang and Singhal (2009) formulate EE as "a theory-based communication strategy for purposefully embedding educational and social issues in the creation, production, processing, and dissemination process of an entertainment program, in order to achieve desired individual, community, institutional, and societal changes among the intended media user populations" (pp. 272–273).

EE has been used to shift the needle of knowledge, attitude, and/or behavior on a broad array of topics (Frank & Falzone, 2021; Singhal et al., 2013), including family planning (Ryerson & Negussie, 2004, 2021), the prevention of cardiovascular diseases (Bouman, 2021), violence against women (Green et al., 2020; Wang et al., 2018), gender inequality (Wang & Singhal, 2021; Chatterjee et al., 2021), child marriage (Obregon et al., 2021), religious tolerance (Gowland et al., 2021), and many other topics.

Much of the literature on EE has focused on fictional formats, particularly dramas, often through the centering of the Sabido Methodology of entertainment-education (Sabido, 2021). But fiction is often rooted in reality, and nonfiction may be far more constructed than the viewer realizes. The documentary is often considered the epitome of nonfiction storytelling (Borum Chattoo, 2020), but it is useful to remember that while establishing his "First Principles of Documentary," John Grierson situated nonfiction film in the tenuous place between truth and fiction, where the author could reshape the world according to her whim and through her ability to make and share stories: "Here we pass from the plain (or fancy) descriptions of natural material, to arrangements, rearrangement, and creative shapings of it" (Grierson & Hardy, 1966, p.146).

The line between fiction and nonfiction is less important for the audience than one might think. This is partly because one of the central aspects of a powerful narrative is the ability of a story to "transport" the audience into the worlds of the story. Green suggests that "transportation can happen whether a story is about actual events (such as a documentary) or whether it is the creation of an author's imagination" (Green, 2021, p. 88). Increased transportation increases audience members' likelihood to adopt story-consistent beliefs, attitudes, and behaviors, so the format matters less while the intention of the storyteller and the transportation of the audience member matter more.

Educational television aimed at youth is a perfect example of a format that blends fiction and nonfiction to transport impressionable audiences into narratives that can shape norms at an important time in their development. Indeed, for more than a half-century, evidence has demonstrated that educational television such as *Sesame Street* can play a profound role in children's cognitive development (see Fisch et al., 1999; Guernsey, 2013; Mares et al., 2013). From Sesame Street to television shows by Bill Nye and Neil deGrasse Tyson, this medium has been used to teach, inspire, and cultivate knowledge across multiple generations of American children. There is every reason to believe that as we look for ways to address the climate crisis, entertainment-education aimed at youth can be particularly effective, both for the primary audience of children and for the adults in their lives.

We saw the opportunity to test a theory—that high-quality, educational television inspired in part by the type of programming created by PBS in the United States could be produced affordably and collaboratively in Africa and distributed for free at scale across multiple commercial markets in Sub-Saharan Africa and beyond, reaching a mass audience. If the project were to draw a significant audience, we anticipated that donors and large NGOs and bilateral institutions would recognize the potential of this approach and begin to fund and/or develop their own projects in this mode. To see why the moment for such a project has come, we must explore trends in media and science education in Sub-Saharan Africa.

Sub-Saharan Screens

Television occupies an unusual place in the hierarchy of development media in Sub-Saharan Africa. For the technological enthusiasts (many of whom might not have significant experience working in the field in Sub-Saharan Africa), their attention will be on the bleeding edge of technologies like artificial intelligence, blockchain, virtual reality, and other digital tools. But across Sub-Saharan Africa, only 29% of individuals use the Internet (World Bank, 2018) with mobile Internet adoption standing at 26% at the end of 2019 (GSMA, 2020). Technological enthusiasts elevate tools that are not always appropriate in a context that finds the bulk of its population further and further isolated by a burgeoning digital divide. For them, television will constitute an old medium.

For those with deep experience in developing media, their attention will still be on the radio, which has consistently enjoyed a place of primacy in this field, particularly in the sphere of entertainment-education, where Sabido-style radio drama and themed talk shows are the default setting for social and behavior change media campaigns. This population may critique television for requiring reliable access to electricity, while radios can run off battery power or hand cranking in some models. For them, television will be too advanced a medium.

But this misses the reality of the increasingly prominent role that television is playing in Sub-Saharan Africa, where television is the fastest-growing legacy medium (Cerbone, 2018). This is largely a function of access to electricity. Though electrification remains a limited resource, it is growing. The World Bank (2021) estimates electrification at 46% of households in 2019 across Sub-Saharan Africa. This is up from 26% in 2000. This growth is a combination of slowly increasing access to grid-tied power and exponentially increasing access to microsolar. For reasons including improved technology, supply chains, and financing schemes, microsolar has expanded rapidly in recent years in Sub-Saharan Africa, allowing previously unpowered communities to leapfrog from no-power directly to off-grid decentralized power. This leads to an "energy ladder" in which these households acquire basic and then more advanced modes of energy and tools of technology, starting with a basic LED light or phone charger and moving toward television (Dominguez et al., 2021). There were nearly 55.34 million TV households in the region in 2015, a number which was expected to rise to almost 75 million by 2021 (Statista, n.d.). According to a different estimate (Dataxis, 2019), among the 215 million households in the Sub-Saharan African region, around 102 million had access to television in 2018. Whatever the exact number may be, the reality is that a previously disconnected population has increasing access to global media. The screens in these newly electrified homes will quickly become a window into a broader world. Whether the stories they receive through that window hurt or help is, in part, dependent on those of us in the business of creating and distributing EE.

Meanwhile, broadcasters are in a difficult position. Audiences clamor for locally produced and salient content, but broadcasters cannot meet this demand with the budgets they are working with. This presents an enormous opportunity for entertainment-education. While historical television stations have demanded that NGOs pay them to play media (often documentary and other didactic content), well-produced entertainment-education creators can sidestep this by offering these gatekeepers content that their audiences want to watch. Organizations such as Impact[ed] (formerly known as the Discovery Learning Alliance), Ubongo, MTV Staying Alive Foundation, and others have been utilizing pro bono or low-cost licensing arrangements with terrestrial broadcasters to amplify their work. By providing their content free of charge to broadcasters, a symbiotic relationship has emerged. This also becomes a measure by which all televisual entertainment-education projects in Sub-Saharan Africa should be judged moving forward: can the producer create something good enough that the broadcasters would acquire and air it with no other incentive other than that they believed they could monetize it on their own? If so, then EE is doing its job. If not, the project has failed to focus on the first E: Entertainment.

We must also note the role that language plays in the growth of television. While radio is often broadcast in local languages, television is generally broadcast in the dominant lingua franca, most often English, French, Kiswahili, or Portuguese. This is because radio stations often have a smaller geographic footprint for their broadcast signal, so local language is essential to compete in a relatively small and fragmented market. Television stations, in contrast, are fewer and often have a much larger broadcast footprint. So, broadcasting in local languages makes less economic and technological sense. For instance, in Uganda, there are more than 40 local languages and dialects. A social and behavior change campaign that utilizes audio will often pick 4–6 of these to translate content into. But of the most watched television stations, all, but one, broadcast primarily in English. Development media veterans will often dismiss television as a medium because of this language aspect. There is no question that information obtained in one's primary language enhances learning and comprehension. But the tradeoff for a creator of entertainment-education where language is concerned is always between breadth and depth. Where cost is a limitation (and when is it not?) and multiple-country distributions are a goal, television offers an opportunity for scale that simply cannot be matched by radio.

Science Education

There is global awareness of the need to enhance Science Technology Engineering and Mathematics (STEM) learning from early years through to post-secondary through education and public engagement. Its importance lies in the very need to create more scientifically literate and engaged citizenship. As the InterAcademy Partnership (IAP) states in its Health Policy position (Ortolani, 2021), "To prosper in this modern age of innovation requires the capacity to grasp the essentials of diverse problems, to recognize meaningful patterns, to retrieve and apply relevant knowledge."

What should this learning look like in practice? Science education should enhance curiosity, wonder, and questioning. It should capitalize on young minds' natural, innate curiosity, giving them a sense of adventure. It should be immersive, inclusive, enabling, and encouraging. There should be a vast opportunity for personal experience, making connections that extend ideas, inquiry, and curiosity while providing experiences that motivate further engagement and learning. It should be a self-propelling, self-sustaining process that results in a snowballing of ideas and connections with learners' reality and community (Rieckmann, 2022). Science learners should be continually creating, communicating, constructing, connecting, concluding, contradicting, collaborating, conferring, creating, and communicating all over again. The science classroom should be a space for explosive ideas, creativity, communication, curiosity, wonder, expression, and of course IMAGINATION. It should be a place that sparks curiosity, keeps learners hooked, and harbors inquisitiveness. A space where the learners are doing, experiencing, exploiting, evaluating, and dissecting. An ideal science education should allow students to interact with and learn from science in an authentic, purposeful way rather than by the simple acceptance of abstract theorem. Science education would also benefit from working with the big ideas: Climate Care, Conservation, Habitat Protection, and more.

However, the teaching of science has not changed in line with the technical advances of the twenty-first century. There is very little evidence within the educational programs of the presentation of big ideas of science from which classroomwide conversations can arise (Harlen, 2010). Science learning has not changed over the generations irrespective of technological advances and worldwide engagement and discussion on many novel scientific phenomena from the applications of nanotechnology in medicine, to the spontaneous formation of mutant variants of disease-causing viruses, to the implementation of RNA vaccine technology in disease prevention, to name but a few recent examples.

This stems from the origins of the teacher's own experience of science education and the long-held belief in the complexity of scientific concepts and phenomena. Often, the method by which teachers prepare scientific content mirrors how they learned about science. Traditional science teaching has demanded an absolute commitment to memorizing many facts, formulae, and abstract theories. Science education is still dominated by rote learning, memorization, and teacher-led lessons that concentrate on repeating keywords or phrases, peppered with closed questions that do not have room for exploring facts and expanding ideas and theory. Didactic presentation of scientific theory robs the learner of the experience or curiosity, of the interconnectedness of science with their immediate environment. It is not a subject area for the fainthearted and cannot be appreciated within the realm of relevant learning that immediately applies to community issues: the community in which the young aspiring African scientist lives.

Successful science education everywhere relies entirely on the nature of its presentation and/or delivery. Successful science education across most of the education structures within the continent of Africa requires an active, dynamic, and continuous overhaul of the teacher development process in a manner that appreciates and recognizes the necessity to inject relevance, storytelling, meaning, and importance into science learning. It is in this context that the *N*Gen* project was formed.

Creating the First Pan African Science Show

With hubs in Tanzania, Uganda, and the United States, Peripheral Vision International is a nonprofit organization that has worked with dozens of other nonprofit organizations to co-create and distribute EE and social and behavior change communication to some of the most remote regions in the developing world. Since 2011, Peripheral Vision International's varied output includes music video programming, television shows that blend journalism and hip hop (Shaker et al., 2019), public service announcements, interactive mobile audio games (Frank et al., 2021), short films, documentary, e-comics, radio shows, and other media that have reached millions of audience members across borders and platforms.

There is a growing awareness of the power and impact of media on a population's knowledge, beliefs, and behavior and a growing understanding that media and communication are what connect, bind, and potentially divide us. When the COVID-19 pandemic hit, it quickly became clear that we were also in the midst of what the World Health Organization called an "infodemic" related to rampant misinformation, conjecture, rumor, and conspiracies (WHO, 2021). To approach the infodemic by creating programming aimed at youth made the most sense for several reasons. Children's beliefs are still taking shape, so it would be an opportunity to shape norms rather than reshape established norms around gender and science. With schools closed and children stuck at home under mandatory lockdown in many places, this audience was more accessible than ever. Because lower-income households in Sub-Saharan Africa have access to fewer screens than higher-income households (Dataxis, 2019), there is a large secondary co-viewing audience of parents and elders. While imported content abounded, there was relatively little youth content produced for or featuring African youth—so this would be an opportunity to provide representation for the largest youth audience in the world to see people like themselves on screen, many for the first time.

When the idea to create *N*Gen* came about in 2020, we wanted the show to focus on science at the very basic, primary school level. We chose the topics for the 13 episodes that comprised the first season advised by educators from around the continent. Most of the teachers working with national curricula deemed science to be an area both neglected by the Sub-Saharan African education systems and one that could transcend boundaries and apply to an abundance of cultures and experiences that our young viewers across Africa would represent.

In this first season topics included the human body (brain, senses), the natural world (bees, wetlands), matter and motion (light, energy). With each one of these topics we emphasized the relevance of science and its ubiquity in everyday life. Science is not an abstract, theoretical, and incomprehensible concept attempting to take form as a medley of symbols on a faded blackboard. Science happens in your kitchen, in your backyard, and when you are sleeping.

A modular format of various small segments ("snackable" for online streaming) emerged that combined to create a half-hour program. The core segments were lessons delivered by teachers and TV presenters. Mainly filmed against a green screen, these were animated with images, B-roll footage, animations, and other lively visuals to bring the lessons to life. They were enhanced by including experiments that kids can do in and around their homes. Interspersed between these lessons were other segments. "Health Tips" were public service announcements on topics like nutrition and masking to prevent COVID-19 transmission. Because children were stuck at home with little physical activity, each episode also featured an exercise routine featuring a lively Tanzanian dancer/exercise instructor named Tadhi. Teachers from across Africa also contributed short videos of themselves giving lessons that we animated and included as the "Africa Teacher Challenge." Social emotional learning and mindfulness were also given a segment in the form of a "Brain Break"—a visualization or mediation exercise aimed at promoting mental health and well-being. Each episode concluded with a "Brain Booster" that tested knowledge acquired over the show.

First launched in September 2020, the reception for *N*Gen* went far beyond our wildest expectations (see an example in Fig. 1). Broadcasters across Africa have shown that their audiences are hungry for engaging, educational, youth-oriented content. The 13 episodes in series 1 were syndicated to 50+ TV networks across Africa through both terrestrial and satellite broadcasts. Though available only in English (because of budgets), broadcasters in Ethiopia and Nigeria translated it into local languages at their own expense.

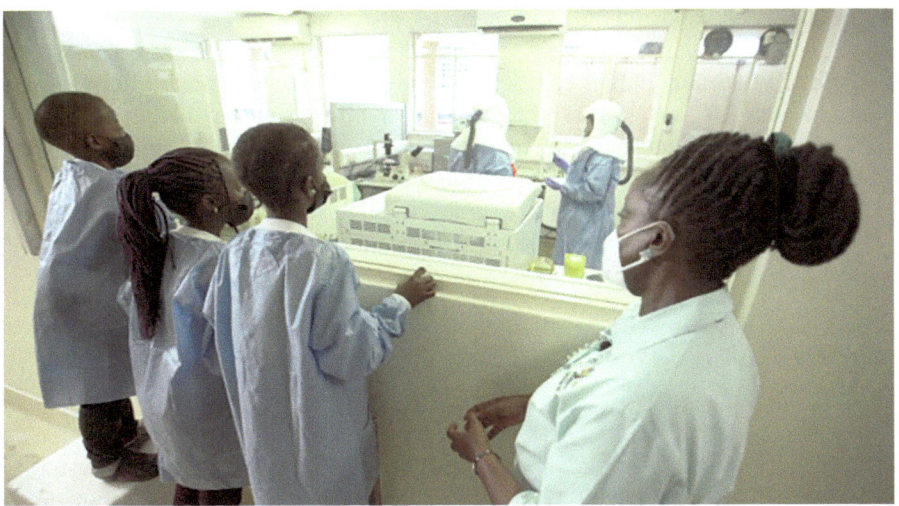

Fig. 1 Adiva, Penzi, and Liam watch with awe as researchers at the Uganda Virus Research Institute handle large quantities of the covid-sars-2 virus, in order to create vaccines. ("To make a vaccine, you have to get something that looks like a virus, and nothing looks like a virus better than the virus itself. So we are multiplying it here, getting good quantities, so that we can use it later to make a vaccine, to protect people from COVID."—Dr. Jennifer Serwanga, Senior Immunologist.) (Photo credit: James Mbiri)

We also found that there was an appetite in Western countries for the show and *N*Gen* became the first program to platform African scientists and educators to a Western audience. It was broadcast in the United States/Caribbean on The Africa Channel, on Common Sense Media's streaming service Sensical, to the African diaspora through the streaming platform AfroLandTV and became available to over 50 million students and 5 million educators across 90 countries via Discovery Education, which is in half the classrooms in the United States. *N*Gen* quickly became the 27th most popular channel among over 650 available channels within Discovery Education in the United States.

But while it was possible to count platforms, the question remained: how many people had actually seen the program? Measuring viewers/television ratings in Sub-Saharan Africa is notoriously difficult due to the large number of national markets and the expense of collecting reliable data. The pandemic has only aggravated this information deficit, with top regional ratings research agencies like Geopoll partially or totally suspending operations in many of the markets that they previously serviced.

But the data we were able to collect was very positive. One snapshot comes from Kenya, where our broadcast partners, Akili Kids TV collected granular data from October 10, 2020, to May 9, 2021, and estimated child viewership at 36,068,045 total views and adult viewership at 34,439,994 total views over the length of the airing of Season 1 (with a combined total of 70,508,039). There were between 600,000–800,000 children and an additional 600,000–800,000 adults watching each episode, depending on the day of the week and time of airing. Unlike US broadcasts, where viewership is very segregated across ages, the high degree of co-viewing between children and adults reinforces the potential of television in Sub-Saharan Africa to transcend these demographics, and that therefore it is more effective to aim younger rather than older when developing EE that can reach a mass audience. These numbers allow us to estimate that across dozens of other broadcasters, each episode of the program would have been seen by *at least* 10 million unique viewers and very likely many more than that.

Shifting to Focus on the Climate Crisis

Television stations clamored for more episodes, and PVI was able to cobble together the budget to produce an additional 13-episode season. Focus group testing in Nigeria with teachers, parents and school children, and broader user feedback led us to shift the format, and the burgeoning climate crises led us to shift the focus.

The research is beginning to show the role that entertainment-education can play in promoting climate-related content. Flora et al. (2014) found measurable changes in high school youths' knowledge, beliefs, involvement, and behavior related to climate science after exposure to an in-school live presentation that they describe as entertainment-education. In testing different frames around climate communication, Jones (2014) found that narrative structures increased the effect of participants toward hero characters, which in turn indirectly influences the persuasiveness of a story. Topp et al. (2019) demonstrated that entertainment-education increased cognitive engagement with climate-related content. Lawson et al. (2019) demonstrated that child-to-parent learning in a US context was capable of boosting climate change concerns in parents. These results were strongest in males and conservative parents who previously displayed the lowest levels of climate concern. Interestingly, daughters seemed incredibly influential in effecting this change. We saw the opportunity to build on this with *N*Gen*.

One of *N*Gen*'s core purposes was also to disrupt gender stereotypes in Africa and beyond. Highlighting the work of female scientists and educators in Africa is an opportunity to inspire millions of children, and girls in particular, as well as their parents to let them know that they are smart, they have value, and they are capable of greatness. We determined that Season 2 would better achieve this by presenting actual African women scientists in their professional settings and engaging directly with topics related not only to their area of specialization but to their career path and how they achieved what they have professionally.

In Season 2 of the show, we also decided to center our lenses on human impacts on our environment, including but not limited to climate change, and ways to mitigate it (see an example in Fig. 2). But how to tell such a large story to primary school children on a continent where conversation about climate change has been largely restricted to esoteric conversations at policy level?

Fig. 2 Linet Kiteresi from the Kenya Marine and Fisheries Research Institute (KMFRI) in Mombasa, Kenya, explains the devastating impacts ocean water pollution on various ecosystems and human health ("Ms. Linet, what do scientists like you do at KMFRI?"—Yana). (Photo credit: Ayaz Rajput)

De Meyer et al. (2020) propose that too much of the coverage of the climate crisis is negative in tone. They suggest that a more effective frame would be to move from issue to action, focusing on the stories of people positively acting to prevent climate change. They point out that rather than knowledge/belief leading to action, it is more effective for action to lead to belief, and that the ideal frame "places 'people taking action' at the heart of each story, fact based or fictional, and regardless of level of interactivity or media format" (p. 10). By providing viewers with this mediated sense of agency, they will look for ways to take action in their own lives. This is also consistent with concepts of "positive deviance," where the focus shifts away from normalizing community problems and toward amplifying problem-solving individuals and practices that may deviate from the social norm in productive ways (Singhal et al., 2013).

The televisual medium allows us to realize this potential to create agency by depicting relatable on-screen personalities taking action. Rather than teach in the traditional sense, *N*Gen* strives to inspire and encourage to inquire, explore, and discover. We decided to initiate conversations about climate change, and other aspects of humans' attitudes and behaviors concerning our planet, by putting children face to face with African scientists and activists in the same settings they are learning about.

Season 2 is for the most part not topic-driven but hero-driven, moving our point of focus from the studio to the field. We researched and selected charismatic science experts from Uganda, Kenya, Tanzania, and Nigeria, where we have a strong production network, as spokespersons for our message. Dr. Perpetra Akite, an entomologist at Makerere University in Kampala, Uganda; Dr. Fiona Wanjiku Moejes, a marine biologist at the Mawazo Institute in Watamu, Kenya; and a Professor of African Architecture and Urban Design, Nmadili Okwumabua in Lagos, Nigeria, are just a few amazing professionals who take the children on adventures to discover the world they live in and interact with. We developed episode briefs, preliminary scripts, and production plans around them.

In one episode, Lorraine (11), Alexis (8), Nimaro (8), and Judah (7) meet Dr. Perpetra Akite at her office at the Zoology Department of Makerere University in Kampala, Uganda (Fig. 3). Dr. Akite ticks some of the boxes of a "typical scientist": she is wearing a white lab coat and has a microscope, and a stack of papers on her somewhat disordered desk. Yet she defies the stereotype of a distant, unsociable scientist. She greets the kids with a warm smile, which throughout the episode hardly leaves her face, welcomes them to her lab, and in the first minute of the interaction ignites the kids' curiosity:

Kids: Good morning.
Dr. Akite: Welcome to the zoology department.
Lorraine: Dr. Akite, what exactly is your job?
Dr. Akite: I am an entomologist.
Kids: Entomologist?
Judah: What is that exactly?
Dr. Akite: An entomologist is somebody who studies insects. We study how they behave. We study their numbers. We study the different kinds of insects that are there. We study how they affect us.
Alexis: Why do we study insects?

Fig. 3 Dr. Perpetra Akite, an Entomologist at Makerere University in Kampala, Uganda, shows four insect-curious kids around her lab. ("Butterflies which are in this box, (…) each of them has a name. An entomologist like myself, we study and even name them."—Dr. Akite.) (Photo credit: James Mbiri)

Dr. Akite: Insects are very important to us. In fact, without insects, we would not be there. Let me show you around.

Dr Akite leads the kids through the maze of glass cabinets, where thousands of specimens are displayed. She opens a case with butterflies from the Papilio family. Each butterfly has a tag with its scientific name and the date it was collected. The kids laugh at the name "Papilio jacksoni."

Judah: Is it Michael Jackson?

Dr. Akite: No. But it was named after someone called Jackson. Entomologists, like myself, we study and even name them.

When, later in the episode, the kids ask Dr. Akite how they could become an entomologist like her, Dr. Akite once again refers to the naming of new species, to further excite the young learners in science:

Dr. Akite: So, your first interest has to be just looking around you because when you look around you, you'll always find an insect. And there's a high chance that you will actually discover a new species someday, and it can also be named after you. For example, I have a moth. It is a little moth. It is now named after me. They used my family name. From Akite, they name it Akitei. Lorraine, if we got a new species of insect and we wanted to name it after you…

Lorraine: A locroach.

Dr. Akite: That must be a new species of cockroach.

Lorraine: Yeah.

Dr. Akite: You now become part of science. Your name remains there.

When Dr. Akite opens another glass case and carefully pulls out a bright orange butterfly, larger than her palm, the kids exclaim a genuine "wow."

Dr. Akite: Here is the biggest African butterfly that we have. It is called the giant African swallowtail. In the 1960s it actually used to occur not far away from Makerere here, but now that butterfly is not there anymore. Do you think you would know why?

Lorraine: Why?

Dr. Akite: Because humans have broken up the forest, and it is a butterfly that needs a very big continuous forest. We say it has gone locally extinct because the forests are so small, they can no longer keep this butterfly.

Judah: If there are no forests in Uganda, what will the butterflies do?

Dr. Akite: If there are no forests in Uganda, all the butterflies that need forest to survive, they will all die. But do you know what will be the biggest problem of that?

All: What?

Dr. Akite: Insects are very important in pollinating the food crops that we eat. So when they also die, then we are going to have problems as well.

She's captured their attention and their concern. The kids ask her to take them to the forest to learn more about how it's changed. In the forest, Dr. Akite comes up with a game: an insect scavenger hunt. The kids, equipped with insect nets, run off into the woods excited to find their little treasures. One by one, they bring what they find, and Dr. Akite names the insects and tells their characteristics, once in a while adding a wow-fact, that further excites the kids about their adventure and builds their respect and awe for insects:

Dr. Akite: You know that butterflies also go very long distances, just like planes. Did you know that? For example, we have butterflies that travel from Europe to Africa. Yes, they look so fragile and small but they have a lot of energy"

She also has an endearing and humorous way of talking about insects in a way that parallels some of the human behaviors that children can relate to:

Dr. Akite: So, (the egg) hatches into a caterpillar, which starts to feed. (...) When it goes into a cocoon, it doesn't eat anymore. Do we also kind of go into cocoons sometimes? At night, we don't want to eat anymore. We sleep. But when we wake up in the morning, we're like caterpillars. We just want breakfast!

The scavenger hunt goes on for hours, the kids enjoy their time in the forest and Dr. Perpetra's teachings. But before they head home, Nimaro voices his concern, which Dr. Perpetra uses as an opportunity to once again emphasize the direct link between human activities and environmental changes:

Nimaro: But some insects are harmful, like the mosquitoes that carry malaria, the locusts that attack crops in Northern Uganda and many other insects.
Dr. Akite: They are part of the big ecosystem. The reason why they have now become a problem to us is because we have gone and destroyed nature. We have gone and built our houses in the swamps where the mosquitoes used to breed. Now they come and breed in our houses. We have gone and taken away some parts of forests, so the insects that live in forests now seem like they're living in our houses. But if we give them a chance to live in their areas, in their habitat, what I call their home, then we would have no problem with them. Do you remember this big butterfly that we saw in the lab, the bright orange butterfly? It was here in Mpanga about 1966, and now we are here in 2021. This butterfly doesn't occur here anymore. We have cut down the forest. The butterflies and other insects are telling us about what we are doing wrong or what we are doing right in the environment. We need all these insects in order for us to survive.
Nimaro: If the insects die, will we also die too?
Dr. Akite: Actually, yes. Now the scientists are very sure.
Lorraine: What can we do to protect insects?
Dr. Akite: The best way of saving the insects is to protect their habitat, just like we protect our homes. To protect insects, it needs all of us. It needs…?
Kids: All of us.

The real heroes are the children who set off on the journeys to meet the experts, engage in their research and advocacy activities, and shower them with questions. Dr. Akite sends a group of young explorers on an insect scavenger hunt into the buzzing and bustling, yet dwindling Mpanga Forest. Dr. Wanjiku Moejes takes them

Fig. 4 Sisters Xamara and Yanna explore the underwater world, guided by marine biologist, Dr. Fiona Wanjiku Moejes off the coast of Kilifi, Kenya. ("And what is the best way to learn about the ocean? It's to get in it!"—Dr Fiona.) (Photo credit: Azim A. Mula)

on a snorkeling safari to a vibrant but fragile coral reef in the Indian Ocean, Professor Okwumabua walks them through a sustainably built house in Nigeria's Delta State to see how it is constructed and the value it brings to its inhabitants and the environment (Fig. 4).

Finding our child- and science-heroes was one of the most important steps in the pre-production process. Because of the gender component of the show, we wanted the featured experts to be predominantly female. We had looked for them via social media, various online networking platforms, and news outlets. When we would approach scientific institutions, many would render our inquiries impossible with bureaucratic requirements, such as multiple permissions from top officials, who (always male) would stall the process. Many were plain unresponsive. At the time of development, *N*Gen* was still an unknown brand with no prominent names attached to it, and not everyone we approached was willing to work with us.

Once we cast the experts, it was time to cast the kids. We did this via schools and educators we collaborated with. The contenders presented a science experiment of their choice, had to talk about one place in the world they wished to travel to, ask one question about a caterpillar, and were invited to touch a cardboard cutout caterpillar. One interesting realization from the casting was that the children who were the best performers at school did not always perform well in front of the camera. Those who performed best were often the kids labeled by their teachers as naughty and underperformers. But they were the ones who asked the most interesting questions, and who dared touch the caterpillar (although it was made out of benign cardboard, most kids were scared to do so!).

Filming with the selected kids and experts in the field was certainly the highlight of the process. In order to generate content for the 13 episodes of Season 2, we spent close to 50 days in the field: in the forests, farms, factories, beaches, oceans, lakes, laboratories, weather stations, schools, national parks, polluted cities, and unspoiled villages. We quickly realized that although the kids were very sensitive to matters concerning our environment, animals, plants, and the planet as a whole, the concept of climate change was not an easy one to get across. Even the featured experts were falling short of explaining it in a succinct, convincing, and—most importantly— child-friendly manner. The filming process required a lot of coaching and second, third, and umpteenth camera takes. As always, the kids would surprise us, and amuse us, with the most unexpected explanations.

In the "Forests" episode filmed in the Fort Portal region in Western Uganda, 8-year-old Elena makes a claim that *"trees give us oxygen and we give them farts"* when trying to explain the carbon cycle. Harriet Nakyesa, a reforestation expert, gives Elena and her friends Jessica, Jacksa, and Elton, a more scientifically accurate explanation, and shows them how to measure the amount of carbon a tree can store. She introduces the term "carbon sequestration" and lets each child take a turn to wrap measuring tape around the trunk (Fig. 5).

Harriet: So, when you're measuring the carbon of the tree, you look at the diameter and the crown of the tree. Can you look up on the top of the tree? You read here. Seventy-three.

Fig. 5 The lead child hero of the Forests episode, Elena, wonders at a piece of "life (which) started from a rock," as forestry expert Mary Ekyaligonza explains the principle of succession. Kyaninga Forest, Uganda. (*"Succession means one generation of plants is being replaced by another when the other one dies. Life starts from a rock. From a rock to the lichens. From lichens to the ferns. From the ferns to the weeds. Then the weeds to the tree."*—Dr Mary.) (Photo credit: James Mbiri)

Elena (voice over): Scientists created a special formula to measure how much good
a tree is doing. Lots of calculations, you see? And the result is...
Harriet: This tree is storing 1386 kilograms of carbon.
Kids: Is that a lot?
Harriet: Yeah, it is a lot, so it has cleaned the environment. So it is a very good tree
because it has stored a lot of carbon.
Kids: Thank you, tree.

Mary Ekyaligonza, who is another forestry scientist accompanying the kids on this
adventure, reminds them of the negative human impact on the environment, and
how that in turn affects us:

Mary: Most of these big trees, which are important to us and clean up the environment to give us all the good services that we need, are disappearing.
Elena: People are cutting them down to make tables and chairs?
Mary: Yes. So every time the population is growing, you find we demand a lot of
forest products, so we cut down the trees. We lose the trees. And in future, you
will find a lot of carbon dioxide in the atmosphere and the earth will not be good
for us to live…When the carbon dioxide is a lot in the atmosphere, it can cause
what we call heating. I don't know if you've heard about climate change.
Elena: I have. It's very dangerous and that humans are causing it.

Yet climate change proves to be a hard notion to both comprehend and explain in
simple yet convincing terms. When contextualizing the documentary-style field
footage in our later scripted and recorded studio segments, we therefore make sure
to focus on something other than terminology, graphs, and cycles. Instead, we
explain the interconnectedness of things, evoking feelings of responsibility and
empathy. We ensure the narrative draws a clear line between our actions today and
our survival as a species tomorrow.

*"I have been around for much longer than anything else. I am going to be around
for much longer. I'd like you to be here as well. I'm stepping in to make sure you
humans don't spoil my plans. So listen up,"* says Earth Auntie, an animated character of the anthropomorphic planet Earth in the closing episode of the season, titled
"Sustainable Living" (see an example in Fig. 6).

The children we filmed seemed most responsive to this framing and approach,
and our hope is that the ones in front of the TV screens will also be moved, and
subsequently motivated to ask questions, to initiate conversations, and—eventually—to be the next generation of African change makers.

Conclusion

It should not be the responsibility of entertainment-education to play a significant
role in averting a climate crisis. It should be the responsibility of lawmakers and
diplomats who create the policies that govern the industries and corporations that
despoil our planet. But we do not live in the world of should. It is an unfortunate

Fig. 6 *N*Gen* Season 2 hosts Nnena and Jiji are joined by a special guest, Ms. Plastic Bottle in the Plastics episode. ("I'm going to need some clever human beings to help me be part of the solution rather than just being the problem. Who's prepared to help me?"—asks Ms. Plastic Bottle.) (Photo credit: June Ndinya)

truth that public opinion is far more likely to shape policy than the reverse. Just as we saw in the battle for civil rights, for gender equality, for marriage equality, and in many other areas, public opinion frequently evolves more quickly than the laws that follow. The soft power of the media often leads and shapes the cultural shift that ultimately results in the hard power of legislation.

Reflecting on the relatively weak agreements that emerged from the COP26 climate summit, US Senator Kristin Gillibrand (WNYC, 2021) stated in an interview: "There is an interest. There is a growing demand. The fact that we have young people all across our country and across the globe who are demanding answers, who are furious with lack of commitment, who are never going to stop and never going to give up. That's going to push politicians to be bolder, to be more aggressive, to work outside their governments, to go farther than their governments are asking."

The need to act on climate change has never been more important. Thus, for actual, effective climate policy, it is necessary that we inform, persuade, and organize a vast citizenry to demand change. Only then can the very real work of policy-making be done. So, for better or for worse, entertainment-education can and must be a part of this approach.

Youth in the same regions where the climate crisis will hit the hardest can and must also be central to the movement. With a median age under 20, the largest youth population in the world is in Africa. The potential of this population is enormous. They are the future not only of the continent but also of the planet. But historically, they have been missing from their own media and the broader discussion around the climate crisis. Meanwhile, Western broadcasters, educators, and mission-driven media platforms are beginning to wake up to the need to include more diverse voices from the global south. This creates an opportunity for entertainment-education to play a pivotal role in platforming powerful voices from within their own communities to cultivate climate consciousness at scale. The *N*Gen* project—one among what we hope to be a growing cohort of efforts—is trying in its small way to help cultivate that movement.

References

Borum Chattoo, C. (2020). *Story movements: How documentaries empower people and inspire social change*. Oxford University Press. https://doi.org/10.1093/oso/9780190943417.001.0001

Bouman, M. (2021). A strange kind of marriage: The challenging journey of entertainment-education collaboration. In L. B. Frank & P. Falzone (Eds.), *Entertainment-education behind the scenes: Case studies for theory and practice* (pp. 61–832). Palgrave Macmillan. https://doi.org/10.1007/978-3-030-63614-2_5

Cerbone, R. (2018). *DTT transforming viewing experience in Africa*. Intelsat. https://www.intelsat.com/resources/blog/digital-television-is-transforming-the-viewing-experience-for-african-households/

Chatterjee, J. S., Pasricha, R., Mitra, R., & Frank, L. B. (2021). Challenging the forcefield: Crafting entertainment-education transmedia campaigns. In L. B. Frank & P. Falzone (Eds.), *Entertainment-education behind the scenes: Case studies for theory and practice* (pp. 265–278). Palgrave Macmillan. https://doi.org/10.1007/978-3-030-63614-2_16

Chikava, E. (2021). *Smallholder farming is a proven path out of poverty, but climate change is changing the rules*. Bill & Melinda Gates Foundation. https://www.gatesfoundation.org/ideas/articles/smallholder-farming-climate-change

CNN International. (2005, January 14). *Then & Now: Dr. Mae Jemison*. http://www.cnn.com/2005/US/01/07/cnn25.tan.jemison/

Dataxis. (2019). *Sub-Saharan Africa market overview 2019: A contrasted growth – Dataxis*. Dataxis. https://dataxis.com/webinars/354512/sub-saharan-africa-market-overview-2019-a-contrasted-growth/

De Meyer, K., Coren, E., McCaffrey, M. S., & Slean, C. (2020). Transforming the stories we tell about climate change: from "issue" to "action". *Environmental Research Letters, 16*(1). https://doi.org/10.1088/1748-9326/abcd5a

Dominguez, C., Orehounig, K., & Carmeliet, J. (2021). Understanding the path towards a clean energy transition and post-electrification patterns of rural households. *Energy for Sustainable Development, 61*, 46–64. https://doi.org/10.1016/j.esd.2021.01.002

Duke University. (2013). *The legacy of Lt. Uhura: Astronaut Mae Jemison on race in space*. Today. duke.edu. https://today.duke.edu/2013/10/maejemison

Fisch, S. M., Truglio, R. T., & Cole, C. F. (1999). The impact of Sesame Street on preschool children: A review and synthesis of 30 years' research. *Media Psychology, 1*(2), 165–190. https://doi.org/10.1207/s1532785xmep0102_5

Fleshman, M. (2012). *Climate change: Africa gets ready | Africa Renewal*. un.org. https://www.un.org/africarenewal/magazine/july-2007/climate-change-africa-gets-ready

Flora, J. A., Saphir, M., Lappé, M., Roser-Renouf, C., Maibach, E. W., & Leiserowitz, A. A. (2014). Evaluation of a national high school entertainment education program: The Alliance for Climate Education. *Climatic Change, 127*(3–4), 419–434. https://doi.org/10.1007/s10584-014-1274-1

Frank, L. B., Sparks, P., Murphy, S. T., Goodfriend, L., & Falzone, P. (2021). The game of life: How playing gamified interactive narratives affects career planning in Cambodia. *International Journal of Communication, 15*(0), 21. https://ijoc.org/index.php/ijoc/article/view/15447

Gerbner, G. (1998). Cultivation analysis: An overview. *Mass Communication and Society, 1*(3-4), 175–194. https://doi.org/10.1080/15205436.1998.9677855

Gowland, S., Colquhoun, A., Nyoi, M. Y. H., & Thawng, V. S. (2021). Using audience research to understand and refine a radio drama in Myanmar tackling social cohesion. In L. B. Frank & P. Falzone (Eds.), *Entertainment-education behind the scenes: Case studies for theory and practice* (pp. 157–174). Palgrave Macmillan. https://doi.org/10.1007/978-3-030-63614-2_10

Green, D. P., Wilke, A. M., & Cooper, J. (2020). Countering violence against women by encouraging disclosure: A mass media experiment in rural Uganda. *Comparative Political Studies, 53*(14), 001041402091227. https://doi.org/10.1177/0010414020912275

Green, M. C. (2021). Transportation into narrative worlds. In L. B. Frank & P. Falzone (Eds.), *Entertainment-education behind the scenes: Case studies for theory and practice* (pp. 87–101). Palgrave Macmillan. https://doi.org/10.1007/978-3-030-63614-2_6

Greenberg, B. S. (1988). Some uncommon television images and the drench hypothesis. *Applied Social Psychology Annual: Television as a Social Issue, 8*, 88–102.

Grierson, J., & Hardy, F. (1966). *Grierson on documentary*. University of California Press.

GSMA. (2020). *Mobile Internet Connectivity 2020 Sub-Saharan Africa Factsheet*. https://www.gsma.com/r/wp-content/uploads/2020/09/Mobile-Internet-Connectivity-SSA-Fact-Sheet.pdf

Guernsey, L. (2013). How kids' television inspires a lifelong love of science. *Smithsonian Magazine*. https://www.smithsonianmag.com/innovation/how-kids-television-inspires-a-lifelong-love-of-science-94268316

Harlen, W. (2010). *Principles and big ideas of science education*. ASE. https://www.ase.org.uk/bigideas

Jones, M. D. (2014). Communicating climate change: Are stories better than "Just the Facts"? *Policy Studies Journal, 42*(4), 644–673. https://doi.org/10.1111/psj.12072

Lawson, D. F., Stevenson, K. T., Peterson, M. N., Carrier, S. J., Strnad, L., & R., & Seekamp, E. (2019). Children can foster climate change concern among their parents. *Nature Climate Change, 9*(6), 458–462. https://doi.org/10.1038/s41558-019-0463-3

Mares, M.-L., & Pan, Z. (2013). Effects of Sesame Street: A meta-analysis of children's learning in 15 countries. *Journal of Applied Developmental Psychology, 34*(3), 140–151. https://doi.org/10.1016/j.appdev.2013.01.001

O'Keeffe, M. (2010). Media and the making of scientists [Doctorate]. In *Dissertations available from ProQuest* (pp. 1–204). https://repository.upenn.edu/dissertations/AAI3447142

Obregon, R., Muhamad, J. W., & Lapsansky, C. (2021). Strengthening integration of communication theory into entertainment-education practice: reflections from the La Peor Novela case study. In L. B. Frank & P. Falzone (Eds.), *Entertainment-education behind the scenes: Case studies for theory and practice* (pp. 137–154). Palgrave Macmillan. https://doi.org/10.1007/978-3-030-63614-2_9

Ortolani, G. (2021). *Science education: Purpose, methods, ideas and teaching resources*. www.interacademies.org. https://www.interacademies.org/news/what-purpose-science-education

Rieckmann, M. (2022). Chapter 2: Learning to transform the world: Key competencies in Education for Sustainable Development. *Issues and Trends in Education for Sustainable Development, 39*–59. https://unesdoc.unesco.org/ark:/48223/pf0000261802

Ryerson, W., & Teffera, N. (2004). Organizing a comprehensive national plan for entertainment-education in Ethiopia. In A. Singhal, M. Cody, E. Rogers, & M. Sabido (Eds.), *Entertainment-education and social change: History, research, and practice* (pp. 177–190). Lawrence Erlbaum Associates.

Ryerson, W. N., & Teffera, N. (2021). The impact of social change communication: Lessons learned from decades of media outreach. In L. B. Frank & P. Falzone (Eds.), *Entertainment-education behind the scenes: Case studies for theory and practice* (pp. 23–38). Palgrave Macmillan. https://doi.org/10.1007/978-3-030-63614-2_3

Sabido, M. (2021). Miguel Sabido's entertainment-education. In L. B. Frank & P. Falzone (Eds.), *Entertainment-education behind the scenes: Case studies for theory and practice* (pp. 15–21). Palgrave Macmillan. https://doi.org/10.1007/978-3-030-63614-2_2

Shaker, L., Falzone, P., Sparks, P., & Kugumikiriza, R. (2019). From the studio to the street: Cultivating democratic norms in Uganda. *International Journal of Communication, 13*(2019), 1612–1630. https://ijoc.org/index.php/ijoc/article/view/10429/2614

Singhal, A., Wang, H., & Rogers, E. M. (2013). The rising tide of entertainment-education in communication campaigns. In R. Rice & C. Atkin (Eds.), *Public Communication Campaigns* (pp. 321–333). Sage.

Statista. (n.d.). *Number of TV households in Sub-Saharan Africa 2016*. Statista. Retrieved January 31, 2022, from https://www.statista.com/statistics/287739/number-of-tv-households-in-sub-saharan-africa/

Topp, K., Thai, M., & Hryciw, D. H. (2019). The role of entertainment in engagement with climate change. *Environmental Education Research, 25*(5), 691–700. https://doi.org/10.1080/1350462 2.2019.1572072

Wang, H., & Singhal, A. (2009). Entertainment-education through digital games. In U. Ritterfeld, M. J. Cody, & P. Vorderer (Eds.), *Serious games: Mechanisms and effects* (pp. 271–292). Routledge.

Wang, H., & Singhal, A. (2021). Mind the gap! Confronting the challenges of translational communication research in entertainment-education. In L. B. Frank & P. Falzone (Eds.), *Entertainment-education behind the scenes: Case studies for theory and practice* (pp. 223–242). Palgrave Macmillan. https://doi.org/10.1007/978-3-030-63614-2_14

Wang, H., Wu, Y., Choi, J. H., & DeMarle, A. (2018). Players as transitional characters: How youth can "Breakaway" from gender-based violence. *Well Played, 8*(1), 27–40. http://press.etc.cmu.edu/index.php/product/well-played-vol-8-no-1/

WNYC. (2021). *Call your senator: Sen Gillibrand on infrastructure, paid family leave, and more | The Brian Lehrer Show*. WNYC. https://www.wnyc.org/story/call-your-senator-sen-gillibrand-111621

World Health Organization. (2021). *An overview of infodemic management during COVID-19, January 2020–May 2021*. www.who.int. https://www.who.int/publications/i/item/9789240035966

WorldBank. (2018). *Individuals using the Internet (% of population) – Sub-Saharan Africa | Data*. data.worldbank.org. https://data.worldbank.org/indicator/IT.NET.USER.ZS?locations=ZG

WorldBank. (2021). *Access to electricity (% of population) – Sub-Saharan Africa | Data*. data.worldbank.org. https://data.worldbank.org/indicator/EG.ELC.ACCS.ZS?locations=ZG

Paul Falzone is a media strategist, producer, and scholar whose career focuses on media and social change. He is the Executive Director of Peripheral Vision International (PVI), a nonprofit organization that researches, creates, and distributes entertainment-education and social and behavior change communication programs in Sub-Saharan Africa and beyond. PVI's output has been seen, heard, read, or watched hundreds of millions of times across the world over the last decade and its projects have been profiled on BBC, CNN, The New Yorker, Al Jazeera, NPR, The Guardian, and dozens of other news outlets. His publications include the recent co-edited collection *Entertainment Education Behind the Scenes: Case Studies for Theory and Practice* (2021, Palgrave Macmillan). He earned an M.A. and a Ph.D. from the Annenberg School for Communication at the University of Pennsylvania.

Joy Kiano is a science specialist, educator, school leader, and teacher of science who is also deeply passionate about the education and welfare of young people. She has a Doctorate in Biochemistry & Molecular Biology, a Master's level Post Graduate Certificate of Education in Science, and has taught widely in both the U.K. and Kenya. She also founded and led Nova Pioneer Girls High School in Nairobi and served as the Learning Initiatives Manager of the first Nova Pioneer Schools in Kenya. As Senior Educational Consultant and Global Ambassador of N*Gen TV Africa, she represents the N*Gen TV science inquiry programs because she believes this unique resource is incredibly important in bringing relevant and engaging science to an African audience, from an African perspective. Dr. Joy is a champion for equitable educational experiences for African learners and strives to make STEM learning accessible, engaging, entertaining, and most of all, informative.

Gosia Lukomska is a creative director, film producer, and educator, who has led the creation of some of the most innovative media initiatives in the field of Social and Behavior Change Communication in Sub-Saharan Africa. Her work has been featured globally in the New Yorker, CNN, BBC, NPR, and Al Jazeera. Prior to her engagement in the development sector, Gosia worked for a cutting-edge marketing agency, Interference Inc. in NYC, where she was responsible for the design and implementation of innovative multi-media productions. Gosia is an occasional guest lecturer at universities around the world and has been involved as an educator with the Urban Arts Partnership, bringing arts-integrated programs to underserved public schools in the New York area. She holds an M.F.A in Integrated Media Arts from Hunter College and an M.S. in Social Psychology from the University of Social Sciences and Humanities in Warsaw, Poland, where she was born and raised.

Rhythm and Glue:
An Entertainment-Education Prototype
for Climate Communication

Emily Coren

Entertainment-education (EE) is an internationally recognized strategy that lever-ages the power of storytelling in entertainment programming for social and behavioral change (Wang & Singhal, 2021). Since its earlier days, EE initiatives have used compelling narratives, inspiring characters, engaging mechanisms, and connection to real-world resources to address public health concerns and complex social issues such as family planning, adult literacy, HIV/AIDS, and domestic violence (Singhal, 2004; Singhal et al., 2013; Singhal & Rogers, 2002). For over 50 years, EE scholars and practitioners have expanded its theory and praxis to more diverse topic areas, entertainment genres, and media platforms to adapt to societal shifts and technological advancement (deFossard, 2008; Frank & Falzone, 2021; Storey & Sood, 2013; Wang & Singhal, 2009). I believe that EE can contribute to a versatile set of tools for the design and implementation of effective climate communication we need now. As articulated in several chapters within this volume, many EE flagship organizations have already shown commitments in the area of sustainable development, environmental health, and climate change (Bish, 2024; Brown, 2024; Garg et al., 2024; Sood et al., 2024) while socially-conscious creative industry professionals and environmental advocacy groups are also devoting more efforts in effective storytelling to facilitate positive climate actions (Falzone et al., 2024; Gurney & N'Diaye, 2024; Hinderfeld et al., 2024; Spiegel & Wang, 2024). My contribution here is to share an EE prototype for climate communication based on my evolving understanding and experiences in both of these worlds. Specifically, this chapter aims to present a prototype for how EE can be adapted to expand the delivery of timely, accurate climate change mitigation and adaptation interventions.

E. Coren (✉)
Stanford University, Stanford, CA, USA
e-mail: emilycoren@gmail.com

© The Author(s) 2024
E. Coren, H. Wang (eds.), *Storytelling to Accelerate Climate Solutions*,
https://doi.org/10.1007/978-3-031-54790-4_7

Background

The accelerating pace of scientific development means that most Americans outside the scientific community will learn the majority of their science after they leave formal schooling (Miller, 2010). However, despite efforts to improve scientific literacy and knowledge of environmental issues, literacy in these areas remains consistently low in the United States (National Science Board, 2020). While formal science education provides a solid base for the public understanding of science, it is insufficient for addressing rapid shifts in information acquisition due to emerging technologies and accelerated changes in scientific knowledge (Pew Research Center, 2015). Developing new mechanisms for the public to encounter opportunities for engaging with such knowledge is critically important, especially with regard to climate change (Cintron-Rodriguez et al., 2021). The cultural integration of science throughout interrelated media forms can build evidence-based discourse around scientific topics. While observing the exponential growth of scientific advancement and the linear cultural integration of scientific work, I became curious about how we could develop better systems to improve the cultural integration of science. I began exploring what mechanisms might help us better achieve these goals.

When I started scoping this work in 2015, it began expansively open-ended. I was curious about how to adapt established EE practices for climate communication. I didn't have a specific narrative goal in mind for the series and was curious about how narrative and scripted media could be used as a tool for advancing climate communication. I spent a year scoping the work through literature reviews and in-depth conversations with leading researchers and practitioners in both EE and climate change communication, and then worked with a team to apply for a National Science Foundation (NSF) Advancing Informal STEM Learning grant (Coren, n.d.). Just as we submitted our application in 2016, the federal government shifted away from climate communication. With a lack of federal support for a climate communication program of this scale at that time, I started to reach out to colleagues in Hollywood to see what could be done to move the work along privately. What I learned in that exploration was that an academic proposal wasn't useful within the media industry and that it needed to be already scripted to advance those conversations. So, I took the NSF proposal and converted it into a scripted narrative with a working title, *Rhythm and Glue*. The title Rhythm and Glue represents the two main characters, Sarah represents "Rhythm" for the music that she loves and Jessica represents "Glue" for her skills at building with whatever odds and ends are available to repair her community. It's also a double meaning in that when we are making sustainability improvements it's easier and more fun with literally a little music and community repair skills.

While I haven't yet succeeded at getting a program on the air, what I did do was to scope the work and build a prototype for a show. This is still useful for climate communication as a community as we continue to expand on applied work. I'm hoping that it is a practical guide for many communities to adapt narrative content for effective climate communication. In the years since I began this exploration, I see so much promise for expansion. I'm hoping to share enough of the process for it to be applicable in any narrative format for any audience segment. I see the screenwriting community uncertain about how to incorporate climate change in narrative media. While I strongly believe that these shows should be written by professional screenwriters, I hope that this prototype creates a functional example to overcome some of the most common challenges. Although not yet fully realized, I'm beginning to see how we might automate EE as a process for providing better public health coverage for communities.

For each topic, I'll give an overview of what the team that I worked on the NSF proposal with learned and then a specific example from the *Rhythm and Glue* prototype. I'll review audience segmentation for character development, story settings, visualizing positive outliers, using an agency for designing the narrative structure, transmedia experience, and program evaluation. Structured as a comedy drama series on television, *Rhythm and Glue* highlights the day-to-day realities of climate change facing residents of the United States, and role models realistic solutions and behaviors that citizens can take to mitigate those effects, for themselves and their communities. The story emphasizes personal and collective efficacy, including layered examples of how climate change is already affecting the health of Americans. Designed as a serial comedy drama, Sarah and Jessica, two female twenty-somethings in Los Angeles humorously grow into adults as they learn to collaborate while sharing their skill sets to build a better city that they both deeply care about. Slapstick, romantic entanglements, and career advancement anchor the story, while real-time audience participation drives engagement with climate topics.

One of the components of this project that sets *Rhythm and Glue* apart from other climate-related content is interactivity with its audience, where audience input shapes the story's conflicts and resolutions. Real-time audience participation facilitates a national conversation about how we can collaborate to mitigate climate change. As an entertainment product, this show provides a model for transitioning sustainable urban transportation internationally and a model for facilitating a national conversation with a variety of stakeholder perspectives. This media platform example connects viewers with a network of nonprofit organizations supporting actionable behaviors that the audience can take locally to mitigate climate change. Content creation is advised by a team of subject experts and iterative research. Audience participation facilitates engagement, building a community base, and increasing the capacity of existing climate mitigation programs. Each season presents an overarching climate change issue. Season One tackles transportation and the current health effects of climate change, with subsequent seasons addressing energy, water, food, and waste (For the full prototype, see Coren, n.d.). This is what we learned....

Character Development

I wrote these character examples as a hybrid of Global Warming's Six Americas audience segmentation tools (Chryst et al., 2018; Maibach et al., 2011) developed by Yale and George Mason University teams and demographic data based on Los Angeles at the time it was written. Global Warming's Six Americas audience segmentation tools break audiences into six groups: Alarmed, Concerned, Cautious,

Disengaged, Doubtful, and Dismissive. The proportion of these groups varies both regionally and over time. The Yale and George Mason University teams continually update and improve these models and you can search for your regional location and see how the trends are changing over time. Proportionally, the Alarmed group has increased in size since I originally wrote this prototype. Additionally, I did my best to highlight STEM (Science, Technology, Engineering, and Math) fields and BIPOC (Black, Indigenous, and People of Color) representation across the character set. Each of the characters features intentionally intersectional social identities. There were early production conversations about relative cast size and production costs that influenced the character choices that are written into this example.

These are the *Rhythm and Glue* character examples:

- *Sarah* a is computer programmer who feels her emotions through music (Figs. 2, 4, and 5). Even typing her fingers on her keyboard and the wheels clicking around on her bicycle provide a soothing rhythm to her. Sarah and Jessica are childhood best friends currently living together as housemates. Sarah is quiet and reluctant to change, she wants consistency. She's a 25-year-old African American asthmatic cyclist. In subsequent seasons, through professional development, she becomes a specialist in computer automation, and conflict resolution training initiated by her conflict with Jose helps her develop stronger interpersonal skills, both skills combined help her grow into a good manager despite her ambivalence to becoming one. She represents the *Concerned* group regarding climate change (as represented by Global Warming's Six Americas). She is convinced that global warming is a serious problem, but while she supports a vigorous national response, she is distinctly not very involved, at least at the beginning of this series.
- *Jessica* is a Latinx registered nurse in her mid-twenties who loves to fix things (Figs. 2, 3, 4, and 5). In her free time, she's perpetually taking old scraps and making new, beautiful things out of them. She is particularly good at loudly expressing frustration with something but then laughing her way through solving it. When she looks back at situations that she thought were problems, she only mentions the good things that she learned from the experience. She's sunny even when things don't work. After becoming increasingly concerned about public health through her experiences with Sarah's asthma and various patients at her medical practice in later seasons, she earned a Master's degree in Public Health and began working at the State Health Department. In Season 1, she learns to bike in a city, representing the transitioning character in Sabido methodology. She is *Cautious* about climate change. She believes that global warming is a problem, but she doesn't view it as a personal threat and doesn't feel a sense of urgency to deal with it.
- *Jose* is a graduate student with an infectious sense of humor who loves cooking and growing food (Figs. 4 and 5). Having grown up on his parent's almond farm, he studied economics as an undergrad to try to find a practical skill. He is now working on his Master's degree in Sustainability at UCLA, trying to figure out how to combine some sort of practical economics with his love of food and being outdoors. In Season 1, he's so overwhelmed by school and work responsibilities that he deflects his remaining energy into video games, tanking his romantic

relationship with Sarah. In Season 3, as an apology to Sarah, he builds her a community garden, so she can spend more time outside while she's home. It works and they move on to organize a CSA delivery program in their neighborhood. Like Sarah, he's *Concerned* about climate change, but not actively doing anything about it.

- *Daniel* is a young professional electrician and physical laborer (Figs. 4 and 5). He's energetic, likes to tinker and build machines, and is skilled at modifying electrical and mechanical systems. He's perpetually moving and good at getting the things around him moving too. He's *Disengaged* with climate change. He hasn't thought much about the issue at all. His electrical work and building skills lead to tinkering with renewable energy in their homes in Season 2: Energy, which later leads him to start a company installing solar panels.

- *Andy*, *Josh*, and *Matt* are casual friends living in another apartment in Sarah and Jessica's complex (Figs. 3 and 5). They are a little like the coffee-drinking/break-room aliens from Men in Black, only they drink beer, not coffee. They are frequently playing drinking games (beer-pong, croquet, barbecuing competitions, etc.) in the courtyard of the apartment complex. They represent the *Doubtful* perspective on climate change with different stages of understanding and acceptance of the problem, and none are actively involved in anything climate change-related. Andy stands out as particularly funny and an effective social organizer.

- *Roger* is the grumpy, misanthropic neighbor in the apartment complex. He provides both conflict and comic relief (Figs. 3 and 5). Roger is retired and belongs to the 1950's American dream. He finds disappointment in what everyone else is doing. He followed the rules he was taught to succeed but the world changed and he's trapped between an ideal that no longer exists and a reality that he doesn't know how to be a part of. Roger likes old cars, sports, and listening to the radio. Conflict arises between him and many of the neighbors. Roger is *Dismissive* of climate change. He's very sure it is not happening and is actively involved in opposing a national effort to reduce greenhouse gas emissions. As a caricature of the position, he's very loud with his opinions, frequently spouting inaccurate information that is sometimes believed by Josh or Matt.

- *Elizabeth "Liz"* is a work friend of Jessica's. She's a Latinx medical assistant, amiable, slight, untidy, and disorganized (Fig. 5). She wears thick, cute dark-rimmed glasses and old comfortable sneakers. She's very empathetic to her patients, but maybe not so patient with Stephanie. She's *Alarmed* about climate change. She's fully convinced of the reality and seriousness of climate change and is already taking individual, consumer, and political action to address it. As a caricature of the Alarmed position, she's annoyingly involved and bossy in social causes. She's always doing marches and suggesting things for everyone else to do. Her initially antagonistic stance toward social activism eventually proves a useful resource to Jessica because she's a well of information. She helps Jessica navigate the hurdles to participation in sustainability topics as Jessica gradually gains interest in them as well.

- *Stephanie* is another friend of Jessica's at work. She's also a medical assistant and is the counterpoint to Liz (Fig. 5). However, detail-oriented Stephanie is particularly good at phlebotomy and untangling administrative/insurance prob-

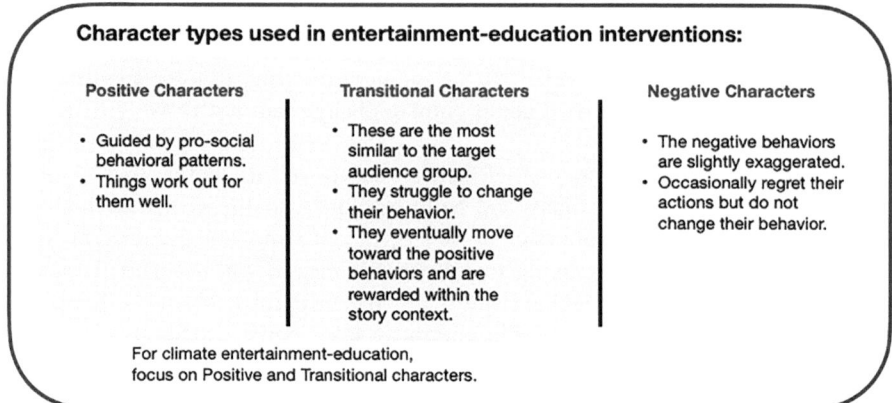

Fig. 1 Character types in entertainment-education programs, modified for climate communication toward positive and transitional characters

lems. Jessica and Liz refer to her for help drawing blood or fixing administrative glitches. She's kind, happy, and overtly materialistic. Tall and a little pudgy, she overcompensates for being self-conscious about her weight by dressing up too much. Stephanie is *Disengaged* with regard to climate change. She likes dancing, drinking, and shoes more than the planet. Conflicts arise between Stephanie and Liz about values and disagreements about consumerism and social activism.

EE characters are often categorized as positive, negative, or transitional characters relative to the intended behavior change. I recommend a modification of that in climate sector entertainment-education to make all characters either positive or transitional, omitting the negative role because we cannot leave people behind in this work (Fig. 1). Story conflicts that begin with negative characters can eventually become transitional characters over time. New story conflicts can be woven in by adding characters that represent new stakeholder groups, emotional, and technical challenges while still allowing the fictional communities to gain cumulative progress. Sarah and Jessica's relationship with Roger begins as antagonistic but gradually warms. They gradually realize that Roger is grouchy because he's lonely and needs social connection. As they develop empathy for him, they find shared goals for their building complex such as adding rooftop solar, even though they have different initial motivations for wanting the solar panels installed with Roger prioritizing energy independence over environmental goals. This recommendation gives both people and characters room to grow.

Story Settings

The stories in *Rhythm and Glue* are set in the present day and focus on current climate actions. Climate adaptation is the largest, fastest set of cultural transformations required in human history. Stories set in the future will lose the detail of optimizing present moment solutions emerging daily. Instead of jumping into a hypothetical future, I encourage you to stay in the present. The residential setting was selected as an apartment complex in Los Angeles with a courtyard supporting intergenerational interactions. The occupational settings of the main characters were selected to showcase a variety of sectors all actively adapting to climate solutions, namely, Jessica works in a healthcare setting, Sarah works in a tech office, Jose works in the agricultural sector, and Daniel works on electrification. The residential setting was selected as a mechanism to illustrate a place where different demographic groups would interact. The medical setting was chosen to illustrate the health impacts of climate change in various subpopulations. As a science communication product, I would answer to the question, "If we optimized narratives to have the largest constructive difference we can make right now?": Set your story here and now in any present-day community, take the climate mitigation and adaptation work already being done, and include it in your story. Collecting local climate goals and building them into storylines by visualizing positive outliers is possible. Collecting crowdsourced geotagged images and stories through an interactive web-based ArcGIS tool called Story Maps (see details in the chapter by Wolf-Jacobs et al., 2024), images of local positive examples of climate engagement sources from community members can be collected and updated in real time. These citizen science-led crowdsourced examples of climate engagement can be used to identify and amplify existing community behaviors that support climate mitigation and adaptation and connect community members to existing programs.

Concept Art

As I got into the story development process, I commissioned concept art through an illustration firm. It was surprising to me how much reference material it took to generate anything close to a representational image set. And, stock art just didn't cut it. I observed lots of details in person that were not visible from search results and web research. A goal of this work is to identify and amplify existing community behaviors that are already in progress and to connect with viewers to those projects and each other for amplifying that work. Adapting to climate change is the fastest global set of cultural changes that humans have ever made. *Positive Deviance* (also referred to as *Positive Outliers* or *Bright Spots*) analyses can help communities to identify existing community solutions to complex challenges and to amplify "what's working?" (Buscell et al., 2014; Pascale et al., 2010; Singhal & Durá, 2017). Automating this process will not just be useful to writers but is potentially useful across all media types. Behavioral solutions may include improvements to electrification infrastructure, transitions to renewable energy sources, local food consumption, transitions to active and public transit, and advocacy for improvements in climate and health policies. There are a lot of possible behavioral interventions for climate, and tailoring communications strategies to local community needs is an area that we have room to improve. Using digital data sources in the systematic identification and understanding of positive outliers in various domains is beginning to be explored (Albanna et al., 2022). Experimentation is also beginning with ArcGIS as a tool to provide meaningful details for concept art, storyboarding, etc. to improve the regional and cultural nuances in real time (Coren & Myers, 2021). Automating this as a process through a GIS tool can help in codifying regional and demographic examples for a more systematic approach to identifying and amplifying the positive outlier examples will be useful across all media types to improve the delivery of accurate community-based reference material. Building a process that moves in real time with communities to identify and amplify positive outliers for climate mitigation and adaptation behaviors will improve our rate of response to environmental public health challenges. Figures 2, 3, 4, and 5 are examples of concept art developed for *Rhythm and Glue.*

Fig. 2 (Left to right: Sarah and Jessica) Sarah and Jessica explore Los Angeles together on bicycles, representing transitions to active transportation with climate and health co-benefits

Fig. 3 (Left to right: Roger, Jessica, Andy, Josh, and Matt) Andy, Josh, and Matt play beer pong in their apartment courtyard. Roger scolds them for the noise and clutter. In Season 1, there are air quality and health plotlines that play off of the social dynamics between Roger and the younger characters within the apartment complex setting

Fig. 4 (Left to right: Sarah, Jose, Daniel, and Jessica) Jose and Daniel join Sarah and Jessica at their apartment to drink and play music in the evenings. The Music Party Scene in Episode 9 is behavioral modeling of social support and community building

Fig. 5 (Left to right: Roger, Josh, Sarah, Jose, Daniel, Andy, Jessica, Liz, Stephanie) Concept art for the main ideas in Season 1: Transportation. This illustration shows each of the characters using different types of multimodal transportation. Jessica, Stephanie, and Liz are out for a walk on a lunch break. Jose is driving for a rideshare. Daniel is riding a cargo bike with a trailer of tools for work. Andy is riding a city bikeshare. Josh is riding an electric bus

Plotlines

EE programs often have a specific focus behavior target. In the formative research stages, the behavioral interventions are defined. For widespread application, I would like to see a systemization of the formative and summative evaluation stages to

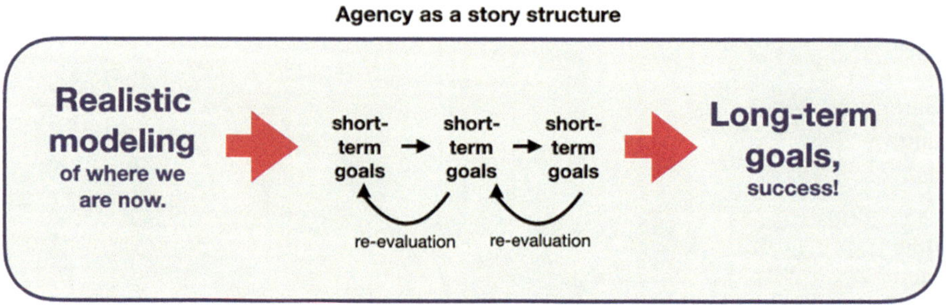

Agency as a story structure

Fig. 6 Agency as a story structure. Stories should use iterative goal setting as a plot tool to socially model agency at both individual and collective levels. Have the characters set short-term and long-term goals and periodically re-evaluate them to reach their long-term goals

improve the rate of development of EE programs. In the climate sector, for each community, multiple parallel behavior change strategies must be sub-targeted for community variability. The trend overall in the story should use an Agency frame, using iterative goal setting to teach characters the skills they need to do at individual and collective levels. A simplified chart of agency as a story structure is in Fig. 6, it can be used as guidelines in narrative development to set character goals.

Story Arcs

These are descriptions of the story arcs for the first five seasons of the Rhythm and Glue prototype.

Season 1: Transportation

Jessica learns to bike in Los Angeles to reduce the amount of time that she's stuck sitting in traffic. Sarah, who is living in the same apartment complex as Jessica and is already an avid bicyclist, teaches her all of the skills that she needs through the transition to active transportation. Sarah and Jose begin the story dating but conflicts arise between them. Jessica starts the season single, meets Daniel when he stops to help her with her bicycle and they begin casually dating. Jose is in school studying agriculture, and goofing off a lot on the side. In the season finale, at a New Year's party, Sarah and Jessica go for a run and Sarah has a scary asthma attack. Jessica saves her life by sharing her inhaler.

Season 2: Energy

Sarah and Jose break up. Sarah pours herself into her work and learns machine automation. Daniel tries to impress Jessica. Daniel and Jessica construct electric artwork together, which leads them to tinker with renewable energy in their homes. A comedy of errors ensues, such as problems with costs, permitting, etc. There are many group discussions about appliance energy efficiency, engineering, and efficient power generation. Jessica fails at biking to work and relapses to using her car, later overcoming it and becoming an advocate for urban biking. While in school Jose learns about climate change affecting food supply and moves from Concerned to Alarmed, he finally takes school seriously. Daniel starts off disengaged with renewable energy but thinks the tinkering is fun; he gets discouraged by repeated failures. As he gains experience and gets plenty of encouragement from Jessica, he gradually becomes more confident and advocates for renewable energy and electrification updates.

Season 3: Food

Jose finishes school and gets a job. Jose builds a community garden to win back Sarah's affection. Sarah and Jose work together to organize an automated Community Supported Agriculture (CSA) delivery with local farms. Jessica sees further public health benefits from improved nutrition, lower air pollution, increased exercise, and fewer insect disease vectors. She gets frustrated that she can't do more to improve public health from her current job and goes back to school for her Master's in Public Health. Daniel goes back to school too, to study business. They see less and less of each other and gradually drift apart. Jessica starts flirting with a new classmate. Jessica and Daniel break up.

Season 4: Water

Watering the garden and increasing city drought and watering restrictions bring up water conservation and find growing efficiency issues for Sarah, Jose, and Jessica. This leads to a larger conversation of municipal water supply being fragmented and the importance of managing it as a society. Jessica is now dating her classmate. They like each other, but they don't have enough interests to stay together. Jessica begins to work in public health. Because of Sarah's success with machine automation, she gets promoted. Sarah gradually gains confidence as a manager. Jessica's art gets recognition. At the gallery opening, Daniel comes and they reunite.

Season 5: Waste

Jessica is offered a residency for her work with recycled objects at the dump. As a result of the residency, she becomes sensitive to the amount of waste in society. Sarah/Jose and Jessica/Daniel are all happy couples; they team up constructively to figure out the engineering challenges. As a result of the project, Sarah and Jessica learned a lot about city planning. Daniel starts a company installing solar panels. Jessica needs help with civics because she's overwhelmed by the art residency, public health clinic and biking, city planning, and compost projects. Daniel is too busy with his new business to help her so she commiserates about it with Sarah, who steps up to the challenge. With Jessica's encouragement, Sarah runs for the local office.

Story Engine

One of my goals in *Rhythm and Glue* is to facilitate a national conversation about transitions to civic sustainability through the microcosm of the show. The pilot sets up the basic story and establishes the characters. All further content has two sections:

1. *Fixed story checkpoints,* mostly revolving around the character's romantic entanglements and Sarah and Jessica's career advancement.
2. *Flexible storylines that incorporate audience participation through transmedia elements* (stories connected across platforms and formats and interlinked with live community programs)**.** These areas give the audience an opportunity to share their successes and frustrations with climate change mitigation activities. These sections mostly relate to social and engineering challenges experienced by show viewers. Through nonprofit partnerships, we planned to increase participation in existing climate change mitigation programs and to identify existing bottlenecks to social-environmental change. By interspersing these two elements, it's an architecture that allows for narrative character development, while still supporting a real-time national dialogue.

Collective Efficacy

Below is a visual example of how plotlines woven together through different character perspectives can increase the visualization of behavioral skill building for a range of skills in parallel. For example, I don't wait to build on health impacts and health interventions in a clinical setting for completing a story modeling active transit skills. I'm building them together within different story settings on ABCD tracks (Fig. 7). I used foreshadowing liberally in the first few seasons of the show to set up health interventions in later seasons. I broke the social–emotional skills into blocks

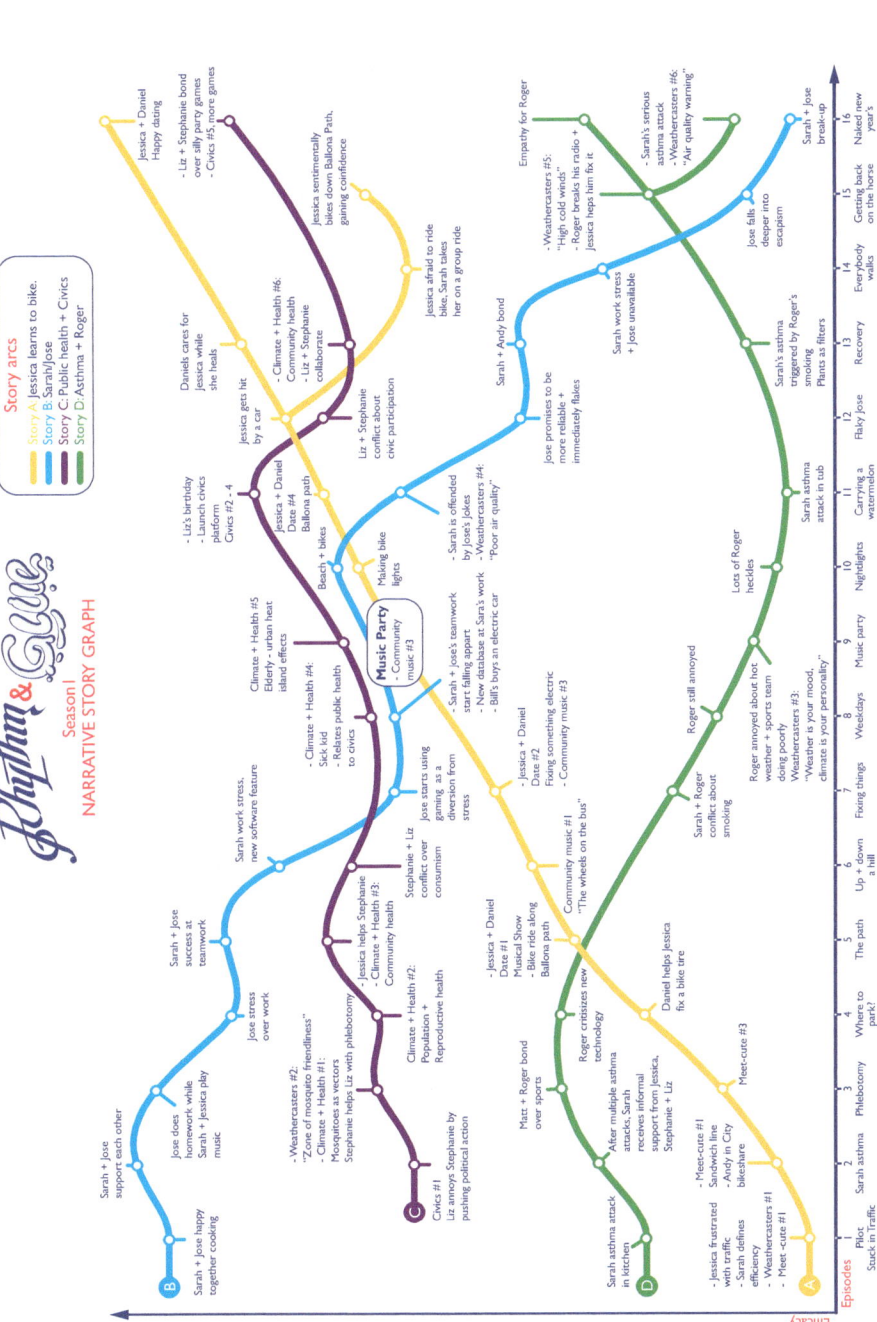

Fig. 7 Here is a detailed example of the Season 1 plotlines in *Rhythm and Glue*. Story A—Jessica gets frustrated with city traffic and, with Sarah's help, learns about active transportation. Sarah, already an avid bicyclist, helps her figure out the nuances of biking in a big city. Along the way, Jessica builds a relationship with Daniel. Story B—Sarah and Jose's relationship implodes as their work stresses increase and they lack effective communication and stress management strategies. Story C—In a medical office, Liz and Stephanie debate overconsumption. Medical cases in the office highlight the relationships between climate and health. Story D—Roger feels lonely and discouraged. His smoking and criticism irritate his community

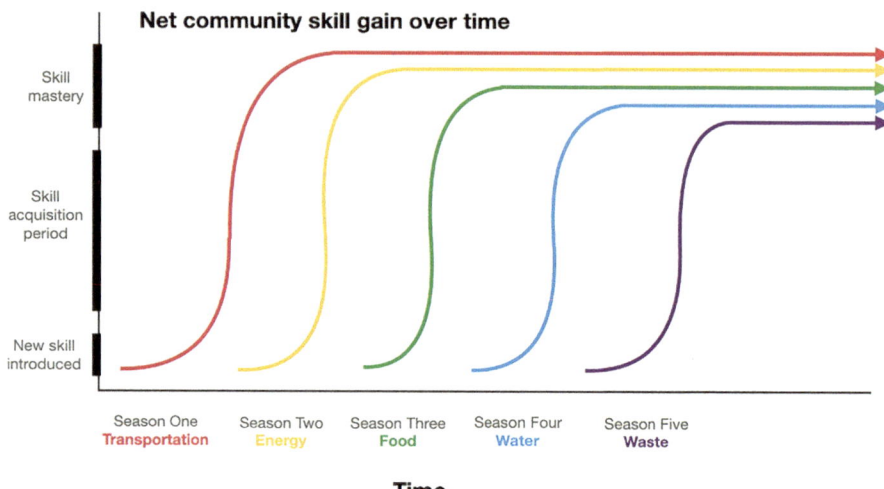

Fig. 8 Over a five-season story arc there is a net community skill gain over time. As the community gradually builds skills, you can see the agency and the skills improve. While there are individual setbacks and frustrations as plot points, the character and community building trends increase cumulatively over the length of the show, visualizing collective agency through narrative

that developed throughout all five seasons. While there are individual setbacks and frustrations as plot points, the character and community building trends increase cumulatively over the length of the show, visualizing collective agency through narrative (Fig. 8).

Season 1 Story Arc

A Story: Jessica is learning active transportation, biking, and dating Daniel. Jessica wants to add a new electronic dimension to her artwork by learning how to solder. When she misses an opportunity through an art museum workshop, due to city traffic, she gets really frustrated. Already an avid bicyclist, Sarah helps Jessica figure out the nuances of biking in a big city. Jessica sees the warmth that Sarah and Jose have together and is appalled by her apparent dating options. She pines for the kind of relationship that Sarah and Jose share, full of shared interests. Roger, their misanthropic neighbor, is jealous of his cat Mini's affection for Jessica and Sarah. Roger perpetually heckles boisterous Andy, Matt, and Josh as they play in the quad. Bicycling leads Jessica to a romance with Daniel. She is pleasantly surprised to find Daniel is a collaborator with a complementary skill set. They both like making things with their hands, and Daniel is able to add the electricity to her art that she was hoping to learn, while she adds purpose and color to his tinkering. Roger gets annoyed by the accumulating bicycles parked in shared areas as he repeatedly trips over them. When biking, Jessica gets hit by a car, she's afraid to get back onto a bicycle. Jessica spends so much time stuck in traffic that she's practically living in her car. A comedic navigation error with a GPS leads her back to a bike path that she fondly remembers with Daniel. She hesitantly regains her confidence in biking, ultimately becoming an advocate for improved bicycle parking in her community by asking for more bike parking at their apartment complex.

B Story: Sarah and Jose share a love of cooking and music. Jose cooks for Sarah, who in turn helps him prepare for the catering that supplements his student income. Sarah's rosy relationship with Jose deteriorates as her stress increases due to work and asthma. Her boss makes a series of ill-conceived choices: selling questionable software, switching databases, and hiring a Chief Security Officer who comically locks down everything in the office. Sarah's chronic asthma is exacerbated by Roger's cigarette smoke blowing into her bedroom window. The coffee machine at Sarah's office breaks repeatedly as a comic gag, and she just can't seem to get a break or even a cup of coffee. Jose himself is overwhelmed by his work driving for Uber, catering, and school assignments. Jose, seeking relaxation from his many commitments, gets pulled into the video games that Josh, Andy, and Matt are playing downstairs. The video games divert his attention away from Sarah and their relationship suffers. Andy provides empathy and restful space for Sarah when she's stressed. The finale is a Naked New Year's party full of wild party games and tons of alcohol. At the party, Sarah and Jessica leave drunk to go running and Sarah has a major asthma attack, but her inhaler is lost. Jessica helps her home narrowly saving her life by sharing her inhaler with her friend. Jose is yet again distracted by video games and not there for Sarah when she needs him and they break up. Jessica consoles Sarah. Season One ends with a circuit overload at the New Year's party causing a power outage. Jessica and Daniel comically open up the wall together to repair the damage, leading us into the Season 2 topic: Energy and power infrastructure.

Community Building

Social skills training is embedded in the design of *Rhythm and Glue* that extends beyond the character and setting design and is globally applicable in building resilient social structures that apply to all genres. Modeling behavioral strategies for community building through media is a way for us to build community resilience by deepening and interconnecting social ties. I'll focus here on building social cohesion and increasing collective efficacy. Climate change threatens psychological health as well as physical well-being (Crimmins et al., 2016; Frumkin et al., 2008; Luber & Lemery, 2015). Building emotional resilience and social cohesion to lower the risk of negative mental health impacts of climate change (Coren & Safer, 2020; Hikichi et al., 2016) becoming a network intervention layer within public health intervention. Rapid anthropogenic change is also increasing population mobility and resettlement globally (Torres & Casey, 2017). Supportive community social ties (i.e., community social cohesion) serve as a source of resilience in the context of protracted and acute disasters (Torres & Casey, 2017). Improved social cohesion is protective of mental health outcomes (Hikichi et al., 2016) and improved social cohesion can be healing post disasters (Wickrama & Wickrama, 2011), therefore establishing and normalizing a set of cultural practices that build and maintain community social ties, improves population resilience, and adaptation to climate change and mental health outcomes. As people move around more frequently because of instability issues, people don't have as long to develop and maintain relationships; therefore, it becomes increasingly important that we be able to build trusting relationships quickly to provide the necessary social services. To improve social cohesion, we need to increase the density of social ties and increase the rate at which we form them. Resilient community structures have an infrastructure that holds its place even when people move away or transfer between project roles. This helps long-term projects persist through administrative and funding changes by providing community anchors. Social structures are dendritic. Improving the frequency of social nodes and the interconnection between them improve both individual and community health. Reconnecting and strengthening social ties may increase resilience and buffer some of the adverse mental health impacts of climate change. Unifying and normalizing a set of behavioral skills for community social support was an intentional strategy built into the story design of *Rhythm and Glue*. The quality of the relationships in the community provided support for each community member as they learned new skills and experienced emotional challenges. Sarah and Jessica's relationship was a key design within the narrative; they are genuinely friends, with narrative representations of people demonstrating loyalty and trust, and doing what they say they're going to do. I want to flag this because many of the shows that I see on television at the moment have a premise of fear and mistrust. Anecdotally, I notice a lot of plotlines creating social rifts just for the entertainment drama of it. There's a way to show community building with plenty of plot and action that still shows empathy and cumulative community building over time. Providing concrete examples of trusting behavior can also support the social justice issues that perpetuate through media as we begin to repair harms from systemic environmental justice policies and air quality-related health disparities such as redlining (EPA, 2022).

In *Rhythm and Glue*, I intentionally started with an emotionally bonded set of characters, Sarah and Jessica. I established at the very beginning of the series that they were close long-term friends. The newer relationships in the show, such as the romantic attachment with Daniel, were built during the onscreen narrative. Sarah and Jose's break-up in the first season leads to emotional growth for Sarah in developing new social-emotional skills and then being able to apply those new skills in professional contexts. These were intentional story choices to provide a range of skill levels at social-emotional skill-building and to provide a context within the story to teach emotional resilience. I'm currently using the nonviolent communication training practice to teach empathy as a skill (Rosenberg, 2015). The basic pattern for the nonviolent communication training is Observation, Feeling, Need, and Request. Observation is what you see or hear. Feelings and Needs are sub-defined sets (Rosenberg, Marshall, 2021). Requests can be for an action, for example, "Please add bike lanes to my street," or for connection, "Would you please reflect on what you heard me say?" I like teaching with this set because the structured pattern of the feelings and needs lists makes emotional communication into a concrete skill for educational purposes. In this context, I'm defining care as an action, not a feeling, as in the act of participating in social care. A strong emphasis was placed on the quality of character engagement by socially modeling pro-social care. Higher network cohesion improves community resiliency. Many climate behavioral interventions are aimed at individual-level behavior change. I include community resiliency overall as a behavioral target, including social cohesion and emotional skill-building.

There are many kinds of network interventions using social network data to accelerate behavior change (Centola, 2010; Hunter et al., 2019; Kim et al., 2015; Valente, 2012). Alteration network interventions (Hunter et al., 2019) support the addition of new members and ties to social networks and are particularly helpful in community resilience support for increasing social cohesion. Practices for building and maintaining social networks will improve community health outcomes as populations move due to an increasing number of direct and indirect stressors from climate change. Social modeling of these social skills provides a unified cultural reservoir for supporting social infrastructure. In recognition of the impacts of climate change on psychological well-being, climate mitigation and adaptation strategies should include a set of shared, unified cultural practices and behaviors that support mental health and community building. As communities experience trauma and migrate, having these cultural skills normalized across a wide range of existing social norms will support both healthier communities now and less fractured communities adjusting to changing environmental conditions. Climate change isn't limited to a single behavioral health intervention but many in parallel (electrification, food consumption patterns, transportation changes). As the climate destabilizes mental health, community building and emotional resilience interventions being culturally normalized might yet become deeply important behavioral interventions.

Suggestions for social support behaviors that can be modeled through media include:

- Investments in enhancing community cohesion such as formal check-ins, buddy systems, and network building (Torres & Casey, 2017). Building and maintaining relationships both physically and remotely.
- Social ties facilitating the exchange of resources (e.g., food, shelter) and information (e.g., early warnings regarding environmental risks) can increase the likelihood of disaster preparedness (Torres & Casey, 2017). For example, people joining or hosting community meals or questions like, "Do you need assistance finding housing? Here, let me show you where to find housing in our community."
- Including people displaced by climate change under international refugee law and providing structured pathways for political and social integration into destination societies (Torres & Casey, 2017).

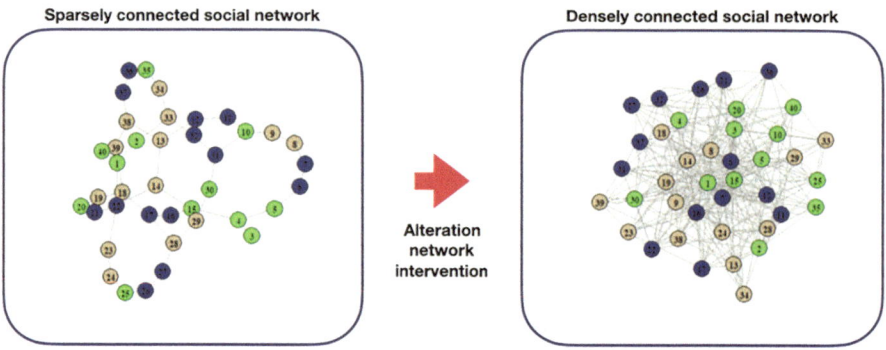

Fig. 9 This is a visual representation of a social system of 40 people moving from a sparsely connected social network to a more densely connected social network. In this illustration, the dots represent people and the gray lines represent the relationships between them

- Providing psychological first aid as an initial disaster response intervention with the goal of promoting safety, stabilizing survivors of disasters, and connecting individuals to help and resources. Psychological first aid assists community members assess the immediate concerns and needs of an individual in the aftermath of a disaster. Increasing the cultural normalization of psychological first aid will help community members assist each other in responding to environmental disasters more effectively. A variety of training resources for psychological first aid is available at the American Psychological Association site (American Psychological Association, 2022) and Coursera (2022).

In social network interventions in public health, the majority of studies focus on the spread of messages across networks. In addition to including that in our future designs for climate communication messaging, I'm proposing that network interventions that build social ties throughout a network improving the density of social support are itself a resilience adaptation (Fig. 9). We need to be teaching community behavioral strategies that improve both the quality of relationships and the density of social connections.

Transmedia Extensions

Transmedia storytelling is a process where key story elements in a fiction set are systematically incorporated across multiple media types, creating a unified and coordinated entertainment experience. Ideally, each story piece makes its own unique contribution to the unfolding of the story (Jenkins, 2010). An example of transmedia storytelling is Star Wars. It starts with three films, then additional films then books, soundtracks, TV shows, games, and toys were produced to extend the story. The same story or threads of it persist across media types. When you connect with the Star Wars franchise universe, you're not watching the same film in movie, book, and audio formats. The story is addictive. The story grows through different times and perspectives across media types, hence transmedia.

As the themes addressed in *East Los High* focused on reproductive health, the transmedia tools connect viewers to live pro-social programs, such as reproductive health services provided in-person through partnerships with Planned Parenthood, as well as safer sex education instructed by characters in pull-out videos such as Cece's vlog (Wang & Singhal, 2016; Wang et al., 2019). The narrative then connected viewers across media types and programs (Wang et al., 2019). Applying this methodology to a climate narrative for what I wrote starts with the main story arc for *Rhythm and Glue* as a television program. Then the story connects to place-based participatory programs as the connective tissue of sorts between them. Videos similar to the character-driven pull-out videos of *East Los High* have characters from *Rhythm and Glue* teaching relevant skills that need more elaboration or space than fit within the show's main narrative. Examples could include Daniel teaching home electrification updates, Jessica teaching active transit skills such as basic bicycle maintenance or how to take your bike on public transit, or Liz and Stephanie teaching emergency health skills for climate and health impacts, such as how to install an air filter on your home to prepare for wildfire season.

The transmedia plan for *Rhythm and Glue* was planned to follow the structure of *East Los High* with a variety of nonprofits assisting with the on-the-ground participation for a range of active transit, electrification, and habitat restoration activities. The first season of the show was planned to focus on transportation, so National partners such as People for Bikes, League of American Bicyclists, and Green for All were combined with local/regional teams such as Los Angeles County Bicycle Coalition, Bike Coalition Philadelphia, Bike Austin, and the Washington Area Bicycle Association. At the time of our NSF submission over 50 organizations had signed on as partners. This was a heavy lift, as the production was a small independent production. There was reluctance on both the part of the media production industry partners to take on a program with this many educational components and there was also reluctance from the nonprofit partners for capacity reasons. While they were all enthusiastic about the premise and goals of the program, the nonprofit partners were universally experiencing their own capacity limitations. Financial support to pay for additional staff time would have been really helpful, though it was beyond the budget that my NSF proposal could cover.

When designing your transmedia interfaces please assess which groups are already doing work for your areas of interest topically and geographically. Listen for what is working for them and listen when your teammates and audiences start to share what's working for them and where they will need capacity support. The content and format should be custom-designed for each intervention matching the intended audience's tone, platform, and language preferences. National-level automation of the transmedia design and evaluation components would help reduce the workload on the production companies and nonprofits who want to participate in this program. It would lighten the load and improve the range and interconnection of participation.

Program Evaluation

Program Evaluation is a key component of an EE initiative. Before a show is written, there's often an extensive research process to plan it out as a communications strategy. Once a show is written, there are focus groups and message testing. There's iterative research that pairs with the entertainment product while it's running and finally, there's a summative evaluation to determine how successful the campaign was at meeting its goals. None of that has a place within the entertainment system. These evaluations are expensive and labor-intensive. If, as a science communication system, we provided these services tailored to the writers and producers, we could streamline this as a process and improve science and public health communication. Evaluation should be coordinated nationally through a centralized system. One of the things that I learned in the pitching process through conversations with entertainment industry leaders was that the depth of formative and summative evaluations, while necessary to make these entertainment-education and health communication projects are too heavy a lift for the production teams themselves. Providing it as a free, streamlined service would improve the adoption of these practices. I note that these methods are constantly improving, so again, centralizing it as a process would improve our collective efficiency in providing an evaluation plan for each show.

Science communication needs to be an iterative design process. I was looking forward to polling the audience for additional directions. I think the key questions to ask the audience are: (1) What do you want to know? (2) How would you like that information delivered? When I first envisioned this work, I hoped to choose a dozen or so different demographic sets of audiences that we'd like to connect with, repeating this entire process with each demographic set and then interweaving the stories like the story crossovers in the Marvel universe demonstrating regional variation examples for sustainability transitions in different locations. It would then have the potential to interconnect different regions' adaptation and mitigation strategies as well as have the space to cross-intergenerational and cultural boundaries.

Engagement works better than deficit communications (Kearns, 2021). Deficit communication is providing people with additional information and expecting them to change their opinions and behaviors. Engagement communication styles are bi-directional, culturally nuanced, and more of an ongoing conversation with your target audiences (Holliman, 2009). There is also discussion within the arts communities that recommends building communities instead of building audiences (Borwick, 2012). Music as auditory communication also fosters community building skills (Higgins, 2012). Fitting science communication into existing industries as functional machinery involves flexibility and staff time for the science communication interface. It's our job to design a system that works for and within the existing social norms of the sectors that we work across. Adding this as a permanent functional role within science communication should be funded permanently as a system update to support "mutual learning by publics and scientists" (McCallie et al., 2009). It should be a system-level design about where to structurally invest and how to coordinate to move information effectively across larger systems.

Future Directions

I hope that future work will build on this prototype. I would like to see the automation of components of the formative research process and image prototyping integrated into community action plans (Wolf-Jacobs et al., 2024). We should intentionally design planned entertainment-education programs for a statistically representative set of target audiences so that each community has representative behavioral modeling of how their community is adapting to climate change. Also, for the United States, an entertainment-education program's evaluation and transmedia components seem to be too heavy to lift for the entertainment partners. If the US science agencies build a transmedia interface and support the formative and summative evaluation then the entertainment programs can "plug in" to the interface and join in where their capacity allows, making participation easier while increasing the accuracy and volume of narrative climate stories. This boundary space is tricky. It will involve deep, trusting partnerships within science agencies to support the infrastructure improvements for science communication and entertainment creatives to be willing to experiment in this shared space.

This chapter provides a toolset and an example for constructing narrative climate stories. Just to recap, characters should be demographically matched to your target audiences, and behavioral interventions should be modeled for multiple parallel targets by identifying existing community positive outliers. Social support behaviors for collective problem solving should be modeled by characters in all plotlines, regardless of the intervention topic. I expect there to be a tremendous amount of creativity and range in the execution of these stories; however, some basic guidelines might facilitate the accuracy of representation so that we can reach shared goals faster as a team. It also is likely to support writers who are interested in social impact stories to get the most impact out of whatever story they would like to tell.

Somewhere between the academic "shoulds" and a tangible plan to build a more robust science communication system, as next steps, I recommend:

- Funding the Action for Climate Empowerment (ACE) work—Scaling this work will require a lot of consistent funding support for staff and programs, building this capacity for public health communication infrastructure is an investment for developing a more adaptive public health communication system that is long overdue.
- A national-level department staffed to coordinate the components of this system and to build a functionally supportive systemization to build and maintain an integrated entertainment-education program as a component of the existing public health communication system.
- I'd establish norms for choosing a representative set of communities and sampling positive outliers of climate adaptation in those regions. By refining our audience segmentation and sampling tools we're also able to improve the visual storytelling of who exactly is doing what within their communities. While the geospatial tools are still nascent for this use, my recommendation from here

would be to use regional and culturally demographically specific characters to build on regionally specific examples of climate action.

- Build partnerships with creatives to create a first set of entertainment-education programs and then deeply evaluate to learn what works and what needs modification.
- While the art products that we generate need to be unique and targeted, the systems that we use to build them can be automated as a process. This would substantially reduce the cost, time, and accuracy barriers.

I think that we need to automate the formative evaluation steps and provide more concrete examples of audience engagement with climate solutions. Due to the increase in technological opportunities, I'm optimistic that we have the technological capacity to automate and coordinate these programs. Despite the scale of the impacts of climate health outcomes, funding this work has been difficult. I'm hopeful that the emerging Action for Climate Empowerment work continues to draw together the capacity to make this work a reality.

We need to be building a coordinated system of messaging across fields originating with community-centered input. There is no one perfect message in climate communication;, we need coordination and accuracy to improve a lot of different kinds of communication assets. As technical criteria, climate communication needs to be bi-directional and designed to meet the communication preferences of the people receiving the messages. In addition to better supporting the groups as communities of practice, we would benefit from a more structured toolset for the coordination of multidisciplinary work. Due to the rapid technological development of geospatial and network intervention tools, we should be pairing those tools with art and communications strategies to develop a systemization of content creation that meets community needs. In Fig. 10, I'm proposing a mechanism to improve the coordination of messaging across regional needs and media types. By "tightening the loops" in this coordination of climate messaging model, we can reduce lag time and improve accuracy across the media landscape.

Technology sector work involves automation that needs to be applied in the civil sector where much of the climate change activism messaging originates. Network analysis can facilitate collective efficacy visualizations, using storytelling to help individuals to see the roles they play over the larger system allowing them to focus on what they're doing so they understand their roles without feeling so overwhelmed thereby improving climate anxiety outcomes. Camera placement should be at the level of individuals, modeling target behaviors. These should be combined with geospatial and network visualizations to show the relative placement of the activities. Time-lapse visualizations can be used to show the relative impacts of policy options. Auditory cues can be placed first modeled within narratives and then repeated in the physical settings where they occur in the stories to reinforce reminders of steps (Fig. 11).

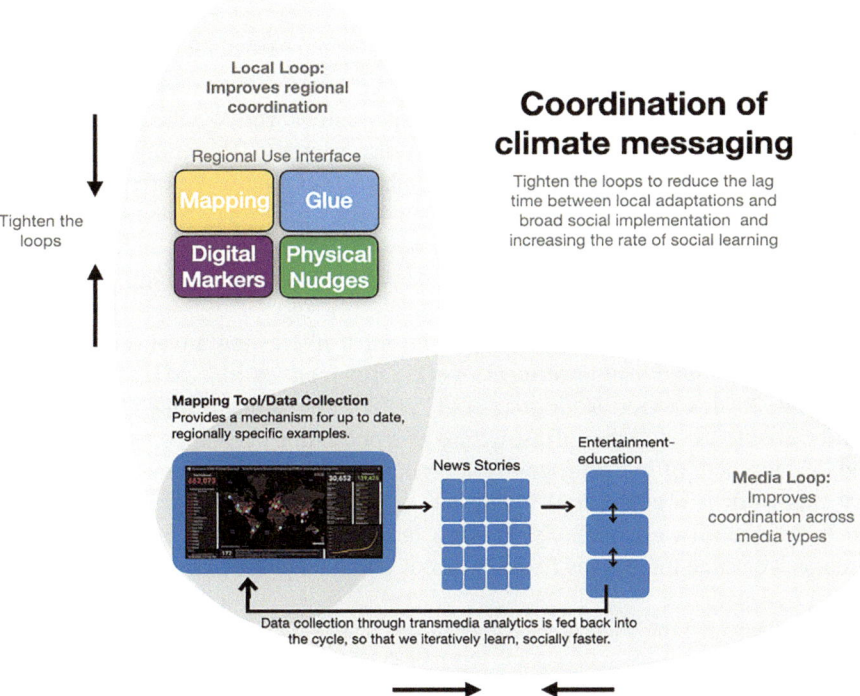

Fig. 10 Coordination of climate messaging first within a region, and then across media types (journalism products, and scripted television)

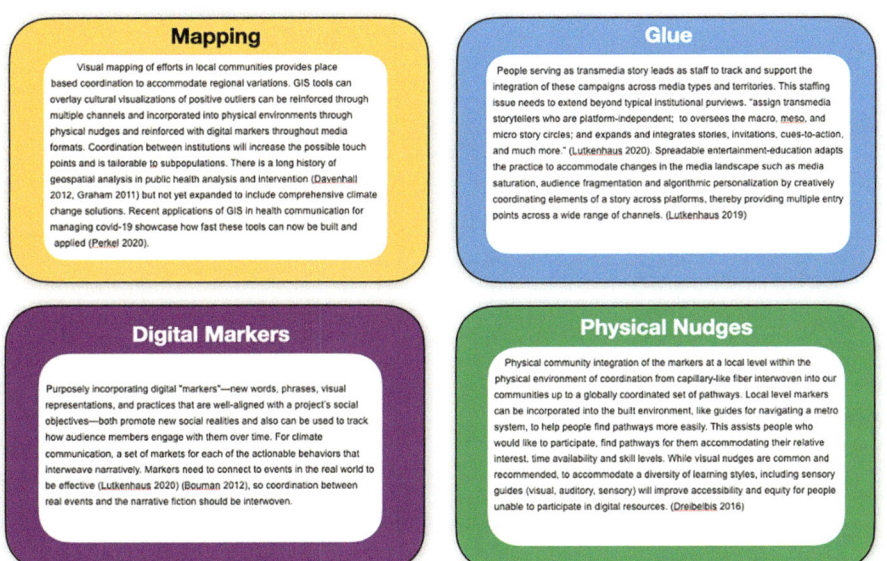

Fig. 11 Defining the mapping, social "glue," digital markers, and physical nudges components of this two-loop system proposal (Bouman et al., 2012; Davenhall & Kinabrew, 2011; Dreibelbis et al., 2016; Graham et al., 2011; Lutkenhaus, 2020; Lutkenhaus et al., 2019)

It is not enough for this work to be done piecemeal. At best that achieves patchy outcomes. Hiring trained artists in coordinated partnerships will take a level of financial support that I have not yet seen in my career. Yet, to achieve the UN sustainable development goals or other global tasks, we must. These jobs must meet or exceed industry-level professional requirements. To build a workforce in science communication we need to advance beyond work for hire and institutional public relations positions, it will mean creating new coordination roles and paying staff better. These are highly skilled jobs and many of them involve multiple competencies across the arts and sciences.

Since writing this prototype, one of the most interesting advancements is the Action for Climate Empowerment work (Bowman et al., 2021; Bowman & Morrison, 2021; Cintron-Rodriguez et al., 2021). I hope that through that work we will be able to develop the infrastructure as a community to support the evaluation and development of many programs applying these ideas. In the meantime, I hope that this work is a useful tool for narrative communities exploring these ideas. I sincerely look forward to all of your ideas and stories. Please stay connected to each other as we refine these tools for all types of narrative practitioners.

Acknowledgments The original NSF proposal was a large team project. I'd like to thank my contributors for their work in developing this project: Martin Storksdieck, Suruchi Sood, Laura Azevedo, Summer Marsh, Robert Diaz LeRoy, Erik Nisbet, Maya Zuckerman, Geoff Harris, Bruce Lewenstein, Susanna Hornig Priest, Bill Ryerson, Arvind Singhal, Jeremy Kagan, Edward Maibach, Frank Niepold, Max Boykoff, Marc Gold, Stephanie Pincetl, Ruth Greenspan Bell, and Hua Wang. Thanks to Tom Valente for assisting me in improving Fig. 9.

References

Albanna, B., Heeks, R., Pawelke, A., Boy, J., Handl, J., & Gluecker, A. (2022). Data-powered positive deviance: Combining traditional and non-traditional data to identify and characterise development-related outperformers. *Development Engineering, 7*, 100090. https://doi.org/10.1016/j.deveng.2021.100090

American Psychological Association. (2022). Understanding psychological first aid. American Psychological Association. https://www.apa.org/practice/programs/dmhi/psychological-first-aid

Bish, J. (2024). Positively life-changing stories today - intergenerational benefits tomorrow. In E. Coren & H. Wang (Eds.), *Storytelling to accelerate climate solutions*. Springer Nature.

Borwick, D. (2012). *Building communities, not audiences: The future of the arts in the United States*. ArtsEngaged.

Bouman, M. P. A., Drossaert, C. H. C., & Pieterse, M. E. (2012). Mark my words: The design of an innovative methodology to detect and analyze interpersonal health conversations in web and social media. *Journal of Technology in Human Services, 30*(3–4), 312–326. https://doi.org/10.1080/15228835.2012.743394

Bowman, T., & Morrison, D. (2021). *Resetting the future—Empowering climate action in the USA*. Changemakers.

Bowman, T. E., Cintron-Rodriguez, I., Crim, H., Damon, T., Dandridge, C., Kretser, J., Morrison, D., Niepold, F., Poppleton, K., Spitzer, W., & Weiland, L. (2021). Building capacity, momentum and a culture of climate action in the United States. *Environmental Research Letters, 16*(4), 041003. https://doi.org/10.1088/1748-9326/abe961

Brown, N. (2024). The power of locally-driven narratives to support and sustain climate action. In E. Coren & H. Wang (Eds.), *Storytelling to accelerate climate solutions*. Springer Nature.

Buscell, P., Lindberg, K., & Singhal, A. (2014). *Inspiring change and saving lives: The positive deviance way*. Plexus Institute.

Centola, D. (2010). The spread of behavior in an online social network experiment. *Science, 329*(5996), 1194–1197. https://doi.org/10.1126/science.1185231

Chryst, B., Marlon, J., van der Linden, S., Leiserowitz, A., Maibach, E., & Roser-Renouf, C. (2018). Global warming's "six Americas short survey": Audience segmentation of climate change views using a four question instrument. *Environmental Communication, 12*(8), 1109–1122. https://doi.org/10.1080/17524032.2018.1508047

Cintron-Rodriguez, I. M., Crim, H. A., Morrison, D. L., Niepold, F., Kretser, J., Spitzer, W., & Bowman, T. (2021). Equitable and empowering participatory policy design strategies to accelerate just climate action. *Journal of Science Policy & Governance, 18*, 2. https://doi.org/10.38126/JSPG180203

Coren, E. (n.d.). Entertainment-education is a public health communication tool for climate solutions. https://www.emilycoren.com/entertainment-education

Coren, E., & Myers, K. (2021). Geospatial Image Collection to Improve the Specificity of Climate Interventions. [Other]. Climatology (Global Change). https://doi.org/10.1002/essoar.10509045.1

Coren, E., & Safer, D. L. (2020). Solutions stories: An innovative strategy for managing negative physical and mental health impacts from extreme weather events. In W. Leal Filho, G. J. Nagy, M. Borga, P. D. Chávez Muñoz, & A. Magnuszewski (Eds.), *Climate change, hazards and adaptation options* (pp. 441–462). Springer International Publishing. https://doi.org/10.1007/978-3-030-37425-9_23

Coursera. (2022). Module 1 introduction – introduction. Psychological First Aid. Retrieved March 19, 2022, from https://www.coursera.org/lecture/psychological-first-aid/module-1-introduction-aRi0t

Crimmins, A., Balbus, J., Gamble, J. L., Beard, C. B., Bell, J. E., Dodgen, D., Eisen, R. J., Fann, N., Hawkins, M. D., Herring, S. C., Jantarasami, L., Mills, D. M., Saha, S., Sarofim, M. C., Trtanj, J., & Ziska, L. (2016). *The impacts of climate change on human health in the United States: A scientific assessment*. U.S. Global Change Research Program. https://doi.org/10.7930/J0R49NQX

Davenhall, W. F., & Kinabrew, C. (2011). GIS in health and human services. In W. Kresse & D. M. Danko (Eds.), *Springer handbook of geographic information* (pp. 557–578). Springer. https://doi.org/10.1007/978-3-540-72680-7_29

deFossard. (2008). *Using Edu-tainment for distance education in community work*. SAGE Publications. https://doi.org/10.4135/9788132108467

Dreibelbis, R., Kroeger, A., Hossain, K., Venkatesh, M., & Ram, P. (2016). Behavior change without behavior change communication: Nudging handwashing among primary school students in Bangladesh. *International Journal of Environmental Research and Public Health, 13*(1), 129. https://doi.org/10.3390/ijerph13010129

EPA. (2022). *Environmental Protection Agency*. EJScreen: Environmental Justice Screening and Mapping Tool. https://www.epa.gov/ejscreen

Falzone, P., Kiano, J., & Lukomska, G. (2024). Let's go! Let's know! N*gen as an EE tool for climate education and agency. In E. Coren & H. Wang (Eds.), *Storytelling to accelerate climate solutions*. Springer Nature.

Frank, L. B., & Falzone, P. (Eds.). (2021). *Entertainment-education behind the scenes: Case studies for theory and practice*. Palgrave Macmillan.

Frumkin, H., Hess, J., Luber, G., Malilay, J., & McGeehin, M. (2008). Climate change: The public health response. *American Journal of Public Health, 98*(3), 435–445. https://doi.org/10.2105/AJPH.2007.119362

Garg, A., Godfrey, A., & Eko, R. (2024). Kembali Ke Hutan (return to the Forest): Using storytelling for youth engagement and climate action in Indonesia. In E. Coren & H. Wang (Eds.), *Storytelling to accelerate climate solutions*. Springer Nature.

Graham, S. R., Carlton, C., Gaede, D., & Jamison, B. (2011). The benefits of using geographic information systems as a community assessment tool. *Public Health Reports, 126*(2), 298–303. https://doi.org/10.1177/003335491112600224

Gurney, C., & N'Diaye, M. (2024). LOLs: Secret weapon against CFCs and CO2? In E. Coren & H. Wang (Eds.), *Storytelling to accelerate climate solutions*. Springer Nature.

Higgins, L. (2012). *Community music: In theory and in practice*. Oxford University Press.

Hikichi, H., Aida, J., Tsuboya, T., Kondo, K., & Kawachi, I. (2016). Can community social cohesion prevent posttraumatic stress disorder in the aftermath of a disaster? A natural experiment from the 2011 Tohoku earthquake and tsunami. *American Journal of Epidemiology, 183*(10), 902–910. https://doi.org/10.1093/aje/kwv335

Hinderfeld, D., Slean, C., & Jacobs, K. (2024). Rewrite the future: Helping Hollywood accelerate climate solutions through storytelling. In E. Coren & H. Wang (Eds.), *Storytelling to accelerate climate solutions*. Springer Nature.

Holliman, R. (Ed.). (2009). *Investigating science communication in the information age: Implications for public engagement and popular media*. Oxford University Press; Open University.

Hunter, R. F., de la Haye, K., Murray, J. M., Badham, J., Valente, T. W., Clarke, M., & Kee, F. (2019). Social network interventions for health behaviours and outcomes: A systematic review and meta-analysis. *PLoS Medicine, 16*(9), e1002890. https://doi.org/10.1371/journal.pmed.1002890

Jenkins, H. (2010). Transmedia storytelling and entertainment: An annotated syllabus. *Continuum, 24*(6), 943–958. https://doi.org/10.1080/10304312.2010.510599

Kearns, F. (2021). *Getting to the heart of science communication: A guide to effective engagement*. Island Press.

Kim, D. A., Hwong, A. R., Stafford, D., Hughes, D. A., O'Malley, A. J., Fowler, J. H., & Christakis, N. A. (2015). Social network targeting to maximise population behaviour change: A cluster randomised controlled trial. *The Lancet, 386*(9989), 145–153. https://doi.org/10.1016/S0140-6736(15)60095-2

Luber, G., & Lemery, J. (2015). *Global climate change and human health: From science to practice*. Jossey-Bass.

Lutkenhaus, R. (2020). Entertainment-Education in the New Media Landscape: Stimulating Creative Engagement in Online Communities for Social and Behavioral Change. https://repub.eur.nl/pub/131186

Lutkenhaus, R. O., Jansz, J., & Bouman, M. P. A. (2019). Toward spreadable entertainment-education: Leveraging social influence in online networks. *Health Promotion International, 1–10*. https://doi.org/10.1093/heapro/daz104

Maibach, E. W., Leiserowitz, A., Roser-Renouf, C., & Mertz, C. K. (2011). Identifying like-minded audiences for global warming public engagement campaigns: An audience segmentation analysis and tool development. *PLoS One, 6*(3), e17571. https://doi.org/10.1371/journal.pone.0017571

McCallie, E., Bell, L., Lohwater, T., Falk, J. H., Lehr, J. L., Lewenstein, B. V., Needham, C., & Wiehe, B. (2009). *Many experts, many audiences: Public engagement with science and informal science education*. A CAISE Inquiry Group Report. Center for Advancement of Informal Science Education (CAISE). http://caise.insci.org/uploads/docs/public_engagement_with_science.pdf

Miller, J. D. (2010). The conceptualization and measurement of civic scientific literacy for the 21st century. In J. Meinwald & J. G. Hildebrand (Eds.), *Science and the educated American: A core component of liberal education* (pp. 241–255). American Academy of Arts and Sciences.

National Science Board, National Science Foundation. (2020). Science and technology: Public attitudes, knowledge, and interest. Science and Engineering Indicators 2020. NSB-2020-7. https://ncses.nsf.gov/pubs/nsb20207/

Pascale, R., Sternin, J., & Sternin, M. (2010). *The power of positive deviance: How unlikely innovators solve the world's toughest problems.* Harvard Business Review Press.

Pew Research Center. (2015). A look at what the public knows and does not know about science. http://www.pewinternet.org/files/2015/09/2015-09-10_science-knowledge_FINAL.pdf. https://www.nonviolentcommunication.com/pdf_files/feelings_needs.pdf

Rosenberg, M. B. (2015). *Nonviolent communication: A language of life* (3rd ed.). PuddleDancer Press.

Rosenberg, M. (2021). Feelings and needs inventory. https://www.nonviolentcommunication.com/pdf_files/feelings_needs.pdf.

Singhal, A., & Durá, L. (2017). Positive Deviance: A Non-Normative Approach to Health and Risk Messaging. *Communication Theory, Health and Risk Communication.* https://doi.org/10.1093/acrefore/9780190228613.013.248

Singhal, A., & Rogers, E. M. (2002). A theoretical agenda for entertainment education. *Communication Theory, 12*(2), 117–135. https://doi.org/10.1111/j.1468-2885.2002.tb00262.x

Singhal, A., Cody, M. J., Rogers, E. M., & Sabido, M. (Eds.). (2004). *Entertainment-education and social change: History, research, and practice.* Erlbaum.

Singhal, A., Wang, H., & Rogers, E. M. (2013). The rising tide of entertainment-education in communication campaigns. In R. Rice & C. Atkin (Eds.), *Public communication campaigns* (pp. 321–333). Sage.

Sood, S., Riley, A., & Birkenstock, L. (2024). Entertainment-education and climate change: Program examples, evidence, and best practices from around the world. In E. Coren & H. Wang (Eds.), *Storytelling to accelerate climate solutions.* Springer Nature.

Spiegel, S., & Wang, H. (2024). Exploring climate science in the metaverse: Interactive storytelling in immersive environments for deep learning and public engagement. In E. Coren & H. Wang (Eds.), *Storytelling to accelerate climate solutions.* Springer Nature.

Storey, D., & Sood, S. (2013). Increasing equity, affirming the power of narrative and expanding dialogue: The evolution of entertainment education over two decades. *Critical Arts, 27*(1), 9–35. https://doi.org/10.1080/02560046.2013.767015

Torres, J. M., & Casey, J. A. (2017). The centrality of social ties to climate migration and mental health. *BMC Public Health, 17*(1), 600. https://doi.org/10.1186/s12889-017-4508-0

Valente, T. W. (2012). Network interventions. *Science, 337*(6090), 49–53. https://doi.org/10.1126/science.1217330

Wang, H., & Singhal, A. (2009). Entertainment-education through digital games. In U. Ritterfeld, M. J. Cody, & P. Vorderer (Eds.), *Serious games: Mechanisms and effects* (pp. 271–292). Routledge.

Wang, H., & Singhal, A. (2016). East Los High: Transmedia Edutainment to Promote the Sexual and Reproductive Health of Young Latina/o Americans. *American Journal of Public Health, 106*(6), 1002–1010. https://doi.org/10.2105/AJPH.2016.303072. Epub 2016 Apr 14. PMID: 27077336; PMCID: PMC4880269.

Wang, H., & Singhal, A. (2021). Theorizing entertainment-education: A complementary perspective to the development of entertainment theory. In P. Vorderer & C. Klimmt (Eds.), *The Oxford Handbook of Entertainment Theory* (pp. 818–838). Oxford University Press. https://doi.org/10.1093/oxfordhb/9780190072216.013.40

Wang, H., Singhal, A. Quist, C., Sachdev, A., & Liu, S. (2019). Aligning the stars in East Los High: How authentic characters and storylines can translate into real-life changes through transmedia edutainment. *SEARCH Journal of Media and Communication Research, 11*(3), 1–22.

Wickrama, K. A. S., & Wickrama, T. (2011). Perceived community participation in tsunami recovery efforts and the mental health of tsunami-affected mothers: Findings from a study in rural Sri Lanka. *International Journal of Social Psychiatry, 57*(5), 518–527. https://doi.org/10.1177/0020764010374426

Wolf-Jacobs, A., Glock-Grueneich, N., & Uchtmann, N. (2024). Mapping out our future: Using geospatial tools and visual aids to achieve climate empowerment in the U.S. In E. Coren & H. Wang (Eds.), *Storytelling to accelerate climate solutions.* Springer Nature.

Emily Coren is a science communicator and an affiliate in the Department of Psychiatry and Behavioral Sciences at Stanford University where she has been working to adapt entertainment-education strategies for health promotion and social change to create more effective climate communication. She has a B.S. in Ecology and Evolutionary Biology and is a certified professional Science Illustrator. She has worked in science communication for almost 20 years, contributing to collections at the Smithsonian Institution's Museum of Natural History, consulting on a World Health Organization clean air campaign, and developing educational content for children's films. In recent years, her work has led to new methods in developing frameworks at a national level, connecting community-led experiences to federal, local, and nonprofit sector programs for climate change communication. She is a member of the National Association of Science Writers and the Society of Environmental Journalists.

Rewrite the Future: Helping Hollywood Accelerate Climate Solutions through Storytelling

Daniel Hinerfeld, Cheryl Slean, and Katy Jacobs

The Natural Resources Defense Council (NRDC) is an international environmental NGO founded in 1970 by lawyers and law students alarmed by ecological destruction. They sought to hold polluters and government agencies accountable and created the first public interest law firm to protect the environment. The organization gradually expanded its toolset beyond the law, first developing scientific and policy capacities and then communication capacities intended to broaden and diversify the environmental movement, lead people from awareness to action, and build political power.

Today, NRDC's top priority is to solve the climate crisis.

Rewrite the Future is a program within NRDC's Communications Department that works with Hollywood professionals to support storytelling about the climate crisis in entertainment film and TV. Its goal is to help shift attitudes and behaviors around the crisis with stories that reflect our climate reality, enable audiences to feel the urgency of the challenge, confront their fear and grief about it, and imagine better futures that inspire hope and action.

The seed of Rewrite the Future was planted in the mid-2000s by the actor, filmmaker, and environmentalist, Robert Redford, a longtime NRDC trustee. At a meeting of the board he hosted at Sundance, Mr. Redford encouraged NRDC to develop the capacity for visual storytelling. In response, NRDC established a documentary unit to make films about the environment. These efforts resulted in documentaries, such as *Acid Test: The Global Challenge of Ocean Acidification* (Bayer & Hinerfeld, 2009), *Stories from the Gulf: Living with Oil Disaster* (Hinerfeld, 2011), *Wild Things* (O'Brien & Hinerfeld, 2014), *Sonic Sea* (Dougherty & Hinerfeld, 2016), and *Our Ocean Planet* (Hinerfeld, 2018). They were well-received and recognized with numerous accolades.

D. Hinerfeld (✉) · C. Slean · K. Jacobs
Natural Resources Defense Council, Inc., New York, NY, USA
e-mail: dhinerfeld@nrdc.org; cherylslean@gmail.com; kjacobs@nrdc.org

E. Coren, H. Wang (eds.), *Storytelling to Accelerate Climate Solutions*,
https://doi.org/10.1007/978-3-031-54790-4_8

Sonic Sea, for example, won two Emmy Awards (Outstanding Nature Documentary and Outstanding Music and Sound) and helped raise awareness about little-known but critically important issues, such as ocean noise pollution (Meulmester, 2017). The NRDC films aired on national television showed up in screenings for policymakers and opinion leaders and proved useful for inspiring targeted policy initiatives, such as ship quieting rules (Bahtiarian, 2019) and congressional action on ocean acidification (Congressional Record, 2009).

However, NRDC recognized that the medium of documentary film, with its relatively small and self-selected audience (Borum Chattoo, 2016), could only go so far in driving change on an issue as fundamental as climate, where the science is well-established but progress has been slow, in part, because public opinion has been fractured by a fossil fuel industry led disinformation campaign (Tabuchi & Friedman, 2021).

According to research such as Yale Climate Change Communication's "Six Americas" study (Leiserowitz et al., 2021), awareness about the urgency of the crisis, its causes and solutions, and the cost and time to fix it vary among Americans, falling short of the political will needed to push through large scale decarbonization. For instance, recent surveys found that over 70% of Americans believe climate change is happening (Leiserowtiz et al., 2018) and 62% say it is impacting their local community (Funk & Kennedy, 2019). However, only 19% think we can solve it (De Pinto et al., 2019). Experts call this disparity the "hope gap"—a recipe for disengagement or fatalism (Upton, 2015). Simply put, people will not work toward a goal they do not think is achievable or cannot envision. To bridge the hope gap, we need stories that show us that a better future is possible and what it might look like.

> We find in our audience research that even the alarmed [those most concerned about climate change] don't really know what they can do individually, or what we can do collectively. We call this loosely 'the hope gap,' and it's a serious problem. Perceived threat without efficacy of response is usually a recipe for disengagement or fatalism. – Anthony Leiserowitz, Director of the Yale Program on Climate Change Communication

The state of public opinion about climate is especially problematic because the societal transformation required to address the challenge is of unprecedented scale and complexity; the time left to achieve that transformation is short (IPCC, 2018, 2021); and the consequences of failure are catastrophic.

Fifty years of entertainment-education research have established the efficacy of public health and social welfare messages in entertainment stories (Frank & Falzone, 2021; Singhal et al., 2004; Singhal & Rogers, 1999; Wang & Singhal, 2021). Based on this body of evidence, we theorized that public attitudes toward the climate crisis might have the potential to shift through the same means and that compelling climate-savvy entertainment may be the fastest way to bring about the cultural change we need.

Theory of Change: Good Stories Can Change Hearts, Minds, and Behavior

For many decades, popular entertainment has played an important role in cultural change. Scripted TV shows and films have helped transform the public discourse and opinions about social issues, such as racism, women in the workplace, immigration, and LGBTQ communities, and have changed behavior around public health issues, such as drunk driving (e.g., Koh & Yatsko, 2017).

How stories achieve this influence is explained by various theories involving the psychological effects of storytelling (Brown, 2015; Moyer-Gusé, 2008; Sood et al., 2017; Wang & Singhal, 2021). The theory of narrative transportation (Green & Brock, 2000; Green, 2021) posits that when we're caught up in a good story we're put into an open and receptive state in which we emotionally identify with and root for characters even though they might be different than we are. We also effortlessly absorb and retain large amounts of information.

The parasocial contact hypothesis (Schiappa et al., 2005) states that exposure to TV and film characters over time builds social affinity similar to how we feel about our friends. We begin to see the world as they do. This can open our minds to new ways of thinking, normalize the unfamiliar, and reduce social group bias and anxiety about change.

Studies of the long-running hit *Will & Grace*, one of the first TV sitcoms with gay lead characters, found large reductions in viewer bias against LGBTQ communities (Schiappa et al., 2006). Medical shows like *ER* and *Grey's Anatomy* that convey health information embedded in emotionally engaging stories have effectively changed behavior (The Henry J. Kaiser Family Foundation, 2004). Primetime television programs such as *Law & Order* have effectively educated viewers about exposure to toxic substances and clarified environmental health policy (Kennedy et al., 2011).

These and other entertainment programs have helped people articulate, process, and understand their (sometimes conflicting) feelings about social developments that challenge the status quo. The more entertaining a story is (i.e., the more thoroughly a viewer is "transported" and identifies with fictional characters), the more potential that story has to bring about social change. These programs have reached large audiences, become societal shorthand, and shifted prevailing attitudes. Legislative and policy wins followed, and the world changed.

Where Is Hollywood on Climate?

In some ways, the climate crisis seems ready-made for film and TV. It is an urgent, high-stakes crisis, rife with human conflict. Heroes and villains abound, and there are trillions of dollars to be made or lost. It is a deeply emotional drama about human survival. And yet storytellers—in Hollywood and beyond—remain curiously quiet about it. There are so few stories involving climate change (or that even *mention* climate change) in mainstream entertainment that a viewer could watch popular film and TV content for weeks and have no idea that humanity has little more than a decade to avert catastrophe (Buckely, 2019; Dembicki, 2019; Lawson, 2019; Littleton, 2019).

When Hollywood *has* taken on climate change, it has tended to do so through a narrow lens—mainly in the genres of disaster (e.g., *The Day After Tomorrow, Deepwater Horizon*), post-apocalypse, and dystopia (e.g., *Mad Max, Blade Runner*), or the "Planet B" genre (e.g., *Interstellar, Midnight Sky*), in which the only hope for humanity is to escape to another world. Nearly absent are stories that reflect today's climate realities, depict characters dealing with the complex emotions and choices that millions of people already struggle with, or allow for the possibility of positive human agency, solutions, and hope.

The Rewrite the Future design phase included months of research, including conversations with writers, directors, producers, executives, and agents, with organizations that support or influence storytellers, and with the Hollywood guilds. That work uncovered an odd dichotomy: a widespread desire among creative professionals to tell climate stories and a culture of resistance to telling such stories based on the perception that many impediments stood in the way. Therefore, an early goal of Rewrite the Future was to identify and analyze those impediments, whether illusory or real, and develop ways to support creative professionals in surmounting them.

Psychological, Creative, and Business Impediments

There is a perception in Hollywood that it is difficult, if not impossible, to tell entertaining stories about climate. Broad awareness of climate change only dates back about 30 years, and for most of that period, the phenomenon has seemed remote and abstract. Many people still have a hard time connecting what they see as impacts on glaciers and polar bears or distant disasters with their daily lives. Yet even as climate awareness has grown with rapidly mounting threats from extreme weather, clear degradation of the natural world, and more robust news reporting, there remains among storytellers an assumption that the issue is primarily scientific, impersonal, and dry—antithetical to entertainment.

Good stories are about *people,* not issues, and many creative professionals have difficulty making the leap from "climate change" writ large to the personal conflicts and transformations that are the grist for drama and comedy. There is, as well, a reflexive association between climate change and disaster/apocalypse, as suggested above.

Another challenge comes from the unconscious climate denial of creative professionals themselves. Writers are people too, and very few people enjoy thinking in detail about existential threats to themselves and their families. Telling a compelling story about climate change requires overcoming the tendency to look away.

Part of Rewrite the Future's job is to inspire writers by reminding them of their unique power to contribute to climate solutions through storytelling. That power is a privilege and can be a salvation. As Greta Thunberg (2018) says, "Once we start to act, hope is everywhere."

For writers, the very act of working through this psychological bottleneck may inspire ideas for character development and plot.

A primary goal of Rewrite the Future is to expand the notion of what climate storytelling can be. This includes highlighting possible angles for compelling climate stories—beyond disaster and apocalypse—that involve characters' everyday lives, livelihoods, hopes, and fears. The provocation for writers is to consider what our present climate reality actually is and how it plays out on a personal level.

Climate change already affects most of us whether or not our house has burned down or been washed away in a flood. We worry about it. We talk about it with our partners and friends. People agonize over whether to have children, they wonder where is the safest place to live, how they can protect their families from wildfire smoke, or whether the ocean is dying. They fear losing their livelihoods in the transition from fossil fuels, resent decades of inaction by older generations, or retreat into denial or nihilism.

This is our present reality, and yet, with few exceptions, we don't see it reflected in scripted TV and film. Neither do we see the dramatic and inspiring struggle and progress by student activists, engineers, scientists, entrepreneurs, and others to save humanity.

Although the subject might seem gigantic and impersonal, the truth is that people made the climate crisis, people are impacted by it, and people are solving it too.

Dorothy Fortenberry, a writer and producer (*The Handmaid's Tale*, *Extrapolations*) and advocate of climate storytelling in TV and film, said the following during a Rewrite the Future panel discussion "Beyond Apocalypse" at the 2021 Sundance Film Festival (NRDCflix, 2021):

> I think any story can and probably should be a climate change story. I think it's sort of like, you know – are walls entertaining? I don't know, but most TV shows have them. Climate change is just as much a part of our reality as walls, so it should be in TV shows as much as walls are. Maybe the episode is about the walls, or maybe they're just there, but they influence everybody's decisions because you have to go around them or through a door … So there's a part of me that feels like if we just made the small adjustment to make it reality, that would be enormous.

Rewrite the Future encourages writers to work with climate themes as they do with other themes: to start with character, relationships, and setting, to focus on the personal and emotional, and to lead with making entertainment rather than "educational entertainment." Because climate impacts and solutions are everywhere, involve everyone, and touch on many aspects of our lives—almost any story can find a natural climate angle. These will be the most effective stories—the ones designed primarily to entertain.

And then there are advertisers. Fossil fuel companies, automakers, and consumer goods conglomerates are perhaps less likely to spend ad dollars on shows that take on climate change and unsustainable consumption. The mere notion that advertisers might use their leverage to quash climate content seems to have had a significant chilling effect.

In addition, some Hollywood professionals are concerned about the politically charged nature of the climate discussion in the United States. Many Americans identify themselves in part by whether they "believe in" climate change or not. They connect the issue with different philosophies about the role of government, the scope of individual liberty, and even the intentions of God. This is potentially problematic for entertainment storytellers because they (and the companies they work for) want their content to reach the largest possible audience and certainly not to alienate millions of potential viewers.

We believe climate stories can be told without triggering identity politics; that advertisers' power to influence content is declining with the rise of subscription-based entertainment platforms, such as Netflix, Apple, and Amazon; and that network creators and executives won't want to miss out on truly compelling content. The climate crisis is arguably the greatest storytelling opportunity of our era.

Finding a Path Forward

Months of conversations with entertainment professionals revealed that many writers, producers, directors, and studio executives are deeply concerned about climate change and recognize the influence their industry could have on society's response to the crisis. They *want* to tell climate stories on screen. Many also know that they need help doing it.

The question we needed to answer was how to provide that help. It would involve gaining the trust of Hollywood professionals, showing them that Rewrite the Future had useful ideas and information to offer, but doing so without infringing on the principle that it is the writer's job to write.

Advocacy groups have long worked in Hollywood, attempting to use the power of film and TV to advance their own objectives. This phenomenon has reached new heights in recent years. Organizations across the spectrum try to integrate their stories into the entertainment content. Although many share core values, their issues are diverse, including public health, immigration, civil rights, women's rights, Muslim and Jewish affairs, LGBTQ communities, gun violence, and the environment. This profusion of advocacy in Hollywood presented another challenge.

The strategy we settled on was that Rewrite the Future would begin by producing, sponsoring, and participating in public discussions about climate storytelling in entertainment. It seemed a good way to spotlight the work of writers already experimenting with climate themes, to find and recruit new Hollywood allies, to identify and find solutions to climate storytelling hurdles, to help create a community of climate storytellers, to illuminate the intersections of climate with other social justice issues, and to establish Rewrite the Future as a trusted resource.

Rewrite Future Public Educational Events

Our moderated panel discussions on climate storytelling topics were initially designed as live events in Los Angeles and included networking and community building. The COVID-19 pandemic necessitated a move to virtual platforms, which has resulted in reaching a larger audience but with less opportunity for live peer interaction. Nevertheless, both formats have been extremely successful in stimulating interest in climate on screen and have helped to address and expand some of the limiting creative assumptions discussed in the last section.

Events have exposed our target audience of content-makers and buyers to our theory of change and have nurtured industry dialogue about entertainment's potentially powerful role in solving the climate crisis. They have grown our network of Hollywood contacts, including serving as an entree to studio executives, content creators, talent agencies, and others who invite us to offer targeted presentations to smaller industry groups. They have also led to opportunities for in-depth climate story consulting with on-air TV shows as well as films and shows in development. We discovered a great appetite for the topics we presented, as reflected in large and engaged audiences and robust follow-up conversations.

"Hollywood Takes on the Climate Crisis"

Our events often feature seasoned Hollywood creatives speaking from their experience integrating climate content into shows and films. For our first public event, we partnered with The Sierra Club, Lear Center Hollywood Health and Society, Center for Cultural Power, and the Good Energy Project to present "Hollywood Takes on the Climate Crisis," a panel discussion and networking opportunity targeted to industry audiences (NRDC, 2019).

In this pre-pandemic networking event, we invited four screenwriters from diverse TV and streaming shows to highlight the wide range of creative approaches to climate content. They included a writer/performer from *It's Always Sunny in Philadelphia*, a writer/producer from *The Handmaid's Tale*, a writer/producer from *Madam Secretary*, and a writer/performer from a comedy web series called *The North Pole* (Healey et al., 2019), unique in the group for having been produced by an environmental justice nonprofit rather than a studio.

We were interested in hearing from writers in the vanguard of scripted climate entertainment about how they had approached the topic creatively through their various genres, characters, settings, and storytelling aims. The variation between shows also extended to different levels of climate engagement, ranging from central events and plots (a climate-exacerbated heatwave in the satirical comedy *It's Always Sunny*, a super typhoon and climate denial in the political drama *Madam Secretary*) to subtle mentions (electric vehicles and winters without snow in the near future dystopia of *The Handmaid's Tale*) to intersectional topics (climate anxiety, food justice, and economic disparity in *The North Pole*'s Oakland, CA).

Each of these writers was passionate about seeing more of our climate reality reflected on screen but was faced with resistance from their showrunners, supervising executives, and writer's rooms, as well as in their own creative process. Alex Maggio of *Madam Secretary* spoke of needing more right-wing characters to inhabit a political climate change story until he found inspiration in a real-life Christian activist and her evangelical preacher father. Since this event, prevailing industry attitudes have changed: One of our panelists, Dorothy Fortenberry, went on to co-executive produce an ambitious AppleTV+ anthology series called *Extrapolations* (Burns et al., 2022), the first show to our knowledge centered around the climate future.

The audience stayed long after the panel ended, continuing the discussion. We were struck for the first time by the entertainment industry's hunger for more dialogue about how to address climate change in their work. We resolved to continue producing events to educate and stimulate discussion among industry audiences.

Sundance Film Festival 2020—Public Launch of Rewrite the Future

Our Rewrite the Future climate in entertainment program was officially launched at the January 2020 Sundance Film Festival, where NRDC hosted film panels, calls to action, and networking events celebrating the power of storytelling to engage the public on environmental issues. Sundance is an important annual gathering for the entertainment industry that attracts significant media attention and is, therefore, an ideal platform to advance our message.

Highlights included NRDC's then-president Gina McCarthy, later National Climate Advisor for the White House, in conversation with comedienne, activist, and NRDC trustee, Julia Louis-Dreyfus; numerous film panels including with Ron Howard on his Camp Fire documentary *Rebuilding Paradise;* interviews on social impact entertainment with activist-artists Eva Longoria and Wilmer Valderrama; and the annual Women in Hollywood reception co-hosted by The Black List screenwriting platform, which led to a partnership on Rewrite the Future's inaugural Climate Storytelling Fellowship (discussed later).

Each event was preceded by a brief overview of our theory of change and a call to action for a better representation of climate reality on our entertainment screens. Our message was well received and, like our other events, resulted in numerous new contacts and potential allies in the industry.

Partner Events

As interest in climate storytelling began to build in Hollywood, allied groups started producing their own workshops and events, inviting Rewrite the Future to partner or participate. For example, we programmed a keynote panel for the first annual

Hollywood Climate Summit (2020), a full-day virtual event produced by our partners, Young Entertainment Activists. Rewrite the Future's panel was the top-viewed segment from the Summit—with over 7000 views on Facebook and an additional 1200 views on Twitch.

In the wake of the sea-change Black Lives Matter protests in the summer of 2020, we partnered with the Producers Guild of America Green initiative (Production Green Guide, 2020) on a panel that featured Reverend Yearwood of HipHop Caucus speaking with TV showrunners about the opportunities for depicting the outsized effects of climate change on frontline communities of color as well as how to draw inspiration from real-world environmental justice leaders who stand up against corporate pollution.

Moving beyond the general climate storytelling panels intended to introduce the subject, at the 2021 virtual Sundance Film Festival, Rewrite the Future produced the first in a series of deeper dives into more focused topic areas, starting with a panel on climate futures called *Beyond Apocalypse: Alternative Climate Futures in Film and TV* (NRDCflix, 2021). Futurist storytelling is one of our target advocacy areas since the vast majority of Hollywood sci-fi depicts apocalypse and dystopia, sometimes with reference to climate change as a cause of the social collapse but more often by some other calamity. Studies have shown that although such representations, if correctly attributed, can attract people's attention to climate change, fear is generally an ineffective tool for motivating personal engagement (O'Neill & Nicholson-Cole, 2009).

To address these issues, the *Beyond Apocalypse* panel proposed that our imagined future narratives should include the healthier society our children might inherit if we succeed in significantly reducing greenhouse gas emissions. Panelist Christiana Figueres, a leading architect of the Paris Climate Agreement and author of *The Future We Choose, Surviving the Climate Crisis*, offered two contrasting future visions: the dystopian future of environmental and social collapse if we do not meet the Paris targets; and the promising future of clean cities, social justice, economic stability, and ecological and human health if we choose the path of equitable decarbonization. The entertainment storytellers on the panel (Sarah Treem, Dorothy Fortenberry, and Rosario Dawson) then reflected on the possibilities for depicting change-making presents and positive futures in TV and films.

The conversation was aimed at entertainment-makers and framed in the language of screenwriting. For example, Sarah Treem described her creative process when devising a 40-years-in-the-future storyline for her critically acclaimed show *The Affair*. While it was important for the writing team to be well-versed in regional climate impacts at the story's setting of Montauk, NY, she related that creative inspiration came primarily from imagining the main character's emotions, especially loss and anxiety (NRDCflix, 2021):

> … what's interesting to think about when incorporating climate change into narratives is that people are very worried about climate change, it's a real anxiety hanging over all of us, and we're all kind of in denial about it because it's too scary to think about. It involves the legacy of our children, it involves the world that we're leaving them, we don't think we have any control over it, and … that kind of anxiety is a real character motivator, like anything else that a character is in denial about or doesn't want to see but is significantly motivating their actions … is something that you could incorporate into character.

The event was well received, with an engaged 500+ live audience at the premiere stream and over 5000 YouTube views at the time of press. The panel has been cited by a number of our industry allies as formative to their thinking on climate storytelling and useful in leveraging further engagement up their studio chains of command.

From our retrospective vantage point, it is clear that Rewrite the Future and our partners' public-facing events have both contributed to, and reflected, a groundswell of interest in climate change content in Hollywood. They contributed to creative decisions by writers and studio executives. In addition, the wide net cast by events and their subsequent coverage in the press successfully piqued the interest of influential industry professionals such as studio executives and TV showrunners who've approached us for a deeper dive into climate storytelling.

Studio Presentations

In response to queries from studios and other industry professionals, we developed a presentation outlining the urgency of the climate crisis and how filmmakers might leverage their uniquely influential product toward accelerating climate solutions without sacrificing their entertainment mandate. It presents an abbreviated version of the theory of change outlined above, including a brief review of narrative persuasion as it applies to entertainment-education and the success of experiments in other issue areas (Fig. 1).

We then ask the next logical question: How might these outcomes be learned from and applied to the climate issue by, say, modeling sustainable behavior change in stories or by depicting the social benefits of transitioning out of the fossil fuel economy in favor of healthier environments and communities? If we make a fossil-free future look attractive on screen, might that help reduce public bias against

STORYTELLING CHANGES HEARTS & MINDS

Narrative Transportation	Character Identification	Parasocial Contact
		Characters join viewers' extended social group
Emotional activation	Investment in outcomes	Trust new people/groups
Open receptivity	Concern-sharing	Normalize the unfamiliar
Info retention	Model values, views	Reduce change anxiety
Raise awareness	and behaviors	Reduce social group bias

Fig. 1 NRDC rewrite the future's theory of change

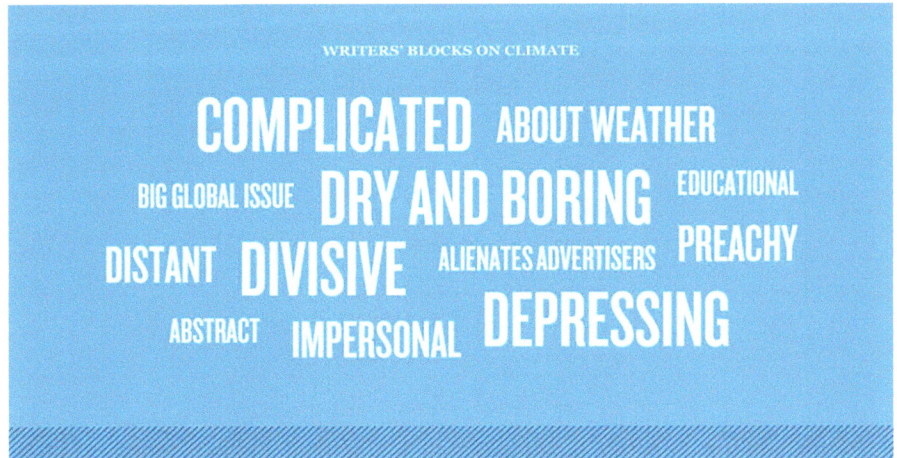

Fig. 2 Word map of commonly stated blocks to climate storytelling

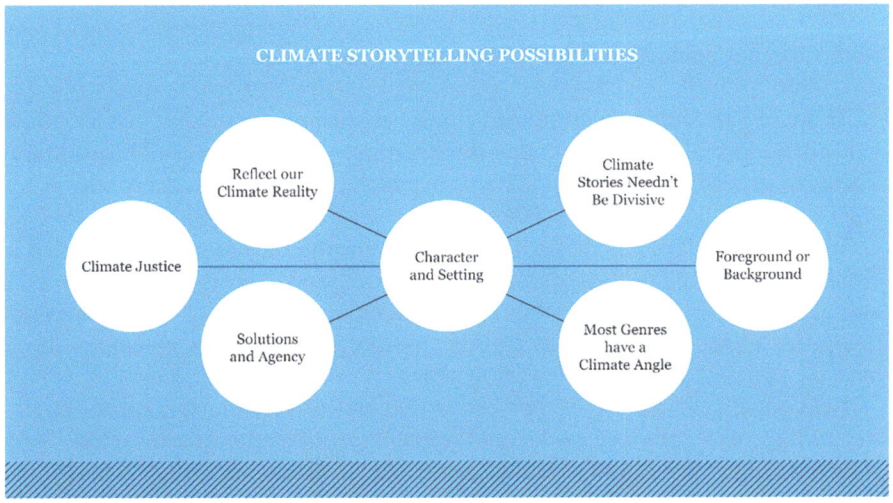

Fig. 3 Chapter headings for climate storytelling angles

status quo nostalgia and resistance to change? This introductory section transitions to climate storytelling tips by sharing our informal industry research findings, including commonly stated concerns about tackling climate in scripted stories (Fig. 2).

The bulk of the deck then explores, with specific examples, some of the many creative opportunities, or "story angles," for engaging with climate in TV and film scripts (Fig. 3). We emphasize the screenwriter's entry point of *character* because

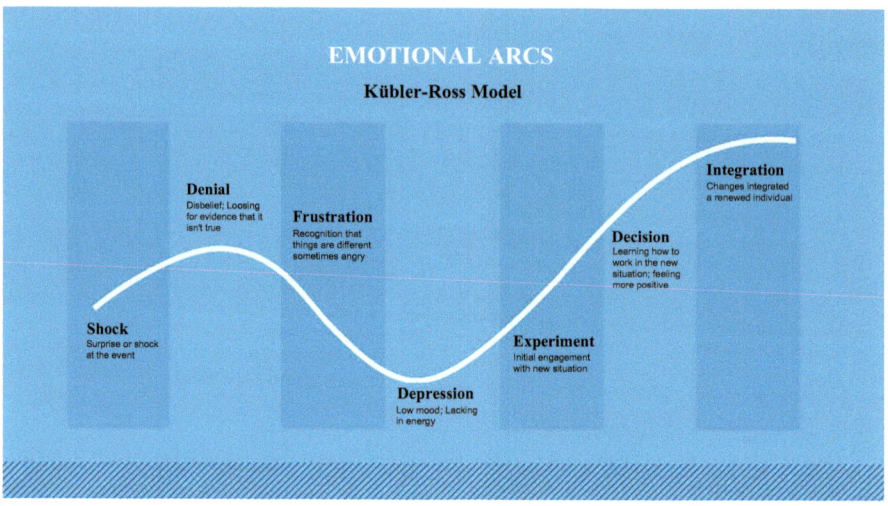

Fig. 4 Kubler-Ross "Stages of Grief": an example of an emotional "arc". (Kübler-Ross E, 1969)

stories are not about issues per se but about people experiencing those issues in their lives.

To address the cultural narrative of climate avoidance, we suggest that a top-level guideline, especially for present-day stories, simply be to reflect our climate reality. There are myriad ways to approach this overarching suggestion and we detail several of them in the subsequent slides.

We start with the climate reality of environmental racism and the fact that all climate impacts are not equal, hitting lower-income and people of color hardest. We also give examples of how frontline communities take control of their destinies with innovative climate resilience and mitigation programs. Stories of empowerment have the potential to inspire viewers to action.

Next comes character development, a rich and diverse entry point for climate storytelling. We explore character, and setting thought-starter questions like "What is the character's job?" and "Where do they live?" discuss climate-affected regions and sustainability jobs from the real world and examples portrayed on film and TV.

"How does the character feel?" begins a section on climate psychology and emotion, an area that stories are particularly well-suited to explore. As Sarah Treem pointed out, culturally relevant stories can engage with our climate reality by capturing how people feel and talk about the crisis—from fear, anxiety, and anger to compassion and engaged response (Fig. 4). The information we present is based on insights and research from climate psychologists.

A solution section includes various actions characters can take in their personal lives (reducing food waste, EVs and bikes, renewable energy, etc.) to real-world system innovations like regenerative farming and eco-industrial design that might inform world-building for stories set in the future. We posit that focusing on "solutions in storytelling" (De Meyer et al., 2021) can inspire ideas for how viewers might act on their concerns about the climate crisis.

We offer ideas based on best practices in climate communication about how to tell climate stories that speak to a range of audiences without being divisive or political, including ideologically diverse characters, settings, and jobs and inspirational spiritual attitudes about nature. We explore a number of environmental examples from different genres of TV and film to show how climate topics are fair game in any present-to-future story, from sitcom to satire, sci-fi and fantasy to procedural drama, and unscripted reality shows. Finally, we make the business case for climate storytelling. Despite the product being entertainment stories, Hollywood is still a business. It is unrealistic to rely on altruism, ethical social activism, or even creative opportunity as a call to action.

A primary business case rests in the cultural relevance of entertainment and the fact that the climate crisis is underrepresented on screen despite having an increasingly dominant presence in many people's lives. The Yale Program on Climate Change Communication reports 72% of Americans are aware of climate change, and 63% on average are worried about it (Marlon et al., 2020), though those data go up when cutting by demographics: 62% of young adults (18–38) "support climate activists" (Ballew et al., 2020b), and Latino (37%) and Black (27%) Americans are more "alarmed" about climate change than white Americans (22%) (Ballew et al., 2020a). These are important segments that studios are eager to court. More content that reflects our climate reality, including the explosion of "green" product choices, could attract new viewers from the climate-concerned. In our current media landscape of streaming providers jockeying for market share, leveraging a competitive edge is the key to success.

We also report on various environmental, social, and governance (ESG) commitments in the industry, where media corporations have long had sustainability programs, but are now starting to step up with more ambitious corporate goals, such as Netflix's recent Net Zero + Nature commitment (Stewart, 2021). Our studio allies have informed us that exposure to what their competitors are doing can serve as a strong incentive to act.

Our basic presentation is adaptable to different applications and audiences. If we're presenting to a group of screenwriters, such as we did at our webinar kicking off the NRDC/Black List Climate Storytelling Fellowship, we will highlight the climate storytelling section and lowlight the business case, and vice versa for presentations at studios and talent agencies.

Expanding Clientele and Building Trust

Our private presentations serve our educational purpose of "expanding Hollywood's understanding of what climate storytelling can be," while also cultivating potential "clients" for the next level of climate engagement, which are our customized story consultations. The presentation format usually includes a discussion period where attendees can reflect on applying the information to their specific slate of projects, and often this will serve as an entree to a relationship on a particular show or film.

The progression from public events to private presentations to writers' room consultations is a deliberate process of building trust in an industry notorious for its "gatekeepers." Before creative executives allow us access to their storytellers, we need to establish our usefulness to the primary business—telling entertaining stories that attract viewers, with a possible beneficial side effect of generating social norm influence. We can't interfere with the former to accomplish the latter. We communicate that by positioning ourselves as experts only in climate impacts, solutions, and the story opportunities that might be found there, being clear to cede expertise on *effective storytelling* to the professionals.

Through this funneling process, we have developed trusted relationships with a number of insider allies who have expressed strong interest in the climate storytelling project. An ally in the executive class is a point of access both up and down the chain—up to content buyers and down to content creators. An ally in the writers' room, particularly a TV showrunner or head writer in a position of authority, is our imprimatur to the writing team.

Rewrite the Future Climate Story Consulting

Entertaining stories speak to universal themes through the specificity of character, setting, event, and detail, and each story's unique milieu will offer different angles for climate and sustainability. The more closely we can work with creatives, the better we can help curate the details that both fit with the story and have the potential to influence our cultural climate narratives.

Outreach Strategy

We have two goals in our consulting strategy that serve to cover a range of projects. For the quickest impact, we aim to encourage climate storylines in the most popular TV shows already running. This is the fastest road to the screen, but because the show premise and characters are already set, we are usually limited to more shallow placements. Usually, we are looking at a single episode of a procedural series, a subplot of an episodic drama or comedy series, or various small but meaningful mentions: one-off lines of dialogue or actions in a scene. The second approach is to work with new shows and films in development that have a high likelihood of being produced. Usually, that means they have established talent attached, but even that is no guarantee, and these projects may have a long road to the screen. However, they offer opportunities for more significant story integrations.

In both cases, we want to intervene in the writing process as early as possible, for example, when the creative team is still determining the plots for the season's episodes. While we feel any accurate depiction of climate reality is useful, we aim for the "highest value" characters (i.e., protagonists) and central storylines, because they will most influence the viewer. But we are also happy to offer consultations on completed scripts to ensure accurate and useful content and opportunities for the insertion of meaningful mentions. In particular, we look to correct the "climate cliches" that reflect unhelpful cultural tropes, such as mocking green behaviors, perpetuating attitudes like "it's too late, we're doomed," or blaming regular people for the crisis instead of the powerful entities most responsible.

Research-Based Process

We rely heavily on our research, which is guided by the story's setting and characters. Is this a story about first responders in Texas? Farmers in India? Federal agents in Washington, DC? We familiarize ourselves with the show or film through available materials such as existing episodes, scripts, treatments or synopses, and then look for promising climate and sustainability angles pertaining to that region and character set. In the research phase, we rely heavily on interviews with NRDC's deep bench of science, legal, and policy experts working in the field to understand and mitigate climate impacts, often with community-based partners.

Our research inevitably unearths a wealth of specific detail that can be mined for stories. We will select, shape, and present this information to the creative teams through writer-friendly research memos as well as creative conversations where options are evaluated and new ideas are unearthed. Climate topics of greatest interest are identified, and we may conduct further research in an iterative, collaborative process with the creative team.

A story consultation can start in a variety of places and proceed by a variety of paths. It can be a one-off script review, an ongoing relationship on a series, or a yearlong engagement with a film through development. The following case studies demonstrate three different pathways for climate story consultation. Because none of these films or shows has been released, we are required by a confidentiality agreement to keep certain information private.

Case Study #1—Family-Friendly Feature FilmFilms in Development

After one of our public events, a small production company specializing in social impact entertainment approached us about a proposed feature film about a teen changing global climate policy through ingenious methods. We felt the story touched promisingly on some of our key communication goals such as contrasting alternative climate futures (NRDCflix, 2021), linking those futures with the choices

Fig. 5 Rewrite the Future typical script review questions

> **Rewrite the Future Script Review Questions**
>
> First, is it entertaining?
>
> Does it get stuck in disaster/dystopia or does it pivot to solutions?
>
> Will it leave viewers despondent/hopeless about climate change or inspire them about prospects for tackling it?
>
> Does it reinforce the importance of human agency in the climate crisis, in other words, conveying that the choices we make now and the actions we take (or fail to take) will determine our climate future, and that our fate is not predetermined?
>
> Does it attribute disaster/dystopia to human made climate change or leave it to the viewer to make that connection?
>
> Does it depict characters experiencing and working through climate emotions like climate anxiety and grief?
>
> Does it reflect that people are already making basic life decisions influenced by climate change, like whether to have children, where to live, how to invest their retirement money?
>
> Does it address environmental inequity and environmental racism – the fact that low -income people and people of color often bear the brunt of environmental challenges, often by design.
>
> Will it alienate some audiences because of its treatment of climate, e.g., center-right audiences?
>
> Does it mock climate deniers?

we are making now, bringing attention to climate impacts on the ocean, and modeling self-efficacy and informed action through a young woman protagonist who teams with diverse friends and allies to make a difference. Moreover, the story offered a co-viewing experience suitable for the whole family. The co-viewing experience is salient because research has shown that children, and especially daughters, can inspire climate concern in their parents, especially fathers and conservative parents (Lawson et al., 2019).

We consulted with the creative team over a year through their development process, including connecting them with NRDC oceans expert Lisa Suatoni, whose knowledge informed important plot points. We reviewed multiple drafts of the story from the outline through several script drafts to identify inaccuracies, suggest new angles, and discuss story choices with an eye to both fact integrity and our mutual goals for audience takeaways (Fig. 5). The screenwriter revised story points based on that feedback.

The producers were keenly interested in making an accessible adventure drama with the potential to inspire audiences of all ages to make informed climate action, and while we all agreed that the film's storytelling was the primary positive influence, we also discussed furthering its impact with an audience engagement campaign upon the film's release. When the creative team felt ready with their script and supporting materials, we continued our support of the project by helping to build an attractive package to go out to studios for financing. We introduced them to an experienced producer who came on as a partner, and we brought it to production company allies who have expressed strong interest in climate stories. We also introduced the project to an Oscar-nominated actress within NRDC's orbit to gauge her interest in taking one of the two lead roles and/or becoming a producer of the film. The film is now being considered for financing, and we will continue to do what we can to help bring it to fruition.

Case Study #2—Scripted TV Series from a Major Studio

The mission of one Rewrite the Future funder is to persuade politically center-right Americans of the urgent need for climate action—an appealing strategy given the fate of proposed federal climate legislation over the last two decades. Rewrite the Future developed a plan with the funder to identify popular TV series that reach a center-right audience, to attempt to gain access to the writers and/or showrunners of those programs, and to help them identify and tell climate change and environmental stories appropriate for their shows. The challenge is to connect with viewers who may be disengaged, doubtful, or even dismissive about the issues. The opportunity is to engage them within a "safe space:" a familiar fictional world populated by characters they already like and trust.

Reliable TV viewership data are not freely available, so identifying TV and streaming series that reach center-right viewers initially posed a problem. We solved it with help from three other Rewrite the Future allies (a funder, a talent agency, and a university) that provided access to viewer data from Nielsen and other sources. That trove of information allowed us to compile a short list of target shows. It also confirmed our hunch that several first-responder, and legal and medical procedural shows, reach large center-right audiences. These shows have a lot of potential for dramatic climate storylines given our new normal of killer heatwaves, superstorms, and year-round wildfire seasons. They also open the door to stories about environmental injustice in urban settings.

We realized we were already acquainted with a writer for one of the shortlist's highest-rated shows. He accepted our request to meet and we made our pitch: We found several compelling, real-world stories about climate change and environmental injustice that fit naturally with the premise and setting of the show. We offered to do the research and produce a story memo, and the writer agreed to share the memo with his colleagues if he thought it added value. We began by conducting interviews with seven NRDC subject matter experts from the fields of climate, environmental justice, clean energy, and water. The interviews helped us identify promising stories specific to the story's location. We then did extensive desk research and several follow-up interviews to flesh them out.

The resulting 4500-word memo is based entirely on real-world events and characters that illustrate the dramatic impacts of climate change on people's health, lives, and livelihoods. These specific details can become points of access for the writers to inspire fictionalized plots and character developments that are both organic to the show and that highlight climate issues and solutions. The memo calls particular attention to the scourge of environmental racism ("the expendability of Black and Brown bodies," as one activist called it) and the fact that the hardships of climate change tend to fall disproportionately on low-income communities and communities of color and to compound other forms of social injustice. Our research also revealed how historically oppressed communities are taking control of their own destinies through community organizing, putting them in the vanguard of the climate fight. The show writer's first response to the memo was, "I am super impressed – there are some terrific and very concrete ideas in here … let me know if you want to plot next steps." At the time of press, the writer had pitched the showrunner and producers one of the stories we proposed, and we were working with him to organize a climate storytelling presentation for the rest of the show's creative staff.

Case Study #3—Episode of a Docu-Series from a Major Studio

Rewrite the Future consults for several streamers and film studios, regularly reviewing fiction and nonfiction content at various stages of development through our particular lens of climate and environmental storytelling. For each review, we engage relevant NRDC subject matter experts to help with fact-checking and to evaluate framing, balance, and representation.

In 2021, we received a request from a major streamer to review an early outline for an upcoming episode of a docuseries, before any footage had been shot. The outline explained the issues to be explored and identified the interview subjects who would appear. The episode focused largely on a country in the Global South, long exploited by industrial nations for its natural resources, including oil and gas, and beset by political corruption and poverty.

Rewrite the Future brought in the energy policy manager for NRDC's International Program to help review the outline. In addition to offering detailed information and resources about the region and its socio-environmental problems, she illuminated the episode's over-representation of Global North interview subjects, under-representation of Global South and NGO subjects, and the omission of subjects from the nation at issue. She explained how this unrepresentative cast would distort the episode's discussion, elevating global market considerations over those of equity and self-determination. She then proposed an alternative cast of subjects from the affected region, offering personal introductions. The filmmakers took her advice, incorporating interviews with the suggested subjects into the episode and reframing the story to center on equity issues. This case illustrates the importance of early intervention—beginning the environmental review as early as possible in the show development process to help catch oversights, inaccuracies, and problems with balance, representation, and framing.

NRDC/Black List Climate Storytelling Fellowship

In service of Rewrite the Future's aim to encourage more and varied climate storytelling in popular entertainment, we developed a first-of-its-kind fellowship program to support new voices in climate storytelling. To complement our work with established Hollywood writers, we sought to reach a broader audience of aspiring screenwriters by partnering with The Black List (n.d.), a respected platform within the entertainment industry for early career writers to showcase their film and TV screenplays, to launch the inaugural NRDC Climate Storytelling Fellowship in April of 2021 (The Black List, 2021). The Redford Center and CAA Foundation, organizations in the entertainment industry focused on using the power of storytelling to create positive change, partnered with Rewrite the Future and The Black List to support the Fellowship.

To qualify for consideration, a script had to include climate in the story meaningfully, involving a major character(s), events, and plot or subplot(s). We encouraged writers to engage with climate themes in ways that reflect the reality of the climate crisis, depict solutions, and imagine a just and equitable future. The Rewrite the Future team also identified a set of story prompts to help writers expand their idea of what climate stories could be.

In June of 2021, Rewrite the Future, The Black List, and the Redford Center hosted a virtual public event to promote the Fellowship: "Climate Screenwriting to Save the World." This event provided an opportunity for the Rewrite the Future team to deliver our Climate Storytelling 101 presentation to a large audience of early-career writers and emphasize the many ways to effectively integrate climate and sustainability topics into screenplays. Following the event, one writer shared, "[...]Turns out I have the perfect project, just going to revise and resubmit... That's why I attended but it's not what I walked away with. To hear climate change spoken of as a given in our day-to-day life — of course it is, but of course it should be in the work we write."

We selected the three fellows from over 240 applicants in September of 2021. Their scripts include a feature film biopic about Rachel Carson that serves as a compelling climate allegory, a feature film mystery/suspense story about the early stages of a global pandemic caused by climate change, and a TV pilot about climate refugees from a ravaged future who come to the present to alter history. In addition to awarding a stipend to support the fellows' revision process, the Rewrite the Future team is providing significant creative feedback to develop the climate angles of their stories through our consultation process outlined above. We also paired each of the Fellows with professional screenwriting mentors Scott Z. Burns (*Contagion*), Naren Shankar (*The Expanse*), and Sarah Treem (*The Affair*), who have tackled climate themes in their own work, to provide feedback and guidance on the Fellows' scripts. At the end of this revision process, we will bring the completed scripts to our allies in the industry who are looking to develop climate content. We intend to make the Climate Storytelling Fellowship an annual program and find ways to bring together this community of climate storytellers throughout the year.

Conclusion

Stories that address the climate crisis are underrepresented in entertainment film and TV by any reasonable measure. Hollywood has been reluctant to tell climate stories for fear of losing advertisers and boring or alienating audiences. It has been creatively stymied by the breadth and complexity of the issue and, likely, by the psychological inhibitions we all experience when confronted by a seemingly overwhelming problem.

Rewrite the Future has been working in Hollywood for 2 years to highlight the void of stories about the climate crisis and to help the creative community tell more varied stories about it. We believe that the ship is beginning to turn. This is evident in the growing demand for Rewrite the Future's consulting services from large entertainment companies, in the strong response to our Climate Storytelling Fellowship, and in the number of writers, executives, and other creative professionals seeking advice and resources. It is also evident in the growing number of TV shows and films beginning to tell climate stories.

We believe that the change is driven by a desire among many Hollywood creators to use their art to help heal the planet. Increasingly, it is also driven by their dawning recognition that the climate crisis is a rich and largely untapped vein of dramatic material. Hollywood *wants* to tell climate stories and is beginning to realize that it *needs* to tell them in order to remain relevant to its audience.

References

Bahtiarian, M. (2019, April). *Quieting ships to protect the marine environment: Workshop final report*. Acentech Project No. 630964. Submitted to transport Canada/Government of Canada. http://voute.bape.gouv.qc.ca/dl/?id=00000167999

Ballew, M., Maibach, E., Kotcher, J. Bergquist, P., Rosenthal, S., Marlon, J., & Leiserowitz, A. (2020a, April). *Which racial/ethnic groups care most about climate change?* https://climatecommunication.yale.edu/publications/race-and-climate-change/

Ballew, M., Marlon, J., Kotcher, J., Maibach, E., Rosenthal, S., Bergquist, P., Gustafson, A., Goldberg, M., & Leiserowitz, A. (2020b, April). *Young adults, across party lines, are more willing to take climate action*. https://climatecommunication.yale.edu/publications/young-adults-climate-activism/

Bayer, T., & Hinerfeld, D. (2009). *Acid test: The global challenge of ocean acidification*. EarthNative Films and Natural Resources Defense Council (NRDC).

Borum Chattoo, C. (2016, September 26). *The state of the documentary field.* https://www.documentary.org/advocacy/state-documentary-field-2016-survey-documentary-industry-members

Brown, W. J. (2015). Examining four processes of audience involvement with media personae: Transportation, parasocial interaction, identification, and worship. *Communication Theory, 25,* 259–283. https://doi.org/10.1111/comt.12053

Buckely, C. (2019, August 14). Why is Hollywood so scared of climate change? *The New York Times.* https://www.nytimes.com/2019/08/14/movies/hollywood-climate-change.html

Burns, S. Z., Ellenberg, M., Fortenberry, D., Jacobs, G., Springer, L. (2022). *Extrapolations.* Media Research/ AppleTV+.

Congressional Record. (2009, October 1). *Proceedings and debates of the 111th* congress (Vol. 155, No. 140). Save The Oceans, comments of Mr. Inslee. https://www.congress.gov/111/crec/2009/10/01/CREC-2009-10-01.pdf

De Meyer, K., Coren, E., McCaffrey, M., & Slean, C. (2021). Transforming the stories we tell about climate change: From "issue" to "action". *Environmental Research, 16,* 015002. https://doi.org/10.1088/1748-9326/abcd5a

De Pinto, J., Backus, F., & Salvanto, A. (2019, September 15). Most Americans say climate change should be addressed now—CBS News poll. *CBS News.* https://www.cbsnews.com/news/cbs-news-poll-most-americans-say-climate-change-should-be-addressed-now-2019-09-15/

Dembicki, G. (2019, July 15). Climate change is everywhere, just not on TV. *Vice.* https://www.vice.com/en/article/wjv3bq/climate-change-is-everywhere-just-not-on-tv

Dougherty, M., & Hinerfeld, D. (2016). *Sonic sea.* Natural Resources Defense Council (NRDC) and Imaginary Forces.

Frank, L. B., & Falzone, P. (Eds.). (2021). *Entertainment-education behind the scenes: Case studies for theory and practice.* Palgrave Macmillan.

Funk, C., & Kennedy, B. (2019, April). *How Americans see climate change and the environment in 7 charts.* Pew Research Center. https://www.pewresearch.org/fact-tank/2020/04/21/how-americans-see-climate-change-and-the-environment-in-7-charts/

Green, M. C. (2021). Transportation into narrative worlds. In L. B. Frank & P. Falzone (Eds.), *Entertainment-education behind the scenes: Case studies for theory and practice* (pp. 87–101). Palgrave Macmillan.

Green, M. C., & Brock, T. C. (2000). The role of transportation in the persuasiveness of public narratives. *Journal of Personality and Social Psychology, 79,* 701–721. https://doi.org/10.1037/00223514.79.5.701

Green Production Guide. (2020). *PGA green: Who's who.* https://www.greenproductionguide.com/about-pga-green/.

Healey, J., Dawson, R., Movement Generation. (2019). *The north pole.* http://www.thenorthpoleshow.com/

Hinerfeld, D. (2011). *Stories from the gulf: Living with the oil disaster.* Natural Resources Defense Council (NRDC).

Hinerfeld, D. (2018). *Our ocean planet.* Natural Resources Defense Council (NRDC).

Hollywood Climate Summit. (2020). *2020 Summit.* https://www.hollywoodclimatesummit.com/2020-summit.

IPCC. (2018). *Global warming of 1.5°C,* special report. https://www.ipcc.ch/sr15/

IPCC. (2021). *Climate change 2021: The physical science basis.* https://www.ipcc.ch/report/sixth-assessment-report-working-group-i/

Kennedy, M. G., Turf, E. E., Wilson-Genderson, M., Wells, K., Huang, G. C., & Beck, V. (2011). Effects of a television drama about environmental exposure to toxic substances. *Public Health Reports, 126*(Suppl 1), 150–159. https://doi.org/10.1177/00333549111260S119

Koh, H., & Yatsko, P. (2017, February). *Jay Winsten and the Designated Driver Campaign.* Harvard Business School Case Study. https://www.hsph.harvard.edu/chc/2017/02/01/jay-winsten-and-the-designated-driver-campaign/

Kübler-Ross, E. (1969). *On death and dying.* Routledge.

Lawson, R. (2019, September 19). *When will Hollywood actually tackle climate change?* Vanity Fair, https://www.vanityfair.com/hollywood/2019/09/climate-change-in-hollywood

Lawson, D. F., Stevenson, K. T., Ptereson, M. N., Carrier, S. J., Strnad, R. L., & Seekamp, E. (2019). Children can foster climate change concern among their parents. *Nature Climate Change., 9*, 458. https://doi.org/10.1038/s41558-019-0463-3

Leiserowitz, A., Maibach, E., Rosenthal, S., Kotcher, J., Ballew, M., Goldberg, M., & Gustafson, A. (2018, December). *Climate change in the American mind*. Yale University and George Mason University/Yale Program on Climate Change Communication.

Leiserowitz, A. Maibach, E., Rosenthal, S., Kotcher, J., Neyens, L., Marlon, J., Carman, J., Lacroix, K., & Goldberg, M. (2021, September). *Global warming's six Americas*. Yale University and George Mason University/Yale Program on Climate Change Communication.

Littleton, C. (2019, September 10). Is Hollywood doing enough to fight climate change? *Variety*. https://variety.com/2019/biz/features/climate-crisis-hollywood-environment-movies-tv-1203329390/

Marlon, J., Howe, P., Mildenberger, M., Leiserowitz, A., & Wang, X. (2020, September). *Yale climate opinion maps 2020*. https://climatecommunication.yale.edu/visualizations-data/ycom-us/

Meulmester, S. (2017, October 5). *"Sonic Sea" wins two Emmy awards*. NRDC. https://www.nrdc.org/media/2017/171016-1

Moyer-Gusé, E. (2008). Toward a theory of entertainment persuasion: Explaining the persuasive effects of entertainment-education messages. *Communication Theory, 18*, 407–425. https://doi.org/10.1111/j.1468-2885.2008.00328.x

NRDC. (2019). *NRDC helps Hollywood take on the climate crisis*. https://www.nrdc.org/sites/default/files/rewritethefuture_onepage_digital.pdf

NRDCflix. (2021, February 2). *Beyond apocalypse: Alternative climate futures in film and TV*. YouTube. https://www.youtube.com/watch?v=wX9V0XZYYjY

O'Brien, M., & Hinerfeld, D. (2014). *Wild things*. Natural Resources Defense Council (NRDC).

O'Neill, S., & Nicholson-Cole, S. (2009). "Fear Won't do it": Promoting positive engagement with climate change through visual and iconic representations. *Science Communication, 30*(3), 355–379. https://doi.org/10.1177/1075547008329201

Schiappa, E., Gregg, P. B., & Hewes, D. E. (2005). The parasocial contact hypothesis. *Communication Monographs, 72*(1), 92–115. https://doi.org/10.1080/0363775052000342544

Schiappa, E., Gregg, P. B., & Hewes, D. E. (2006). Can one TV show make a difference? *Will & Grace* and the parasocial contact hypothesis. *Journal of Homosexuality, 51*(4), 15–37.

Singhal, A., & Rogers, E. M. (1999). *Entertainment-education: A communication strategy for social change*. Lawrence Erlbaum.

Singhal, A., Cody, M. J., Rogers, E. M., & Sabido, M. (2004). *Entertainment-education and social change: History, research, and practice*. Lawrence Erlbaum.

Sood, S., Riley, A. H., & Alarcon, K. C. (2017). Entertainment-education and health and risk messaging. In J. F. Nussbaum (Ed.), *Oxford research encyclopedia of communication*. Oxford University Press. https://doi.org/10.1093/acrefore/9780190228613.013.245

Stewart, E. (2021, March 30). *NetZero + Nature: Our commitment to the environment*. https://about.netflix.com/en/news/net-zero-nature-our-climate-commitment

Tabuchi, H., & Friedman, L. (2021, October 27). Oil executives to face congress on climate disinformation. *The New York Times*. https://www.nytimes.com/2021/10/27/climate/oil-congress-climate-disinformation.html

The Black List. (2021). *NRDC climate storytelling fellowship*. https://blcklst.com/partnerships/opportunities/86

The Black List. (n.d.). https://blcklst.com

The Henry J. Kaiser Family Foundation. (2004). *Entertainment education and health in the United States*. Issue Brief, https://www.kff.org/wp-content/uploads/2013/01/entertainment-education-and-health-in-the-united-states-issue-brief.pdf

Thunberg, G. (2018, November). *School strike for climate—Save the world by change the rules*. TEDxStockholm. https://www.ted.com/talks/greta_thunberg_school_strike_for_climate_save_the_world_by_changing_the_rules/transcript?language=en

Upton, J. (2015, March 28). *Media contributing to "hope gap" on climate change*. Climate Central. https://www.climatecentral.org/news/media-hope-gap-on-climate-change-18822

Wang, H., & Singhal, A. (2021). Theorizing entertainment-education: A complementary perspective to the development of entertainment theory. In P. Vorderer & C. Klimmt (Eds.), *The Oxford handbook of entertainment theory* (pp. 819–838). Oxford University Press. https://doi.org/10.1093/oxfordhb/9780190072216.013.42

Daniel Hinerfeld is an Emmy-winning filmmaker and award-winning journalist who has worked at the NRDC since 2003. Daniel founded NRDC's documentary film unit and co-founded Rewrite the Future, an initiative that helps Hollywood tell stories about the climate crisis. Daniel previously worked for NPR (National Public Radio) as a reporter, producer, and senior editor. He co-created The Tavis Smiley Show and the nationally syndicated political discussion show, Left, Right & Center. He has written for the Washington Post, the Los Angeles Times, and Rolling Stone, among other publications.

Cheryl Slean is a climate communicator, playwright, and filmmaker with a diverse academic background in physics, literature, and sustainability. She's taught filmmaking and creative writing at several universities and started a production company specializing in narrative strategies for educational media. She co-founded Rewrite the Future, a climate storytelling in entertainment initiative at the NRDC, and she currently works as Manager of Sustainability and Storytelling at Netflix.

Katy Jacobs is the director of Entertainment Partnerships at the NRDC and works with partners across the entertainment industry to use pop culture and storytelling to elevate environmental issues and create impact. Katy led NRDC's inaugural Sundance Film Festival activation in 2020. In 2021, she developed and launched a first-of-its-kind climate storytelling fellowship in partnership with The Black List, CAA Foundation, and The Redford Center. Prior to the NRDC, Katy worked in documentary film as a producer and impact consultant. She also served as the director of Public Programs, Festivals, and Exhibits at the Paley Center for Media where she produced the first PaleyFest NY in partnership with the NYC Mayor's Office of Media & Entertainment.

LOLs: Secret Weapon Against CFCs and CO2?

Celia Gurney and Mamoudou N'Diaye

You see this line? It's global temperatures, okay? And it's going UP! It's going up!!! If this was going down, you'd be right, and WE'D be the dumbasses! But it's going UP! So YOU'RE wrong, and YOU'RE the dumbass! Okay?! Even if you flip the chart upside down—IT'S STILL GOING UP!!!

–Ronny Chieng, Correspondent on The Daily Show with Trevor Noah (2021)

Climate change is predominantly regarded as a serious and boring domain, associated with things that don't light the average person's fire, such as Doom and Gloom™ (Clayton et al., 2015; O'Neill et al., 2013) and granular data on things such as rainfall rates and ocean temperatures.[1] It's not something you'd want to bring up immediately at a party, per se—unless you're trying to clear a six-foot radius around yourself so you can let one rip in peace (the original social distancing). Some people still don't even accept that it's happening, hence Ronny Chieng's rant above, but even those who do accept that global warming is happening often end up feeling helpless due to the massive scale of the problem (Borum & Feldman, 2020) and unmotivated due to its diffuse, slow-moving, and future-concentrated consequences.

All this leads to a lot of hesitation around making comedy about climate change à la Chieng. *Is there anything* funny *about climate change*, people think? *Is it even* appropriate *to make jokes about climate change? Will we offend the coral reefs and risk being canceled by rising sea levels? Will climate change comedy actually make*

---

[1] No offense to the scientists and weather geeks out there who love granular data—consider yourselves above average! Yeah…we'll go with that….

C. Gurney (✉)
Los Angeles, United States
e-mail: climatecomedyprojects@gmail.com

M. N'Diaye
Brooklyn, NY, United States
e-mail: egodeathproductionco@gmail.com

E. Coren, H. Wang (eds.), *Storytelling to Accelerate Climate Solutions*,
https://doi.org/10.1007/978-3-031-54790-4_9

people laugh, *or will it be kind of like when Ken at the office makes "jokes" during his slide deck presentation and we all sort of go "heh, heh, heh," in a bored, mechanical way with dead eyes and deader spirits?!*[2]

The answer is yes, Yes, YES, and we'll explain, but first, allow us to introduce ourselves. We are two comedy writers and performers who are passionate about advancing climate justice. Climate justice, for those who may not know, is a response to the climate crisis that acknowledges (1) climate change is a product of the extractive capitalism cemented by colonialism and slavery, and (2) that it exacerbates social justice issues today—especially racial justice issues (Heglar, 2019; NAACP and the Clean Air Task Force, 2017; Yearwood Jr., 2020; Fig. 1 provides a good visual). Therefore, the only "solution" to the climate crisis (in quotes because the best we can hope to do is mitigate) is one that confronts these issues in tandem with the earth sciences part. Got that? Ok, cool. Now who are these two people talking your ear off at the party??[3]

- One of us, Mamoudou N'Diaye, is a comedian, TV/film writer, DJ, baby abolitionist, full-size mental health advocate, racial justice and climate justice creative consultant, and former seventh-grade teacher who has written for Netflix,

Fig. 1 Climate justice is social justice. (Image used with permission from Freedom to Breathe)

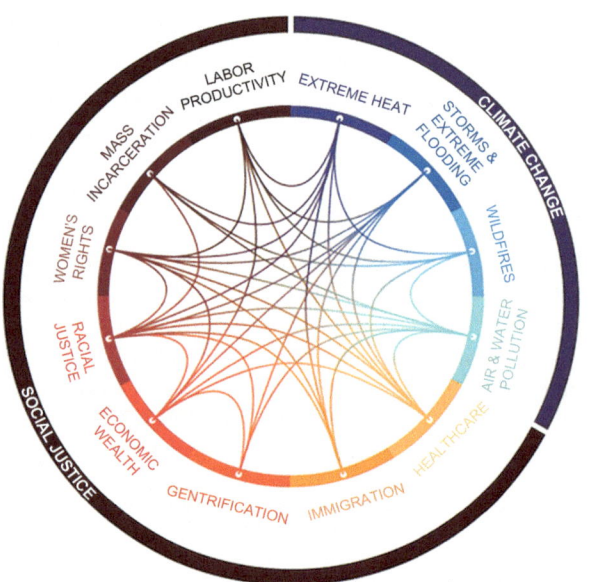

---

[2] For more dead-eye content, you may want to check out the podcast Dead Eyes, which is about an actor getting fired from Band of Brothers because he had "dead eyes."

[3] "I already know who you are," you think bitterly. "I read your bios at the top of the article." Whatever, killjoy—we didn't know they were gonna be there until we'd already gotten attached to this flow of ideas.

ABC, Hulu, Apple, and Amazon with a degree in Cognitive Behavioral Neuroscience from The College of Wooster.
- The other one, Celia Gurney, is an Upright Citizens Brigade-trained improviser and video producer who spent 3 years at Climate Nexus, a climate change communications nonprofit, translating climate science and politics for the digital space using her B.A. in Environmental Studies from the University of Washington, Seattle.

In short, we've always had really clear, straightforward/focused career aspirations that our boomer parents could tooooOOooootally understand and had faith in from day 1.[4] Anyway, we felt it was important to tell you who we are so you'll understand why this chapter feels different from the rest of this book. We're just two comedians, standing in front of you, asking you to accept that climate justice is already starting to permeate the comedy space,[5] and that its full permeation of that space is a key tool in the Achieving-a-Just-Transition-Before-We-All-Burn-to-a-Crisp Toolbox.[6] By the end of this chapter, we want you to have a sense of (1) the many different ways people are combining climate communication and comedy, (2) how you might combine them yourself, (3) where the climate comedy field is heading, and (4) what it needs to get there. So we're gonna walk the talk and discuss all this stuff using—you guessed it—comedy. And maybe a troupe of cannibalistic child mimes. Read on to find out!!!

Literature Review

Entertainment-education (EE) is a type of public health programming that combines—yep, you guessed it—entertainment (about 80%) with education (about 20%) (Coren et al., 2020). The intention behind EE is to put theory into practice to solve societal problems (Wang & Singhal, 2021). A common component of EE is transmedia, which is media outside of the show itself that reinforces the information and resources discussed in the show—for example, social media posts and online Q&As with characters (Coren et al., 2020).

One key figure in the history of EE is Dr. Albert Bandura of Stanford, who cooked up a little something called social learning theory. It was a lentil-based dish blasting more flavors at your palate than Goldfish themselves…okay fine, it was actually the idea that watching and trying to emulate others is one of the primary ways humans learn (UC Berkeley, 2021a, b). (Are you happy now??!!![7]) In a famous study, he had children watch videos of adults beating up an inflatable Bobo the

---

[4] Our parents refused to be reached for comment.

[5] Boombox over our head playing "Careless Whisper."

[6] Available for a very limited time on earth. Come get yours today—and make sure to use it by 2030!

[7] Michelle Branch shoutout.

clown doll. The children then mimicked that behavior when left alone with Bobo[8] (Goode, 2021). This showed that rewards and punishments were not the only way to shape behavior (Goode, 2021) and that an onscreen example of a given behavior could be just as influential as an in-person one (Friedman, 2013).

Then, in the 1970s, Mexican TV executive Miguel Sabido—widely recognized today as the father of EE—started blending Bandura's theory into a delicious smoothie. The other ingredients, as Sabido details in his chapter of *Entertainment-Education Behind the Scenes* (Frank & Falzone, 2021), were "Wilbur Schramm's theory of communication, Eric Bentley's drama theory" and Sabido's own theory of tone[9] (Sabido, 2021, p. 18). Sabido had observed the huge behavior-modeling impact of a Peruvian telenovela, *Simplemente María* (Wang & Singhal, 2021). He then created a telenovela called *Ven Conmigo*, in which multiple characters signed up for literacy classes that were also offered by the Mexican government in real life (Desmon, 2018). Enrollment in the real-life classes grew by 1100% in the year after the show aired (Coren et al., 2020), with more than 500,000 people signing up (Sabido, 2021).

Sabido went on to create shows that talked about family planning, which gave women a blueprint for asserting more control over their own bodies (Friedman, 2013) and are thought to have decreased Mexico's population growth rate by about one-third in less than a decade (Desmon, 2018). Sabido and his team also started collaborating with TV writers and producers in other countries. Soon there were TV shows in China and India featuring characters with HIV/AIDS, a show in Kenya that communicated the dangers of female genital mutilation, and a Bolivian radio drama about alcoholism and relationship violence, to name a few (Friedman, 2013). A recent example of EE in the USA is Hulu's Latinx-created and Latinx-written series *East Los High* (Wang & Singhal, 2016), which aired from 2013 to 2017 (IMDb, n.d.-a). The show aimed to reduce pregnancy among Latinx youth via safe sex modeling by Latinx characters and reinforcement by transmedia that directed people to reproductive health resources (Coren et al., 2020).

With increasing recognition that climate justice is a public health issue, there's a growing cohort of so-called boundary organizations bringing storytellers—many of them comedic—into collaboration with climate movement professionals.[10] These include Storyline Partners, Define American (Define American, 2022), and Participant Media (Borum, 2021), Population Media Center (Ryerson & Negussie, 2021), the Skoll Center for SIE (Schwartz, 2020), the Center for Media & Social Impact (Borum & Feldman, 2020), Albert and Doc Society (York St. John University,

---

[8] Wherever Bobo is now, we hope he's at peace, and not at all connected to those reports of clowns beckoning children into the woods in 2016.

[9] We unfortunately do not have time to dive into all those theories, but we can give you a little theory-of-tone teaser. Apparently in *The Devil Wears Prada*, Meryl Streep placed her energy in the "'nodes' behind the eyes," which produced an "intelligent and lucid" tone (Sabido, 2021, p. 17). Don't you wanna know more??!!!

[10] The ultimate goal is marriage, and the tactics consist entirely of a Bachelor(ette)-type reality show called Methane Me or CFC U At The Altar.

2021), Johns Hopkins Center for Communication Programs, the Norman Lear Center's Hollywood, Health & Society program, NRDC's Rewrite the Future program (NRDC, 2022), The Center for Cultural Power, and The Good Energy Project. While Sabido's model of entertainment-education necessitates close coordination with governments and the private sector, as well as "a solid, viable hypothesis that can be quantitatively and qualitatively tested"[11] (Sabido, 2021, p. 20), boundary organizations' work typically results in a looser configuration of the two Es.[12] This is referred to as social impact entertainment, cultural strategy, or narrative strategy (Borum, 2021; Borum & Feldman, 2020; Ryerson & Negussie, 2021; Schwartz, 2020). This development reflects the way EE practitioners and social justice activists have both adapted to the "post-millennial participatory media age," as Caty Borum wrote in her chapter of *Entertainment-Education Behind the Scenes* (2021, p. 44). In the time of streaming services and social media, it's much harder to get everyone's eyes on the same thing at once—so that's no longer the goal. Instead, the goal is to create and maintain a "wide constellation of entertainment narratives that accumulate" to more accurately represent marginalized groups (Borum, 2021, p. 44) and move public opinion about social justice and public health issues.

TL;DR:[13] This climate change comedy combo isn't some obscure, irrelevant rabbit hole that one person wrote about for their dissertation before ultimately going on to work as a corporate lawyer. It's a current iteration of a decades-long tradition of using storytelling for social change. Universities, nonprofits, "big greens," and production companies are all in on it—and as you'll see in the rest of this chapter, it's gathering steam.

Fine, But Why Comedy?

Listen, we didn't come here to prove that comedy is a worthwhile climate change communication strategy. Other people have already shown it empirically! That's right, folks—REAL SCIENTISTS are out there conducting experiments where they get to, for example, show some participants traditional news clips and others clips from *The Daily Show*. That means those scientists "have to" watch *The Daily Show* at length to find the clips they want to use.[14]

However, even though we don't have anything to prove, we want to make sure we're all on the same page before we proceed. So we're going to go over some of the theory, research, and case studies that show comedy is a worthwhile communication strategy, even for good ol' Doom & Gloom™ climate change. And don't just

---

[11] Okayyy, someone's showing off his knowledge of science....

[12] More knitwear than textile. Upscale knitwear, though. Missoni.

[13] For those not in the know: TL;DR is Internet slang for "too long, didn't read." It is used to introduce a brief summary of the preceding text. You're welcome for bringing you into the twenty-first century.

[14] Well-played.

take our word for it—feel free to follow the trail of breadcrumbs (our citations) back to Hansel and Gretel's house (the original sources).[15] It's a stunning, turn-of-the-century stone Tudor with all the original chimneys (meaning: the original sources are fascinating and explain their findings better than we ever could).[16]

Borum and Feldman's Framework

As non-academics working on this chapter in between network TV scripts (Mamoudou) and sad little TikTok videos (Celia), we had neither the time (Mamoudou) nor the level of executive functioning (Celia) to reinvent the wheel here. Fortunately, in their book *A Comedian and an Activist Walk into a Bar: The Serious Role of Comedy in Social Justice*, Caty Borum and Lauren Feldman (2020) shared a meticulously researched framework (synthesizing 200+ studies) explaining why comedy is a valuable tool for advancing social justice. Even more fortunately, they were kind enough to allow us to share a SparkNotes version of their framework here, seasoned with a bit of information we found via other sources.

So, without further ado, here is Borum and Feldman's (2020) framework (plus some other stuff), all regurgitated in caveman speak.

So, Why Comedy?!? Well, It's

- *Your favorite applesauce…*with a dash of truth-flavored penicillin, a Trojan horse, a spoonful of sugar, etc. See, people don't generally think of late-night comedy shows as educational programs. But when they watch them, they usually end up learning something about serious issues—and devote more attention to those issues than they otherwise would (Baum, 2005, as cited in Borum & Feldman, 2020). Some would even say[17] the teams behind late-night comedy shows are performing an important public service by doing most of the hard work to understand the news. It takes a lot less energy to watch a funny presentation than it does to do all that information gathering, analysis, and creative thinking yourself (Borum & Feldman, 2020). Plus, a 2017 study found that people actually learn more about climate change and sustainability topics when comedy is involved (Gravey et al., 2017). Figure 2 displays the interaction between comedy and climate crisis information.

---

[15]Yes, birds did eat Hansel and Gretel's breadcrumbs, but fortunately for you, our breadcrumbs are meticulously formatted in APA and saved to the cloud.

[16]All knowledge of architectural styles courtesy of the House cards from The Game of Life (early 2000s version).

[17]We. We would say.

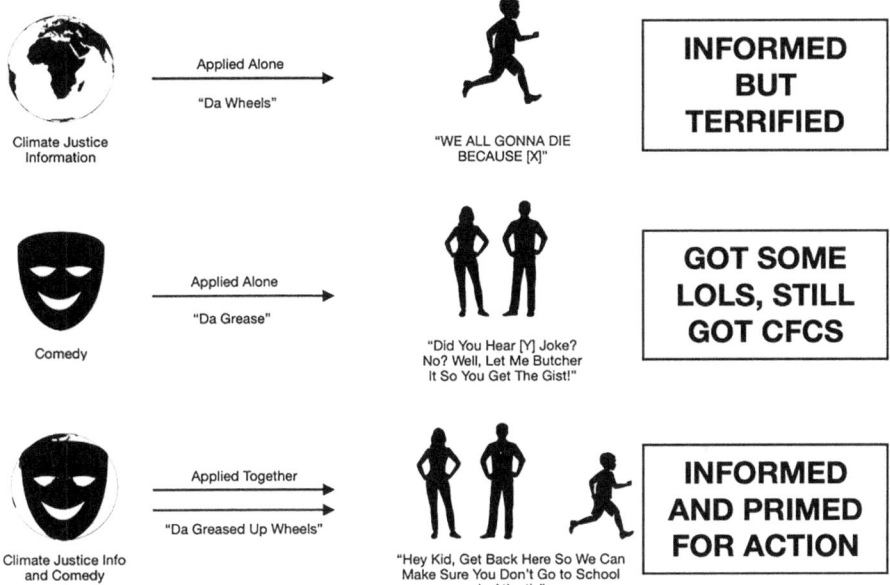

Fig. 2 Interactions between comedy and climate crisis information

- *A Facilitator of Parasocial Relationships*. Parasocial relationships form when people feel they know a character or an actor personally because of all the time they've spent watching and/or reading about them[18] (Dibble et al., 2015; Horton & Wohl, 1956). When it comes to reducing prejudice against marginalized groups, parasocial contact may be the next-best thing to face-to-face contact (Kim & Harwood, 2020). For example, a 2006 study of college students found that those who watched *Will & Grace* more frequently harbored less prejudice against lesbians and gay men. This correlation was especially strong for the students who didn't spend much time with lesbians and gay men IRL (Schiappa et al., 2006). Positive parasocial relationships with comedy characters (vs. characters from dramas) may be more powerful in this regard (Zhao, 2016). With regard to the climate justice movement, parasocial contact with the marginalized groups most affected by climate change could move less directly affected people towards action.
- *Keeping Skeptics Busy*. People don't have as much brain space to counter-argue—or push back on ideas they disagree with—when they're focusing on getting a joke or immersed in a funny story (Nabi et al., 2007). When participants in a 2008 experiment listed their thoughts in response to funny and serious versions of political messages, those who were exposed to the funny versions gener-

---

[18] Was anyone else in a problematic parasocial triad with Ariana Grande and Pete Davidson in 2018? No??

ated fewer negative thoughts about the messages (Young, 2008). Narrative comedy, meanwhile, disarms by putting great storytelling and compelling characters front and center rather than overt persuasive messages, so audiences aren't as primed to analyze (and potentially push back on) the content they consume (Green, 2021). Narrative comedy may also reduce counter-arguing via transportation, which is when a viewer is so immersed in a story that they forget about reality (Green, 2021). Because we all know that when you get a break from reality, you do not cut it short by arguing that such-and-such character "could never afford that apartment."[19]

- *Straight-Talking.* Comedians can tell it like it is. There's a time and a place to appear impartial—and per John Oliver's 2014 segment (Last Week Tonight with John Oliver, 2014b), it is NOT when marginalized people are getting pummeled by climate changes that 97% of scientists agree are due to human activities (NASA, 2021). Compared to traditional media outlets, satirical news shows have a lot more freedom to add commentary, transparently share their own analyses, call out hypocrisy, and speak truth to power (Berlant & Ngai, 2017; Borum & Feldman, 2020; Martin, 2007). Perhaps best of all, they can swear! While not an end all be all, comedians shine a light on the issue—kinda like a clown holding a flashlight to an issue so politicians, special interest groups, and real fixers can get in there with their toolset and get to the real work.
- *Magnetic and Sticky.* Research has repeatedly shown comedy's power to get our attention (Borum & Feldman, 2020), and a recent study found that people are more likely to remember things they learn from humorously presented news clips than seriously presented ones (Coronel et al., 2021; Yount, 2021). Scientists don't completely agree on why comedy is memorable, but we'll let you go down that rabbit hole on your own time[20] (Borum & Feldman, 2020). Anyway, apparently corporations have seen all this research, because commercials. Have you seen Allstate's Mayhem series? Pemco's Northwest Profile series?![21] (A regional classic! Seriously—run, don't walk.) Comedy runs so rampant in commercials that "SNL can't even do commercial parodies anymore," according to an improv teacher one of us had once.[22] So why (are the public and voluntary sectors still largely) so serious, to quote Heath Ledger's Joker?[23]
- *Easily Amplified Because People Share It.* A study of tweets about social issues found that funny tweets are more likely to get retweets than serious ones, and a 2021 study found that viewers of news clips are more likely to share funny clips than serious ones (Coronel et al., 2021). But enough about studies—why don't we mix things up by getting an analysis in here?! According to the authors of a 2013 analysis of a campaign to reduce unintended pregnancies in Iowans under

---

[19] DAD.

[20] And sweetie?? Have fun!!!!

[21] Admittedly these are kind of deep cuts.

[22] No idea who that teacher was, but we thank them for being the cornerstone of this chapter.

[23] Turns out *The Dark Knight* was actually about climate change communications.

30, humor was so key in getting women to share the campaign that all campaigns should consider using it[24] (Borum & Feldman, 2020; Campo et al., 2012; Venzke, 2012). Additionally, to top things off with some anecdotal evidence, one participant in CU Boulder's Stand Up for Climate Change live comedy show observed that comedy helps people talk about the climate crisis without sounding like Debbie Downer all the time (Boykoff, 2019).

- *Enhancing Political Will.* According to a 2018 study (Nabi et al., 2018), college students' climate policy advocacy increased after emotional experiences. Being moved to laughter is an emotional experience, and usually a more fun one than being moved to tears...unless you have a "high need for affect," which means you enjoy feeling extreme emotions in general, including sadness.[25] (Green, 2021). At least a couple pieces of legislation can even thank comedy for helping them get passed: 1) the 9/11 First Responders Health Bill, which Jon Stewart advocated for on *The Daily Show*, and 2) the Sexual Assault Survivors' Bill of Rights Act, which got a boost from a Funny or Die video called "Even Supervillains Think Our Sexual Assault Laws Are Insane" (Funny or Die, 2016; Hill & Holbert, 2017; Kim, 2019). In front of the right eyes, this shit works!
- *Taught by Big-Name Institutions.* An increasing number of colleges and universities offer comedy writing and/or performance degree programs, including Columbia College Chicago, Emerson College, and DePaul University (Columbia College Chicago, 2021; DePaul University, 2001–2021; Emerson College, 2021). (Obviously not all of the students who get these degrees will become comedy writers or performers—only the ones who are as funny or funnier than us. So two, maybe three people). Others offer comedy minors, like USC and NYU, or significant course and extracurricular options, like UCLA, Harvard, Indiana University-Bloomington, and UC San Diego (FOOSH Improv UCSD, n.d.; Harvard, 2021; The Comedy Attic, 2021; Tisch School of the Arts, 2021; UCLA School of Theater, Film & Television Professional Programs, 2021; University of Southern California, 2021). UNC Chapel Hill even has a famous, long-running class called Gram-O-Rama in which students write and perform sketch comedy about grammar rules (Moger, 2017). The University of Colorado at Boulder's Inside the Greenhouse program puts on an annual Stand Up for Climate Change comedy show and runs an annual comedy video competition.
- *Used by Boundary Organizations.* You already heard about boundary organizations in the literature review, but here is a more in-depth explanation of one. American University's School of Communication houses the Center for Media & Social Impact (CMSI), a creative innovation laboratory and research center (CMSI, 2021b) that produces reports on social change comedy (CMSI, 2021c) and facilitates collaborations between diverse comedians and social justice professionals (CMSI, 2021a, c). Its Comedy ThinkTanks program "curates a room

---

[24] This is a particularly useful recommendation for Celia, who has been trying to get an obscure awareness campaign off the ground for years: the Campaign for Seals: We Bark, Too™.

[25] If that's you, you probably have fond memories of reading *Bridge to Terabithia* as a child.

of professional comedy writers and performers to collaboratively co-create with social justice groups to create original comedy—sketches, program ideas, campaigns" (CMSI, 2021a). Meanwhile, the Yes, And Laughter Lab mentors diverse comedy writers leading up to an event where they get to pitch their work to "the entertainment industry, social justice organizations, philanthropists and activists who can help bring their work into the comedy marketplace—and into movements for social change" (Yes, And Laughter Lab, 2022). (Fun fact: Mamoudou was a winner in the inaugural class of 2019!) CMSI is directed by the same Caty Borum who co-wrote *A Comedian and An Activist Walk into a Bar* (Center for Media & Social Impact, 2021a, b, c, d), and it received $1.1 million in grants in 2020[26] (American University, 2020).

- *Suitable for Any Topic (When Used Responsibly).* Comedy is about having a unique perspective. That's a filter you can put on literally anything,[27] and comedians are doing it constantly. Tig Notaro's breakout special was about getting diagnosed with breast cancer (Notaro, 2013). Yedoye Travis hosted Dark Tank, a podcast where he and guests debated potential solutions to racism, including "mass hauntings" and outlawing sunscreen (Travis, n.d.). Hell, haven't you ever heard an amazing roast joke about a dead person at their funeral?! If you're worried climate change and comedy won't mix, then WE'RE worried…that you've never seen comedy. Of course, there are rules about how to and who should joke about certain topics—a comedian code of ethics, if you will. For instance, if someone's going to be the butt of a joke, it should be a person in power; it's called "punching up" and has a history dating back to court jesters roasting their kings and queens.[28] See Fig. 3 for examples of punching up vs. punching down in climate justice comedy.

For your convenience, we created a mnemonic device with the first letters of each of the above bullet points so you can remember all those good reasons to use comedy for the climate. That mnemonic device is "Yaks Meet Us…" It's a fun first half of a sentence that you can complete with "in St. Louis," "Where We Are," "At the Mall; It's Goin' Down," or whatever feels right to you. And again, we do encourage you to check out the original, much more detailed, much more eloquently written framework as conceived and explained by Borum and Feldman (2020).

Research Questions and Hypotheses

In this chapter, we asked two important research questions:

- RQ1: Who is making climate change comedy today and in what formats?

---

[26] Really, *really* rich people believe in this work, you guys!

[27] Take that, Instagram!!!!

[28] Only back then, if the jokes were bad, jesters got killed. Today, they start their own podcast, which is a death of its own now that we say it out loud….

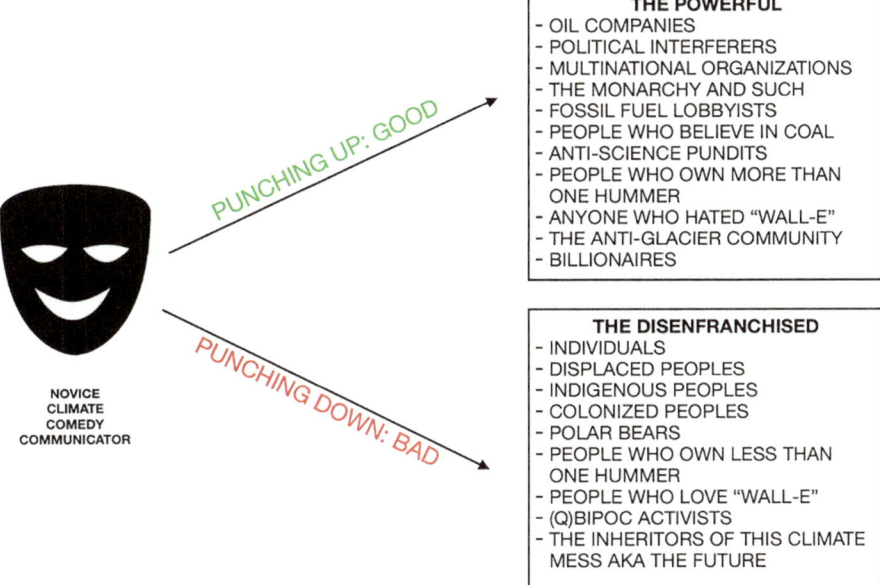

THE POWERFUL
- OIL COMPANIES
- POLITICAL INTERFERERS
- MULTINATIONAL ORGANIZATIONS
- THE MONARCHY AND SUCH
- FOSSIL FUEL LOBBYISTS
- PEOPLE WHO BELIEVE IN COAL
- ANTI-SCIENCE PUNDITS
- PEOPLE WHO OWN MORE THAN
  ONE HUMMER
- ANYONE WHO HATED "WALL-E"
- THE ANTI-GLACIER COMMUNITY
- BILLIONAIRES

THE DISENFRANCHISED
- INDIVIDUALS
- DISPLACED PEOPLES
- INDIGENOUS PEOPLES
- COLONIZED PEOPLES
- POLAR BEARS
- PEOPLE WHO OWN LESS THAN
  ONE HUMMER
- PEOPLE WHO LOVE "WALL-E"
- (Q)BIPOC ACTIVISTS
- THE INHERITORS OF THIS CLIMATE
  MESS AKA THE FUTURE

NOVICE
CLIMATE
COMEDY
COMMUNICATOR

PUNCHING UP: GOOD

PUNCHING DOWN: BAD

Fig. 3 How to (and how not to) haha the climate crisis

- RQ2: What will it take to more fully harness the power of climate change comedy in order to improve the efficacy of the climate justice movement, invite in more allies, and build a more inclusive clean energy economy?

Based on our personal experiences in multiple American cities and on the Internet, we proposed the following hypotheses:

- H1: A bunch of dope creators are making climate change comedy in a variety of formats, from sketches and live shows to explainer videos and more.
- H2: In order to more fully harness the power of comedy to improve the efficacy of the climate justice movement, invite in more allies, and build a more inclusive clean energy economy, boundary organizations will need to work with writers and producers to consistently and thoughtfully integrate climate change into comedy programming. This will mean improving and increasing climate change coverage by late-night comedy shows (and perhaps giving the issue its own late-night show), as well as in half-hour comedies, stand-up specials, and other projects. Things also need to work in the other direction, which is to say comedy needs to become a more accessible, acceptable tool for teachers, organizers, scientists, and other non-entertainers in the climate justice movement.

Methodology

Sample

We asked a group of about 20 climate crisis-communicating comedians (biological sex nonwithstanding) for qualitative comedological data/content to provide contextual real-world examples of laughter measuring art option (or LMAO) style entertainment as a vehicle for Matrix-esque uploads of information about the current climate crisis, its history, and our future.

"Comedian"[29] (or "comic") was defined as an artist who produced chortles, chuckles, lols, laughs, snickers, hahas, lmfaos, guffaws, giggles, "NPR-style nods of assent," titters, crying laughing emojis, tee-hees, and/or hoots and hollers from a captive[30] audience, digital or live. "Roflmaos" were excluded from this study as it is not 2007. NFTs were not included in this study because frankly the authors still don't get what they are.

"Climate crisis communication" was defined as "climate crisis communication." It really was not that deep; if subjects discussed the impending climate crisis, its causes, effects, histories, impacts, advocates, and detractors using facts, figures, stats, data sets, infographics, PowerPoints, Windows Movie Maker, emo lyrics on their Facebook wall, tweets (or whatever that emerald heir is calling them these days), fleets (RIP), Tumblrs, YouTube, Instagram, Tik, Tok, TikTok, or TokTik, they were considered climate crisis communicators for the purposes of this study.[31]

Data Collection

Responses were compiled using a radical new quantum electron microscopic nuclear magnetic resonance spectroscopic acquisition software designed to optimize a streamlining of data collection. This software is called "Google Forms."

Respondents responded to the form and the authors kinda just took the hot parts and dropped them in this mf.[32] This form was intended to produce the best results by way of the authors not having to make up nice things to say about them, but the authors were ready to let the compliments fly if this methodology was considered "challenging" by any of the respondents.[33]

---

[29] We didn't accept "comedienne" or "comicstress" because well that's just stupid. Abolish gender.

[30] Not literally captive, although what a weird kidnapper if it was real, right?

[31] Ordained by Albert Gore.

[32] You say "lazy," we say "you're just jealous."

[33] Okay, well actually, in the end a lot of respondents didn't live up to their name, by which we mean they did not *respond* to the form. The authors' plot to put together a book chapter without doing any actual work was foiled. They are working through their resentment in therapy.

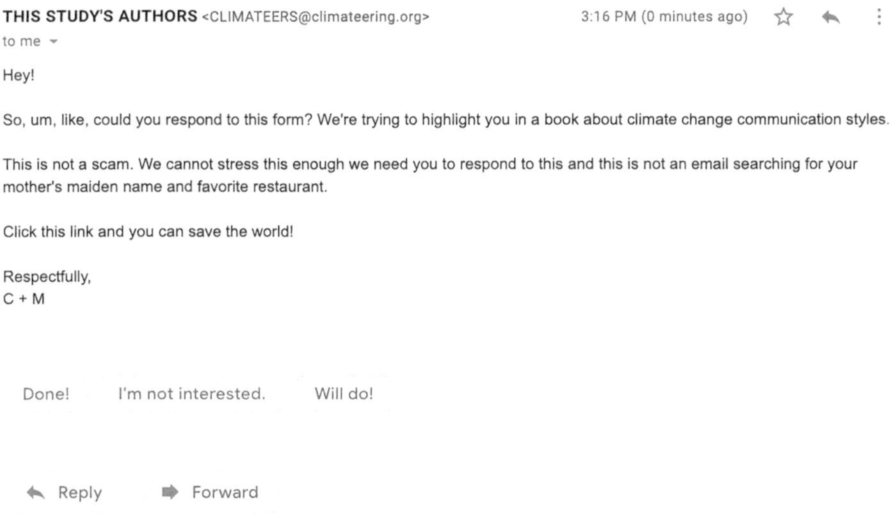

THIS STUDY'S AUTHORS <CLIMATEERS@climateering.org>   3:16 PM (0 minutes ago)   ☆   ↰   ⋮

THIS STUDY'S AUTHORS <CLIMATEERS@climateering.org>   3:16 PM (0 minutes ago)   ☆   ↰   ⋮

to me ▾

Hey!

So, um, like, could you respond to this form? We're trying to highlight you in a book about climate change communication styles.

This is not a scam. We cannot stress this enough we need you to respond to this and this is not an email searching for your mother's maiden name and favorite restaurant.

Click this link and you can save the world!

Respectfully,
C + M

Done!   I'm not interested.   Will do!

↰ Reply   ➡ Forward

Fig. 4 A desperate attempt at data collection

We sent all respondents an email with a Google Form and a promise of no money. No really. That's it. Figure 4 shows what we followed up with a few days later.

The measurements we used were a scale from "exists" to "doesn't exist." If the respondent communicated information about the climate crisis using comedy, it exists. If the respondent doesn't communicate information about the climate crisis using comedy, it was weird they got the email in the first place.

Results

We got…some community responses. Like, not a lot, but not enough. The rest we added from Web research that we conducted ourselves, and it took frickin' forever (fund your local climate change comedy communication chapter writers aka us), but the good news is H1 was supported.

We focus here on creators who have not yet reached maximum visibility in the American media landscape, and/or who, to our knowledge, have not yet been highlighted in bird's-eye-view-style academic literature about climate change comedy. If you wanna split hairs, some of them might be better described as climate change communicators *using* comedy. Many other creators can be found in books like *A Comedian and an Activist Walk into a Bar: The Serious Role of Comedy in Social Justice* by Caty Borum and Lauren Feldman (2020); *Creative (Climate) Communications* by Maxwell Boykoff (2019); *Entertainment-Education Behind the Scenes*, edited by Lauren B. Frank and Paul Falzone (2021)—and more are popping up every day. We hope that the list below can serve as inspiration, a block of case studies, and a directory[34] for those who are, want to become, want to collaborate with, or simply appreciate climate change comedy practitioners.

Please also note that our list is shamefully US-centric (and within that, coast- and city-centric) and English language-centric, and by no means exhaustive. Making the list was like fighting the Hydra from Greek mythology: In researching one person, we learned about two (or three, or six) others we should have been profiling, too. And there are tons more out there. So if you have created or know of any climate change comedy projects you think we should include in future editions of this book // shout from the rooftops, please email us at climatecomedyprojects@gmail.com with the subject line FYI: (Title of Project). At the very least, we will watch and, if we feel so inspired, share on social media.[35] WHEW, are we long-winded or what?! Let's get to this list of creators already!

Scripted Episodic Comedy

Dallas Goldtooth A member of the Dakota and Dine tribes (Yardley, 2016), Dallas Goldtooth is also essentially a one-man boundary organization, which we didn't know was possible. He grew up observing his dad run the environmental justice organization Indigenous Environmental Network and then later joined its staff as a Keep It in the Ground campaigner (Funes, 2021; Yardley, 2016). Now, he plays the spirit of a long-dead Indigenous man on the FX show *Reservation Dogs*. It's pretty unusual for someone to cut their teeth in the environmental advocacy world and then transition over to Hollywood, but in Goldtooth's case, it makes perfect sense. Before joining the show, he had regularly made crowds laugh as an emcee at pipeline protests (Yardley, 2016). He had also started a sketch group called The 1491s with his stepbrother and three friends—one of whom was Sterlin Harjo, who co-created *Reservation Dogs* with Taika Waititi (IMDb, n.d.-b; Yardley, 2016). The show repeatedly touches on the LANDBACK movement, which calls for public lands to be returned to Indigenous people and is part of the climate mitigation puzzle (Funes, 2021; Wozniacka, 2021). You can find Goldtooth on Instagram, Twitter, and TikTok: @dallasgoldtooth.

Movement Generation Movement Generation provides training and support for organizers, and in the name of that mission, the organization has put sustained efforts into comedy. Its Green Collar Comedy YouTube videos include one where Boots Riley plays the earth (Movement Generation, 2015), as well as several co-created by The Other 98% that substitute "frack" for another F-word we all know and love. Movement Generation also executive produced the web series *The North*

---

[34] Minus actual emails and phone numbers. At most, you'll know how to DM people on social media.

[35] Fan mail addressed to the authors will also be accepted, but only the extremely fawning, complementary kind.

Pole, which weaves storylines about wildfires and green startups into a show about longtime North Oakland residents trying to keep their community intact in the face of gentrification, deportation threats, and more (The North Pole, 2019a, b, c). Notable cameos include Mistah F.A.B., Zumbi, W. Kamau Bell, and Rosario Dawson, who's also an executive producer (The North Pole, 2019a; The North Pole Show, 2017). You can really go down a rabbit hole reading about the show's actors and creative team (Reyna Amaya, Santiago Rosas, Donte Clark, Favianna Rodriguez, Josh Healey, Yvan Iturriaga, Darren Colston…), which is why it took Celia about 4 hours to write this paragraph (The North Pole, 2019a). Read more at movement-generation.org and follow along on Twitter: @MoveGen or on Instagram: @movementgeneration.

Layel Camargo The Impact Producer for season 2 of the climate justice comedy web series *The North Pole Show* (described under Movement Generation above), Layel Camargo, is a cultural strategist, land steward, filmmaker, and artist. They are a transgender and nonbinary person who is also a descendent of the Yaqui tribe and Mayo tribes of the Sonoran Desert. They graduated from UC Santa Cruz (The Center for Cultural Power, 2021) with dual degrees in Feminist Studies and Legal Studies (Camargo, n.d.) and in 2021 produced and host "Did We Go Too Far," a podcast with Movement Generation (Vasquez & Camargo, 2021). As the Ecological Arts and Culture Manager at The Center for Cultural Power (The Center for Cultural Power, 2021), they created "Climate Woke," a national campaign to center BIPOC voices in climate justice, alongside Favianna Rodriguez and The Center for Cultural Power (MACRO, 2022). Due to wanting to shape a new world, they co-founded "Shelterwood Collective" (Camargo, n.d.). This land-based organization teaches land stewardship, creative envisioning and healing for long-term survival. Layel was a Transformative Justice practitioner for 6 years and still finds ways to bring their lessons in alternatives to the carceral system to all their work. They were named on the Grist 2020 Fixers List (Grist, 2020), as well as celebrated by Yerba Buena Center for the Arts list of people to watch out for in 2019 (Yerba Buena Center for the Arts, 2019). In January 2022, they spoke on a Sundance panel hosted by Favianna Rodriguez called Pass the Mic: Centering Communities of Color in Climate Storytelling, which also featured documentary filmmaker Sabrina Schmidt Gordon and Varshini Prakash of Sunrise Movement (MACRO, 2022). They support and advise on The Good Energy Project's forthcoming Climate Storytelling Playbook (MACRO, 2022). Find Layel on Instagram and Twitter: @LayelCamargo.

Stand-Up

Hip Hop Caucus and The Center for Media and Social Impact In August 2019, Hip Hop Caucus participated in the CMSI's Comedy ThinkTanks program, which functions such as a traditional TV writers' room, but mixes social justice advocates and other experts in with the comedy writers (Borum, 2020). Here's what Hip Hop

Caucus' Think 100% page said about the film that emerged from the week-long collaboration:

Ain't Your Mama's Heat Wave is a stand-up comedy special from the frontlines of the climate crisis. It's filmed in the St. Paul's district of Norfolk, VA, a Black public housing community that is being redeveloped because of climate flooding, sea level rise, and a legacy of racist urban policies. The city of Norfolk, which is below sea level and sinking, is grappling with the climate crisis and racial injustice.

Four Black millennial stand-up comedians [Kristen Sivills, Aminah Imani, Clark Jones and this chapter's very own Mamoudou N'Diaye] take the stage to 'make the climate crisis funny' in front of a St. Paul's audience who are at risk for a Hurricane Katrina-like disaster and who are currently being displaced from their homes. Things are not so funny when it's clear that climate threats can mean life or death. But, in the Black American tradition of struggle, resilience, and triumph in the face of existential threat, the joy of comedy, music, and art informs and empowers. (Think 100%, 2021)

Follow these two orgs on Instagram and Twitter: @HipHopCaucus and @ CMSImpact. They're both on YouTube under their full names, and you can read more at hiphopcaucus.org and cmsimpact.org.

Her-icane   Jenny Gorelick and Kate Villa started this showcase of women and non-binary comedians in 2017 to raise money for Hurricane Maria relief efforts (K. Villa, personal communication, February 5, 2022). Then, they kept it going, intent on supporting Puerto Rico's recovery even when the media had stopped talking about it, and fundraising for survivors of subsequent climate change-fueled disasters (Villa, 2022). Organizations supported included Hispanic Federation, RAICES Texas, and California Community Foundation's Wildfire Relief Fund (McCarthy, 2019; Villa, 2022). Comedians such as Alex Song-Xia (*Rick and Morty*, *The Tonight Show Starring Jimmy Fallon*), Mariah Smith (Keeping Up with the Kontinuity Errors), and Alison Leiby (*The Marvelous Mrs. Maisel*; "Oh God, An Hour About Abortion") graced the stage (Caveat, 2021; Her-icane, n.d.; Smith, n.d.; Song-Xia, n.d.). Her-icane was also selected as a semi-finalist for the Yes, And Laughter Lab (Villa, 2022; Yes, And Laughter Lab, 2019). A May 2019 Time Out blurb about the show said, "considering its consistently terrific lineups and the mounting havoc wreaked by climate change, we can safely say there's no end in sight" (Time Out, 2019). Guess that blurb hadn't heard about the coronavirus. Follow @villafied and @jennycestquoi on Instagram.

Sketch

The Juice Media   Founded by historian and satirist Giordano Nanni and based in Melbourne, The Juice Media is known for its Honest Government Ads videos, which imagine a world in which governments tell it like it is and use a lot of profanity (The Juice Media, 2019b). Every expletive-studded dose of reality is delivered in a relentlessly upbeat tone by Nanni's partner, Lucy Cahill, who does the voiceovers for two "Australian" government reps played by Zoë Amanda Wilson and Ellen

Burbidge (The Juice Media, 2019b). In one video, Wilson's rep explains, with a huge smile on her face, that scientists call our current stage of climate change "We're F**ked"…which is a change from the previous stage, which scientists called "Listen to Us or We Might Be F**ked" (The Juice Media, 2019a). In another, Burbidge's beaming rep asks and answers questions like, "Does our EV policy include incentives to help you afford an EV? No! That would make us like governments that actually *encourage* EVs, like Norway!" (The Juice Media, 2021). The contrast between the words and their delivery really underscores just how much governments try to distract and placate their constituents while making no meaningful progress on climate change. The Juice Media YouTube channel, thejuicemedia, has over 800K followers (The Juice Media, n.d.). Find full-length vids there and follow @thejuicemedia on Instagram, Twitter, and TikTok for more.

Global Heartwarming Giovanni Fusetti is a scientist, clowning instructor, and permaculture designer who founded Hèlikos International Theatre Creation School in Florence, Italy (Hèlikos, n.d.). (Okay polymath!!) Tejopala Rawls is a freelance climate change professional who has worked on energy efficiency policy and faith group engagement in Australia, among other things, and who also does stand-up comedy (Rawls, n.d.). In 2015, the two came together to launch Global Heartwarming (Global Heartwarming, 2015b), a website featuring five short comedy videos about climate change for environmental campaigning organizations to use leading up to COP 21 (Global Heartwarming, 2015a). One of those videos, The Summit (alternately titled The Negotiation), shows a bunch of climate change negotiators standing on a beach throwing around fancy but ultimately meaningless words such as "unilateral agreement" and "methodological consistency" as the tide rises around them (Global Heartwarming, 2015c). Check it out on the Global Heartwarming YouTube channel.

Monty Hempel Shortly after Kellyanne Conway's infamous coining of the phrase "alternative facts" (Swaine, 2017), an environmental studies professor and documentary filmmaker named Monty Hempel produced a short satirical video called *Alternate Science (Vol. 1)*. In the video, Hempel himself plays Dr. Theodore Droop, a "black carbon climatologist" who claims to have discovered a planetary cooling mechanism called "thermal erectile dysfunction." Puns and NSFW[36] diagrams ensue (Boykoff et al., 2017a). Hempel, who was the Hedco Chair in Environmental Studies and the Director of the Center for Environmental Studies at the University of Redlands, sadly passed away in 2019 (University of Redlands, 2019). Nevertheless, his contribution to this field shows how comedy gives people a pass to behave in unexpected ways, which can allow for more frank expression.

---

[36] For those less online: Not Safe for Work.

Ads

Climate Ad Project The Climate Ad Project doesn't always make funny videos, but when it does, they're good. Highlights from their collection include a Mars tourism ad and a clip comparing carbon offsets to "murder offsets." This team of multidisciplinary individuals (originally brought together by, of all websites, *Twitter*[37]) believes climate change deserves as many funny ads as Allstate and Amazon Alexa. The Climate Ad Project was originally the idea of NASA climate scientist Peter Kalmus, who seems to know a lot more about effective science communication than Leonardo DiCaprio's character in *Don't Look Up* (Climate Ad Project, 2021). Check out the Climate Ad Project YouTube channel, visit climateadproject.org, and follow @ClimateAd on Twitter and @climateadproject on Instagram for more.

The Potential Energy Coalition × JOAN In 2019, the Potential Energy Coalition partnered with creative agency JOAN to produce a campaign called Save Florida Man (Chapman, 2020; JOAN, 2022). The campaign was inspired by the Internet's fascination with frequent headlines about one "Florida man" or another doing something unusual, like trying to bring a kitten into a nightclub (Leibowitz, 2012). It centered around a video starring real Florida Man Robby Stratton, who made headlines in 2018 for bringing a live alligator on a beer run. In the video, Robby explains that sea level rise is encroaching on Florida Man's natural habitat and that soon enough, newspapers might be completely bereft of Florida Man hijinx. Now, we don't actually think Florida Man is an endangered species, and we definitely don't want any more alligators to experience this kind of harassment. But we *do* want more funny videos that approach climate change from creative angles and garner millions of impressions (JOAN, 2022). With a Chief Creative Officer who writes for McSweeney's and Reductress on the side, we know the Potential Energy Coalition will deliver—and JOAN's past work speaks for itself (McSweeney's, n.d.; Reductress, n.d.). Visit potentialenergycoalition.org and joancreative.com, @potentialenergycoalition on Instagram, and @JoanCreative on Instagram and Twitter for more.

Celia Gurney AKA one of the authors of this chapter, is an Upright Citizens Brigade-trained improviser who spent 3 years immersed in climate change communications at Climate Nexus. There, she produced a comedic video series about individual climate action called #9for2019 (Climate Nexus, 2019). Think mockumentary-style videos where "random people" get asked questions such as "Why do you only date people who live car-free?" and then answer things like, "They have better butts." The project test-drove an expanded suite of digital marketing tools, with the goal of generating sign-ups for a partner organization, The Environmental Voter Project. The #9for2019 website received over 4000 visits during the promotional period and recruited 20+ new "climate voters." At Climate

---

[37] The audience gasps.

Nexus, Celia also contributed to a Black History Month series on climate justice leaders and Freedom to Breathe, a collection of events with on-the-ground climate justice advocates (Freedom to Breathe, 2018; Nexus Media News, n.d.). More recently, she created comedic content for Shut The Fossils Up and Yellow Dot Studios' campaign to protect New York State's climate law from fossil fuel interference. Find out what else she's up to at celiagurney.com.

Multiple Formats

Mamoudou N'Diaye AKA the other co-author of this chapter (pay attention) is a Brooklyn-based comedian who accidentally made his way into climate crisis communication by way of his racial justice work. He was featured in the Think 100% documentary *Ain't Your Mama's Heatwave* as the host of the comedy portion of the documentary about the climate crisis effect on the city of Norfolk, VA, and then became a writer for it (Think 100%, 2021). Prior to that, he/me/I have worked with Rollie Williams on his New York-based climate justice late-night show *An Inconvenient Talk Show*, where he/me/I was the DJ and occasionally a correspondent, delivering out-of-the-box solutions to the climate crisis via PowerPoint.[38] In 2019, I/myself/me performed at The Center for Cultural Power's (TCCP) Climate Woke Summit and then partnered with them where I wrote sketches/web series based on climate justice with TCCP one called *Law and Order: Climate Justice Unit.* On top of that, I've written too many treatments about the integration of climate justice into a TV format and I'm just happy to be nominated at this point. Not a fan of social media so I'll just leave this here: mamoudoundiaye.com.

Inside the Greenhouse This interdisciplinary program at the University of Colorado, Boulder, shows students that creative, funny climate change communication is possible and helps them produce some of their own. Among the offerings are two classes ("Creative Climate Communication" and "Climate Change and Film"); a science film internship program with its own film festival; a comedy show aptly titled "Stand Up for Climate Change"; and an annual short comedy video competition open to the public, in which several people/groups on this list have placed over the years (Boykoff et al., 2017b, 2018, 2020, 2021; CIRES, 2019). We have to hand it to Inside the Greenhouse for its massively successful efforts to incubate and incentivize the creation of climate change comedy in many different formats and from many different sources. We also have to pay them $1,000,000[39] for helping us find a lot of the creators on this list. Follow @everydayclimate on Instagram and @ITG_Boulder on Twitter, and for cryin' out loud, just go to insidethegreenhouse.org already.

---

[38] Often intersecting with topics surrounding race, reassigning the burden of blame for the current crisis, and adding Greta Thunberg to the Marvel Cinematic Universe. Better than Hawkeye.

[39] In Monopoly money.

Chuck Nice New York-based comedian Chuck Nice has long participated in science comedy via StarTalk, the podcast he hosts with Neil deGrasse Tyson (StarTalk, 2022), but in 2019, he founded the Shhh, It's Real Campaign, which uses entertainment to inspire young people to act on climate (Shhh, It's Real!, n.d.). That same year, he hosted a climate change-themed variety show at NYC Climate Week (New York Comedy Club, 2019) that showcased several stand-up comics, at least one musician, and pre-recorded sketches that Nice himself wrote in collaboration with his assistant. One of the funniest aspects of his sketches is the way he incorporates children (Climate Comedy, 2021). You don't have to look any further than the climate crisis to see that kids often act more rationally and responsibly than adults—and Nice plays with that dynamic to great comedic effect. His limited-run podcast, *Pod Zero*, is all about the climate crisis and features interviews with Peggy Shepard of WE ACT for Environmental Justice, comedian Roy Wood, Jr. of The Daily Show (Nice, 2021a), and climate educator Dr. Eugene Cordero of San Jose State (Nice, 2021b). All the follows: @ShhhItsReal and @chucknicecomic on social media, and don't forget the Chuck Nice YouTube channel.

Climate Town Climate Town's funny explainer videos about things like the invention of the concept of carbon footprints by oil companies and fast fashion will teach you about climate change "in a way that doesn't make you want to eat a cyanide pill" (Williams, n.d.). The series is produced by documentary filmmaker Benjamin Boult, Daily Show writer Nicole Conlan, and Rollie Williams, who also hosts (Boult, n.d.; Conlan, n.d.). It's a natural continuation for Williams, who, in the Before Times, regularly dressed up as Al Gore and hosted something called *An Inconvenient Talk Show* at Caveat[40] bar/theater in Manhattan (Caveat, n.d.). He'd give a humorous PowerPoint presentation on some aspect of the climate crisis—for example, the story of the Keeling Curve. Then a comedian would come on and do a 10-minute set, and then "Gore" would interview a climate scientist (e.g., Geoffrey Supran) or climate science communicator (e.g., Kendra Pierre-Louis). For the grand finale, improvisers impersonating various celebrities would play a climate change-themed, Jeopardy-style game. On multiple occasions, *An Inconvenient Talk Show* moved its audience to act: once, by allowing the Sunrise Movement to recruit volunteers at a show, and later by doing a pandemic Zoom show that doubled as a fundraiser for WE ACT for Environmental Justice. If you like this paragraph's energy, check out the Climate Town YouTube channel, @climatetown on all social media, and The Climate Denier's Playbook, the podcast Williams and Conlan started in 2023.

---

[40] Rare sincere footnote: Caveat has science comedy shows coming out of its ears. Get your tickets immediately at www.caveat.nyc.

Explainer Videos

Politically Aweh A project of Bouncing Biscuit Studios with distribution by premier investigative news outlet Daily Maverick, Politically Aweh, creates fast-paced, funny, high-production value videos about South African news and politics (Schiffrin, 2019). One of those videos won first place in the 2020 Inside the Greenhouse Comedy & Climate Change Video Competition (Boykoff et al., 2020)! Titled *Climate Change in South Africa: How Bad Can It Be?*, it was the first in a three-part series featuring Zipho Majova as the host and frequent cutaways—to barrages of news clips, Nelly's Hot in Herre music video, and a GIF of Troy from Community (Politically Aweh, n.d.). More recent videos include one about South Africa's renewable energy future (Politically Aweh, 2020) and another about the Karpowership energy deal (Politically Aweh, 2021). Stephen Horn, the post-production freelancer and filmmaker who developed Politically Aweh, describes the project as "South Africa's answer to 'Last Week Tonight with John Oliver.'" which might sound a little cocky—if it wasn't spot on (Horn, n.d.). Check out @politicallyaweh on social media and the Politically Aweh YouTube channel.

ClimateAdam ClimateAdam is a YouTube channel dedicated to explaining complex ideas around climate change in playful and accessible ways (right down our alley, yeah?). But you wouldn't know this creator's got a British accent so for some reason it just means more and hits harder. Anyway, the ClimateAdam channel, founded in 2014, is run by Dr. Adam Levy (read that with a capital DOCTOR), a climate scientist from Oxford who makes sketch comedy-infused explainer videos. Their channel has received over half a million views (Levy, n.d.-b), with videos shared by Huffington Post, Scientific American and Upworthy (Levy, n.d.-a). ClimateAdam has been awarded funding from multiple sources, including the Royal Society of Chemistry and the Climate Communications Project (Levy, n.d.-a). In 2019, the channel was awarded second place in the Inside the Greenhouse Comedy & Climate Change Video Competition (Boykoff et al., 2019). (We can only assume any awards ceremony was hosted by Captain Planet.) Levy aims at a lot of the current climate conversations about all these damn heat waves, IPCC reports, and even the coronavirus effect on climate in a way that feels soothing—because their voice goes down like a smooth cuppa at your mate's flat in Battersea. You can hear that voice by visiting @climate_adam on Instagram or @ClimateAdam on Twitter and TikTok.

All About Climate Roshan Salgado D'Arcy's channel functions as a kind of climate denier Mythbusters. In response to the common climate claim "one degree of global warming isn't that much," for example, he tactfully converts and contextualizes: The amount of energy absorbed by the oceans due to global warming is equal to seven billion Hiroshimas since 1871—or more than one Hiroshima every second from 1871 to 2017 (Salgado D'Arcy, 2021). Screenshots of studies referenced build credibility with viewers, as does his master's degree in Climate Change, Development

& Policy from the University of Sussex and several years working in science television production in the UK (BBC, Channel 4 and Plimsoll Productions) (Salgado D'Arcy, 2022). While Salgado D'Arcy's use of comedy is sparing, it's also effective, combining with his upbeat delivery to keep his 1.18K subscribers engaged. All About Climate on YouTube, people.

Just Have a Think Dave Borlace's dry sense of humor sneaks up on you in his 10- to 30-minute explainer videos, which often explore earth systems and renewable energy technologies. For example, in an April 2020 video titled "Is the Gulf Stream collapsing?" he describes *The Day After Tomorrow* as "a great action movie starring Dennis Quaid and a very young Jake Gyllenhaal, which, in the best traditions of Hollywood blockbusters, provided two hours of fabulous entertainment, but bore very little resemblance to real-world events or possibilities" (Borlace, 2020). See also the following description of his YouTube channel: "The channel is not a debating forum about whether Human Induced Climate Change is a real phenomenon or not. If that's what you're after then I can highly recommend chat forums on social media, where people on both sides of the argument go round and round in circles achieving precisely nothing at all" (Borlace, n.d.). Borlace created Just Have a Think in 2018 after reading Arctic scientist Peter Wadhams' book *A Farewell to Ice* (Buckley, 2020). Today, his channel has over 370,000 subscribers (Borlace, n.d.). You could become one of them…

Simon Clark Simon Clark recently finished his PhD in Atmospheric Physics and Dynamics—and you can experience this triumph vicariously in the PhD series of videos on his YouTube channel (Clark, n.d.). Or, you can click immediately over to the videos about nuclear power, the Kyoto Protocol, and how much carbon we can pump into the atmosphere before sh*t gets really serious. Spoiler alert: We don't have much of that allowance left, and sh*t is already really bad (Clark, 2021). Clark's editing choices, like opening a video about nuclear with footage from The Simpsons, definitely succeed in harnessing comedy's memorability (Clark, 2020). Warning: You WILL click on the video titled "Why weathermen were illegal wizards for 97 years." Follow @simonoxfphys on social media or just go to the famed YouTube channel!!!

Podcasters

Thimali Kodikara, Mothers of Invention Thimali Kodikara is the series producer and a host of Mothers of Invention, a podcast about intersectional feminist climate change solutions (Robinson et al., 2018–2020; Schlossberg, 2019). Her co-hosts? Comedian Maeve Higgins and former Irish president (also Ireland's first woman president) Mary Robinson. Casual. Kodikara works on not only an editorial strategy, but also a social justice strategy for the show, which she explains isn't typical for podcasts (York St. John University, 2021). She and the rest of the Mothers of

Invention team keep conversational chemistry fresh by playing around with hosting configurations and format. In one episode, they had climate activist Xiye Bastida come on to guest co-host with comedian Pooja Reddy (Bastida & Reddy, 2020). In another, they envisioned a climate-justiced future in which the first Indigenous woman president of the USA had just been elected and Maeve was the longtime partner of Michael B. Jordan (Robinson et al., 2020). Kodikara is big on this type of thing, advising people to create "a giant mood board" of what they want to see down the road (York St. John University, 2021). "What is the point of being artists if we can't do exactly that: visualize the future from outside the boxes everyone else is functioning in?" she said in an interview. "We are very, very powerful in that regard" (York St. John University, 2021). Feeling inspired yet? Follow @oneloudbellow on Instagram, @apathysuckseggs on Twitter, @mothersinvent on both, and listen to Mothers of Invention at www.mothersofinvention.online or wherever you get your podcasts.

Dave Powell and Oliver Hayes, Sustainababble Hosted by Dave and Ol, two friends who both work at environmental organizations in the UK, Sustainababble is a weekly comedy podcast about climate change and nature. They firmly believe that just because everything is really bad and getting worse, that doesn't mean there aren't crumbs of hilarity to be found—even if it's only laughing at the dark sh*t some companies and businesses get up to. Episodes vary from interviews with environmental professors and activists to Dave and Ol attempting to understand something they're told is simple but rarely is—like whether there are 60 harvests left (Powell & Hayes, 2021), or whether it's okay to own a cat[41] (Powell & Hayes, 2018). They pride themselves on their human, no-BS approach to talking about the climate and nature crisis—and having a wry, but comforting chuckle from time to time. They also wrote most of this paragraph, which is why it has a charming "across the pond" vibe.[42] Follow @thebabblewagon for updates and find episodes at www.sustainababble.fish or wherever you normally listen to podcasts.

Journalists

Gizmodo/Earther Brooklyn is part of New York City. So why do you have to write "Brooklyn, NY," instead of "New York, NY" when you're sending mail there? Earther's relationship to Gizmodo is kind of like that, meaning: We're confused about who to address this paragraph to. Gizmodo/Earther, please get in touch. Anyway, the people who write for this website, whatever it's called, regularly find funny ways to talk about the climate crisis. With headlines such as "French Car Ads

---

[41] Wait, IS it okay to own a cat?! *Authors rush to episode*.

[42] Sadly enough, we replaced "awfulness" with "dark sh*t" so it would sound more consistent with the rest of our chapter.

Will Soon Be Required by Law to Tell You Not to Drive a Car" (Ropek, 2022), "Huntington Beach High School May Want to Change Its Mascot After This Oil Spill" (Noor, 2021), and "New York City Plan to Close Streets is Weak as Hell" (Funes, 2020), you'd be forgiven for thinking you were clicking on a piece from *The Onion*. Except *The Onion* doesn't consistently highlight the climate justice angle, or, ya know, write true stuff. Follow @EARTH3R on Twitter for regular climate snark.

Social Media

Mary Annaïse Heglar, Head Greentroller If you have Twitter and think about climate change regularly, you've probably come across Mary Annaïse Heglar. She writes essays (and tweets) about climate justice and climate grief, and she gets to the heart of things so expertly that you just want to retweet all the way down her profile. She co-hosts an intersectional climate change podcast with journalist Amy Westervelt called Hot Take (Heglar & Westervelt, 2019–2021), and she also trolls fossil fuel companies on Twitter to disrupt and bring attention to their greenwashing tactics. As Kate Yoder reported in Grist (Yoder, 2020), sometimes this "greentrolling" can be pretty funny—like when BP suggested people calculate their carbon footprints and Heglar replied, "Bitch what's yours???" (Heglar, 2019). To a recent Shell post about EVs, she replied, "This you?" and linked to an Earther article about Shell's CEO getting roasted at a climate conference (Heglar, 2021). A highlight of the greentrolling movement seems to have been in November 2020, when Shell asked what people are "willing to change to help reduce emissions." It was a great alley-oop for hordes of Twitter users, with nearly 8000 people quote tweeting it and many flexing their comedy writing muscles (Shell, 2020). To that we say: like and subscribe. Or more specifically, follow @MaryHeglar on Twitter and Instagram, and jump on the greentrolling bandwagon.

Climate Meme Accounts Many of the best climate change memes around used to come from @climemechange on Instagram. Bursting onto the scene circa July 2018, the account built a 125K-strong following in just a few short years. Among other things, it once juxtaposed the statement "wind turbines are such an eyesore," with an image of their alternative: hundreds of pump jacks littering a desert landscape. Unfortunately its old posts are no longer visible, (Climemechange, n.d.) but we hope the account will return from this dormant period someday with fresh content. Though we both know its creator personally, we'll never reveal their identity—not even if you trap us in bathtubs full of all the deer ticks that have shifted their range north due to climate change. Mwahahaha! The @climate_memes Instagram account, meanwhile, is a little less mysterious: It's run by Climatepedia at UC Irvine. Click over there for memes featuring Squidward and vintage Lindsay Lohan, plus a clip of Doja Cat talking about cow farts (Climate Memes, n.d.).

Alex Engelberg and Other TikTokers Alex Engelberg (@alexengelberg) is a TikTok star who frequently combines music and comedy, to the delight of app users (Engelberg, n.d.-b). In 2021, he posted a video that started out as a corny song about ways to go green, but when he started to list his tips for being energy smart, they weren't your typical individual actions. We won't spoil the ending—you should just go watch it—but we will say it was impressively succinct commentary. It also has at least 2.6 million views on TikTok alone. Engelberg doesn't typically post about climate change, and he isn't the only TikToker who's made a funny, one-off climate change video that went viral, but that's exactly why this vid makes us so happy. It's evidence that climate change is increasingly filtering into the Internet's collective consciousness and that at least 2.6 million people recently thought about it for 31 seconds (Engelberg, n.d.-a).

Influencers

Levi and Leah Levi Hildebrand is not just extremely photogenic[43] and good at what he does, but he also specializes in helping others tell their stories about the effects of climate change and other environmental research. He loves it so much that he even has his own production company that he founded in 2017 to focus on this work called One Island Media (Hildebrand, n.d.). Finding his lane as a creative consultant and collaborating with companies such as Blinkist, Klean Kanteen, and Sunski Sunglasses (Hildebrand, 2022), Levi now shares a YouTube channel with his wife Leah, and the two of them leverage it to combat climate change by promoting recycling, minimalism, and a zero-waste lifestyle and by testing sustainably sourced items for their 135K+ followers (Hildebrand & Tidey, n.d.). Those two good-natured, attractive Canadians will almost make you want to move in with them in Victoria, B.C.

Kurtis Baute Here's the thing about Kurtis. The man makes fire videos. His whole thing should be our whole thing; he believes in science for a better world! He believes that in order to take on big issues such as climate change the world needs to be more science literate. He believes this so much that rather than convince you through speaking to you about the atmosphere, the man sealed himself in a DIY biodome to talk about how people interact with the atmosphere (Baute, 2018). Y'all wanna know just how much plastic there is in the world? Kurtis tried to spend 24 hours not touching any plastic to show us just how prevalent it is and where it ends up (Baute, 2019). Not only that but he knows the importance of demystifying conspiracy theories such as flat earthers, tackling convenient consumerism's effect on the environment, getting out there to protest pipelines, and, like any good community-minded person, spotlighting other climate communicators (Baute, n.d.) because, in the words of the famous Wildcats, we're all in this together.

---

[43] Must be nice….

Scientific Papers

Scientists Yes folks, things have gotten so dire that even scientists themselves are getting in the comedy game. In 2018, four of them—Guillaume Chapron, Harold Levrel, Yves Meinard, and Franck Courchamp—published an article in the journal *Trends in Ecology & Evolution* called "A Final Warning to Planet Earth." It was a satirical response to the "warning to humanity" a large group of scientists had put out the year before, flipping the script and demanding that Earth—not humanity—change its behavior to resolve the whole environmental catastrophe thing (Chapron et al., 2018). Of all the canary-in-the-coal-mine moments we've observed while working on this crisis, scientists going all Jonathan-Swift's-A-Modest-Proposal on us is honestly one of the scariest—and definitely the most likely to remind us of high school English class. Hats off to Guillaume & Company.

Discussion

Well damn, it's almost as if the authors have been right the whole damn time.

Major Findings

This study sought to determine whether the causes, effects, histories, perils, and solutions to the earth's climate crisis can be and have been effectively communicated via comedic pathways. And it identified more than 25 distinct entities doing that, so…science…win? Comedy win? Scimedy win? You get it; these findings clearly indicate that there are multiple people and groups who are actively out in the streets using their platform and senses of humor to inform the masses of the evil patterns of Big Oil, the mythic carbon footprint, the convenience economy, political interference in offsetting climate research and action, etc. They successfully disseminate pertinent and urgent information in a captivating and charming way. So, to the millions of people running around asking, "Where are all the climate comedians?" we say "Ummmmm…please stop asking that. They're here, in this chapter. Duh."

There is value in talking about climate change and justice in entertainment. The type of work that our respondents—and non-respondents, those f**kers—are doing needs more visibility. More visibility makes these issues easier to talk about with clear information, an easier way into a hard conversation, and connects us all under the banner of "YO IF WE DON'T HANDLE THIS WE ALL GONNA DIE!"[44]

This study has examined how comedic artists are using their platforms to elevate the conversation surrounding the climate crisis and has gone some way to collect

---

[44] Can we get someone to fly that banner over Rockaway Beach this summer?

25+ different options for multiple different tastes of comedy in one place. Phrases get lazily bandied about when the climate crisis is pitched as an area for comedic focus: It's "too dark to talk about" and "no one is talking about it meaningfully." Well, all of these climate comedy creators have been able to make space and take time to craft a way into the conversation using humor, and research shows how laughing can break tension. Why not use the thermal energy from that broken tension and transform it into actionable energy?[45]

And that brings us to H2, which, in case you don't remember, was our hypothesis about what it will take to more fully harness the power of climate change comedy in order to improve the efficacy of the climate justice movement, invite in more allies, and build a more inclusive clean energy economy. More good news: Everything we hypothesized in H2 was supported, and we're about to elaborate on that for…wait, let us check…the rest of this chapter!

People learn about war, genocide, racism, sexism, homophobia, transphobia, capitalist exploitation, etc., through mass media avenues such as *The Daily Show with Trevor Noah, Full Frontal with Sam Bee, and Last Week Tonight with John Oliver*. Sometimes those shows cover climate change, and they've been doing so more effectively for a lot longer than traditional news media (just google "Colbert Report climate" to see for yourself). However, the climate crisis has no singular voice in late night to propagate information and highlight the intersections of climate justice with many other types. And it deserves one. More about that in the "Future Work" section.

Notwithstanding, this is not all the voices out here making a splash about climate as the sea levels rise all around the world. Most of these voices highlighted in this chapter are in the continental USA and the climate crisis is global, often affecting communities that do not benefit from the hyper-focused lens of media attention on America and its (for lack of a better term) shenaniganry. Research shows the Global South and other states ravaged by settler colonialism and exploitation of working-class indigenous people are unfairly and disproportionately negatively affected by the climate crisis in ways that get invisibilized by the powers that be behind the world's camera (Brändlin, 2019). BIPOC, immigrant, queer, native, poor, and disabled voices must be allowed at the table and not spoken for by other people.

Limitations

Building off of the point above, a major limitation of this study was limited resources to be able to travel to other parts of the world or have access to people potentially using comedy to tell their story. Even Greta got a boat, Springer! Give the authors of this chapter a raft with Wi-Fi capability in the future.[46]

---

[45] This is science, right?

[46] A Hermione-style time turner wouldn't hurt, either.

Another limitation of this study was the definition of humor. Humor cannot be defined by just two artist/activists.[47] Comedy is hyper-subjective; not everyone can deliver *30 Rock* levels of joke volume with *An Inconvenient Truth* level of PowerPoint prowess and even so, that's not everyone's cup of tea. Armed with the resources and connections at their disposal, the authors did their best to present a wide array of senses of humor, but they acknowledge that they mostly used word of mouth to find the respondents in this study, and they're are aware that this chapter is just the tip[48] of the tip of the iceberg.

Future Work

Assuming we have a future,[49] here are some steps we'd love to take to use the accessibility of comedy to further the message of the climate crisis without leaning into the dark, dreary, depressing facts, and figures but rather laughing together *with* the dark, dreary, depressing facts, and figures.[50]

All issue-driven comedy should shine a light on the issues and not make light of them. It's punching up like Mario, not punching down like if Mario gave up plumbing and became a horrible open mic comedian. The respondents to this study all are shining examples of highlighting the issue at hand. The climate crisis is not only here but it has been couch surfing on the earth for decades, and it's about time for us to address the elephant in the room; the climate crisis needs to get a job and not spend its entire day watching *Judge Judy*—okay I think we lost the plot here. The climate crisis is here, and we cannot pretend that a climate apocalypse is our future when we see its effects of it in our present (Well, 2022).

Al Roker, Lord of the Weatherfolk, will not be leading us to the Climate Promised Land, nor will Al Gore.[51] It will take all of us to commit to change. In order to commit to change, one must know what is the problem and sometimes the best person to communicate what the problem isn't a dry, jargon-heavy specialist. That's where comedy comes in. So…

Let Comedy Sit at the Comms Table Comedy needs to be taken more seriously as a communication strategy. We're not saying "Hey, let's treat comedians with the exact same weight that we do our elected officials!" However, what we are saying is that we need to engage the populace on the ins-and-outs of the climate crisis in a way that is personal, personalized, and unifies people around the goals outlined in climate solutions. Those changemakers can find agents in comedians who are

---

[47] 3? Maybe. Also, "activist" is generous in Celia's case, though she might get there someday.

[48] If you looked at this footnote, you're immature :/

[49] Hopefully you're reading this from the future, in which case can you send a message backwards through time to let us know how we did?

[50] Cut to us at a dinner party with the facts and figures.

[51] Is it in the contract that if your name is "Al" you have to do something with the weather?

actively using research to get the correct, pertinent, and all-inclusive messages surrounding the climate crisis to the people through their humor. It is the stance of this chapter that these artists and others, if given space, resources, and more eyes, can draw more people toward climate action and, simultaneously, demystify climate justice by adding more information to replace overwhelming doom, gloom, and anger at billionaire exoduses to low orbit (or further). Hopefully, this chapter brings even one new fan to these people but supporting them where they're at the writing of this chapter isn't the only way to use their talents.

Clone John Oliver, But Make It Climate The climate crisis affects us all. Some more than others, but without a planet, where the f**k will we tweet opinions from the toilet from? Most of these hosts come from the Jon Stewart Academy of Hahas Over Boohoos, and in September 2021, a lot of them participated in Climate Night.[52] But on the whole, none has been more impactful than Liverpool's own John Oliver with *Last Week Tonight.* John Oliver's show finds a way to take an issue that might be a bit off the mainstream beaten path and weave it into the current cultural conversation. With topics ranging from India's elections (Last Week Tonight with John Oliver, 2014c) to FIFA Corruption (Last Week Tonight with John Oliver, 2014a), John talks about it all including issues that overlap with the climate crisis like floods (Last Week Tonight with John Oliver, 2017b) and the Paris Agreement (Last Week Tonight with John Oliver, 2017a). His show's deep-dive format gives complex, intersectional issues more breathing room. Its once-a-week schedule allows things (the issues themselves, as well as his writers' draft scripts) to marinate and potentially develop further before airtime.

Finesse Those Calls to Action John Oliver is also particularly good at calls to action. Many of his hashtags and charity ventures have brought awareness, visibility, and resources to vulnerable communities through the power of comedy (Borum & Feldman, 2020, p. 65; Ohlheiser, 2015; Simerman, 2015). This man started a fake megachurch, thousands of people donated to it, and he sent all the money to Doctors Without Borders (Ohlheiser, 2015). Why did that work? We haven't done any peer-reviewed studies on it, but maybe because he focused on one very specific ask, embedded humor *in the action itself* (the megachurch was called Our Lady of Perpetual Exemption), and consolidated individual actions into a more impactful collective action. You don't have to be John Oliver to do something in that ballpark. Think of Her-icane fundraising for disaster relief and Sunrise signing up volunteers at An Inconvenient Talk Show. Watch and learn, climate comedy people—and imagine the mobilization that could happen if an Oliver-style late-night show was completely dedicated to the climate crisis' effects, globally and locally.

---

[52] Delivering what it promised, Climate Night was an event where everyone made climate jokes *on the same night.* A helpful Gizmodo article (Kahn & Taft, 2021) ranked the hosts' performances and kept track of "hot burns." Samantha Bee's tackled a level 2 climate topic, extreme precipitation: "Even if we stopped clogging our sewage systems with rat kings of human grossness, our sewage system would still be in trouble because it wasn't designed for our changing climate."

Give the John Oliver Clone Some Friends Why stop at one content perspective about climate? Our results section presents compelling and diverse POVs on the climate crisis. Centralizing one late-night show isn't the answer, because no one show can encompass all points of view. We need a mutual highlighting of all the different people doing this work, a focus on the most vulnerable communities, and a seat at the table and more importantly an equal voice for those communities. This requires the TV/film industry and the climate change think tanks/organizations to invest in voices outside of the same seven people of no color who do not live with the effects of the crisis (pollution, poor air quality, famine, sea level rises, constant flooding, residing in hurricane disaster zones, etc.) providing a solitary voice. Guess what, TV Exec? Everyone in your target demo needs this information about the climate crisis so invest, invest, invest! Invest in all voices from queer people, to BIPOC folks, to people living outside of "the West" and people living on reservations, from older conservationists to younger ones—especially that last part.

Nurture New Voices Like You Work for American Idol One way to invest in new voices: support and create more programs that welcome people into the social impact entertainment space and provide networking opportunities. The IllumiNative Producers Program (Illumi*Native*, n.d.), the Yes, And Laughter Lab, and Inside the Greenhouse are good examples. After-school writing and performance programs for underserved youth, like Story Pirates Changemakers and Superhero Clubhouse's Big Green Theater, are also key (Story Pirates, n.d.; Superhero Clubhouse, n.d.). And those of us who know about those programs need to make sure everyone else does, too. (ATTN comedians: share info!) We need projects like a movie about New York's Black food justice advocates (Thomas et al., 2019), a stand-up special featuring comics from Miami's climate-gentrifying Little Haiti (PBS, 2022), and a half-hour comedy about farmers in the Brazilian Amazon transitioning to chop-and-mulch agriculture thanks to a fictionalized version of Projeto Tipitamba (Embrapa, 2020). If any of that's going to happen, people from those communities need to be at the helm, making decisions from conception to launch. So let's keep building out the infrastructure that gets them to the table and makes the table a safe place to be.

Remember: There's Life Beyond Satire When most people think of climate change comedy, they probably think of satire: late-night satirical news shows, *Don't Look Up*, articles from *The Onion* like "Optimistic Researchers Say There Still Time to Head Off Climate Change Before It Starts Killing Rich People" (The Onion, 2021). Satire is a key piece of the puzzle (it punches up with great force!), but it's not a one-size-fits-all answer to engaging people. Research has shown it's good at getting people who care less about climate change to sit up and pay attention (Anderson & Becker, 2018), increasing their certainty that it's happening (Brewer & McKnight, 2017). On the other hand, the people who are getting lampooned might mistake you for one of them (it's more common than you'd think), and your words could end up reinforcing their views (Lamarre et al., 2009; Garrett et al., 2019). They might also dig their heels in more in response to being the butt of the joke (Borum & Feldman, 2020, p. 65). That is, if they even tune in at all (Mitchell et al., 2014, as cited in Borum & Feldman, 2020, p. 65). Also, sometimes people are just looking for more of an escape from reality, or they could use some hope (Osnes et al., 2019), or they just wanna see vampires make out or whatever. So if we want this movement to have maximum comedy reach, we need a whole charcuterie board of different offerings. That means scripted episodic TV, stand-up, sketch comedy, comedic documentaries (Borum & Feldman, 2020, p. 79), musical comedy, reality TV, podcasts hosted by comedians, live shows that convert investigative journalism into stand-up or scientist monologues into improv (Borum & Green-Barber, 2018; Thank You, Robot, 2019), funny journal articles, and everything else we can dream up. Climate stuff doesn't always have to be front and center, it just has to be part of each project's world in a meaningful way. In fact, it would probably be wise for some projects to avoid the word "climate" altogether and stick to "clean air and water" language, too (Boykoff, 2019, p. 38).

Make That Mood Board for the Just Transition Humanity's current mood re: the climate crisis is not good, which makes sense. There's a lot to grieve. We know what we don't want, and we can see it very clearly because it's coming, but while we keep an eye on that, we also need to rally around a clear vision of what we DO want.

And that is climate justice and a complete overhaul of our relationship with this planet, with each other, with materialism, with capitalism, with food, fashion, technology, etc. There are a lot of experts who, all together, know the details of how we can get there and what it will look like. Comedy programming (and other art) can help the rest of us wrap our brains around that. So let's make that mood board like Thimali Kodikara advised. How will this new paradigm improve the lives of families in 2065, for example? A sitcom could show us. What kinds of health problems will completely disappear from hospitals? We bring you: Season 53 of Grey's Anatomy. How universal and robust will mutual aid be? How will island communities find a way forward? How will we dismantle white supremacy and capitalism? What new classes will be required in schools? Will every city block have a huge community garden in the middle, shared across all the backyards? Which countries will former fossil fuel execs have tried to flee to, only to be found and brought to justice, and what will that justice look like? Will we even have countries?!

Shared visualization actually helps make change. For example, in 2014, conversations about clean water and alternatives to extractive industry united an Appalachian community at an art festival. The next time "land men" tried to purchase their land to sell to energy companies, more residents were prepared to say no, and they organized for the long term (Helicon, 2018). The future they had in mind was more appealing than what the land men offered. To quote Mary Annaïse Heglar quoting writer Toni Cade Bambara: "The role of the artist is to make revolution irresistible" (2020).

Invest in The Future, Even If They're Roasting You Youth are inheriting the world. What's scarier? They are making fun of us while doing it. Even spookier?!

They're good at it and they're almost always right. Even more Hocus-y and Pocus-y?!![53] Not enough people are taking them seriously.[54]

Times change. Media changes. The way we consume media changes. While the youth audience seems to be a perpetually vexing market for older generations to understand in order to platform, one immutable truth is that they will be here when older generations will not be. TV and film may or may not be the best vehicle to get that crucial climate information to that population but you know what? They're already doing it without any of our help.

Millions of teens use social media daily to share information. They already have chosen the climate crisis as a priority threat amidst all the other social traumas they dissect, lampoon, roast, sing about, rap about, dance about, throw soup cans at, etc. Systemically they don't have power, but their influence has immense worth when it comes to communication regarding adapting to the current crisis, cutting down emissions, etc. I mean, these mfs were using TikTok to get tickets to rallies so actual attendees would show up to an empty stadium (Lorenz et al., 2020). You understand how petty, nay, how *powerful* that is? Don't you want that power on your side?! Infuse that youthful raw talent and motivation with the messaging and creativity and we got us one hell of an effective communication method that informs the community when they're actually babies and not just acting like one.[55] Additionally, Gen Z should be hired to run all transmedia for the entertainment-education people (callback!!![56]).

Make Comedy a More Accessible Tool We need more guidance and support for non-comedians who want to integrate comedy into their climate justice work. The vast majority of organizers, teachers, health professionals, social workers, nonprofits, communicators, etc., do not have access to boundary organizations who can advise them and/or connect them to entertainment professionals. They're on their own, and they might be too nervous to just start throwing spaghetti at the wall, jokes-wise, until they "find their brand."

So what's the answer? Can we somehow "Oprah's Favorite Things" boundary organizations? ("And you get a boundary organization! And YOU get a boundary organization!") Sounds great, but expensive and time-consuming. Maybe we could make people check out the special humor issue of the journal *Environmental Education Research* that came out in summer 2022—the call for papers alone cited enough research on humor in education about "thorny issues" to make your head spin (Chandler et al., 2022). Maybe we could make everyone check out The Good Energy Project's screenwriter playbook (The Good Energy Project, n.d.). Or maybe we could distribute PDFs with comedy communication tips across the internet, but

---

[53] The authors were going to reference the Disney Channel Original Movie *Twitches* here but chose to not do it. Wait, they just did.

[54] Except us because they scare us.

[55] Insert GOP Senator's name here. Seriously, just pick one, it'll probably work.

[56] High-level comedy technique. Only the greats can swing it.

require a password and not give it to the bad guys?? Respectfully requesting that you conduct research into that yourself, because these authors are TIRED. We've worked hard to alert you to dozens of inspiring examples of climate comedy projects in the results section, so go forth and check those out ASAP. We're clocking out, dammit!

Conclusion

Look at y'all, you made it here! Either you had a good time or you had an *amazing* time! Or maybe you also just took a nap. Regardless, the aim of this experiment was to ask if comedy is an effective way to Trojan horse climate crisis education to the world.[57] That hypothesis was supported by all the talented and wonderful comedic climate communicators that we rustled up. While it is a massive undertaking to inform the world of all the ins-and-outs surrounding the crisis, our investigation found that many people laced up their Gore-Tex boots and are already tackling this inconvenient (and impending) truth.

Comedy is subjective. And when the subject of said comedy is a global crisis powered by decades of exponentially rising CO_2 levels, apathetic politicians with toothless promises, and corporations facing no accountability and attempting to shift the blame to the people?? Well, it's easy to get overwhelmed. But the artists who dive in, sift through the data, find the truth, and make it their mission, nay, their civic duty, nay on that nay, their PURPOSE to shine a comedic light on this hefty hefty hefty dilemma deserve a bigger platform, more space and time to share, and a tip of the hat for their contributions. Facts and figures are not everyone's love language, but we all can use a laugh. Something that tempers the social anxiety surrounding climate change with something soft, sweet like a spoonful of sugar, which, as a famous English child endangerer once said, helps the medicine go down (Poppins, 1965).

Additional deeper investigations into all the nooks and crannies of the comedy, TV, and film would build upon this experiment and illuminate more alternative methods of climate communication from even more people. Tackling the climate crisis as a fully backed primetime initiative, agreed upon by all networks, targeting the demos of each outlet would be a way to hit as many targets as possible. Peppa Pig, Rachel Maddow, Zendaya, Larry David, and RuPaul all talking about the Green New Deal? Can you imagine?! It's really an achievable goal. One step in the right direction would be admiring and hiring the brilliant communicators discovered in this chapter—and keeping an eye out for the new ones popping up constantly, Whac-A-Mole style. In the words of our wise editors, which we're too lazy to paraphrase: "Let's keep deepening engagement through a range of comedy at the local, national and international levels." Boom.

---

[57] Equal to or greater than Greta Thunberg with a gun? You decide.

Thank You

…to all the creators we mentioned, as well as everyone who provided feedback on drafts of this chapter—especially Caty Borum, Lauren Feldman, Rollie Williams, Bartees Cox, Patrick David Chandler, and of course our two editors, Hua Wang and Emily Coren. We could not have done it without you! (Well, we COULD have, but it would have been worse. Radical honesty.)

References

American University. (2020, January 16). *AU's Center for Media & Social Impact receives $1.1 million.* https://www.american.edu/media/news/20200116-cmsi-2020-grants.cfm

Anderson, A. A., & Becker, A. B. (2018). Not just funny after all: Sarcasm as a catalyst for public engagement with climate change. *Science Communication, 40*, 524. https://doi.org/10.1177/1075547018786560

Bastida, X., & Reddy, P. (Hosts). (2020, October 28). We may be small but our impact is huge [Audio podcast episode]. In *Mothers of Invention*. Doc Society.

Baum, M. A. (2005). *Soft news goes to war: Public opinion and American foreign policy in the new media age.* Princeton University Press.

Baute, K. [Kurtis Baute]. (2018, November 2). *I sealed myself in a jar: Part one* [Video]. YouTube. https://www.youtube.com/watch?v=PoKvPkwP4mM&t=181s

Baute, K. [Kurtis Baute]. (2019, February 4). *I went 24hrs without touching plastic. This is what happened* [Video]. YouTube. https://www.youtube.com/watch?v=XgD9ItnBQsA

Baute, K. [Kurtis Baute]. (n.d.). *Videos* [YouTube channel]. YouTube. https://www.youtube.com/user/ScopeofScience

Berlant, L., & Ngai, S. (2017). Comedy has issues. *Critical Inquiry, 43*(2), 233–249. https://doi.org/10.1086/689666

Borlace, D. [Just Have a Think]. (2020, April 19). *Is the Gulf Stream collapsing?* [Video]. YouTube. https://www.youtube.com/watch?v=6Yz8nZbZPE8

Borlace, D. [Just Have a Think]. (n.d.). *About* [YouTube channel]. YouTube. https://www.youtube.com/c/JustHaveaThink/about

Borum, C. (2020). *Comedy for racial justice in the climate crisis: Leveraging creativity and building community power in Ain't Your Mama's Heat Wave.* Center for Media & Social Impact. https://cmsimpact.org/wp-content/uploads/2016/08/CMSI_HHC_web.pdf

Borum, C. (2021). Entertainment-education as social justice activism in the United States: Narrative strategy in the participatory media era. In L. Frank & P. Falzone (Eds.), *Entertainment-education behind the scenes: Case studies for theory and practice* (pp. 39–59). Palgrave Macmillan. https://doi.org/10.1007/978-3-030-63614-2_4

Borum, C., & Feldman, L. (2020). *A comedian and an activist walk into a bar: The serious role of comedy in social justice.* University of California Press.

Borum, C., & Green-Barber, L. (2018). An investigative journalist and a comedian walk into a bar: The role of comedy in public engagement with environmental journalism. *Journalism, 22*(1), 196–214. https://doi.org/10.1177/1464884918763526

Boult, B. (n.d.). *Benjamin Boult is a documentary filmmaker, producer, and cinematographer based in New York City.* https://www.benjaminboult.com/about

Boykoff, M. (2019). *Creative (climate) communications: Productive pathways for science, policy and society.* Cambridge University Press. https://doi.org/10.1017/9781108164047

Boykoff, M., Osnes, B., Safran, R., & Pezzullo, P. (2017a). *2017 Second place winner, alternate science (Vol. 1) (USA)*. Inside the Greenhouse at University of Colorado Boulder. https://insidethegreenhouse.org/media/2017-second-place-winner-alternate-science-vol-1-usa

Boykoff, M., Osnes, B., Safran, R., & Pezzullo, P. (2017b). *Winners announced! 2017 comedy & climate change video competition*. Inside the Greenhouse at University of Colorado Boulder. https://insidethegreenhouse.org/news/winners-announced

Boykoff, M., Osnes, B., Safran, R., & Pezzullo, P. (2018). *2019 ITG comedy & climate change short video competition*. Inside the Greenhouse at University of Colorado Boulder. https://insidethegreenhouse.org/news/2019-itg-comedy-climate-change-short-video-competition

Boykoff, M., Osnes, B., Safran, R., & Pezzullo, P. (2019). *Comedy & climate change video competition: Announcing the 2019 winners!* Inside the Greenhouse at University of Colorado Boulder. https://insidethegreenhouse.org/node/3934

Boykoff, M., Osnes, B., Safran, R., & Pezzullo, P. (2020). *Comedy & climate change video competition: Announcing the 2020 winners!* Inside the Greenhouse at University of Colorado Boulder. https://insidethegreenhouse.org/news/comedy-climate-change-video-competition-0

Boykoff, M., Osnes, B., Safran, R., & Pezzullo, P. (2021). *Who we are*. Inside the Greenhouse at University of Colorado Boulder. https://insidethegreenhouse.org/who-we-are

Brändlin, A. (2019, August 28). *The global injustice of the climate crisis*. Deutsche Welle. https://www.dw.com/en/the-global-injustice-of-the-climate-crisis-food-insecurity-carbon-emissions-nutrients-a-49966854/a-49966854

Brewer, P. R., & McKnight, J. (2017). "A statistically representative climate change debate": Satirical television news, scientific consensus, and public perceptions of global warming. *Atlantic Journal of Communication, 25*(3), 166–180. https://doi.org/10.1080/1545687 0.2017.1324453

Buckley, M. [Inside Ideas]. (2020, August 4). *How we build a better world. Just Have a Think Dave Borlace* [Video]. YouTube. https://www.youtube.com/watch?v=uCP1dHJjQ6M&t=593s

Camargo, L. (n.d.). *Layel Camargo [They/Them]* [LinkedIn page]. LinkedIn. Retrieved January 14, 2022, from https://www.linkedin.com/in/layel-camargo-they-them-37450233/

Campo, S., Askelson, N., Spies, E., & Losch, M. (2012). *Avoiding the Stork: A statewide social marketing campaign to reduce unintended pregnancy*. Retrieved January 14, 2021, from https://www.researchgate.net/publication/266815491_Avoiding_the_Stork_A_statewide_social_marketing_campaign_to_reduce_unintended_pregnancy

Caveat. (2021). *Alison Leiby: Oh god, an hour about abortion*. https://www.caveat.nyc/event/oh-god-an-hour-about-abortion-9-28-2021

Caveat. (n.d.). *An inconvenient talk show*. https://www.caveat.nyc/series/an-inconvenient-talk-show

Center for Media & Social Impact. (2021a). *Comedy ThinkTanks*. https://cmsimpact.org/program/comedy-think-tanks/

Center for Media & Social Impact. (2021b). *Spotlighting storytelling and social change*. https://cmsimpact.org

Center for Media & Social Impact. (2021c). *Reports*. https://cmsimpact.org/report-list/entertainment-popular-culture

Center for Media & Social Impact. (2021d). *Yes, And Laughter Lab: Uplifting diverse comedy for social change*. https://cmsimpact.org/report/yes-laughter-lab-uplifting-diverse-comedy-social-change/p/comedy-can-change-world-introducing-yes-laughter-lab/

Chandler, P., Dillon, J., & Russell, C. (2022). *CFP for EER SI: Humour & environmental education*. Lakehead University. https://www.lakeheadu.ca/users/R/crussell/cfp-for-eer-si-humour-environmental-education

Chapman, M. (2020, October 13). *Work – Save Florida Man*. Potential Energy Coalition. https://potentialenergycoalition.org/work-save-florida-man/

Chapron, G., Levrel, H., Meinard, Y., & Courchamp, F. (2018). A final warning to planet Earth. *Trends in Ecology & Evolution, 33*(9), 651–652. https://doi.org/10.1016/j.tree.2017.12.010

CIRES. (2019). *Stand up for climate change*. https://cires.colorado.edu/events/stand-climate-change

Clark, S. [Simon Clark]. (2020, September 9). *Why nuclear power will (and won't) stop climate change* [Video]. YouTube. https://www.youtube.com/watch?v=k13jZ9qHJ5U&t=331s

Clark, S. [Simon Clark]. (2021, February 25). *The climate is lost when we fill this cube* [Video]. YouTube. https://www.youtube.com/watch?v=jUQrNt72_Os&t=122s

Clark, S. [Simon Clark]. (n.d.). *Videos* [YouTube channel]. YouTube. https://www.youtube.com/c/SimonOxfPhys/featured

Clayton, S., Devine-Wright, P., Stern, P. C., et al. (2015). Psychological research and global climate change. *Nature Climate Change, 5*(7), 640–646. https://doi.org/10.1038/nclimate2622

Climate Ad Project. (2021). *About.* https://climateadproject.org/about/

Climate Comedy. (2021, July 19). *Chuck Nice opens stand up for climate change 2021* [Video]. YouTube. https://www.youtube.com/watch?v=sVkfL3Gsml0

Climate Memes. [@climate_memes]. (n.d.). *Climate memes* [Instagram profile]. Retrieved January 23, 2022, from https://instagram.com/climate_memes

Climate Nexus. [Climate Nexus]. (2019). #9for2019: *Individual climate action* [YouTube playlist]. YouTube. https://www.youtube.com/playlist?list=PLAzsC5SmJmagI76fU-J3EU4eqJaqHy3vQ

Climemechange. [@climemechange]. (n.d.). *Climate change memes* [Instagram profile]. Retrieved January 8, 2022, from https://instagram.com/climemechange

Columbia College Chicago. (2021). *Comedy writing and performance (BA).* https://www.colum.edu/academics/programs/comedy-writing-and-performance

Conlan, N. (n.d.). *Hi, my name is Nicole.* http://www.nicoleconlan.com

Coren, E., Safer, D., & Gurney, C. [Celia Gurney]. (2020, October 20). *Accelerating public health solutions for climate change using entertainment education methods* [Video]. YouTube. https://youtube.com/watch?v=nXuqiouVfFg&t=300s

Coronel, J. C., O'Donnell, M. B., Pandey, P., Delli Carpini, M. X., & Falk, E. B. (2021). Political humor, sharing, and remembering: Insights from neuroimaging. *Journal of Communication, 71*(1), 129–161. https://doi.org/10.1093/joc/jqaa041

Define American. (2022). *Changing the narrative about immigrants through innovative pop culture strategies.* https://defineamerican.com/hollywood

DePaul University. (2001–2021). *Comedy arts.* https://www.depaul.edu/academics/undergraduate/majors/Pages/comedy-arts.aspx

Desmon, S. (2018, April 30). *A conversation with Miguel Sabido, entertainment education pioneer.* Johns Hopkins Center for Communication Programs. https://ccp.jhu.edu/2018/04/30/miguel-sabido-entertainment-education-pioneer/

Dibble, J. L., Hartmann, T., & Rosaen, S. F. (2015). Parasocial interaction and parasocial relationship: Conceptual clarification and a critical assessment of measures. *Human Communication Research, 42*(1), 21. https://doi.org/10.1111/hcre.12063

Embrapa. (2020, June 16). *ECLAC ranks Embrapa project among top transformative for sustainability.* https://www.embrapa.br/en/busca-de-noticias/-/noticia/53363766/eclac-ranks-embrapa-project-among-top-transformative-for-sustainability

Emerson College. (2021). *Comedic arts (BFA).* https://www.emerson.edu/programs/comedic-arts-bfa

Engelberg, A. [@alexengelberg]. (n.d.-a). *Check out these quick tips to conserve energy and fight climate change* [Video]. TikTok. https://vm.tiktok.com/TTPdrL9de5/

Engelberg, A. [@alexengelberg]. (n.d.-b). *Videos* [TikTok profile]. TikTok. https://vm.tiktok.com/TTPdrLB6yv/

FOOSH Improv. [@fooshimprov]. (n.d.). *FOOSH Improv* [Instagram profile]. Retrieved March 18, 2024 from https://instagram.com/fooshimprov

Frank, L., & Falzone, P. (Eds.). (2021). *Entertainment-education behind the scenes: Case studies for theory and practice.* Palgrave Macmillan.

Freedom to Breathe. (2018). *Freedom to Breathe.* https://www.freedomtobreathe.org

Friedman, P. (Director). (2013, August 9). *Poor Consuelo Conquers the World* [Film]. Belladonna Productions & Escape Pictures.

Funes, Y. (2020, May 2). *New York City plan to close streets is weak as hell*. Gizmodo. https://gizmodo.com/new-york-city-plan-to-close-streets-is-weak-as-hell-1843203952

Funes, Y. (2021, September 27). *The hidden climate messages in "Reservation Dogs"*. Atmos. https://atmos.earth/reservation-dogs-dallas-goldtooth-climate-justice/

Funny or Die. [Funny or Die]. (2016, February 24). *Even supervillains think our sexual assault laws are insane* [Video]. YouTube. https://youtube.com/watch?v=bTJSHoR-cQ0

Garrett, R. K., Bond, R., & Poulson, S. (2019, August 16). *Too many people think satirical news is real*. The Conversation. https://theconversation.com/amp/too-many-people-think-satirical-news-is-real-121666

Global Heartwarming. (2015a). *About*. http://globalheartwarming.org/phone/about.html

Global Heartwarming. (2015b). *Cast & crew*. http://globalheartwarming.org/phone/cast%2D%2D-crew.html

Global Heartwarming. (2015c, November 27). *The Negotiation* [Video]. YouTube. https://www.youtube.com/watch?v=GWdVQ2yVugQ&t=2s

Goode, E. (2021, July 29). Albert Bandura, leading psychologist of aggression, dies at 95. *The New York Times*. https://www.nytimes.com/2021/07/29/science/albert-bandura-dead.html

Gravey, V., Hargreaves, T., Lorenzoni, I., & Seyfang, G. (2017). Theoretical theatre: Harnessing the power of comedy to teach social science theory. *Journal of Contemporary European Research, 13*(3), 1319–1336. https://www.jcer.net/index.php/jcer/article/view/824

Green, M. C. (2021). Transportation into narrative worlds. In L. Frank & P. Falzone (Eds.), *Entertainment-education behind the scenes: Case studies for theory and practice* (pp. 87–101). Palgrave Macmillan. https://doi.org/10.1007/978-3-030-63614-2_6

Grist. (2020). *Meet the fixers*. https://grist.org/grist-50/2020/

Harvard. (2021). *Advanced TV writing: the half-hour comedy*. https://online-learning.harvard.edu/course/advanced-tv-writing-half-hour-comedy?delta=0

Heglar, M. A. (2019, August 13). *Climate change isn't racist—People are*. ZORA. https://zora.medium.com/climate-change-isnt-racist-people-are-c586b9380965

Heglar, M. A. (2020, April 1). *We can't tackle climate change without you*. WIRED. https://www.wired.com/story/what-you-can-do-solve-climate-change/

Heglar, M. A. [@MaryHeglar]. (2021, November 30). *This you?* [Tweet]. Twitter. https://twitter.com/maryheglar/status/1465671194041008131?s=21

Heglar, M. A., & Westervelt, A. (Hosts). (2019–2021). *Hot Take* [Audio podcast]. Critical Frequency.

Helicon Collaborative. (2018, February 14). *Farther, faster, together: How arts and culture can accelerate environmental progress*. ArtPlace America. https://www.artplaceamerica.org/view/pdf?f=/sites/default/files/public/pictures/environment.pdf

Hèlikos. (n.d.). *Giovanni Fusetti*. http://www.helikos.com/pages/blog.php?lang=en

Her-icane. [@hericanecomedy]. (n.d.). *Posts* [Instagram profile]. https://instagram.com/hericanecomedy

Hildebrand, L. (2022). *Youtube*. https://www.levihildebrand.com/youtubesponsorship

Hildebrand, L. (n.d.). *Levi Hildebrand* [LinkedIn page]. Retrieved January 14, 2022, from https://www.linkedin.com/in/levi-hildebrand-22357b36/?originalSubdomain=ca

Hildebrand, L., & Tidey, L. [Levi & Leah]. (n.d.). *Videos* [YouTube channel]. YouTube. https://www.youtube.com/c/LeviHildebrand/featured

Hill, M. R., & Holbert, R. L. (2017). Jon Stewart and the 9/11 first responders health bill. In C. M. Madere (Ed.), *Viewpoints on media effects: Pseudo-reality and its influence on media consumers*. Lexington Books.

Horn, S. (n.d.). *Howzit! I'm Stephen*. http://www.stephenhorn.co.za

Horton, D., & Wohl, R. R. (1956). Mass communication and para-social interaction: Observations on intimacy at a distance. *Interpersonal and Biological Processes, 19*(3), 215–229. https://doi.org/10.1080/00332747.1956.11023049

IllumiNative. (n.d.). *The future is Indigenous*. https://illuminatives.org/the-future-is-indigenous/

IMDb. (n.d.-a). *East Los High*. https://www.imdb.com/title/tt2312036/

IMDb. (n.d.-b). *Reservation Dogs*. https://www.imdb.com/title/tt13623580/

JOAN. (2022). *Save Florida Man*. https://www.joancreative.com/work/save-florida-man

Kahn, B., & Taft, M. (2021, September 23). *Late night climate comedy segments, ranked*. Gizmodo. https://gizmodo.com/late-night-climate-comedy-segments-ranked-1847730858

Kim, C. (2019). *The battle over extending the September 11th Victim Compensation Fund, explained*. Vox. www.vox.com/2019/6/20/18691670/jon-stewart-9-11-september-11th-victim-compensation-fund-explained

Kim, C., & Harwood, J. (2020). Parasocial contact's effects on relations between minority groups in a multiracial context. *International Journal of Communication, 14*, 364–385. Retrieved January 19, 2022 from https://ijoc.org/index.php/ijoc/article/download/13010/2912

Lamarre, H. L., Landreville, K. D., & Beam, M. (2009). The Irony of satire: Political ideology and the motivation to see what you want to see in The Colbert Report. *The International Journal of Press/Politics, 14*(2), 212–231. https://doi.org/10.1177/1940161208330904

Last Week Tonight with John Oliver. [LastWeekTonight]. (2014a, June 9). *FIFA and the World Cup: Last Week Tonight with John Oliver (HBO)* [Video]. YouTube. https://www.youtube.com/watch?v=DlJEt2KU33I

Last Week Tonight with John Oliver. [LastWeekTonight]. (2014b, May 12). *Climate change debate: Last Week Tonight with John Oliver (HBO)* [Video]. YouTube. https://youtu.be/cjuGCJJUGsg

Last Week Tonight with John Oliver. [LastWeekTonight]. (2014c, May 19). *India election update: Last Week Tonight with John Oliver (HBO)* [Video]. YouTube. https://www.youtube.com/watch?v=8YQ_HGvrHEU

Last Week Tonight with John Oliver. [LastWeekTonight]. (2017a, June 5). *Paris Agreement: Last Week Tonight with John Oliver (HBO)* [Video]. YouTube. https://www.youtube.com/watch?v=5scez5dqtAc

Last Week Tonight with John Oliver. [LastWeekTonight]. (2017b, October 30). *Floods: Last Week Tonight with John Oliver (HBO)* [Video]. YouTube. https://www.youtube.com/watch?v=pf1t7cs9dkc

Leibowitz, B. (2012, June 12). Man with cat denied entry by Fla. strip club, arrested. *CBS News*. https://www.cbsnews.com/news/man-with-cat-denied-entry-by-fla-strip-club-arrested/

Levy, A. (n.d.-a). *DR ADAM LEVY: ClimateAdam*. https://climateadam.co.uk. Accessed January 11, 2022.

Levy, A. [ClimateAdam]. (n.d.-b). *About* [YouTube channel]. YouTube. https://www.youtube.com/c/ClimateAdam/about

Lorenz, T., Browning, K., & Frenkel, S. (2020, June 21). TikTok teens and K-Pop stans say they sank Trump rally. *The New York Times*. https://www.nytimes.com/2020/06/21/style/tiktok-trump-rally-tulsa.html

MACRO. [StayMACRO]. (2022, January 21). *MACRO X SUNDANCE 2022 | PASS THE MIC: CENTERING COMMUNITIES OF COLOR IN CLIMATE STORYTELLING* [Video]. YouTube. https://www.youtube.com/watch?v=rprY8CKZX_o

Martin, R. (2007). *The psychology of humor: An integrative approach*. Academic Press. https://doi.org/10.1016/C2016-0-03294-1

McCarthy, S. L. (2019, January 31). 5 comedy shows to catch in N.Y.C. this weekend. *The New York Times*. https://www.nytimes.com/2019/01/31/arts/comedy-in-nyc-this-week.html

McSweeney's. (n.d.) *Articles by Casey Rand*. https://www.mcsweeneys.net/authors/casey-rand

Mitchell, A., Gottfried, J., Kiley, J., & Matsa, K. E. (2014, October 21). *Political polarization & media habits*. Pew Research Center. https://www.pewresearch.org/journalism/2014/10/21/political-polarization-media-habits/

Moger, A. (2017, December 8). *Blending comedy and academia*. UNC-Chapel Hill. https://www.unc.edu/discover/blending-comedy-academia/

Movement Generation. (2015, August 3). *Green Collar Comedy* [YouTube playlist]. YouTube. https://www.youtube.com/playlist?list=PLgfdVMERG3OOPsZfPMOx690x39ysjbUHV

NAACP and the Clean Air Task Force. (2017). *Fumes across the fence-line: The health impacts of air pollution from oil & gas facilities on African American communities*. https://naacp.org/resources/fumes-across-fence-line-health-impacts-air-pollution-oil-gas-facilities-african-american

Nabi, R. L., Moyer-Gusé, E., & Byrne, S. (2007). All joking aside: a serious investigation into the persuasive effect of funny social issue messages. *Communication Monographs, 74*(1), 29–54. https://doi.org/10.1080/03637750701196896

Nabi, R. L., Gustafson, A., & Jensen, R. (2018). Framing climate change: Exploring the role of emotion in generation advocacy behavior. *Science Communication, 40*(4), 442–468. https://doi.org/10.1177/1075547018776019

NASA. (2021). *Scientific consensus: Earth's climate is warming.* https://climate.nasa.gov/scientific-consensus/. Accessed August 18, 2021.

New York Comedy Club. (2019). *136 degrees funny: Because climate is no laughing matter.* https://newyorkcomedyclub.com/events/136o-funny-because-climate-is-no-laughing-matter-new-york-comedy-club. Accessed January 8, 2022.

Nexus Media News. (n.d.). *Black History Month.* https://nexusmedianews.com/tagged/black-history-month/

Nice, C. (Host). (2021a, February 23). Climate education [Audio podcast episode]. In *Pod Zero.* Shhh, It's Real.

Nice, C. (Host). (2021b, January 26). Environmental justice [Audio podcast episode]. In *Pod Zero.* Shhh, It's Real.

Noor, D. (2021, October 5). *Huntington Beach High School may want to change its mascot after this oil spill.* https://gizmodo.com/huntington-beach-high-school-may-want-to-change-its-mas-1847797297

Notaro, T. (2013, July 16). *Live* [Audio comedy album]. Secretly Canadian. https://open.spotify.com/album/6ttCxEGqI0tX85k80YPYNu

NRDC. (2022). *Rewrite the Future: NRDC helps Hollywood take on the climate crisis.* https://www.nrdc.org/RewriteTheFuture

O'Neill, S. J., Boykoff, M., Niemeyer, S., & Day, S. A. (2013). On the use of imagery for climate change engagement. *Global Environmental Change, 23*(2), 413–421. https://doi.org/10.1016/j.gloenvcha.2012.11.006

Ohlheiser, A. (2015, August 24). John Oliver has received "thousands" of donations for his televangelism ministry. *The Washington Post.* https://www.washingtonpost.com/news/acts-of-faith/wp/2015/08/24/john-oliver-has-received-thousands-of-donations-for-budding-televangelism-ministry

Osnes, B., Boykoff, M., & Chandler, P. (2019). Good-natured comedy to enrich climate communication. *Comedy Studies, 10*(2), 224–236. https://doi.org/10.1080/2040610X.2019.1623513

PBS. (2022). *Freedom to Breathe Ep. 2: Climate gentrification in the US.* Peril and Promise. https://www.pbs.org/video/freedom-to-breathe-ep-2-climate-gentrification-in-the-us-odjsmc/

Politically Aweh. (2020, November 13). *Same Old Eskom or Green New Deal? South Africa's renewable energy future explained* [Video]. YouTube. https://youtu.be/sijx84RliFo

Politically Aweh. (2021, June 18). *There's Something Fishy About the Karpowership Deal (ft. KG Mokgadi & Bearleii Sober)* [Video]. YouTube. https://youtu.be/8d55Dqf9WbE

Politically Aweh. (n.d.). *Climate change in South Africa mini series* [YouTube playlist]. YouTube. https://m.youtube.com/playlist?list=PLG8zV9d7X947TiV6Ov5PfTsgatCNT9i0l

Powell, D., & Hayes, O. (2018, March 12). #104: CATS [Audio podcast episode]. In *Sustainababble.* http://www.sustainababble.fish/?p=652

Powell, D., & Hayes, O. (2021, July 25). #221: SOIL [Audio podcast episode]. In *Sustainababble.* http://www.sustainababble.fish/?p=1352

Rawls, T. (n.d.). *Tejopala Rawls* [Linkedin page]. LinkedIn. Retrieved January 14, 2022, from https://www.linkedin.com/in/tejopala-rawls-71a32989/?originalSubdomain=au

Reductress. (n.d.). *Author archives for Casey Rand.* https://reductress.com/author/caseyrand/

Robinson, M., Higgins, M., & Kodikara, T. (Hosts). (2018–2020). *Mothers of Invention* [Audio podcast]. Doc Society.

Robinson, M., Higgins, M., & Kodikara, T. (Hosts). (2020, November 4). Brave enough to imagine [Audio podcast episode]. In *Mothers of Invention.* Doc Society.

Ropek, L. (2022, January 6). *French car ads will soon be required by law to tell you not to drive a car.* Gizmodo. https://gizmodo.com/french-car-ads-will-soon-be-required-by-law-to-tell-you-1848310770

Ryerson, W. N., & Negussie, T. (2021). The impact of social change communication: Lessons learned from decades of media outreach. In L. Frank & P. Falzone (Eds.), *Entertainment-education behind the scenes: Case studies for theory and practice* (pp. 135–189). Palgrave Macmillan. https://doi.org/10.1007/978-3-030-63614-2_3

Sabido, M. (2021). Miguel Sabido's entertainment education. In L. Frank & P. Falzone (Eds.), *Entertainment-education behind the scenes: Case studies for theory and practice* (pp. 39–59). Palgrave Macmillan. https://doi.org/10.1007/978-3-030-63614-2_2

Salgado D'Arcy, R. [All About Climate]. (2021, April 20). *"It's only 1°C" – How significant is global warming really?* [Video]. YouTube. https://www.youtube.com/watch?v=Mam6JTXLqkA&t=159s

Salgado D'Arcy, R. (2022). *Roshan Salgado D'Arcy* [LinkedIn page]. LinkedIn. Retrieved January 10, 2022, from https://uk.linkedin.com/in/roshan-salgado-d-arcy-657a64125

Schiappa, E., Gregg, P., & Hewes, D. (2006). Can one TV show make a difference? Will & Grace and the parasocial contact hypothesis. *Journal of Homosexuality, 51*(4), 15–37. https://doi.org/10.1300/J082v51n04_02

Schiffrin, A. (2019, May 1). South Africa's Daily Maverick exemplifies the travails facing Global Muckrakers. *Columbia Journalism Review*. Retrieved September 27, 2021, from https://www.cjr.org/watchdog/south-africa-daily-maverick.php

Schlossberg, T. (2019, November 10). Telling stories to battle climate change, with a little humor thrown in. *The New York Times*. https://www.nytimes.com/2019/11/10/climate/mary-robinson-maeve-higgins-thimali-kodikara-podcast.amp.html

Schwartz, T. (2020). *The State of SIE: Welcome note*. The State of SIE. https://www.thestateofsie.com/teri-schwartz-welcome-note/

Shell. [@Shell]. (2020, November 2). *What are you willing to change to help reduce emissions? #EnergyDebate* [Tweet]. Twitter. https://twitter.com/shell/status/1323184318735360001?s=21

Shhh, It's Real! [@ShhhItsReal]. (n.d.). *Tweets* [Twitter profile]. Retrieved January 8, 2022 from https://twitter.com/shhhitsreal?s=21

Simerman, J. (2015, September 17). HBO comedy host John Oliver gives Orleans Parish Public Defender's Office a crowdfunding boost. *The New Orleans Advocate*. https://www.nola.com/news/article_b3b8a8f0-803b-5311-adff-6cab33a2d85b.html

Smith, M. (n.d.). *Bio*. https://www.mariahmariahmariah.com/bio

Song-Xia, A. [@alexsongxia]. (n.d.). *Posts* [Instagram profile]. https://instagram.com/alexsongxia

StarTalk. (2022). *About us*. https://www.startalkradio.net/about-us/

Story Pirates Changemakers. (n.d.). *Story Pirates Changemakers*. https://storypirateschangemakers.org/about-1

Superhero Clubhouse. (n.d.). *Big Green Theater*. http://www.superheroclubhouse.org/bgt/

Swaine, J. (2017, January 23). Donald Trump's team defends "alternative facts" after widespread protests. *The Guardian*. https://www.theguardian.com/us-news/2017/jan/22/donald-trump-kellyanne-conway-inauguration-alternative-facts

Thank You, Robot. (2019, February 2). *Science exclamation point*. Caveat. https://www.caveat.nyc/event/science-exclamation-point-2-2-2019

The Center for Cultural Power. (2021). *Layel Camargo: Ecological arts and culture manager*. https://www.culturalpower.org/people/Layel%20Camargo

The Comedy Attic. (2021). *Upcoming shows*. https://www.comedyattic.com

The Daily Show with Trevor Noah. [The Daily Show with Trevor Noah]. (2021, August 30). *Global temperatures are going UP* [Video]. YouTube. https://www.youtube.com/watch?v=Kpya7-slQzE

The Good Energy Project. (n.d.). *Flipping the script on climate change*. https://www.goodenergystories.com

The Juice Media. [thejuicemedia]. (2019a, September 3). *Honest Government Ad | We're F**ked* [Video]. YouTube. https://www.youtube.com/watch?v=cOmdkN6MOwU

The Juice Media. (2019b). *The Juice Media: About*. https://www.thejuicemedia.com/about/

The Juice Media. [thejuicemedia]. (2021, April 24). *Honest Government Ad | Electric vehicles* [Video]. YouTube. https://www.youtube.com/watch?v=fLflYkgnNBY

The Juice Media. [thejuicemedia]. (n.d.). *Videos* [YouTube channel]. YouTube. https://www.youtube.com/channel/UCKRw8GAAtm27q4R3Q0kst_g

The North Pole. (2019a). *About.* http://www.thenorthpoleshow.com/about/

The North Pole. (2019b). *Episodes.* http://www.thenorthpoleshow.com/season-1/

The North Pole. (2019c). *Episodes.* http://www.thenorthpoleshow.com/season-2/

The North Pole Show. [The North Pole Show]. (2017, September 12). *The North Pole S1, Ep 3: "Drought"* [Video]. YouTube. https://www.youtube.com/watch?v=D8Ie6e4LXCg

The Onion. (2021, September 3). *Optimistic researchers say there still time to head off climate change before it starts killing rich people.* https://www.theonion.com/optimistic-researchers-say-there-still-time-to-head-off-1847614847

Think 100%, a project of Hip Hop Caucus. (2021). *Ain't Your Mama's Heat Wave.* https://think100climate.com/films/aint-your-mamas-heat-wave/

Thomas, M., Jain-Conti, S., Lee, M., Gurney, C., Cox, B., & Agnew, O. (2019, February 25). The food revolution will not be televised. *Nexus Media News.* https://nexusmedianews.com/the-food-revolution-will-not-be-televised-d4a0a9e1737d/

Time Out. (2019, May 22). *Her-icane: Women+ stand up comedy for disaster relief.* https://www.timeout.com/newyork/comedy/her-icane-women-stand-up-comedy-for-disaster-relief

Tisch School of the Arts. (2021). *Minor in comedy writing.* Retrieved August 30, 2021, from https://tisch.nyu.edu/special-programs/minors/minor-in-comedy-writing

Travis, Y. (n.d.). *Dark Tank with Yedoye Travis* [Audio podcast]. Forever Dog/Brain Machine.

UCLA School of Theater, Film & Television Professional Programs. (2021). *Writing for late night comedy online.* https://professionalprograms.tft.ucla.edu/writing-for-late-night-comedy

University of California, Berkeley. (2021a). *Cognitive constructivism.* https://gsi.berkeley.edu/gsi-guide-contents/learning-theory-research/cognitive-constructivism/

University of California, Berkeley. (2021b). *How social learning theory works.* https://hr.berkeley.edu/how-social-learning-theory-works

University of Redlands. (2019). *In memoriam: Monty Hempel (1950–2019).* https://www.redlands.edu/bulldog-blog/2019/december-2019/in-memoriam-monty-hempel-19502019/

University of Southern California. (2021). *Comedy minor.* https://catalogue.usc.edu/preview_program.php?catoid=14&poid=16743&hl=comedy&returnto=search#

Vasquez, T., & Camargo, L. (Hosts). (2021). *Did We Go Too Far?* [Audio podcast]. We Rise Production Society.

Venzke, D. (2012). *Avoiding unexpected deliveries.* https://now.uiowa.edu/2012/09/avoiding-unexpected-deliveries

Villa, K. (2022, February 4). *Personal communication.*

Wang, H., & Singhal, A. (2016). East Los High: Transmedia edutainment to promote the sexual and reproductive health of young Latina/o Americans. *American Journal of Public Health, 106*(6), 1002–1010.

Wang, H., & Singhal, A. (2021). Mind the gap! Confronting the challenges of translational communication research in entertainment-education. In L. Frank & P. Falzone (Eds.), *Entertainment-education behind the scenes: Case studies for theory and practice* (pp. 223–242). Palgrave Macmillan. https://doi.org/10.1007/978-3-030-63614-2_14

Well, E. (2022, January 3). This isn't the California I married. *The New York Times.* https://www.nytimes.com/2022/01/03/magazine/california-widfires.html

Williams, R. [Climate Town]. (n.d.). *About* [YouTube channel]. YouTube. https://www.youtube.com/c/ClimateTown/about

Wozniacka, G. (2021, November 5). Indigenous leaders call for landback reforms and climate justice in "required reading". *Yes Magazine.* https://www.yesmagazine.org/environment/2021/11/05/indigenous-authors-land-back-climate-justice

Yardley, W. (2016, September 13). An environmental activist who uses comedy to help stop oil pipelines. *Los Angeles Times.* https://www.latimes.com/nation/la-na-sej-dallas-goldtooth-activist-comedian-20160902-snap-story.html

Yearwood, L., Jr. (2020, June 22). *Climate justice is racial justice, racial justice is climate justice.* Shondaland.com. https://www.shondaland.com/act/a32905536/environmental-justice-racial-justice-marginalizedcommunities/

Yerba Buena Center for the Arts. (2019). *YBCA 100*. https://ybca.org/ybca-100/2019-honorees/

Yes, And Laughter Lab. (2019). *2019 Yes, And... Laughter Lab winners*. https://yesandlaughterlab.com/2019-winners/

Yes, And Laughter Lab. (2022). *Comedy can change the world*. https://yesandlaughterlab.com/

Yoder, K. (2020, November 19). Greentrolling: A "maniacal plan" to bring down Big Oil. *Grist*. https://grist.org/energy/greentrolling-a-maniacal-plan-to-bring-down-big-oil/amp/

York St. John University. [YSJU Events]. (2021, March 18). *Thimali Kodikara: Education and the frontlines of social justice* [Video]. YouTube. https://www.youtube.com/watch?v=I5pcIqRYMs0

Young, D. G. (2008). The privileged role of the late-night joke: Exploring humor's role in disrupting argument scrutiny. *Media Psychology, 11*(1), 119–142. https://doi.org/10.1080/15213260701837073

Yount, A. (2021, January 6). *Delivering the news with humor makes young adults more likely to remember and share*. Annenberg School for Communication, University of Pennsylvania. https://www.asc.upenn.edu/news-events/news/delivering-news-humor-makes-young-adults-more-likely-remember-and-share

Zhao, L. (2016). *Parasocial relationships with transgender characters and attitudes toward transgender individuals* (Publication No. 533) [Doctoral dissertation, Syracuse University]. SURFACE at Syracuse University. https://surface.syr.edu/etd/553

Celia Gurney is an Upright Citizens Brigade-trained writer and performer who spent 3 years immersed in climate change communications at Climate Nexus. In 2019, she produced a comedic video series about climate action called #9for2019, which had the goal of generating sign-ups for the Environmental Voter Project. At Climate Nexus, Celia also contributed to a Black History Month series on climate justice leaders as well as Freedom to Breathe, a collection of events across the country featuring on-the-ground climate justice advocates. More recently, she created comedic content for Shut The Fossils Up and Yellow Dot Studios' campaign to protect New York State's climate law from fossil fuel interference.

Mamoudou N'Diaye is a Brooklyn-based comedian who accidentally made his way into climate crisis communication by way of racial justice work. He was featured in the Think 100% documentary *Ain't Your Mama's Heat Wave* as the host of the comedy portion of the documentary about the climate crisis effect on the city of Norfolk, VA, and then became a writer for it. Prior to that, he worked on the New York-based climate justice late-night show *An Inconvenient Talk Show*, where he was the DJ and occasional correspondent, delivering out-of-the-box solutions to the climate crisis via PowerPoint. In 2019, he performed at The Center for Cultural Power's Climate Woke Summit and partnered with them, writing sketches/web series based on climate justice like *Law and Order: Climate Justice Unit*. On top of that, he has written way too many treatments about the integration of climate justice into a TV format.

Climate Fiction to Inspire Green Actions: A Tale of Two Authors

Denise Baden and Jeremy Brown

The flourishing genre of "climate fiction" (cli-fi) is increasingly prevalent across the arts as anxieties rise about the fate of the planet. The phrase itself was not visible until first used by Dan Bloom in 2015 (Svoboda, 2016). But cli-fi, as a trend, was already noticeable in the 1990s with the significant rise in fictional books depicting human-caused climate change as a focal point of the narrative. Toward the end of the 2000s, dozens of cli-fi novels had entered circulation as the impacts of climate crisis became more obvious (Schneider-Mayerson, 2017; Trexler, 2015)—*Solar, The Year of the Flood, The Wind-Up Girl, A Children's Bible* and *Clade*, to name just a few best-sellers. Similarly, Svoboda (2016) provides an overview of 60 movies which fall within the genre of cli-fi. Notable examples of climate films include *The Day After Tomorrow*, entering cinemas in 2004, dramatizing the collapse of ocean currents (Lowe et al., 2006), and most recently, *Don't Look Up* (2021), a dark comedy about an apocalyptic comet which parodies much of the world's indifference to the climate crisis. At the time of writing, the film had already broken records on Netflix (Lewis, 2022). Similarly, many works of literary cli-fi are receiving a positive reception. For example, *Florida* (2018), an anthology of short stories by Lauren Groff, was a finalist of the National Book Awards (Schneider-Mayerson et al., 2020).

However, empirical evidence is mixed over whether climate fiction can bring about the desired changes in behavior needed to make a meaningful contribution to "saving the planet." For example, a study by Schneider-Mayerson (2018, p. 494) found that many of the new behaviors reported by novel readers did little to "contribute to climate mitigation, collective adaptation, or the pursuit of climate justice" with very few political actions. For example, reading *Flight Behavior* merely caused one participant to start using recycled shopping bags—a necessary but insufficient response to the scale of the climate emergency. *The Day After Tomorrow*

D. Baden (✉) · J. Brown
University of Southampton, Southampton, UK
e-mail: D.A.Baden@soton.ac.uk; j.d.brown@soton.ac.uk

© The Author(s) 2024   203
E. Coren, H. Wang (eds.), *Storytelling to Accelerate Climate Solutions*,
https://doi.org/10.1007/978-3-031-54790-4_10

meanwhile, left viewers "feeling" motivated, but somewhat unable to respond, given the film fell short of depicting adequate solutions to the climate crisis (Lowe et al., 2006).

Furthermore, climate dystopias, while attempting to raise awareness of the need for change, may have unintended consequences, such as defensive reactions on the personal front, such as avoidance, denial, and/or on the political front such as hostility against immigrants. Another concern is that even if climate fiction increases awareness of the hazards of unmitigated climate change, this awareness may result in eco-anxiety, rather than effective action (Baden, 2019). However, the outlook is more promising for stories that depict positive solutions. For example, Baden (2019) recruited 91 participants to read and comment on a selection of four short stories with an environmental theme. The results showed that fictional stories with a solution focus were more effective than catastrophic stories at motivating intentions to carry out ecofriendly behavior.

The authors of this chapter come from different backgrounds, with different objectives and genres. For example, compared to Denise Baden's *Habitat Man*, the stories of Jeremy Brown begin with a greater degree of dramatic or "scary" suspense about the threat of monsters, villains, and potential disasters. Yet the authors agree that increasing a sense of agency is important to showcase the kinds of behavior that might avert the futures we fear and bring about a happy ending. As creators, we share the belief that raising the alarm without increasing a sense of being able to do something about the issue often simply leads to eco-anxiety. Helping to move readers from high eco-anxiety and low agency to a sense of being able to make a difference (see Table 1) is an important driver for our work.

A particular aim of Denise Baden is also to reach the disengaged (low eco-anxiety/low agency), and her novel *Habitat Man* is targeted at a mainstream audience who might avoid "green" literature but would enjoy a rom-com. In the following section, Denise describes how she drew upon psychological theories of behavior to inspire green actions in readers. Techniques utilized include highlighting the benefits of green behavior to the user, presenting green actions as the norm, showing solutions, and tapping into more spiritual aspects relating to humanity's relationship with nature. At the same time, romance, humor, and mystery are used to keep readers engaged with the plot and balance the tips on wildlife and climate mitigation that readers can adopt themselves.

In *The Renegades* series, aimed at younger readers, the solutions promoted are not always immediately achievable for the reader, but Jeremy Brown highlights that

Table 1 Objective of moving toward low eco-anxiety and high agency

	Low agency	High agency
High eco-anxiety	Pointless fretting, passive despair, nightmares, pessimism; risk of alienating others	Driven action at expense of mental health, risk of burnout and aliening others
Low eco-anxiety	Lack of awareness, avoidance; if not part of solution then part of problem; mindless complicity	Calm, effective action

the aim is to plant ideas and values that may provide seeds of future actions in the long term, while modeling some behaviors that readers can engage in, such as political protest.

The common approach that we share in this chapter is around the need to create role models that generate this positive sense of agency, which can be radical, or focus more on gentle changes in behavior. In the remainder of the chapter, we illustrate this point through our experiences in writing, or co-creating, our own works of eco-fiction. In doing so, we seek to share our wisdom with creative writers embarking on a similar journey into climate fiction. We conclude our chapter by showcasing an ongoing initiative that promotes fiction to inspire green actions: the free series of Green Stories Writing Competitions set up by Denise Baden in 2018.

Denise Baden

Motivation to Write Habitat Man

When you work as an academic in the world of sustainability, it is easy to be extremely busy, while achieving very little. We share among ourselves latest news, findings, and articles about climate change problems and solutions and publish academic articles that only those who care about climate change read and teach courses on sustainability that only students who already care about the topic choose. It's a sad fact that we spend most of our time talking in our own echo chambers. In desperation, I turned to fiction as a way to reach a wider audience—those who wouldn't dream of watching a climate change documentary. A perfect framework for a book came to me in the form of a visit from a local ecologist in 2019 who'd retired early to become a green garden consultant, advising people how to make their gardens more wildlife friendly. It immediately occurred to me that this would make a lovely TV series—a bit like "All Creatures Great and Small"—each episode he could visit different gardens with character arcs in the form of a love story and mystery. I realized the best route to this was to first write a book which could then be adapted if successful. In Sept 2021, I published *Habitat Man* (D. A. Baden, 2021)—an eco-themed rom-com. The blurb is below:

> Tim is fifty, single and in a job he hates. Inspired by a life-coaching session, he sheds his old life to become Habitat Man, giving advice on how to turn gardens into habitats for wildlife. His first client is the lovely Lori. Tim is smitten, but first he has to win round Ethan her teenage son. Tim loves his new life until he digs up more than he bargained for, and uncovers a skeleton, one that threatens to bring out the skeletons in his cupboard too. Only Jo, Tim's long-time best friend knows his secret, but can she be trusted?

I began by thinking of what I wanted readers to do as a result of reading my novel, and then considered how I could inspire such behaviors. In the next section, I present extracts from my book *Habitat Man* which illustrates how I drew upon psychological theories of behavior change to promote certain behaviors. I also discuss the steps I took to avoid the trap of info-dumping or preaching. The age-old premise of "show don't tell" came in very useful, as did the use of humor to disguise the green message.

Theory of Planned Behavior

One of the best known and most widely used theory of behavior is Ajzen's (1985) Theory of Planned Behavior (Fig. 1). Ajzen posits that behavior is a product of:

1. Attitudes toward the behavior, which in turn are informed by awareness and knowledge.
2. Social norms—do the people who matter to you approve or disapprove of this behavior?
3. Perceived behavioral control—how able do you feel to engage in this behavior?

A similar concept to "perceived behavioral control" is self-efficacy (Bandura, 1997). Self-efficacy can be broken down into three aspects: (1) *self-efficacy expectancy*, i.e., perceptions of competence/capability in carrying out the behavior, (2) *outcome expectancy:* belief that the behavior will result in the desired outcome, and (3) *outcome value*: the outcome of the behavior is desirable.

For example, according to this model, recycling behavior will be affected by attitudes toward recycling, which in turn is affected by awareness of issues related to recycling. The normative aspect would be whether the people close to us will disapprove if we don't recycle, and whether there are recycling facilities available (perceived behavioral control).

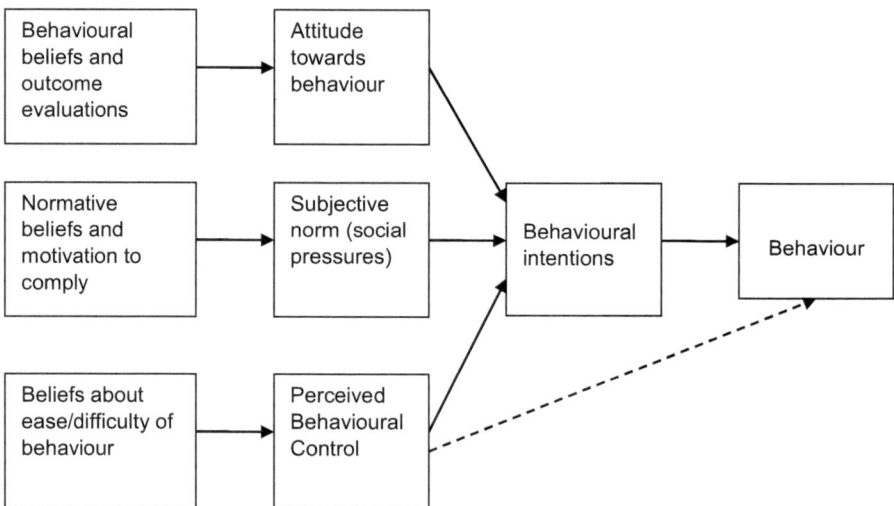

Fig. 1 Theory of planned behavior (Ajzen, 1985)

Research indicates that the most effective drivers of behavior are social norms and perceived behavioral control. Countless studies into sustainable behavior report an attitude-behavior gap. For example, despite most people claiming positive attitudes toward green products, and intentions to purchase them, most purchase decisions are based on price and quality (Miller & Levine, 2019; Sheeran, 2002; Sheeran et al. 2003). Indeed, it is normative influences and feelings of being able to make a difference that drive behavior in reality. This is an issue as currently almost all climate change communications, including cli-fi, focus on raising awareness of the issues.

Those of us who are engaged in our own personal efforts to mitigate against climate change may have been inspired to do so by documentaries showing the effects of plastic on ocean creatures, orangutans wondering homeless in decimated rainforests, frightening statistics of rising temperatures, and desertification and acidification. Not unnaturally, we take what inspired us and try to share it more widely. However, many, when faced with such bleak prospects, switch off, go into denial, or succumb to paralyzing eco-anxiety, which, unless accompanied by effective action, simply adds mental health issues to the growing list of impacts of climate change (Baden, 2019).

One way to protect against this is to focus on solutions, especially those that people may not be aware of. In *Habitat Man*, there were plenty of opportunities for my character to hold forth about the terrifying statistics relating to biodiversity and extinction, but I decided to skip the alarm and focus almost entirely on the solutions. Possibly the only time the issue of biodiversity loss is directly mentioned is on his first visit as follows:

> I rubbed the grey soil between my fingers… I dug another spadeful, still looking. A lone woodlouse, what had happened to the rest? Where were the worms, the springtails, the beetles, the millipedes? These were the building blocks of the food chain. I felt a jolt of anxiety in my gut. I'd not dug in soil for twenty years. Statistics were one thing, but the soil was barren. Where has all the life gone?

The above extract focuses initially on <u>attitudes</u>—raising awareness of the issues and consequences of barren soil. The next paragraph moves quickly onto <u>perceived behavioral control</u>—ways to solve this issue that are possible for readers to do and importantly solutions that readers may not already be aware of. Please note Florence is a dog!

> Florence trotted into the garden and squatted, gazing serenely into the distance.
> 'Do you use a wormer on Florence?'
> 'Yes?'
> 'That will kill all worms it comes in contact with.' I held some unmoving soil in my hand. 'How about flea treatment?'
> 'Is that bad?'
> 'Pets rarely get fleas in winter so maybe give it a rest till summer.'
> 'The vet recommended it each month just in case.'
> 'It's a good income stream for them, but one typical flea treatment has enough pesticide to kill millions of insects, not just fleas, but bees too, and it gets into the water supply, affecting aquatic organisms and amphibians.'
> 'OMG, I feel like a murderer now.'
> 'Look for less toxic alternatives. If they don't work you can always do a stronger treatment if problems occur.'

That is probably the longest info dump in the whole novel and the plot quickly moves onto the relationships developing between the characters. But most readers, even if they don't own a pet themselves, will know someone who does, and few are aware of the toxic effects of pet treatments. This focus on solutions that continue throughout the book, and no issue is presented without corresponding information on what anyone can do to address it, thus tapping into readers' feelings of behavioral control—they know what can be done and how to do it.

However, if readers wanted a book on green solutions or wildlife gardening, they'd have got one, and my aim was to reach a wider audience who wouldn't choose such material, so in each garden there is something to keep the reader hooked. In the gardens with the love interest, Lori, the readers are kept engaged by seeing the romance play out. Similarly, in the chapter called "The Polyamorist," I cover the topic of home composting, but the reader is hooked, waiting for Dawn the polyamorist to make a move. In the scenes with the wizard, Tim promotes the idea of a pond to attract frogs and bats, but the reader turns the pages in anticipation of some fun magic or wizardry. In Daisy, the Feng Shui gardener's garden, there's the mystery of the body in the garden unfolding to keep the reader's interest.

Having a body naturally results in the need for a coffin and a burial, enabling me to promote the idea of green funerals. An inquest into the death provides a natural opportunity for a witness from the Natural Death Center to talk about the environmental benefits of shallow burial in a willow coffin, and avoiding toxic embalming fluids. Following the rule of "show, don't tell," this is followed up by a natural burial scene, so we experience an alternative approach for ourselves:

> The intermittent sound of fiddles gave way to a proper tune, and gradually the chatter subsided and everyone looked towards Andrew and Katie. They brought their fiddling to a graceful close and we stood before the curtain of flowers and willow. Brian's voice still taking to Paul was discordant in the sudden quiet, and he bumbled to a halt.
>
> The music of the garden took over from the fiddles. Undeterred by the crowd, a tiny brown wren, tail cocked in the air, trilled its liquid song from the new willow fence. Nearby, a chiff-chaff chanted the repetitive call of its name. A queen bumblebee burred, her legs loaded with balls of pollen for her hungry offspring. A brimstone butterfly fluttered by, investigating the flowers on the willow bower, its bright yellow wings a flash of sunshine.
>
> A roar of a plane flying overhead reminded us that we weren't in the deep countryside, but in a suburban small garden, underneath the flight path from the airport a few miles down the road. When the plane had passed, Fern nodded at Andrew and Daisy and together they carefully lifted the willow curtain down from the branches and walked it to the end of the garden.
>
> I steeled myself to look. But it wasn't the deep, dark, rectangular coffin-shaped hole I'd pictured in my head. The hole in the ground was just as I'd left it, pond-shaped and three feet deep, except now Grandad, as I thought of him, was laid out in his baggy trousers and a colourful knitted jumper in the willow coffin, surrounded by the bones of his wife. I exhaled with relief. This was absolutely right. The shallow pond-shaped hole was like nature's opening arms welcoming them back to the earth.

Another climate change solution I wanted to promote was food choices, especially the use of insects as a greener alternative to meat. The main barriers here are negative attitudes toward insects as food, so simply raising the idea makes a start toward normalizing the concept. I used the device of the comic side kick to develop a sub plot. Jo, the lead character's best friend, devises a recipe generator that randomly picks seasonal ingredients to create recipes. Most challenging is the joker column which can include anything from edible insects to nettles. This provides humor and the opportunity to promote sustainable food choices. Fiction is now driving reality as a student inspired by the idea is working on making it a reality (see random recipe generator at Baden, n.d.).

Some sections focused on tapping into the normative drivers of behavior. The hero Tim leaves his job in finance in the city where the culture is one of competition and material values, because they turn down his proposal for costing for nature accounting system. He sets up as a wildlife gardening consultant—Habitat Man—and in the process comes across a variety of other characters who all represent different motivations toward sustainable behaviors. One character is a Buddhist who rejects material values and another is a pagan—the Wizard of Woolston who wants a habitat for bat and frogs. This allows me to show the overlap between the love of nature, fundamental to many pagans and alternative religions, and practices that promote biodiversity. Tim also visits the back garden of a psychologist who researches in trends toward non-consumption, which provides an opportunity to suggest that conspicuous consumption is out of date.

> I turned to Eric. 'How does your neighbour like your garden?'
> 'Now they want to copy us.'
> 'It proves my research, I think.' Eleanor set down a bowl of steaming pasta in front of me.
> 'What's your research?' Lori asked.
> 'It was psychology of consumption, wasn't it?' I said.
> 'Now I look at the new tendency to non-consumption.'
> 'Like the Share Shop?'
> 'Not quite. For example, our neighbour tells me they also want upcycled decking, but before, she'd insist on new. And it's because the meaning has changed you see. The story behind new decking might be rainforest destruction, air pollution, climate change, it's not a nice story, but the story behind your decking is a lovely story. We don't just buy the product, we buy the story.'

The above extracts all show in various ways how the drivers of behavior, as delineated by Ajzen's Theory of Planned Behavior (1985): attitudes (informed by awareness), social norms, and perceived behavioral control, were utilized to inspire greener behaviors in readers.

Rational Choice Model

Another common model of behavior is the rational choice model (Becker, 1976), which posits that people will make decisions based upon the costs and benefits to themselves. Thus, in *Habitat Man*, I often chose to emphasize the pleasurable aspects of the desired action, rather than the green aspects. For example, ponds are one of the most effective ways to increase biodiversity in back gardens, but instead of presenting it in those terms, which may make readers think they *should* install a pond for the greater good, I include a scene that allows readers to experience the delights of a pond through a child's eyes. To keep readers hooked, the back story about a lost child that has been hinted at throughout the book is revealed in more detail.

Similarly, Tim, aka Habitat Man, visits a family whose members are very proud of their composting toilet and keen to highlight the green benefits, but in the next extract, the focus is on the more pleasurable aspects, and again readers are kept engaged by the gradual revealing of the back story of the child. When asked if he has children, Tim finds an excuse to leave by asking to use the composting toilet the family has in their garden. This allows the reader to experience a composting toilet vicariously through the character. The extreme decline of amphibians is hinted at rather than stressed:

> 'Erm…' I put the wine down and shook my head. 'I think I will have a go at your composting toilet.'
> Eric smiled delighted. The boys jumped up to follow me, but he held them back.
> 'Leave him in peace. It's his first time.'
> I left the sounds of chattering children and walked outside. A few slow soft drops of rain remained, then petered out when the sun emerged, setting the raindrops sparkling against the vegetation. I walked down the garden to the hut and went in. I sat down. It was perfectly quiet except for the distant sound of a wood pigeon. It smelled of forests and fresh air. The feeling of calm and sanctuary in the toilet reminded me of Daisy's garden. That sense of perfect harmony between art and nature, soothing to the senses and the spirit. No harsh lights, whirr of fans, smell of urine overlaid with air freshener. Instead, daylight streamed in through the small window, which I now saw had a picture set into the glass, a frog on a lily pad amidst dragonflies and bulrushes. The sun caught the stained-glass window and brought the scene suddenly to life creating an almost religious experience. The elusive frog so sensitive to water pollution, safe here where our waste was used to nurture life. A benign, quiet smell of wood shavings. I heard the characteristic chirp of a grasshopper and smiled. Danny would love that, and the pond too.
> This time, I didn't push the memories away. I thought back to the decision I'd faced, try again to see him and risk a jail sentence and destabilising a new family, or walk away. In the sanctuary of the composting toilet, at last I forgave myself.

I must admit, I wondered if I'd gone too far when I read the last line in the following review on Amazon:

> What I loved most about Habitat Man was the gentle, non-preachy way in which it nudges the reader to think green and love nature, whilst being sucked into an enthralling drama. Yes, there are 'radical lifestyle changes for the benefit of all' messages, but they are peeking out from behind the willow hedges, rather than stuffed down the reader's throat. Emerging from the pages is a satisfying triumph of good over evil, of love over anger, and of green solutions over wanton consumerism. There's also a slightly weird fixation with pooing outside!

Jeremy Brown

Backstory of the Renegades: Defenders of the Planet

The basic job description of most superheroes is to "save the world," but in the age of the climate crisis, who on Earth is going to "save the planet"? This is a question that's been at the front of my mind since 2017, when a friend shared her idea of developing a comic book about climate change, out of her passion for the Marvel universe. I jumped at the earliest opportunity to bring the idea to life, because it strongly resonated with my love of art and undergraduate learning about the power of storytelling.

In 2018, I started sketching ideas for the plot and the three main characters—Katelyn, Leon, and Mo, otherwise known as "The Renegades." But it soon became obvious that the comic would only become a reality with the help of professional artists, scriptwriters, and friends. Thankfully that's where Katy, David, Libby, Ellenor, Jonny, and others came to the rescue, and within 18 months, we had completed three 96-page graphic novels in *The Renegades* series—all edited, printed, and sold by our publishers at Dorling Kindersley (Fig. 2).

In the following section, I draw on both my personal story of producing *The Renegades: Defenders of the Planet* (Brown et al., 2020, 2021a, 2021b), and learning from my academic studies, to offer the following three recommendations—firstly, to design characters that model both achievable and aspirational behaviors; secondly, to break outside of our echo chambers; thirdly, to play around with careful metaphors. My advice here is intended for any creative writers that plan to enter the exciting universe of climate fiction (in general). But as I'm a PhD student specializing in comics, that's where I will begin my discussion.

The Role of Cartoons, Comics, and Comedy

I went down the route of comics and graphic novels because their visual nature strongly appealed to my artistic side—especially having enjoyed painting and drawing as a teenager. But as revealed in the word "comic" itself, part of what sets apart

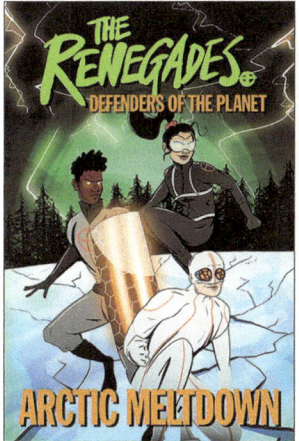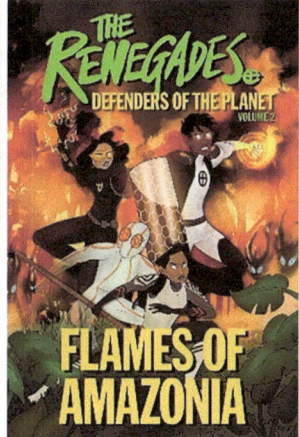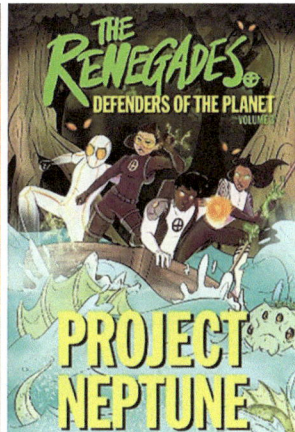

Fig. 2 The three graphic novels in the series, *The Renegades: Defenders of the Planet* (Brown et al., 2020, 2021a, 2021b). © 2020–21 Jeremy Brown, and Dorling Kindersley Limited

this form of storytelling from others, such as photography, is their humor (Bouchard et al., 2018). The academic literature similarly underlines the psychological importance of comedy, especially among children, with one study illustrating how humorous cartoons about earthquake risks eased children's fears of disaster in Iran (Sharpe & Izadkhah, 2014). However, the relevance of comic books even holds among young adults, with evidence from the USA revealing that 23% of 18- to 34-year-olds were reading comic books once a month (or more) during 2018, while 48% had done so at least "occasionally" (Yougov, 2018; cited by Statistica, 2021).

Superhero movies often enjoy a larger audience than printed comics, even across several age groups (National Tracking Poll, 2019; cited by Statistica, 2021b). But many films (from *Captain America* to *Spiderman*) were originally adapted from comics, and compared to films with actors, comics have the ability to imagine endless possibilities, without breaking the bank. Similar is true for animated movies and TV shows: *Not bounded by the physical laws governing 3-D space, animated programs can feature casts of hundreds and take place in any geographic or historical time frame* (Arnold et al., 2004, p. xiii). One of the most notable examples is *Captain Planet*—an animated TV series from the 1990s, starring a fun superhero created to inspire kids to go green. The story follows the adventures of five young "Planeteers," recruited from across five continents to save the Earth from destruction. To aid their mission, the Spirit of Gaia endows the Planeteers with the powers of Water, Fire, Earth, Wind, and Heart. When combined, they usher the arrival of the Captain Planet himself, who aids their fight against polluting villains (TBS 1990, cited by King, 1994).

On reflection, engaging with comedy and superhero stories would have been a much healthier way to cope with a traumatic period of eco-anxiety that I experienced in 2018. It was partly triggered by watching *Before the Flood*—a gripping documentary emphasizing the serious moral injustices of the climate crisis. Within my mind, I've now moved on from the worst of the eco-anxiety, especially since I'm less exposed to the dangers of extreme weather than many communities on the frontline. But in my body, I can still feel the side effects of the trauma, even as I write this paragraph. My story was complicated by several stressful mistakes I made at the time, but I can only wonder whether some of the trauma would have been avoided if instead I'd watched cli-fi comedies, such as *Don't Look Up*. With hindsight, it also would have been much better if my first instinct had been to reach out to trusted friends, like-minded activists, and professional therapists. And if you are experiencing eco-anxiety yourself, I would especially recommend the resources provided by the Climate Psychology Alliance (2022). But in my own case, I'm thankful that I eventually managed to channel my difficult experience into writing about eco-anxiety in the comic series. It was a therapeutic process that gave meaning to tough times.

In the next section, I will focus on a series of top tips from my comic experience that are relatable to climate fiction in general.

Tip 1: Model Actions That Are Both Achievable and Aspirational

When writing *The Renegades: Arctic Meltdown* (Brown et al., 2020), one of the environmental dilemmas we faced was about how our comic heroes would travel between the various and distant ends of the Earth where our stories were set: Eastern Africa, Bangladesh, Brazil, Texas, Britain, and the Canadian Arctic. In a typical twentieth-century story, we might expect the characters to catch a plane. But to be true role models for the twenty-first century, we felt it was important to invent an alternative mode of transport. Initially, we came up with the idea of a solar-powered sailing boat (Fig. 3). While this was suitably sustainable, it didn't resolve the "need for speed" demanded by our heroes when suddenly alerted to the far-off locations of our villains and monsters. As a result, the team later came up with the futuristic gizmo of a solar-powered "jet." (I like to think it's what Luke Skywalker would fly if he was watching his carbon footprint!).

This raises the question of whether our fictional solar jet has potential to inspire something similar in real life. The paper by De Meyer et al. (2020) underlines the psychological importance of featuring characters that act out behaviors that are easily "doable" or "meaningful" to readers. Otherwise, readers are often left confused over what they can specifically do about the climate crisis themselves. This finding is reinforced by Baden's (2019) experiments which show that positive solutions that can be easily replicated by readers tend to be more effective at producing eco-friendly intentions. Unsurprisingly, none of our readers have so far told us they plan to buy a multi-million pound electric aircraft! But as social scientists and authors, do we also need to consider the potential for a "legacy" in the long term?

Fig. 3 The solar sailing ship in *Arctic Meltdown* (Brown et al., 2020)

Aspirational Legacy

The importance of legacy is clear in the pioneering works of early science fiction. For instance, Hunley (1995) observes that at the age of 17, Robert Goddard was resting in a cherry tree at home while reading the *Journey from the Earth to the Moon* by Jules Verne and *War of the Worlds* by H.G. Wells. These stories stirred Goddard to imagine a technology that would have potential to travel to Mars—a highly enlightening experience for Goddard, which he commemorated by returning to the tree on the anniversary of his revelation (Hunley, 1995). As an adult, Goddard eventually went on to become the pioneer of the liquid-propelled rocket.

Some scholars of eco-psychology could argue that, strictly speaking, over 99% of readers of Jules Verne and H.G. Wells were *not* inspired to invent a pioneering technology. But it is unlikely their true aim was to motivate all readers to become world-class engineers themselves. Perhaps a more universal objective was to bring about a more widespread change in attitudes about risky technologies. Indeed, James (2016) remarks that Wells managed to foretell the future of atomic weapons, planes, and tanks, thereby harnessing the power of creative writing to help either bring about a better world or warn of potential dystopias. Wells was also a highly political author (James, 2016), unafraid to draw parallels between the Martian invaders and the ecocide and racism of imperial expansion:

> We men, the creatures who inhabit this Earth, must be to them [The Martians] at least as alien and lowly as are the monkeys and lemurs to us... And before we judge them too harshly, we must remember what ruthless and utter destruction our own species has wrought, not only upon animals, such as the vanished bison and dodo, but upon its own inferior races (Wells, 2003, p. 42–43).

Wells' efforts to influence the "attitudes" of readers may have been a necessary first step toward real action such as readers' actions to resist racism and dystopian technologies. Following in the tradition of Wells, our *Arctic Meltdown* comic intended to raise awareness of the threat of methane from thawing permafrost, at the same time as modeling radical political behavior that could act to reduce the threat. In all honesty, most of our readers (especially children) are unlikely to take part in non-violent direct action as a result of our comic's story—in this case, a scene where the protagonists occupied the villain's airport, inspired by the real activists of the Zone à Defendre, near Nantes in France (Jordan, 2019).

Achievable Behaviors

Yet we still attempted to balance this radicalism with general climate solutions which can be more easily imitated by the majority of teen readers, for example, fictional solar gadgets that could, in theory, inspire readers to campaign for their schools to install solar cells on their rooftops. It is likely that modeling these types of solutions would be perceived to be more "doable." "Doable" in this context, equates to Ajzen's concept of perceived behavioral control or Bandura's concept of self-efficacy as discussed earlier. Thus, presenting solutions that are "doable" would be likely to have a greater impact on readers' behavior.

Tip 2: Break Free from the "Echo Chamber"

When the time came to plan our comic's sequel (*Flames of Amazonia)* (Brown et al., 2021a), I was keen to feature characters that modeled the benefits of plant-based diets. This would come naturally to the character of Leon, our vegan climate activist. But to be more easily achievable to most readers, we suggested that not everyone has to strictly follow Leon's example of cutting out animal products completely. In another words, eating the occasional cheese omelet would not threaten the Amazon, as long as our heroes (and readers) stayed clear of beefburgers that originated from deforested land.

At the same time, I became increasingly conscious of the risks of demonizing low-income workers in the global cattle industry, particularly after listening to a lecturer's (at King's College London) warning of the perils of political polarization. Indeed, it is very difficult to escape the reality of the so-called "Culture Wars" that have erupted since 2016. As a result, I invented the character of Jack (Fig. 4)—a rugged cattle rancher from rural Texas—whom we presented not as a villain but as a hero. There was of course a need for the other characters to question Jack's choices, to the point of eventually persuading him to devote at least part of his

Fig. 4 Jack, the cattle rancher character in *Flames of Amazonia* (Brown et al., 2021b)

farmland to growing a plant-based alternative. But the point is that Jack became "part of the team," as well as being someone who could challenge Leon whenever he verged into being too sanctimonious.

After publication, I was grateful to see that several (but not all) of the online reviewers agreed that we had succeeded in our efforts to prevent the narrative from being too "preachy." Importantly, I suspect this resulted from how we had reached out to people outside of our echo chambers. In our case, this meant having conversations on Zoom with agricultural scientists from the typically Republican state of Texas. These conversations were particularly fruitful in the task of fleshing out the character of Jack—since they were people that had actually spent time with cattle ranchers in real life. Ideally, we would have likewise had video calls with real communities that had inspired the character of Alma (a teenage activist from an indigenous community living in the Amazon rainforest). But this proved to be too difficult within the constraints of our tight deadline. Conducting other forms of online research was therefore our best alternative effort at breaking free from another dominant echo chamber that cli-fi writers need to be aware of: the highly industrialized world of the Global North.

Tip 3: Play Around with Careful Metaphors

My final nugget of wisdom is around how best to balance entertaining metaphors with educational science. Here I speak again from experience, because when embarking on *The Renegades* comics, I was all too aware that I came from a scientific background. I have a creative side; but my earlier years were focused more on landscape painting and cartoon drawings, rather than creative writing. Therefore, I'm guessing the honest advice of anyone reading my first attempts at fictional storylines would have been "don't give up the day job, mate." Except I had already given up my day job! (A long story for another time.) But thankfully Katy and David soon came to the rescue. Essentially, I recruited them to bring to life my initial ideas for fictional characters and sketchy story arcs, with the end product being a professional script that could be shared with the publishers.

As well as adding emotional and literary depth to the fictional characters, the rest of the team's input as experienced creative writers was crucial to introducing vivid and engaging metaphors. For example, in our third comic in the series, *Project Neptune* (Brown et al., 2021b), Katy suggested portraying the melting ice as a metaphor for the tragic disintegration of her parents' health—a symbolic representation for how personal grief can often feel as overwhelming as a melting glacier crashing into the ocean. Another example is how we depicted the saltwater trees of the Bangladesh coast as "mangrove creatures" that could breathe, crawl, swim, or fly, as mythical animals. Beyond the benefit of being more visually playful, this metaphor meant that the gradual process of mangrove conservation could instead happen at the dramatic pace of the narrative. The consequences of the heroes' actions could therefore play out within days or hours, at a speed fast enough to chase the villains.

However, such fictional 'shortcuts' and metaphors can dilute the educational messages about how the audience can act themselves, so there may be trade-offs between exciting fantasy and effectiveness at inspiring relevant environmental action. This was evident in *The Day After Tomorrow*, in which the film dramatically exaggerated the speed at which the climate crisis could disrupt the world's ocean currents, culminating in a thick ice sheet enveloping the Statue of Liberty. The result was that several viewers encountered problems in separating science from fiction. As expressed by one participant of an academic study: "you didn't know exactly where the truth ended and Hollywood started" (Lowe et al., 2006, p. 447).

We faced a similar dilemma with *The Renegades* in how our villains played an exaggerated role in the physics of coastal flooding. But to minimize any confusion between real science and fantasy fiction, we included two pages of non-fiction at the end of the book. These pages gave us the space to explain certain scientific details about sea level rise. This was important because the enjoyment of the narrative would have been diluted if we had crammed too much science into the story. Another tactic was to harness our book launch events as opportunities to explain real-life solutions to climate change in more detail. This was challenging for our launch of *Project Neptune*, given that Southampton isn't exactly well known for its tropical mangrove forests! But linking back to my first tip, I still feel there is potential to explain how the mangrove creatures can be metaphors for achievable solutions on a local level, such as the protection of saltmarshes in the UK.

To sum up, I'd recommend to any writers of climate fiction that you feature entertaining role models that act out solutions that are sometimes metaphorical and always aspirational. But it's important that your protagonists balance their heroic aspiration with solutions that are still achievable to readers of many different cultural and political identities. I hope the experience and knowledge I have shared here has been useful. Good luck with all your writing adventures—our planet needs all the help it can get!

Conclusion

Our work approached the topic of the climate through different genres and styles, yet both adhered to the guiding principles that fiction should demonstrate the risks of climate disaster, while highlighting the possibility of hope and presenting solutions. Progressive climate fiction highlights both radical actions individuals can take, such as non-violent direct action, alongside more gentle climate activism such as sustainable gardening.

Appendix 1 lists some positive examples of climate fiction and Table 2 demonstrates some of the variety of approaches that can be undertaken.

Both *Ministry for the Future* and *The Carbon Diaries* are hybrid, showing a mixture of policy solutions and disasters, yet both finish on a note of hope. *The Carbon Diaries* has a happy ending on the personal front in the romance tradition, yet the climate issues remain. It also shows that the carbon rationing introduced

Table 2 Mapping examples of climate fiction onto a framework of suspense and agency

	Gentle agency (e.g., sustainable gardening)	Radical agency (e.g., non-violent direct action)
Medium/low suspense	*Habitat Man* by Denise Baden *Beauty and the Bin* by Joanne O'Connell *Amara and the Bats* by Emma Reynolds *No More Fairy Tales: Stories to save our Planet* edited by D.A. Baden	*Burning Sunlight* by Anthea Simons *The Stone Wētā* by Octavia Cade *Visco by David Fell*
High suspense (risk of a climate disaster, but with happy ending)	*Captain Planet* by Barbara Pyle and Ted Turner *The Last Bear* by Hannah gold *A Psalm for the Wild-Built* by Becky Chambers *The Carbon Diaries* by Saci Lloyd	*Green Rising* by Lauren James *The Renegades* comics by Jeremy Brown and team *The Ministry for the Future* by Kim Stanley Robinson *Fairhaven by Steve Willis and Jan Lee*
Dystopian endings (solutions that fail to prevent disaster)	Take extra care when writing these types of stories!	

suddenly and drastically on an unprepared population in the story does operate as a solution, but also implicitly, one can see that personal carbon allowances, if applied earlier, might have prevented the worst of the disaster at a fraction of the cost.

In his Ph.D., Jeremy Brown is planning to apply these insights by exploring how comic style stories may have the potential to entertain and empower young people to imagine a brighter future for our planet. In doing so, the hope is that the audience will be inspired to imitate fictional superheroes that model a range of both gentle and radical responses to the climate crisis.

New, developing sub-genres such as "solarpunk" and "hopepunk" lead the way in this effort to resist a dystopian ending to our planet's story. Hopepunk was defined in 2017 by fantasy author Alexandra Rowland as, "standing up and fighting for what you believe in. It's about DEMANDING a better, kinder world, and truly believing that we can get there if we care about each other as hard as we possibly can, with every drop of power in our little hearts." We believe this emerging type of storytelling has an exciting future ahead.

Green Stories Project, Denise Baden

In 2018, I set up the Green Stories Writing competitions (Green Stories Project, n.d.). These are a free series of competitions that anyone can enter and have covered a range of formats from stage plays and radio plays, to flash fiction, screenplays to full-length novels, even interactive fiction (Green Stories Previous Competition Successes, n.d.). The aim is to create a cultural body of work that would engage the mainstream audience, not just in green issues, but informing them and hopefully inspiring them to adopt the kinds of green solutions that would be effective.

On the website, we present transformative solutions that could enable a genuine shift toward a truly sustainable society (Green Stories Project, n.d.). For example, changing our metric of success from the GDP, which measures mostly consumption, to a metric, such as the well-being index or Happy Planet Index, would shift attention toward sustainable policies rather than short-term economic gains at the expense of a healthy eco-system (Hoekstra, 2019). Similarly, shifting from a culture of ownership—buy, use, dispose—to an access-based culture or sharing economy allows greater equity and access to resources at much reduced resource costs and planetary impact (Baden & Frei, 2021). For example, almost all of what we own isn't being used. It's estimated for example that the typical drill is used for an average of 8 minutes a year (Skjelviket al., 2017). A library of things or shared shed would allow everyone access at much reduced environmental impact (Baden, Peattie, & Adekunle, 2019). Similarly, ideas such as personal carbon allowances or carbon credit cards would drive sustainability innovations across the board and enable a more equitable and rapid transition to a net zero economy (Fuso Nerini, Fawcett, Parag, & Ekins, 2021).

Twenty competitions have been run since its inception with thousands of writers entering and engaging in these solutions, so the green stories competitions were successful on that front, resulting in several publications: *Visco* by David Fell (2022) and *Fairhaven* by Steve Willis and Jan Lee (2024). However, they have been less successful in terms of authors writing these green solutions into their stories. Inevitably free competitions will attract many poor entries, but it was disappointing that not even ten percent met the green stories criteria:

1. Story telling ability—the story should be engaging
2. Incorporate green solutions, e.g., by

   – Showcasing what a green/sustainable society might look like and/or how we might get there
   – Smuggling green solutions into a story aimed at mainstream readers, which can be quite subtle
   – Using positive role models

We found that even among the well-written stories, many had green themes but few included solutions that readers could engage with. For example, several showed a heroine heading off to the rainforests to take on evil loggers. However, this is an issue that most are already aware of. On its own, awareness does nothing to address the problem. Much more effective, I believe, would be a story that did not have an explicitly green theme but did show positive role models of characters enjoying plant-based diets, or upcycling, or campaigning against companies that engaged in deforestation etc.—behaviors that will make a difference.

Social marketing approaches aim to apply marketing strategies to promote desired behaviors rather than products (Lee & Kotler, 2019). Thus, just as some books/films product place products, we encourage writers to "product place" sustainable attitudes behaviors products and policies. It may be more effective at engaging a mainstream audience if the story *isn't* specifically about climate change. Any kind of genre—rom com, crime drama, legal drama, children's book, sci-fi, etc.—can showcase sustainable technologies, practices, products, or ideas in the background. BAFTA's "Albert" project trains production companies' editorial staff on how to do this through "planet placement."

We incorporated this technique in our anthology, *No More Fairy Tales: Stories to Save Our Planet,* where we teamed experienced writers with climate experts to create 24 short stories each with climate solutions at their heart. One story—a whodunnit—set in a citizens' assembly has been adapted as an interactive play *Murder in the Citizens' Jury*, which engages the audience in choosing what should be done (see Green Stories Project, n.d.).

Another acceptable approach could be to focus on characters. Currently, characters in fiction who are green/ethical are often portrayed as priggish or aggressive, and we'd like to see attractive characters behaving in sustainable ways. Green Stories partnered with BAFTA for the #ClimateCharacters, which is a series of fun social media posts comparing the carbon consumption of various fictional characters. For example, James Bond with his walk-in wardrobe of luxury suits and single-use sports cars kills bad guys with a much higher carbon footprint than Jack Reacher who travels by bus and shops in thrift shops! (See website Green Stories Project, n.d.). We want to start a conversation that asks if it is okay, in a time of climate and biodiversity crisis, that many of the top series and films have characters whose lifestyles are destroying our beautiful planet?

Check out upcoming competitions, projects, and events via the website https://www.greenstories.org.uk/

Appendix 1

Online Resources

Green Stories Project. https://www.greenstories.org.uk/
Role modeling. https://theconversation.com/positive-fictional-role-models-the-missing-ingredient-to-fight-the-climate-crisis-177684
Heated. Substack newsletter. https://heated.world/
Inkcap Journal. https://www.inkcapjournal.co.uk/
Lights On newsletter. https://lightson.substack.com/
Down to Earth by The Guardian newsletter. https://www.theguardian.com/info/ng-interactive/2021/oct/20/sign-up-for-down-to-earth-the-best-way-to-make-sense-of-the-biggest-environment-stories
How to Save a Planet podcast. https://gimletmedia.com/shows/howtosaveaplanet

Recommendations for Positive Climate Fiction

Adult Fiction

Baden, D.A (2021). *Habitat Man.* Habitat Press.
Chambers, B. (2021). *A Psalm for the Wild-Built.* Tor.
Cade, O. (2020). *The Stone Wētā.* Paper Road Press.
Fell, D. (2022). *Visco.* Habitat Press.
Wigmore, R. (2021). *Foxhunt.* Queen of Swords.
Walker, L. (2018). *Melt.* Lacuna Publishing.
Miller, S. (2018). *Blackfish city.* Ecco.
Woodrow, L. (2020). *470.* Melliodora Publishing.
Willis, S. and Lee, J. (2024). *Fairhaven.* Habitat Press.

Childrens Fiction

James, L. (2021). *Green Rising.* Walker Books
O'Connell, J. (2020). *Beauty and the Bin.* Pan Macmillan.
Reynolds, E. (2021). *Amara and the Bats.* Simon & Schuster.
Gold, H. (2021). *The Last Bear.* HarperCollins.
Miyazaki, H. (1996), *Princess Mononoke.* Studio Ghibli
Sutcliffe, W. (2021). *The Summer we Turned Green.* Bloomsbury.
Smith, M. (2021). *If Not Us.* Text Publishing.
TBS Productions, Inc. and DIC Enterprises, Inc. (1990). *Captain planet and the planeteers: The power is yours.* Public Relations kit.
Okogwu, T. (2022). *Onyeka and the Academy of the Sun.* Simon & Schuster.
Vachharajani, B. (2019). *A Cloud called Bhura: Climate Champions to the Rescue.* Speaking Tiger.
Simmons, A. (2021). *Burning Sunlight.* Anderson Press.

References

Ajzen, I. (1985). From intentions to actions: A theory of planned behavior. In J. Kuhl & J. Beckmann (Eds.), *Action control: From cognition to behavior* (pp. 11–39). Springer.
Arnold, D. L., Koeingsberger, K. M., Chow, V. W., Broderick, M., Mullen, M., Rushkoff, D., et al. (2004). *Leaving Springfield: The Simpsons and the possibility of oppositional culture.* Wayne State University Press.
Baden, D. A. (n.d.). *Random Recipe generator.* https://www.dabaden.com/habitat-man/random-recipe-generator/
Baden, D. (2019). Solution focused stories are more effective than catastrophic stories in motivating pro-environmental intentions. *Ecopsychology, 11*(4), 254–263.
Baden, D., & Frei, R. (2021). Product returns: An opportunity to shift towards an access-based economy? *Sustainability, 14*(1), 410.

Baden, D., Peattie, K., & Adekunle, O. (2019). *Sustainable business model innovation: The potential of libraries of things*. Paper presented at the EURAM, Lisbon.

Bandura, A. (1997). *Self-efficacy: The exercise of control*. W.H. Freeman.

Becker, G. S. (1976). *The economic approach to human behavior*. University of Chicago Press.

Bouchard, F., Sansoulet, J., Fritz, M., Malenfant-Lepage, J., Nieuwendam, A., Paquette, M., & Tanski, G. (2018). "Frozen-Ground Cartoons": Permafrost comics as an innovative tool for polar outreach, education, and engagement. *Polar Record, 54*(5–6), 366–372.

Brown, J. D., Jakeway, K., Mererid-Jones, E., Reed, E., & Selby, D. (2020). *The Renegades: Arctic Meltdown*. Dorling Kindersley. https://www.waterstones.com/book/the-renegades/jeremy-brown/katy-jakeway/9780241457832. Accessed 21/12/21.

Brown, J. D., Jakeway, K., Mererid-Jones, E., Reed, E., & Selby, D. (2021a). *The Renegades: Flames of Amazonia*. Dorling Kindersley. www.dk.com/uk/book/9780241490662-the-renegades-flames-of-amazonia/. Accessed 07/01/22.

Brown, J. D., Jakeway, K., Mererid-Jones, E., Reed, E., & Selby, D. (2021b). *The Renegades: Project Neptune*. Dorling Kindersley. www.dk.com/uk/book/9780241535356-the-renegades-project-neptune/. Accessed 07/01/22.

Climate Psychology Alliance. (2022), https://www.climatepsychologyalliance.org. Accessed 12/01/22.

De Meyer, K., Coren, E., McCaffrey, M., & Slean, C. (2020). Transforming the stories we tell about climate change: From 'issue' to 'action'. *Environmental Research Letters, 16*(1), 015002.

Fell, D. (2022) Visco. Habitat Press.

Fuso Nerini, F., Fawcett, T., Parag, Y., & Ekins, P. (2021). Personal carbon allowances revisited. *Nature Sustainability, 4*(12), 1025–1031.

Green Stories Previous Competition Successes (n.d.). https://www.greenstories.org.uk/previous-competitions

Green Stories Project. (n.d.). https://www.greenstories.org.uk

Green Stories Project. (n.d.). *Story ideas and resources*. https://www.greenstories.org.uk/story-ideasresources

Green Stories Project (n.d.). *Theatre in Education*. https://www.greenstories.org.uk/theatre-in-education

Green Stories Project (n.d.) *Climate Characters*. https://www.greenstories.org.uk/climatecharacters

Hoekstra, R. (2019). *Replacing GDP by 2030: Towards a common language for the well-being and sustainability community*. Cambridge University Press.

Hunley, J. (1995). The enigma of Robert H. Goddard. *Technology and Culture, 36*(2), 327–350.

James, S. J. (2016). Science journals: The worlds of HG wells. *Nature, 537*(7619), 162–164.

Jordan, J. (2019). *For the love of winning: An open letter to extinction rebellion*. Roar, https://www.roarmag.org/essays/open-letter-extinction-rebellion-zad/. Accessed 07/01/22.

Lee, N. R., & Kotler, P. (2019). *Social marketing: Behavior change for social good*. Sage Publications.

Lowe, T., Brown, K., Dessai, S., de França Doria, M., Haynes, K., & Vincent, K. (2006). Does tomorrow ever come? Disaster narrative and public perceptions of climate change. *Public Understanding of Science, 15*(4), 435–457.

Miller, M. D., & Levine, T. R. (2019). Persuasion. In *An integrated approach to communication theory and research* (pp. 261–276). Routledge.

National Tracking Poll. (2019). https://morningconsult.com/wp-content/uploads/2019/02/190242_crosstabs_HOLLYWOOD_Adults_v3_ML.pdf. Accessed 29/10/21.

Schneider-Mayerson, M. (2017). Climate change fiction. In *American literature in transition: 2000–2010* (pp. 309–321). Cambridge University Press.

Schneider-Mayerson, M. (2018). The influence of climate fiction: An empirical survey of readers. *Environmental Humanities, 10*(2), 473–500.

Schneider-Mayerson, M., Gustafson, A., Leiserowitz, A., Goldberg, M. H., Rosenthal, S. A., & Ballew, M. (2020). Environmental literature as persuasion: An experimental test of the effects of reading climate fiction. *Environmental Communication, 17*, 1–16.

Sharpe, J., & Izadkhah, Y. O. (2014). Use of comic strips in teaching earthquakes to kindergarten children. *Disaster Prevention and Management, 23*(2), 138–156. https://doi.org/10.1108/dpm-05-2013-0083

Sheeran, P. (2002). Intention—behavior relations: A conceptual and empirical review. *European Review of Social Psychology, 12*(1), 1–36. https://doi.org/10.1080/14792772143000003

Sheeran, P., Trafimow, D., & Armitage, C. J. (2003). Predicting behaviour from perceived behavioural control: Tests of the accuracy assumption of the theory of planned behaviour. *British Journal of Social Psychology, 42*(3), 393–410. https://doi.org/10.1348/014466603322438224

Skjelvik, J. M., Erlandsen, A. M., & Haavardsholm, O. (2017). *Environmental impacts and potential of the sharing economy* (Vol. 2017554). Nordic Council of Ministers.

Statistica. (2021). *Frequency of reading comics in the U.S. 2018, by age group*, https://www.statista.com/statistics/943127/comic-book-reading-frequency-by-age-us/. Accessed 29/10/21.

Svoboda, M. (2016). Cli-fi on the screen (s): Patterns in the representations of climate change in fictional films. *Wiley Interdisciplinary Reviews: Climate Change, 7*(1), 43–64.

TBS Productions, Inc. and DIC Enterprises, Inc. (1990). *Captain planet and the planeteers: The power is yours*. Public Relations kit.

Trexler, A. (2015). *Anthropocene fictions: The novel in a time of climate change*. University of Virginia Press.

Willis, S. and Lee, J. (2024) Fairhaven. Habitat Press.

Wells, H. G. (Ed.). (2003). *The war of the worlds*. Broadview Press.

Yougov. (2018). *Stan Lee, Fieldwork Dates:13th – 14th November 2018*. today.yougov.com. Accessed 11/01/22.

Denise Baden   is a Professor of Sustainable Practice at the University of Southampton. Following a degree in Politics and Economics, she worked in the publishing industry in sales, then returned to academia in 2002 to obtain a Ph.D. in psychology. Since then, Denise has been a researcher and lecturer primarily in sustainable practice and business ethics. Her primary interest is in how to promote green practices to the mainstream. Denise set up the University of Southampton Green Group in 2001 and represents the sustainable implementation group at the University. Denise has published numerous publications in the academic realm on positive role models and the role of fiction in promoting pro-environmental behaviors. Denise set up the free series of Green Stories Writing competitions in 2018 which have run 20 free competitions. She has written three screenplays, two plays, one musical. Her first novel *Habitat Man* was published in Sept 2021, followed by an anthology—*No More Fairy Tales: Stories to Save Our Planet*. Denise is listed on the Forbes list of Climate Leaders Changing the Film and TV industry.

Jeremy Brown   is studying for a Ph.D. in climate communication under the supervision of Denise Baden. After an integrated master's in Environmental Sciences, he specialized in comic books while studying for an M.A. in "Climate Change: History, Society and Culture" at King's College London. While completing his M.A., Jeremy managed a team of artists and writers to create a series of superhero comics about the climate crisis—*The Renegades: Defenders of the Planet*. The three titles were published by Dorling Kindersly between autumn 2020 and 2021. His Ph.D., based at the University of Southampton, is currently focused on exploring the potential for climate fiction comics to inspire readers to imitate the fictional role models. Jeremy is also active within local politics and is keen to encourage climate activists to find creative ways to engage with the democratic process.

Visuals as a Catalyst for Climate Science Communication

Kalliopi Monoyios, Kirsten Carlson, Taina Litwak, Tania Marien, and Fiona Martin

If someone told you that your city produced 54 million tons of carbon dioxide gas each year in emissions, you might be impressed. But then you would realize you really have no idea what 54 million tons of gas looks like—is that a lot? A little? Is that concerning? And you're suddenly lost. In 2012, a data visualization firm called Real World Visuals and the Environmental Defense Fund recognized this as an opportunity to experiment with how they could make this very statistic, the annual CO_2 emissions for New York City, instantly understood by audiences "who don't know they need to know" (Real World Visuals, 2017).

Their brilliant solution was to translate the volume of a single metric ton of carbon dioxide gas into a large blue sphere measuring 33 feet across. They then animated the streets of New York as these spheres accumulate at the rate of one per

Kirsten Carlson, Taina Litwak, Tania Marien and Fiona Martin contributed equally with all other contributors.

K. Monoyios (✉)
Visible Science LLC, Denver, CO, USA
e-mail: studio@kalliopimonoyios.com

K. Carlson
Fathom It Studios, Stuttgart, Germany
e-mail: kc@kirstencarlson.net

T. Litwak
United States Department of Agriculture, Darnestown, MD, USA
e-mail: taina@litwakillustration.com

T. Marien
Talaterra Inc., Riverside, CA, USA
e-mail: tania@talaterra.com

F. Martin
Visualizing Science® LLC, Kirkland, WA, USA
e-mail: fiona@visualizingscience.com

E. Coren, H. Wang (eds.), *Storytelling to Accelerate Climate Solutions*,
https://doi.org/10.1007/978-3-031-54790-4_11

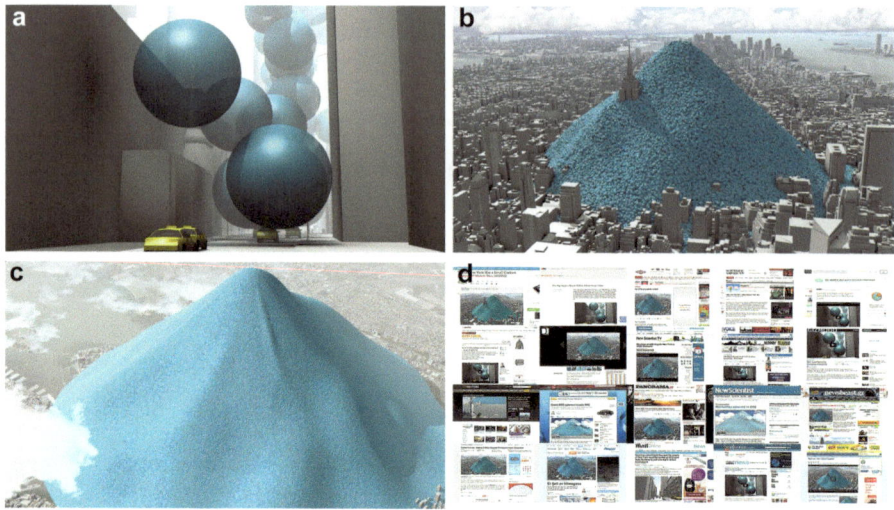

Fig. 1 Stills from the animated sequence "Carbon Emissions in New York in 2010" (courtesy of Real World Visuals). (**a**) Each 33-foot sphere represents 1 metric ton of CO_2 emissions. (**b**) Daily emissions. (**c**) Annual emissions. (**d**) Collage of some of the >100 articles and blogs featuring the images and animation (Real World Visuals, 2012). (Images used with permission)

0.58 s. A single day's emissions form a pile of spheres roughly the height of the Empire State Building. Eventually, 1 year's worth of carbon emissions forms an imposing mountain of blue balls covering Manhattan Island. Since its publication, the animation has been cited in over 100 articles and blogs (Fig. 1) and viewed over 400,000 times on YouTube (Real World Visuals, 2012).

Carbon Visuals and the Environmental Defense Fund could have simply shared the numbers with the world: 54,349,650 metric tons of carbon dioxide added to the atmosphere by New York City in 2010—that's 148,903 tons a day, 6204 tons an hour, and 1.72 tons a second. But they would have lost most of us who aren't accustomed to dealing with large numbers and/or who don't have a working concept of how much gas weighs (beyond "very little"). Their great insight was that we are visual animals. Making the invisible visible was all that was needed to get an important point across about the untenable rate of greenhouse gas emissions coming from New York. And we can look to the overwhelmingly positive media response and high engagement numbers on YouTube as evidence of the potential for great visuals to carry science messages far and wide (Real World Visuals, 2012).

The success of Carbon Visuals' animation was no accident. Their creatives have backgrounds in both art and science. They are able to delve into the details of the science without intimidation and then zoom out to apply their understanding of how people *see* things to create visuals that are intuitive and striking. They are quintessential examples of what we call *visual science communicators*.

In this chapter, we aim to draw back the curtain on what visual science communicators can bring to the table on climate change communications. We break down the many fronts on which visual collaborations can effectively push climate science forward. With visuals that have the power to transcend language and cultural barriers, learning differences, and knowledge or skill gaps (between highly specialized scientists and your average Weather Channel watcher, for example), we have the opportunity to broaden and deepen engagement as quickly as possible—improving understanding and increasing the rate of cultural uptake of mitigation and adaptation skills. We are here, with our sleeves rolled up, ready to work with scientists, policymakers, and communications teams on the critical work of climate communication.

From Science Illustration to Visual Science Communication

Science illustration is a natural collaboration between two disciplines—science and art—with common roots in observation. In the Age of Exploration, scientists routinely documented their discoveries, mapped out their knowledge, and drew elaborate illustrations of exotic species to share their stories with people back home. In the twentieth century, our increasingly specialized world siloed scientists apart from artists. Exacerbating this gap was fine art's turn away from realism toward abstraction and modernism in the nineteenth century where concepts and processes eclipsed technical skill (Remington & Pontis, 2021). Despite this increasing trend of siloed disciplines, an enthusiastic group of practitioners known as science illustrators persevered, bridging the gap between science and art.

Science illustrators comprise a specialized group with advanced training in both science and art. Their work has been, until recently, primarily descriptive with most of their efforts spent documenting the anatomy of plants and animals. Traditional science illustration work goes back well beyond the Renaissance and scientists/artists like Leonardo da Vinci (d. 1519), human anatomists like Andre Vesalius (d. 1564), and botanical artists like Leonhart Fuchs (d.1566), to the exacting botanical paintings of Xu Xi (China, d. 975) and the painters of the Tomb of Nebamun (Egypt, 1350 BCE). You'll recognize contemporary practitioners' work in medical settings (patient pamphlets and doctors' offices), in resources like birding books, specialty science journals, and popular science magazines.

As science advances into understanding more and more of what cannot be seen easily, or seen at all, traditional scientific illustration is changing too. It has become a valuable method of conveying complex concepts like minute cellular and molecular structures and biochemical mechanisms and processes (see, for example, Jennifer Fairman's spectacular depictions of SARS-CoV-2 done for the Johns Hopkins medical community at the height of the pandemic (Fairman Studios, 2020) and Mesa Schumacher's spread for National Geographic on viruses in general (Mesa Studios, 2021)). At the other extreme of the scale of scientific exploration, it is an invaluable tool for describing the vastness of space and our understanding of the physics of the universe (see space artists such as Mark Garlick (n.d.) and Lynette Cook (1998)). Of critical importance to us at this moment in history, it can also be used to illustrate

Fig. 2 In this sweeping depiction of how rising sea levels will affect the Red Sea coastline, visual science communicator Xavier Pita literally brings a solution-minded approach to the fore by illustrating two scenarios—how the coastline might change if we invest in infrastructure adaptations (*front*) and what we might face if we do nothing (*back*). In this way, Pita unpacks the suite of changes that may or may not unfold in the complex interplay of climate change and mitigation efforts. (Illustration by Xavier Pita/King Abdullah University of Science and Technology (KAUST, 2021). Image used with permission)

complex systems and how they work, as in Xavier Pita's masterfully executed tableau titled "Sea Change in the Red Sea; Adapting to Climate Change" (Fig. 2).

In these cases, what a select few scientists can view directly with highly specialized equipment or understand via complex calculations is made accessible to much larger audiences with images that feel intuitive and real. Creating these illustrations requires a significant understanding of the science and the ability to translate and crystallize complex information into an image that tells the story and can be readily understood.

Reflecting this change in how science has advanced, the skill set of science illustrators has also changed in stride to stay relevant. Consequently, many science illustrators have adopted the title "Visual Science Communicator" to more accurately reflect the broad array of visual art tools and science communication skills they have at their disposal. Visual science communicators typically have formal training in both science and art, which results in a unique combination of skills: a high level of understanding of the science and the ability to distill and communicate that information visually. They excel at telling engaging stories through well-designed graphics and illustrations that appeal to wide and varied audiences. These are the people who are critical to establishing a robust catalog of imagery that accurately and effectively conveys the seriousness of our climate situation and spurs policymakers and the public to act swiftly and decisively.

Why Are Visuals More Effective Than Words Alone?

A striking characteristic of human memory is that pictures are remembered better than words (Grady et al., 1998, p. 2703).

If a story does not sustain our attention, then the brain will look for something else more interesting to do (Zak, 2015, p. 6).

With the proliferation of the Internet, smartphones, and ubiquitous online publishing, information is more accessible than ever. Our attention has become a valuable commodity, with advertisers finding more ways to monopolize it and monetize it (Iyengar & Massey, 2018). As a result, accurate information has a tougher slog to break through the never-ending scrolling we engage in, and people increasingly ask the question "why should I care?" or "what can I possibly do about it?" when confronted with challenging scenarios. We need tools to compete effectively, engage viewers, and inspire action. Visuals—particularly those with a narrative bent that engage our emotions and trigger the neurotransmitter oxytocin—are a powerful tool to harness in this pursuit (Zak, 2015).

Research supports the assertion that visuals add to the impact of scientific endeavors by improving dissemination and deepening engagement (Ibrahim et al., 2017; Zaelzer, 2020). Aside from the anecdotal evidence members of the general public can easily conjure (e.g., school textbooks chock full of explanatory figures, museum displays employing heavy use of two-dimensional and three-dimensional visuals, and a nearly endless supply of social media feeds populated with eye-catching imagery), studies suggest that when put into the world with imagery, original scientific papers enjoy more engagement on platforms like Twitter than papers that are tweeted without (Ibrahim et al., 2017; Koo et al., 2019). Simply put, using effective imagery translates to *increased visibility* for scientists' research. The value in increased visibility need hardly be argued among researchers, but it begs the question: How does this ultimately translate to a better informed and engaged public, particularly on the vexing problem of implementing climate change solutions?

For one, images are processed differently than words and seem to engage the memory centers of the brain more directly (Grady et al., 1998). Furthermore, imagery that is novel and/or triggers emotions can heighten attention and memory in ways that words cannot (Grady et al., 1998; Zaelzer, 2020; Zak, 2015). After all, there's only one way to write the word "water" but near-infinite ways to depict it visually. In this way, images can serve as a universal communication tool reaching beyond language and education barriers (Zaelzer, 2020). They can communicate large data sets and real-world impacts "at a glance" (e.g., weather patterns), allowing easier comparison of findings. And they have the power, as with infographics, to mitigate information overload by visualizing findings and solutions in more succinct and appealing ways (Houser, 2020).

When paired with storytelling principles, images can invite viewers along for a ride, increasing their sense of empowerment and belonging. This often entails utilizing artistic techniques that evoke feelings in a viewer, ideally allowing the transfer of information in nonthreatening ways, such as by adding culturally sensitive, emotionally moving, and/or entertaining content (Hassenzahl et al., 2013; Houser, 2020). Neuroscientific studies have shown that areas of the brain involved in processing visual, esthetic, and emotional stimuli—the prefrontal cortex, orbitofrontal cortex, amygdala, and insula (the limbic system)—are closely linked to learning and memory, as well as decision-making under social pressure (Zaelzer, 2020). All of these aspects give visual science communication in government, NGOs, news media, pop culture, youth education, and informal learning environments an outsized role in educating the public and spurring societal change.

Invisible Visualizers?

Increasingly, there is a trend in academic circles to train scientists and educators to be better science communicators (Thorp, 2021). Organizations like the Alan Alda Center for Science Communication and conferences like ComSciCon do an admirable job of training thousands of scientists each year in this endeavor. Typically, these programs focus on written communication, but conversations about visual communication do occasionally seep into their curricula. This training for academics has great value, as scientists benefit from an increased understanding of the importance of communication in everything from grant writing to reporting results, but also an appreciation of the skill set that is required to be a good communicator. However, this cannot be a replacement for collaborating with trained science communicators.

As H. Holden Thorp points out in a recent editorial for *Science,* "not every scientist wants to take time away from research to be a voice for science… most scientists prefer to persuade by performing meticulous, credible work" (Thorp, 2021, para 3). Though he is referring specifically to science writers, the same holds true for visual science communication—expecting every scientist and researcher to also be a graphics expert is unrealistic. For most people then, teaming up with professionals who specialize in communicating with the public is the best use of their time and funds. The time is ripe to recognize that science research and science communication are different, equally skilled, technical specialties. We must build teams of professionals with the broad array of skills necessary to disseminate research, make it relevant, and make it actionable to critical masses of people around the globe. These teams absolutely should include specialists such as visual science communicators.

Astonishingly, an outsized portion of people looking to increase their reach via visual science communication (often referred to as "SciComm") have no idea where to find professionals in the field or how to go about productive collaborations. In a pinch, academics equip graduate students with Adobe Photoshop or Illustrator and ask them to create figures with little or no training in visual arts. Likewise, publishers may tap social media editors and graphic designers who may well have good design skills, but lack sufficient understanding of the science. Despite the rise in popularity of the "graphical abstract" in a number of peer-reviewed medical journals, numerous guidelines available to authors on the nuts and bolts of creating graphical abstracts fail to mention that there are visual science communicators trained to do this work. Contributing to the problem with advocating for the use of trained visual science communicators in science outreach efforts are the multiple terms we use to describe what we do: "science illustrator," "data visualizer," "information architect," etc., all describe various specialties that can all fall under the umbrella of visual science communication. Adopting the overarching umbrella term as an industry standard with subspecialties such as "data visualizer," "infographics specialist," etc., contained within will only help others conceptualize what we can offer.

Finding a visual science communications professional is easy if you know where to look. There are several professional associations of visual science communicators in the United States and Europe, each with hundreds of skilled and accomplished members. These include the Guild of Natural Science Illustrators (GNSI), the Association of Medical Illustrators (AMI), and the Association Européenne des Illustrateurs Médicaux et Scientifiques (AEIMS), among others. More specialized organizations cater to subfields such as the International Association of Astronomical Artists, the American Society of Botanical Artists, and the UK-based Society of Botanical Artists. There are also numerous small companies in the USA and elsewhere that specialize in science-based visual storytelling. Figures in published works often include a credit line or acknowledgment of the artist (or at least they should) allowing one to find reputable illustrators that way. Resources such as the AMI's Medical Illustration and Animation Sourcebook (2019) and GNSI's (n.d.) online gallery of practicing visual science communicators offer a convenient concentration of professionals who advertise their portfolios to potential clients in the science and medical illustration realms. Additionally, medical illustration programs and the numerous certificate programs in science illustration can serve as valuable resources (see the AMI's list of Graduate, Undergraduate and Certificate Programs for training medical illustrators at the Association of Medical Illustrators (2000) and the GNSI's expansive list including single courses and certificate programs for people interested in broader science illustration on their website's education page and via their monthly newsletter (2019)). Often, emailing the instructor of a science illustration program can open the door to many talented and eager professionals, some fresh out of school and others with years of experience.

Though we highly recommend seeking out visual science communicators to collaborate with, there are other models that exist. Recognizing the very real constraints of funding, companies like BioRender are attempting to create tools that assemble icons and spot images in automated ways to create figures for research papers and presentations—a sort of DIY tool for researchers on a budget. ClimateVisuals.org has built a curated library of photographs depicting positive and inspiring actions by communities worldwide as they adapt to their new climate realities. These photographs are available for use by climate communicators free of charge, many under Creative Commons licenses. But though these types of resources have real benefits, they do not yet approach the utility of having visual science communicators embedded in outreach and communications efforts across research, news media, public health, and educational institutions.

PART I: What Visual Science Communicators Bring to the SciComm Table

I work with some of the best science communicators in the world, and I see how hard they have endeavored to hone their craft. This is a profession and a full-time job - not something that can be picked up in a workshop. —H. Holden Thorp, Editor-in-Chief, Science journals (Thorp, 2021, para 4).

Drawing is really hard (Colbert & Munroe, 2014, 01:35).

Research Publications and Academic Training

There are many situations in which academics working on climate research could take better advantage of the power of visual science communication. Poster sessions—where research results are displayed visually with text and graphics—have been integral parts of scientific conferences for decades. Yet the concept of submitting a graphical abstract, also known as a visual abstract, to be shared in online settings where scientists convene is relatively new (Ibrahim et al., 2017). Despite the widespread adoption of graphical abstracts in at least 75 medical journals (Ramos & Concepcion, 2020), a quick survey of author guidelines for the top journals publishing *climate science* reveals no mention of graphical abstracts to promote engagement or dissemination of critical research.

In the current landscape, without access to staff visual science communicators, researchers' graduate students, lab techs, and postdocs—who are often not interested in or able to provide satisfactory graphics for their papers—may resort to repurposing old publication graphics. If a repurposed diagram is not laser-focused on communicating the author's intent, it fails to inform quickly and seamlessly and risks losing the viewer altogether. In teaching, this is particularly important, but nowhere is it more important than on social media, where scientific conversations and informal learning are increasingly happening (Ramos & Concepcion, 2020).

Ibrahim et al. (2017) found that adding quality graphics to Twitter posts about scientific research leads to higher impact factors for published papers and greater reach in the scientific community, the news media, and beyond. It follows that dedicating resources to at least one visual science communicator in an academic department, or even one position shared across closely related departments (where budget constraints are real), is well worth the investment.

External impacts are not the only benefits of working with a good visual science communicator. Working with someone who brings a different set of skills and perspectives to the workspace can allow researchers to see their work in new ways and even open new lines of inquiry (Schwabish, 2021). Good visual storytelling can make potentially inaccessible research sing, inspiring students and colleagues to pursue new lines of research that suddenly spark their interest. It can also reveal patterns in the data, clarify thinking, and/or expose areas in which further research is needed. In one example, Kalliopi Monoyios, working as a staff illustrator at the University of Chicago, was given video footage of African lungfish engaging in a walking behavior and was tasked with finding an elegant way to show the novel fin locomotion for publication in two-dimensional media. By illustrating stills from the video footage (Fig. 3), Monoyios created a stacked time-lapse depiction of the walking movement that highlighted pivot points around which the lungfish was propelling itself forward. This visual prompt in turn sparked new questions for the researcher as she pondered whether future experiments could quantify how much force was exerted with each "step." It is well accepted that having diverse perspectives in a boardroom leads to better business outcomes; the same holds true for scientific laboratories.

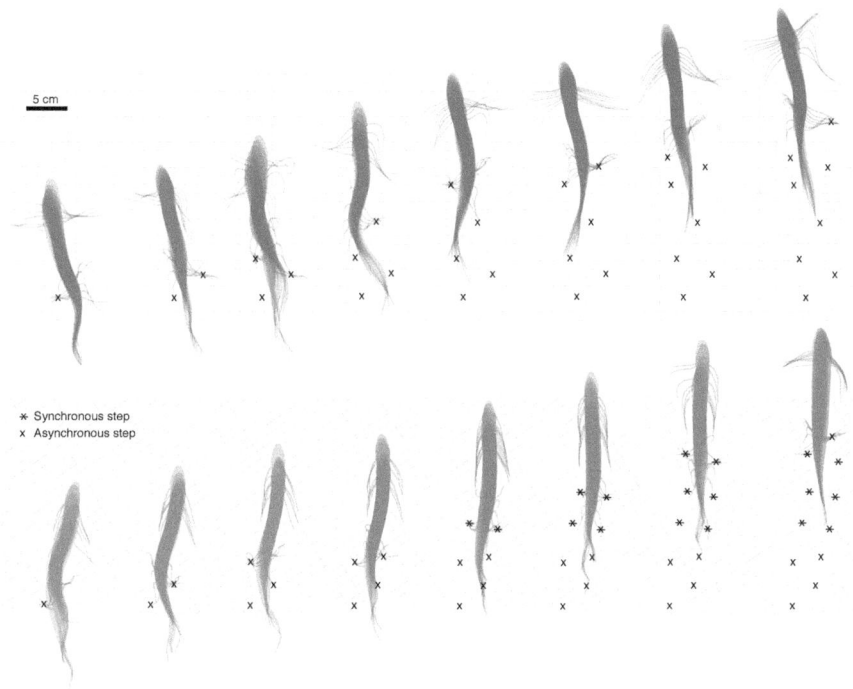

Fig. 3 Lungfish walking. Working with a departmental scientific illustrator led the lead researcher, Heather King, to consider designing force plate experiments she hadn't previously considered. (Illustration by Kalliopi Monoyios (King et al., 2011). Image used with permission)

Public Outreach by Governmental and Nongovernmental Organizations

Ignoring the chance to improve our visual communication is inexcusable given the urgency of climate change… (McMahon et al., 2016).

It is important that government agencies—with their vast collection of scientific teams doing critical research on climate change and solutions—use effective visual storytelling to engage with the public. Likewise, nongovernmental organizations (NGOs) such as nonprofits, social movements, and citizen science groups can amplify their efforts to combat climate change complacency with quality support from visual science communicators. According to the Intergovernmental Panel on Climate Change report (Intergovernmental Panel on Climate Change, 2021), the earth will warm between 2 °C and 6 °C (3.6–10.8 °F) over the next century. In many regions, warming has already surpassed 1.5 °C above preindustrial levels. The US National Intelligence Council Report (National Intelligence Council, 2021) predicts the impacts of climate change—rising temperatures, extreme weather, droughts, food insecurity, health risks, and conflict—will accelerate trends of massive migration and global instability within the next 20 years. Obviously, getting these types of academic findings seen and understood on a personal, even emotional level, by the largest possible audience is critical. We have little time to act if we are to assist governments and NGOs in the swift, decisive action that is warranted.

In the United States, 13 federal agencies, including the Environmental Protection Agency (EPA), the National Science Foundation (NSF), the National Oceanic and Atmospheric Association (NOAA), the Department of Defense (DOD), and the Department of Agriculture (USDA), worked on the *2018 Climate Change Report*. All these agencies have communications departments staffed with quality journalists and "visual information specialists." They are not staffed with "scientific illustrators" (a different US government job description series), and there is no official classification for visual science communicators. Without a formal category for specialists combining science degrees with visual communication skills, federal agencies risk hiring candidates who may unintentionally misconstrue the science and/or fall short of motivating decision-makers and the public to take action. Clearly, there is a need to create staffing structures within the communications sections of government agencies that specify an appropriate skill set for producing scientific graphic communication.

Furthermore, "existing communications teams are too dependent on stock images," says Taina Litwak, who at the time of this writing, is the sole staff scientific illustrator we are aware of for the United States Department of Agriculture (USDA), a collection of 29 units employing over 100,000 people. While stock photos can sometimes approximate what you are trying to say or establish an overall tone, rarely do they convey the context and nuance that make a scientific revelation stand out. And context and nuance are critical elements of effective graphics, according to Jen Christiansen, Senior Graphics Editor at *Scientific American* (Schwabish, 2021).

We also see missed opportunities and heavy reliance on stock imagery in information-heavy emails from top NGOs such as Friends of the Earth Action,

Center for Biological Diversity, National Audubon Society, Climate Hawks Vote, the Climate Reality Project, the Nature Conservancy, Sierra Club, and 350.org. They are largely devoid of visual storytelling and tend to feature a sympathetic photo or two (often featuring the poster child of climate change imagery: polar bears) as a decorative afterthought to the salient information packed into the prose. These NGOs are working hard to make the world aware of the urgency of climate change and are instigating and advocating for widespread action. Yet, they could use visual storytelling to much greater effect.

Not all climate advocates are reliant on stock photography, however. Some government agencies are already harnessing the power of visual science communication professionals either as full-time staff or, more commonly, independent contractors. NOAA's Climate Program Office (CPO) has a staff of highly trained data visualizers and mapping specialists who contract with freelance visual science communicators regularly to assist in rendering climate science illustrations and designing reports for the public and decision-makers. CPO's Climate.gov (2022) *News and Features* regularly posts climate updates paired with striking visualizations on their website as well as on social media. The Climate Explorer (n.d.), as part of the U.S. Climate Resilience Toolkit, provides an innovative interactive experience where users can "get a feel for future conditions" with visual data tailored to their own city or county. The NSF has contracted with visual science communicators such as Nicolle R. Fuller of Sayo Studio to produce some wonderfully detailed, narrative graphics for their program called "The Future of Work at the Human–Technology Frontier" (Fig. 4). In another example, the USDA's staff scientific illustrator, Taina Litwak, rendered the "Farming and Eating Insects" poster (Fig. 5) to communicate how using insects as a protein source makes sense from an environmental perspective.

An NGO doing something similar with a talented team of graphics editors is Climate Central (n.d.). Their initiative titled "Picturing Our Future" presents interactive depictions of climate change in which viewers can manipulate photos to see the effects of various temperature increases on cities around the world. It's extremely well done and not something that could have been accomplished with stock photography.

There are numerous online resources for climate change information and shareable research presentations such as the United Nations' Say It with Science (n.d.), Yale University's Program on Climate Change Communication (n.d.), Climate Outreach's Climate Visuals (n.d.), and the Climate Advocacy Lab (n.d.). Unfortunately, they are not always great resources for *visual storytelling*. Though they do offer bountiful information, they could all benefit from the addition of high-quality, targeted conceptual illustrations and data visualizations. Several do offer figures sourced from previously published papers, data digested into colored maps and graphs, and wonderful galleries of individual photographs, but many papers and presentations they offer are heavy on dry text, tables, and bar charts. What they are lacking are memorable, accessible visuals, particularly ones that can be picked up and shared easily over social and news media. This is a particularly interesting omission given that there seems to be a general understanding that drawing

Fig. 4 This very detailed visual, "The Future of Work at the Human–Technology Frontier," was created for the National Science Foundation by Nicolle R. Fuller during her work as a contractor there. It's a good example of the depth and breadth of a story that can be told with quality narrative scientific illustration. (Illustration by Nicolle R. Fuller (National Science Foundation, 2021). Image used with permission)

connections to people's lives and livelihoods while providing them with positive examples of how we can (and are already) adapting is critical in getting individuals to change (Markowitz et al., 2014).

Mainstream News Coverage of the Climate Crisis

Climate scientists' level of engagement with mainstream media is high compared with other disciplines (Entradas et al., 2019). So, it is even more important that they use every tool available to communicate with audiences with varying levels of interest and scientific literacy. Encouragingly, news outlets with larger budgets are creating print and online interactive narrative pieces that foster emotional engagement through impressive graphics and powerful storytelling, using tools such as ArcGIS StoryMaps. These are prime examples of what can be accomplished when visual elements are central to our science communication efforts.

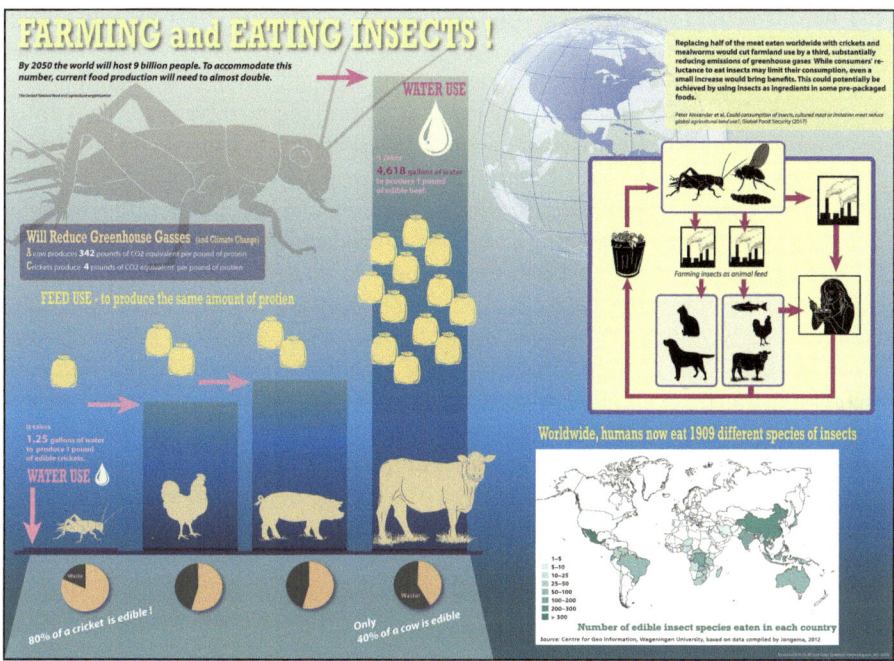

Fig. 5 To the best of our knowledge, the USDA, a collection of 29 agencies employing over 100,000 people, has one staff scientific illustrator position. She produced this poster, "Farming and Eating Insects," to promote a more sustainable alternative to traditional animal protein sources. This poster is used for public educational presentations by USDA staff. (Illustration by Taina Litwak. Image used with permission)

The staff of *The Washington Post* won the "2020 Pulitzer Prize for Explanatory Reporting" for the series 2 °C: *Beyond the Limit* (WashPostPR, 2020). Based on the authors' analyses of global data sets and nearly 170 years of temperature records, the 10-article series mapped every place that has already warmed by 2 °C (3.6 °F)—the threshold that international climate negotiators say the earth collectively must never reach. Using photojournalism, interactive illustrations, and intuitive maps, they successfully drive home the point that extreme climate change is already a life-altering reality across 10% of the earth's surface.

In 2019, the Norwegian Broadcasting Company (NRK) produced an interactive story package called *Chasing Climate Change* (Norwegian Broadcasting Company, 2020). In a country of just five million people, it drew one million page views and won awards for its digital storytelling. It is optimized for mobile, rich with compelling photographs, and contains minimal text with scientific details tucked into pop-up features for those who are interested in engaging more deeply with the content. They followed it with a second piece titled *Velkommen til Oslo i år 2100!* (Støstad, 2020). Set in the year 2100, it is a continuously scrolling illustrated work centered around an engaging cartoon woman who accompanies the reader throughout the story. As she uses her yellow umbrella creatively to float, fly, and stay dry through the narrative, the viewer is exposed to how livelihoods in Norway will change toward the end of the century due to climate alterations. The Society for News Design recently awarded the program "2020 Best In Show for Medium-Sized Newsrooms," and it was featured by the Global Investigative Network in "Climate Storytelling Impact: Lessons from Norway's Public Broadcaster" by Cherilyn Ireton on July 26, 2021. One of the judges commented, "Bringing disparate datasets of the most daunting and complex topic of our time and presenting it in a relatable, personal, and non-overwhelming fashion is a huge achievement."

The New York Times (NYT), too, has produced many successful experiments in visual science communication. In April 2021, they published a story titled, *Bad Future, Better Future: A Guide for Kids, and Everyone Else, About Climate Change—and What We Can Do About It*, by Julia Rosen (Rosen, 2021). Through a richly illustrated, continuously scrolling experience, illustrator Yuliya Parshina-Kottas draws young readers into the story of how we got to this point in history and what we might do about it. Simple but fresh interactive features (e.g., some images unexpectedly advance to the right instead of scrolling down) underscore the experience of being led through a narrative as though on a physical journey with the protagonist, wrist in hand, being pulled here and there as the story unfolds.

Likewise, in July 2021, while much of the population of the United States sat inside due to poor air quality, Nadja Popovich (data and graphics reporter on the NYT's Climate Desk) and Josh Katz (NYT graphic editor) utilized data from NOAA's Global System Laboratory to create a riveting online graphic visualizing the hotspots and dilutions of near-surface wildfire smoke, again for *The New York Times* (Popovich and Katz, 2021). On a darkened map of the United States and southern Canada, bright orange flames leap up in at least seven states and two provinces as winds spread the reddish-purple smoke across the continent. Though we were not able to access metrics on this particular image, we might assume Popovich and Katz's animated map was at the very least served up to the 5.33 million digital subscribers of *The New York Times* (Statista, 2022). This is a potent example of how much more visceral, concise, and memorable a well-executed stand-alone image can be than any of the articles that would have accompanied it.

Formal Learning Environments

Anyone who has jokingly (or seriously) referred to themselves as a "student of life" understands intuitively that we never stop learning. Most of us begin learning, however, in formal environments, defined as structured learning facilitated by a teacher inside a traditional classroom. These are environments where we expect to encounter visuals, and indeed, a robust tradition of visual science communication already exists in K-12, technical, undergraduate, and graduate curricula.

Classroom Textbooks

> When the polar bear is the most visible mascot of climate change, it does the rest of us a disservice by making the issue seem remote and distant (Hayhoe, 2017, para 5).

From illustrations to infographics, textbooks remain a lasting and effective communication medium in formal classroom environments, despite new models of publishing. And given that "inoculation"—reaching audiences early and first with accurate climate information—is a better way to combat misinformation than trying to "debunk" retroactively (Lewandowsky, 2021), the staying power of textbooks should be encouraging. However, in a recent survey of image allocation in college biology textbooks over the last 50 years, Jennifer Landin, Teaching Associate Professor in the Department of Biological Sciences at North Carolina State University, found that while the use of visuals about climate change has increased over time, the textbooks she surveyed are only featuring two climate change images on average (Fig. 6), a number that seems comically low (Ansari & Landin, 2022). Her research reveals that discussion of the topic usually includes photographs of species facing extinction due to climate change (e.g., polar bears and butterflies) instead of visual interpretations of data or solution-oriented imagery (think renewable energy farms). As such, students could be forgiven for believing that climate change doesn't affect them—our communication to date has focused on distant species or far-away places that most will never see in real life, rather than direct ways in which climate change will impact their personal lives and communities. Landin also noted that coverage of climate change topics—including infographics describing changes in temperature, CO_2 levels, and species migration—has not increased proportionately in response to the amount of data available.

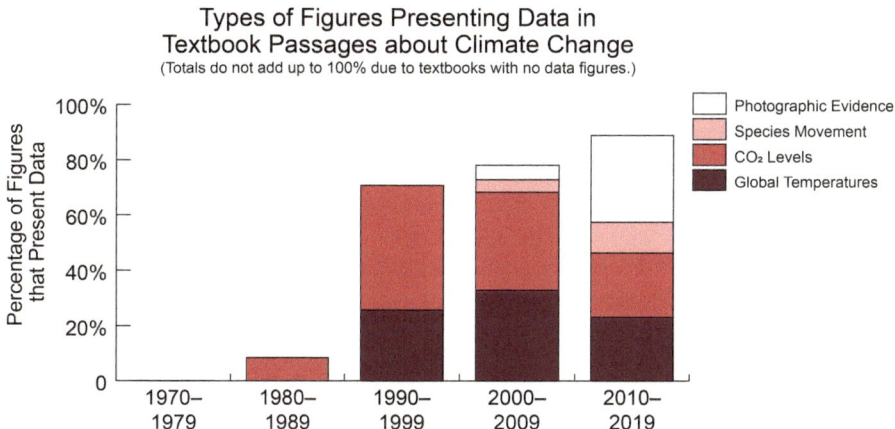

Fig. 6 Sorted by decade, Jennifer Landin's survey of historical textbooks reveals the paltry number of visuals dedicated to explaining climate change, despite the enormous amount of supporting data we have amassed over the last 30 years (Ansari & Landin, 2022). (Image used with permission)

How can we do better? Clearly, textbook publishers need to engage visual science communicators and their authors in increased climate science information. However, a major constraint in textbook publishing is the enormous amount of work that goes into each edition, forcing new editions to come out in a punctuated fashion, often with years passing between revisions. As textbooks are increasingly accessed online, this lack of responsiveness should abate. Until then, more responsive modes of teaching will need to pick up the slack in climate science communication.

Lesson Plans Utilizing Visual Science Communication

In addition to her work quantifying the number and type of illustrations included in textbooks, Landin touts the merits of reintroducing art education into science curricula (Landin, 2015). She cites the benefits gained by looking intently at an object for a long period of time—you actually see more the longer you look—and notes that previous generations of scientists were required to know how to draw, precisely so they could "learn to observe" properly. Despite this, drawing has been largely eliminated from science curricula at this point, save for a few efforts like those being led by Landin.

One example of a lesson plan utilizing visual science communication comes from three educators at the Monterey Bay Aquarium Research Institute (MBARI), who hope to bring scientific data directly to educators and their classrooms. In 2016, they created a teaching module titled "What's the Bigger Picture?" Available online through MBARI's website (Chierici et al., 2016), the lesson plan leads students through an exercise in creating graphs that visualize global climate change data, while using art to illuminate the context (Chierici et al., 2016). By imitating scientist-artist Jill Pelto's (n.d.) innovative illustration style, students transform their informative but staid line graphs into dynamic scenes that tell the story behind the data they depict (Fig. 7).

Books and journals published by the National Science Teaching Association (NSTA) are another rich source of information and lesson plan inspiration. Their publications address learning for all age groups and settings: *Science and Children* (elementary school), *Science Scope* (middle school), *The Science Teacher* (high school), *Journal of College Science Teaching* (postsecondary), and *Connected Science Learning* (informal). The July/August 2021 issue of *Science Scope is of special relevance to this chapter,* available on the NSTA website. The entire issue is dedicated to visual literacy and offers middle school teachers guidance on how to use graphs better in classrooms; how to integrate cross-cutting concepts from the Next Generation Science Standards; how to apply the "Drawing to Learn" strategy to encourage student understanding; and how to use "Big Data" to learn about the history of earth.

Fig. 7 This student piece is an example of the fruits of the visual science communication exercises in the MBARI lesson plan (Chierici et al., 2016). Here, the student illustrates the effects of climate change that are being seen in coral reefs. The graph that was referenced for this (Upton, 2016) shows that the water temperatures near the Great Barrier Reef have been rising significantly since the beginning of the 1900s. Since corals are only able to withstand limited temperature ranges, even the smallest rise in temperature can cause extreme stress and result in coral bleaching. The left side of the image shows cooler water temperatures and healthier, more colorful coral, whereas the right side of the image shows the death of the coral as the water temperatures get warmer. (Illustration by Caralyn Rexroad. Image used with permission)

Informal Learning Environments

In contrast to formal learning, informal learning accounts for how Americans learn most of their science post-schooling (Falk & Dierking, 2019). Informal learning environments include museums, zoos, botanical gardens, libraries, community centers, and themed events like wildflower festivals and BioBlitz. Informal learning also occurs through the Internet, television, radio, podcasts, and countless other everyday experiences, from childhood well into adulthood. As such, informal learning environments provide critical spaces for the public to learn about new ideas and to engage with these ideas in their own way (National Research Council, 2009). The Committee on Learning Science in Informal Environments recommends informal science opportunities be developed through collaborations between community members and educators, because such partnerships lead to *inclusive* science learning (National Research Council, 2009). Visuals created for science communication can and should be present in all of these scenarios.

In the following sections, we focus on print media aimed at youth audiences, though in many cases they can be appreciated by all ages. Our emphasis is on particularly interesting collaborations that bring science into these informal learning spheres. Then, because we can't possibly document all the fronts on which visual science communication is currently being incorporated into experiences for the general public, we highlight a few possibilities for informal science-art partnerships in climate communication.

Children's Literature

A successful arena for science communication where scientists and visual science communicators already work well together is in children's picture books. Overall, consumer sales of children's books increased 9% from 2020 to 2021, and this section of the book industry is on track to experience the strongest year of sales it has had since 2014 (Green, 2021). Additionally, book publishers are turning their attention to comic books and graphic novels for emerging readers (ages 4–8) in response to the growing popularity of this same category with middle school students. At the 2021 American Booksellers Association Children's Institute, it was reported that manga and comic sales were up 17% in 2021. Of great promise for visual science communicators is that publishers see an opportunity to tell many kinds of stories through this genre (Alverson, 2021).

A prime example of researchers using picture books in their outreach programs comes from the NSF's Long Term Ecological Research (LTER) Network. Since 1980, 26 LTER sites have been documenting and analyzing environmental change around the world. As part of their outreach efforts, the program has created the Schoolyard Book Series for children, highlighting findings at 15 of the LTER sites (LTER Network, 2020). With an emphasis on beautiful and accurate illustrations, scientifically reviewed content, and a narrative to encourage children to engage with the science featured at each site, this collection is an admirable example of how scientists can directly influence the richness of our education landscape.

One of the books in the LTER series, *Sea Secrets: Tiny Clues to a Big Mystery* (Fig. 8), is written by Mary Cerullo and Beth Simmons and illustrated by visual science communicator Kirsten Carlson of Fathom It Studios. It uses a visual nonfiction narrative to invite young readers to explore ocean ecosystem shifts among three different species and their food source, krill, from California to Antarctica via the food web.

Given that NSF requires an outreach component for all of its grants, we would like to see more successful proposals combining narrative, science illustration, and cutting-edge science in future research projects supported by NSF.

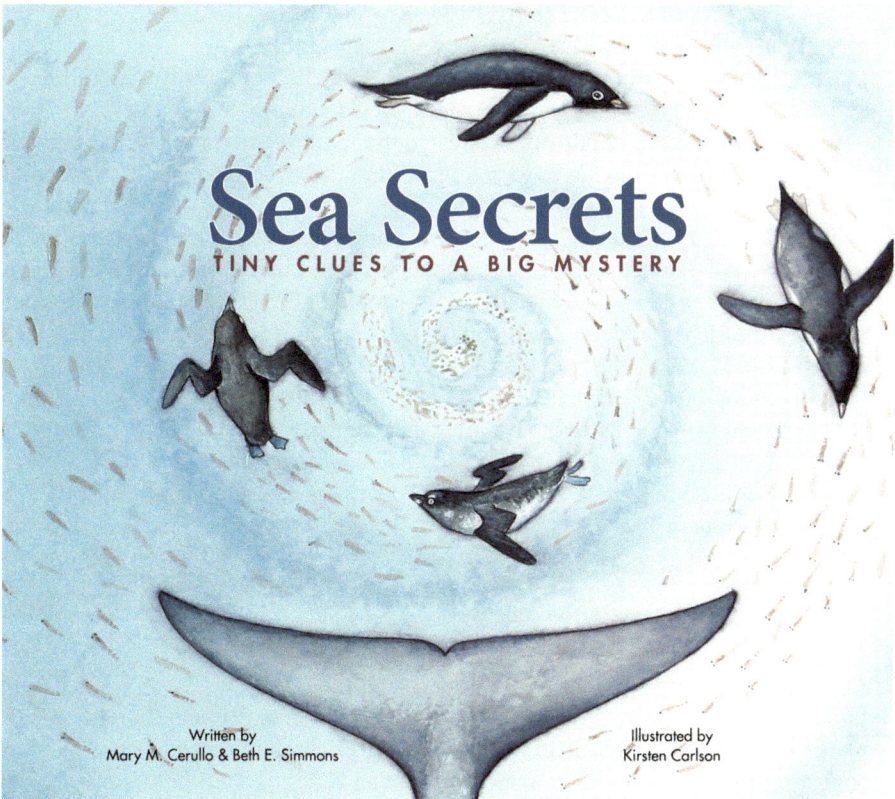

Fig. 8 The cover of *Sea Secrets: Tiny Clues to a Big Mystery*, illustrated by Kirsten Carlson (Cerullo & Simmons, 2015). (Image used with permission)

Children's Magazines

Magazines have a leg up on textbooks and picture books in that they are published more frequently and can theoretically be more current with the frontier of our scientific knowledge. They have the potential to be a critical tool in our dissemination of climate science progress.

The National Wildlife Federation's *Ranger Rick* magazine has been enchanting children for over 50 years with spectacular photography showcasing the wonder of the natural world alongside editorial-style illustrations that embellish their stories. But *MUSE*, a magazine for ages 9–14 published by Cricket Media, combines science and art in a way that more closely resembles the ethos of visual science communicators. In February 2017, *MUSE* dedicated an entire issue to climate change (Cricket Media, 2017). The following year, they published an issue designed to "teach young readers how to gain an accurate understanding of a visually represented data set, as well as how to detect faulty infographics." In this important issue, they broached many of the topics we discuss in this chapter with articles such as

"Making Facts Plain to See: The Art of Data Visualization" and "Secrets of Visual Storytelling." In these articles, they seek out and highlight the breadth of forms visual science communication can take, from Jill Pelto's innovative adornment of graphs mentioned earlier to Florence Nightingales' famous rose diagram (Thompson, 2016) that gave visual weight to the different causes of mortality in the Crimean War, ultimately revealing that most soldiers didn't die of combat, but of preventable diseases.

It should be noted that magazines aimed at educators are another frontier for visual science communicators. The popular environmental education magazine *Green Teacher* boasts an audience of 15,000 readers, 72% of whom are classroom teachers, librarians, and outdoor educators (Green Teacher, 2022). *STEM ED* Magazine (2021) is another prime example; their objective is "to support and inspire as many educators as possible to enhance STEM learning." These magazines' audiences and their reliance on photography would suggest they have an opportunity to collaborate more with professional visual science communicators to incorporate illustrations, infographics, and other visual learning activities that could then be passed on to students.

Outlets with Broad Appeal for All Ages

In the section on lesson plans, we broached the topic of how visual science communication exercises can aid learning in K-12 classrooms. But children are not the only group that can benefit from using drawing as a conduit to understanding and connecting more fully with the world around them.

Scientists' field and lab notebooks have long been places to record observations, reflections, sketches, diagrams, and data tables. Though they are not generally used for outward communication, they are a critical tool for observation and discovery. Nature journals, a close cousin of scientists' lab notebooks, provide a way for the general public to engage in this same exercise of careful, deliberate observation. Starting a nature journal does not require any previous art experience, nor does it use fancy or expensive supplies. It is a place to record what Dirnberger et al. (2005) call *raw knowledge*: newly acquired knowledge that will be processed and refined over time. Journals help individuals become more familiar with their surroundings regardless of age or education and develop positive attitudes toward the natural world. Nature journaling can even be used successfully in urban environments to combat the "extinction of experience" which Lyn Baldwin, Associate Professor at Thompson Rivers University in British Columbia, refers to as the ever-increasing divide between people and nature (Baldwin, 2017). Luckily, teachers and outdoor educators are increasingly familiar with nature journaling thanks to the growing popularity of global events such as John Muir Laws' Wild Wonder Nature Journaling Conference (Laws, 2017) and International Nature Journaling Week (n.d.).

In addition to the popular science magazines, we mentioned in the section on mainstream media and news, graphic novels are another great frontier for visual science communication. Thanks to publishers' growing appetite for books in this genre, we are beginning to see titles that expand beyond memoirs and fantasy into serious science communication works. Clifford V. Johnson's *The Dialogues: Conversations About the Nature of the Universe*, Maris Wicks' *Primates: The Fearless Science of Jane Goodall, Dian Fossey, and Biruté Galdikas,* and Michael Keller and Nicolle Rager Fuller's adaptation of Darwin's *On the Origin of Species* are successful examples of nonfiction graphic novels intended for an adult audience. Likewise, titles like the *Max Axiom, Super Scientist!* Series, and *The Manga Guide* series cater to younger readers with similarly serious aims of science education.

PART II: *What Makes Effective Visuals*

> Identifying "good quality" science communication is not just a matter of weeding out or getting rid of the "fake news," as we often hear (Collver & Bucchi, 2021, p. 23).

Visual science communicators dedicate their careers to thinking about strategies and considerations to take into account when distilling complex ideas into intuitive visuals that people connect with. Whether you are a practitioner or are collaborating with one, it is useful to be aware of and think critically about the elements that contribute to the best communication efforts so that we may always be learning and adapting to the cultural shifts that unfold at an ever-accelerating rate. As Jordan Collver and Massimiano Bucchi rightly point out in their pithy Lifeology course *Style in Science Communication*, there is no single formula that will produce a winning graphic every time (Collver & Bucchi, 2021). Rather, each communication attempt represents a complex interplay between various considerations, the most important of which we attempt to collect in the sections that follow.

Visual science communication has a language (Fig. 9), in this case comprised of visual elements, some of which are broadly recognized across cultures and others that are more effective with people who have specialized knowledge or training. Knowing which elements to use entails defining a specific goal that can generally be answered by questions like "Who am I trying to reach?" and "How do I want people to react?" Answering these critical questions will go a long way toward narrowing down the myriad options there are in creating an effective visual. Social psychologist Jonathan Haidt's research indicates that "intuitions come first, strategic reasoning second," (Haidt, 2012), so it's important to remember that visual science communication is more than just information transfer. Effective climate visuals must engage people, give them a reason to change their behavior, and leave them feeling empowered to do so.

When done well, graphics serve as a universal, accessible language, translating complex scientific concepts into infographics or illustrations that can be understood "at a glance" by anyone and anywhere. They provide a way for abstract or distant concepts to be visible, quantifiable, and relatable. They may employ narrative, emotion, analogies, and even humor to disarm people and open their minds to the possibility of change. In our age of information overload, novel, targeted images cut

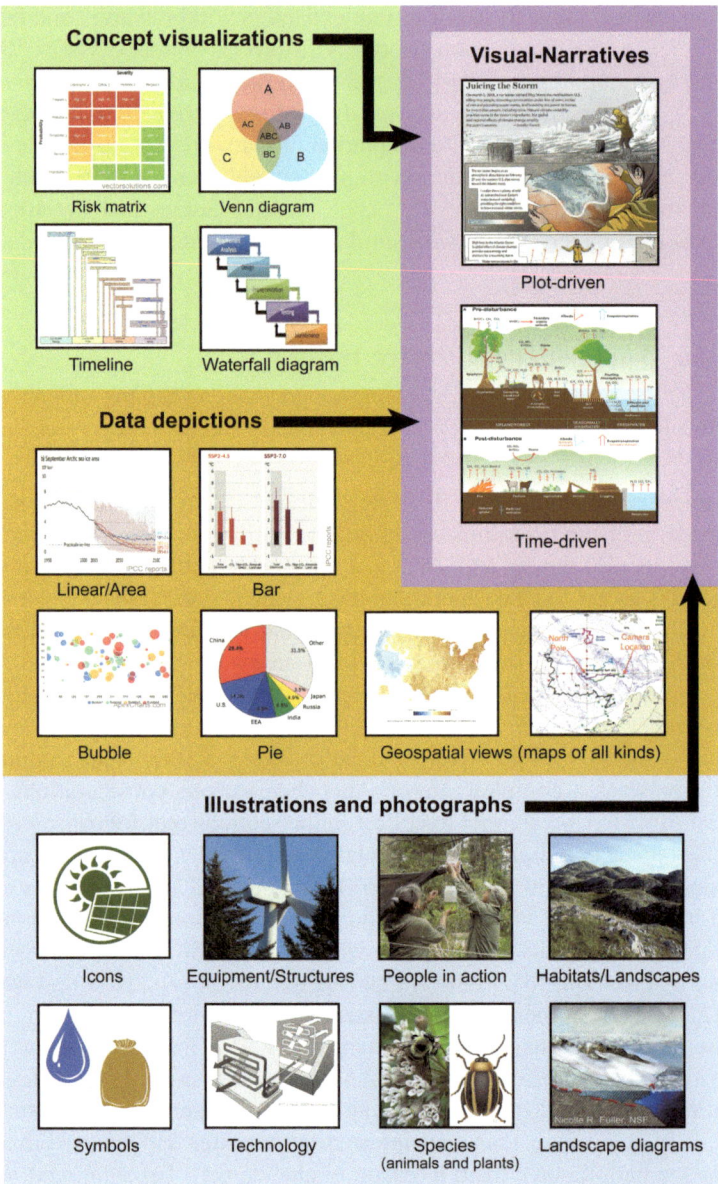

Fig. 9 A wide array of elements can be used to create visual narratives. The most successful graphics often use several types of visual elements in one piece. Concept visualizations (*top*) can arrange data in easy-to-digest, visual formats, where relationships become clearer than numbers or words can communicate alone. Similarly, data depictions (*middle*) simplify and compare complex data sets (e.g., weather patterns), using universal visual elements we can understand at a glance. Illustrations and photographs (*bottom*) can range in detail from graphic icons to realistic illustrations to photographs. This is just a subset—there are many other visual tools not shown here. (Figure arranged by Taina Litwak and Emily Coren; thumbnail images are copyrighted and included with permission from individual artists/publishers. Image used with permission)

through the noise, relieving "climate fatigue," offering solutions, and inspiring citizens to get involved (Houser, 2020). They provide a window into the world of scientists, increasing transparency and building trust with local communities. Ultimately, our best research revelations are fruitless if we do not utilize our full tool kit as we strive to increase climate science literacy and resilience.

Judicious Use of Photography

> Imparting urgency and concern is just a matter of showing people how to connect the dots among the issues they already care about, and how those issues are affected by—and in many cases are threatened by—a changing climate (Hayhoe, 2017, para 24).

In Part I, we lament the missed opportunities for meaningful communication presented by poorly chosen stock imagery. However, the practical appeal of stock imagery is real. Faced with limited or nonexistent budgets, the low cost of a licensed image is hard to beat. Additionally, licensing terms and pricing on stock image aggregators are crystal clear, providing time savings and avoiding steep learning curves that can present themselves when working with independent image creators for the first time. If stock imagery is the only feasible option, previously mentioned collections such as ClimateVisuals.org are an excellent resource for images highlighting communities modeling positive climate adaptations. What they avoid, and what anyone perusing stock imagery for climate communications should resist, is the temptation to feature only signature species and faraway places being altered by climate change. These generic images can contribute to "climate fatigue," causing people to disconnect from the problem and any possible solutions (Markowitz et al., 2014). Better images center communities around the world modeling positive climate adaptations they are making locally (Fig. 10).

Cultural Connections and Two-Way Engagement

> Science is a search for evidence, but science communication must be a search for meaning (ElShafie, 2018, p. 1213).

Diverse audiences—with varying levels of trust in science, let alone scientific literacy—will interpret information in different ways. Making sense of climate concepts, or meaning-making, is subject to cultural norms, group identities, values, and past experiences of each individual or community (Leiserowitz et al., 2021). A shift in scientific communication to recognize this fact is currently underway. Dr. Maja Horst, Professor of Science Communication at University of Copenhagen, suggests reframing science communication as an aspect of culture (Horst & Davies, 2021). The resulting science communication is positioned not just to teach knowledge from the top down (the "deficit model"), but instead to foster "two-way engagement," coordinating with existing community values and serving community needs. The deficit model assumes that scientific literacy is a knowledge problem and that

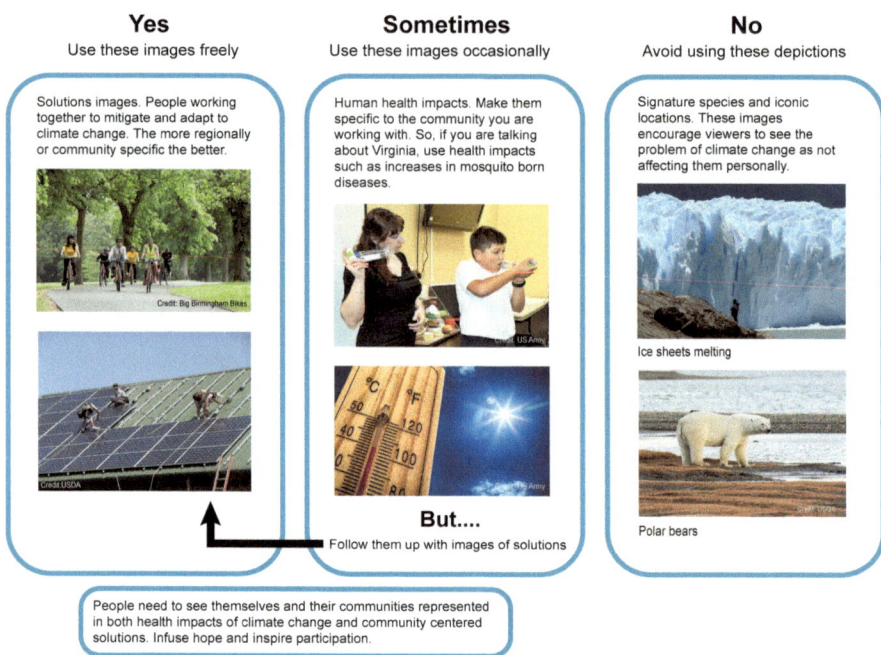

Fig. 10 Imagery chosen with care and consideration of specific communities can increase communication quality and connection with an intended audience. (Graphic by Emily Coren and Taina Litwak. Image used with permission)

sharing more information will enable critical thinking and better decision-making. However, simply sharing knowledge is not enough to change behavior.

Katharine Hayhoe's work embraces this approach. Hayhoe, the Chief Scientist for The Nature Conservancy and a Distinguished Professor and Endowed Chair at Texas Tech University, has made impressive inroads sharing climate science with audiences who would be politically turned off by phrases such as "global warming" and "climate change." In an article for *Foreign Policy* titled "Yeah, the Weather Has Been Weird" (Hayhoe, 2017), she shares an anecdote in which she was careful not to say the word "climate" followed by "change" in a routine lecture on climate science to a politically right-leaning audience in the United States. With this tiny alteration, she went on to present the climate science data in a way that was relatable and nonthreatening and received no pushback at all. Afterward, a woman approached her and said, "You know those people who are always talking about global warming? I don't agree with them at all. But this? *This makes sense.*" What is her secret? By her own admission, she is dishing out grim news. But she understands the issues that matter to her audiences and follows her stark dose of reality with common sense actions these communities can take immediately: "Make smart water choices, plan ahead, and prepare for a water-scarce future." Her audiences leave feeling empowered, not defeated, and she is widely praised for her science communication efforts as a result.

What would a similar approach look like in visual communications? Being culturally responsive begins with understanding that people don't just receive information passively; instead, it is filtered through their own lens of life experiences and communities of trusted peers. To fully achieve this goal, representation matters, among people of color and marginalized groups as well as in conservative and liberal-leaning sections of society. People are more likely to listen to leaders and scientists who look like them, reflecting the demographics of their local community and circle of trust (Houser, 2020; Marsh, 2020; McIntyre, 2021; Zaelzer, 2020). For this reason, we should all be on the lookout for ways to understand and involve the communities we are communicating with. Then, logically, our images should reflect those communities so that people can see themselves in them. They see themselves as part of the solution and know what to do next.

Revealing the Process of Science

> Scientists get tempted to pretend that they know something with certainty. You have to be willing to embrace the idea that science is uncertain and why you love it anyway (McIntyre, 2021).

Those of us trained in science understand what a powerful tool it is to help us make sense of our world. We trust the process of inquiry, discovery, and building consensus. We understand that what we think we know today may change as new, better data clarify concepts and processes. We are comfortable with the fact that science is self-correcting—as we gather new information, we form new conclusions (or become more certain of the conclusions we'd previously drawn). But large sectors of the public don't understand science this way—they don't see it as a process but rather a set of facts. They mistake scientists' certainty in the *process* of science (we know this thing because our best data support it) with a misconception that scientists think they know it all and are never wrong. So when experts talk about "global warming" and citizens are experiencing an arctic blast with record low temperatures in February, it's no wonder scientists get labeled as out of touch and untrustworthy.

Every science communicator has an opportunity to underscore the process of science with each story they tell, and visual science communicators are no different. Being transparent about the nature of research and unknowns can help build trust. Images showing scientists working in the field and data visualizations including known and unknown variables (e.g., white/gray areas on a map) help keep the communication open, honest, and more complete. Additionally, in framing our narratives, we can look for opportunities to explain why recommendations change when new findings emerge, such as with the CDC's changes in mask-wearing recommendations during the COVID-19 pandemic.

The Role of Emotion, Empathy, and Humor

Scientific communication in academic circles often favors professionalism and a high degree of specialization. Experts create utilitarian graphs and tables to explain their findings; these displays fare well enough in academic circles and peer-reviewed journals, but are often unintelligible or uninteresting to the general public. Communication that is emotionally moving, in tune with local communities, and solution-oriented might be more effective—consistent with the two-way engagement model mentioned previously (Markowitz et al., 2014; Zaelzer, 2020).

The power of visual storytelling to engage the reader or viewer on an emotional, "gut" level is well documented. Numerous studies and academic papers have weighed in on the impact of visual narratives on fostering emotional engagement (Bracken et al., 2014; Zak, 2015). Art ranging from fine art to editorial art, realistic representations, and data visualizations has the power to convey subtle emotions that connect to human experiences through the deliberate application of color, tone, rendering style, and composition (Hassenzahl et al., 2013). Our evolutionary instincts lead us to gravitate toward the esthetic, the unusual, and the compelling (Dutton, 2009), no matter how our cultures or subcultures define them. The skillful visual science communicator learns about the community they are attempting to communicate with and takes care to catch attention, arouse curiosity, and heighten learning by picking up on and establishing meaningful connecting points.

W. Sean Chamberlin, writer and oceanography professor at Fullerton College, knows that creating emotional connections can heighten engagement and retention. With 25 years of experience teaching, he saw a need for editorial figures students could relate to on an emotional level that would help them connect better with the terminology describing the earth's seven spheres. In June 2021, he hired visual science communicator Fiona Martin of Visualizing Science® LLC to create oceanography e-textbook figures with novel perspectives, geared toward young college students in urban Los Angeles (Fig. 11). Her resulting figure is colorful and dynamic and takes care to relate the seven biospheres to humans by placing the "anthrosphere" centrally, suggesting it has a disproportionate impact on the other six natural domains. There is no text in the figure itself, inviting students to interpret the figure on their own. This is significantly different from the approach taken by traditional illustrations of this concept that center on equally weighted geospatial views of the earth. Importantly, this framing is justified by evidence that suggests images are more memorable, and consequently better understood, when they include people and familiar objects or symbols (Borkin, 2014).

Lastly, a little humor goes a long way in an age of information overload and existential crises. Fear, distrust, or skepticism about climate science can evaporate when we have a shared appreciation for hilarious cartoons, memes, or deadpan humor. Science comics such as the wildly popular *xkcd* and the YouTube channels MinutePhysics and MinuteEarth have been leading the way for over a decade in this realm and have amassed millions of followers each. Another website and Facebook page, I Fucking Love Science, boasts 23 million followers, sharing science images with a dose of humor and encouraging social interaction and a sense of community

Fig. 11 A novel perspective of the earth's seven spheres. *Clockwise from top left:* heliosphere, atmosphere, cryosphere, hydrosphere, biosphere, and geosphere. *Center:* The anthrosphere, representing humans, has had a disproportionate impact on the rest of the earth's systems. (Illustration by Fiona Martin (Chamberlin, 2022). Image used with permission)

(Horst & Davies, 2021; Marsh, 2020). Beyond these examples, successful science comic strips such as Karen Romano Young's *Antarctic Log* abound and are an excellent example of how visual science communicators can seek out individual communities and highlight climate actions being taken today (Fig. 12).

Visual Analogies and Metaphors

> In the charged atmosphere of climate politics, analogy enlivens deadening data (Houser, 2020).

Using metaphors and comparisons to everyday objects makes science relatable and interesting, especially for abstract concepts we can't see with our own eyes such as public health threats or carbon emissions. Visual science communicators regularly use metaphors from everyday life to help get a point across. For example, the Swiss cheese diagram, originally devised by James T. Reason in his book *Human Error* (1990), explains how an accumulation of errors can lead to adverse events. It has since been used widely by safety analysts in industry. Virologist Ian Mackay revived

Fig. 12 *Antarctic Log* is a science comic written by Karen Romano Young. It is aimed at middle-school-aged kids and features actionable messages, often with a climate change focus. (Illustration by Karen Romano Young (Romano Young, 2015). Image used with permission)

it recently to explain how single interventions against the spread of COVID-19 are not as effective as combining several layers of protection (Fig. 13). Each intervention (layer) is imperfect and has holes. But by combining personal and shared responsibilities, there is a better chance of success. Incidentally, the mouse represents the potential for misinformation, which can erode protections (Mackay, 2021; Roberts, 2020). The figure is simple enough that it could be understood by children and adults and is a brilliant example of the power of metaphor in communicating complex ideas.

"Carbon Emissions in New York" by Real World Visuals cited at the beginning of this chapter is another powerful example of the "concretization" of an abstract concept. "The problem is that some people are very cut off from quantitative information. You put numbers and graphs in front of people and they bounce straight off," said Real World Visuals' creative director, Adam Nieman. "Our [goal] is to make the cause of climate change visible because very few other people are approaching it like that" (Hahn, 2022, para 11). Interestingly, their approach centers around assessing whether intended audiences fall into the categories of "push" or

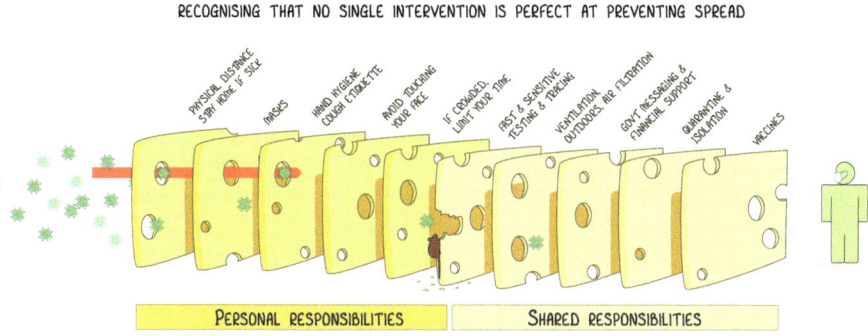

Fig. 13 Swiss Cheese Respiratory Virus Pandemic Defense (version 4.3). (Illustration by Ian M. Mackay (Mackay, 2021). Image used with permission)

"pull" audiences. Whereas "pull" audiences come already interested in the topic and are able to navigate more complex graphics because of their own intrinsic motivation to learn more, "push" audiences need more coaxing. By creating graphics that use analogies to "draw on our wider experience of the world," they are able to appeal to the coveted "push" audiences.

Addressing Information Overload

To a nonscientific audience and students, the sheer amount of climate information received from the news and other media can be overwhelming. Too much confusing, negative, seemingly conflicting evidence of climate change can leave people in a state of paralysis or defeat, not knowing what to do (Houser, 2020). In a Climate Outreach study (Corner et al., 2015), researchers found that participants were more likely to respond to emotional images of climate impacts, especially local events, and clearly, this has been the strategy of environmental activists for decades. However, we also know intense emotion can lead to disillusionment and feelings of being overwhelmed; ultimately, this is a recipe for inaction. To combat feelings of despondency, Climate Outreach suggests following images of climate impacts with images of climate solutions, giving viewers actions they can take to prevent or mitigate disasters (Corner et al., 2015; Markowitz et al., 2014). At least on the topic of climate change, it would seem hope is a better motivator than fear.

In October of 2001, *Scientific American* published the print article "Drowning in New Orleans" by Mark Fischetti (Fischetti, 2008), which turned out to be eerily prophetic of the disastrous consequences that followed Hurricanes Katrina in 2005 and Ida in 2021. The article begins with an ominous warning in large blue text, *"A major hurricane could swamp New Orleans under 20 feet of water, killing thousands. Human activities along the Mississippi River have dramatically increased the risk, and now only massive reengineering of southeastern Louisiana can save the city."* Then a paragraph into the text, *"New Orleans is a disaster waiting to happen."* It is not hard to see how such a dire prediction could be overwhelming enough to shut people down. The opening figure describes the rate of land loss due to human-made levees that restrict the flow of sediments into the marshlands. An inset cutaway of New Orleans shows the city sits in a "bowl" below sea level, ringed with levees. Solutions are offered, but only in text sections of text below each impact. And the price tag on the suite of infrastructure improvements that might mitigate the risk is stated as a whopping $14 billion. To residents of a city with a median income of $27,133 in 2000, this number must have seemed impossibly large. And the cool, academic, and distant visual depictions of their city may have done little to connect with how the people of New Orleans experience their city.

In the case of this article, it's worth considering whether it might have been more effective to visualize what a climate-resilient New Orleans would look like on a human scale in addition to the warnings about impacts and the call for large-scale structural changes. A hopeful depiction of what an actively adapting city could be— with manageable action items that policymakers and ordinary citizens could take— may have done more to nudge residents toward action.

Naturally, we are not suggesting that altering the approach of one article could have stopped the disasters that unfolded with Katrina and Ida. But when applied across all the media we have to communicate climate resilience, this sensibility may help communities "see" solutions and see themselves in them, ultimately getting people to make them happen (Corner et al., 2015). In short, climate solutions— including smart infrastructure, green technology, sustainable food production, and positive public health measures—should outnumber disaster imagery. And importantly, though there is room to dream about a future of resilient cities, we must remember to follow the examples set by skilled communicators like Katharine Hayhoe and provide common sense steps that people can start with today: conserve, prepare, and adapt.

Measuring Impact

The great irony here is that we spend a lot of time talking about science and empiricism, yet our ability to reflect on our own effectiveness has been stunted by reduced job opportunities and limited funding. Currently, our best way to measure the impact of visual science communication is through proxy: the number of page views an article gets, the number of shares and likes, whether people stay on a given page, and how long they linger. These metrics are regularly used as estimates of an online *article's* success and, in turn, are assumed to extend to any graphics that accompany it (J. Christiansen, personal communication, December 8, 2021; A. Montañez, personal communication, January 3, 2022).

Another way to gauge a particular graphic's success is when it gets picked up by policymakers or mainstream media outlets. A graphic by Amanda Montañez called The Costs of Climate Change was printed in a *Scientific American* article titled "Here's How Much Climate Change Could Cost the US" in 2018 (Thompson). Shortly after, it was used by Congressman Jimmy Panetta (CA) in a presentation on the costs of climate change, earning it a coveted spot in the success column. Additionally, the frequency with which an image is requested for reuse can be a direct indication of success (J. Christiansen, personal communication, December 8, 2021), but it still provides no perspective on how large numbers of people understand or respond to it; at best, it tells us another person or small group of people think it is informative enough to use it in their own publication.

Beyond these measures, there are few metrics being systematically collected that give us direct, measurable feedback on how our visualizations are performing. Visual science communicators rely on accepted design principles, powerful visual narratives (Zak, 2015), and human-recognizable objects and colors (Borkin, 2014), for example, to help make images clear and memorable and inspire the cooperative, adaptive behaviors we seek to spark. There is evidence that certain types of imagery evoke feelings of salience (climate change is important) and others evoke self-efficacy (I can do something about this), but rarely do images accomplish both (O'Neill et al., 2013). Aside from infrequent insights such as these, rarely do we get direct confirmation that our use of these tools successfully hits our intended mark.

Visual science communicators whose work is produced for museums and other institutions that regularly poll visitors to look for signs of learning (Diamond et al., 2009) may be one exception to this rule. Prior to digital publishing, Jen Christiansen, Senior Graphics Editor at *Scientific American*, regularly held focus groups that allowed her to get feedback and connect with people she was creating graphics for. However, she has not held one in years and cites this as a lost opportunity to understand how people are receiving the graphics we pump out (Schwabish, 2021). We have no way to close the learning loop for the rest of us working in visual science communication. By and large, visual science communicators are left to their own devices to intuit what works and what does not with indirect or incomplete metrics to guide us.

The concept of A/B testing, in which two versions of an article are published online and served randomly to viewers to see which performs better, is well within our technological capability. In fact, it is regularly employed for headlines. Yet publications such as *Scientific American* that have a vested interest in this type of insight do not currently have the capability to apply this testing to their images (J. Christiansen, personal communication, December 8, 2021). Essentially, they are at the whim of the platforms they use to manage their publishing, and the add-on services that try to fill in the feature gaps. One would imagine that if enough users requested this type of functionality, publishing platforms might respond accordingly.

Ultimately, understanding and refining our own effectiveness is an exciting opportunity for all of us with a vested interest in visual science communication. We have a suite of creative tools at our fingertips and ideas galore—we only need partners willing to iterate, test, and track performance with the communities we serve.

Adopting a Collaborative Mindset

> Rather than keeping disciplines and tasks divided, artists and scientists should collaborate in a more organic fashion. In some cases, the artist's role is to break down conventional methods so that science can make leaps. They can also help scientists make information more compelling so that it generates awareness, cultivates support and helps the general population feel that they are a part of scientific progress (Sabraw, 2021, p. 8).

Scientists have a long track record of collaborating with peers on research projects. Encouragingly, many are also interested in distributing their work in nontraditional ways, by collaborating with professionals in fields outside academia (Allen et al., 2018). Successful collaborations of this sort have resulted in animations and two-dimensional art, social media content, multimedia, storytelling, and programming at national parks too numerous to name (Allen et al., 2018; Cheng et al., 2018; ElShafie, 2018; Harrower et al., 2018).

From 2011 to 2016, The Scientific American Blog Network hosted *Symbiartic*, a blog written by Glendon Mellow, Kalliopi Monoyios, and Katie McKissick that documented this growing field of "Science-art," or "SciArt." They documented hundreds of scientists and artists using art as a medium to spark conversations and educate ever more general audiences on various aspects of science—and that was just the tip of the iceberg. As these science-art collaborations showed, the possibilities for partnerships between scientists and informal science educators are exciting and near endless.

Science communicators looking to expand their reach can share their work in existing public programs or they can become involved in detailed projects such as the design of exhibits, museums, and nature centers (Alpert, 2018). Potential partners may also be found at gardens, zoos, aquariums, and through independent professionals working in related and sometimes entirely different fields. When looking for these collaborations, we encourage communication teams to be open to working with content developers, cultural interpreters, artistic directors (Rudenko, 2018), environmental educators, authors (Armstrong et al., 2018), game designers (Kipnis, 2018), advocates (Inches, 2021), food system experts (Kiss the Ground, n.d.), natural resource specialists, consultants, and visual science communicators. In addition, people can look to organizations such as SciArt Initiative (n.d.) that are creating formal ways for scientists and artists to collaborate through innovative residencies like *The Bridge*.

The search for new partners takes time, but we have much to gain from this investment of energy. Those who develop these cross-disciplinary partnerships have the opportunity to change public perceptions about who scientists are, increase interest and trust in STEM disciplines, and reframe what an individual thinks about a particular topic (Merson et al., 2018). All of this has the potential to influence policy decisions and enact change.

Inspiring Action

Adoption of climate solutions will happen faster if our climate science communication inspires citizens worldwide to act. Activism takes many forms and may mean writing letters and calling on government representatives to vote a certain way. It can mean organizing postcard campaigns to other voters, signing and delivering petitions, organizing and/or showing up for rallies, volunteering for local conservation groups' physical efforts, or making small personal changes in daily choices such as eating less meat, taking public transportation, or buying an electric car. In the book *Empowering Climate Action* (Bowman & Morrison, 2021), numerous authors repeatedly stress the need for massive climate change education and "bottom-up" public engagement/activism. Katharine Hayhoe points out that even the most vocally opposed politicians understand the gravity of climate change but won't act until their constituency demands it (Hayhoe, 2021). If visual science communicators are consistently included in science communication teams, our stories will move people and inspire them to take action. We trust decision-makers to take it from there.

Conclusion

Effective communication of climate change solutions will require a suite of adaptations. Science communication efforts need to take into account accessibility, cultural sensitivity, representation, emotional engagement, and relevance. We must also find ways to overcome information overload and "climate fatigue," maintain transparency in climate research, and demystify the scientific method. This will involve building trust with local communities, building bridges between academics and the public, and convincing decision-makers that funding is needed to support climate mitigation. Visuals serve as a powerful universal language that can interpret science, transcend barriers, invoke empathy, and inspire citizens to implement practical solutions. Our communication efforts can improve by systematically incorporating well-designed, compelling diagrams, illustrations, infographics, data visualizations, graphic novels, and more into climate science outreach at every level. To do so, researchers, publishers, news outlets, and communications teams need to recognize the importance of collaborating with professionally trained visual science communicators and create space in their budgets to support them. This small shift in priorities and funding allocations will dramatically improve the reach and rate of climate mitigation and adaptation efforts.

References

Allen, L., Char, C., Hristov, N., Wright, T., & Merson, M. (2018). Beyond the brown bag: Designing effective professional development for informal educators. *Integrative and Comparative Biology, 58*(1), 77–84. https://doi.org/10.1093/icb/icy026

Alpert, C. L. (2018). So you want to share your science . . . Connecting to the world of informal science learning. *Integrative and Comparative Biology, 58*(1), 85–93. https://doi.org/10.1093/icb/icy008

Alverson, B. (2021, February 26). *Comics formats go younger.* PublishersWeekly.Com. Retrieved September 9, 2021, from https://www.publishersweekly.com/pw/by-topic/childrens/childrens-industry-news/article/85687-comics-formats-go-younger.html

Ansari, R. A., & Landin, J. M. (2022). *Coverage of climate change in introductory biology textbooks, 1970–2019.* PLoS ONE 17(12): e0278532. https://doi.org/10.1371/journal.pone.0278532

Armstrong, A. K., Krasny, M. E., & Schuldt, J. P. (2018). *Communicating climate change: A guide for educators.* Comstock Publishing.

Association of Medical Illustrators. (2000). *Education.* AMI. Retrieved April 29, 2022, from https://ami.org/medical-illustration/enter-the-profession/education

Association of Medical Illustrators. (2019). *Medical Illustration and Animation Sourcebook.* Retrieved April 29, 2022 from https://www.medillsb.com/

Baldwin, L. K. (2017). Drawing care: The illustrated journal's "path to place". *Journal of Teaching in Travel & Tourism, 18*(1), 75–93. https://doi.org/10.1080/15313220.2017.1404723

Borkin, M. A. (2014). Perception, cognition, and effectiveness of visualizations with applications in science and engineering [Doctoral dissertation, Harvard University]. https://dash.harvard.edu/handle/1/12274335

Bowman, T., & Morrison, D. (Eds.) (2021). *Empowering climate action in the United States.* Changemakers.

Bracken, B. K., Alexander, V., Zak, P. J., Romero, V., & Barraza, J. A. (2014). Physiological synchronization is associated with narrative emotionality and subsequent behavioral response. In *Foundations of augmented cognition. Advancing human performance and decision-making through adaptive systems* (pp. 3–13). https://doi.org/10.1007/978-3-319-07527-3_1

Cerullo, M. M., & Simmons, B. E. (2015). *Sea secrets: Tiny clues to a big mystery.* Taylor Trade Publishing.

Chamberlin, W. S. (2022). *Our world ocean: Understanding the most important ecosystem on earth.* Blue Planet Publishing.

Cheng, H., Dove, N. C., Mena, J. M., Perez, T., & Ul-Hasan, S. (2018). The Biota Project: A case study of a multimedia, grassroots approach to scientific communication for engaging diverse audiences. *Integrative and Comparative Biology, 58*(6), 1294–1303. https://doi.org/10.1093/icb/icy091

Chierici, J., Couchon, K., & FitzGerald, N. (2016). *What's the Bigger Picture? MBARI.* Retrieved April 29, 2022, from https://www.mbari.org/what-is-the-bigger-picture/

Climate Advocacy Lab. (n.d.). *About.* https://climateadvocylab.org/about

Climate Central. (n.d.). *Picturing our future.* https://picturing.climatecentral.org

Climate Visuals. (n.d.). *Climate visuals, a project of climate outreach.* https://climatevisuals.org

Climate.gov. (2022, April 27). *News and features.* Retrieved April 29, 2022, from https://www.climate.gov/

Colbert, S., & Munroe, R. (2014, September 3). "What If?" author Randall Munroe talks about using science to answer bizarre hypothetical questions. *The Colbert Report.* Other, Comedy Central. Retrieved December 8, 2021, from https://www.cc.com/video/ewijdy/the-colbert-report-randall-munroe

Collver, J., & Bucchi, M. (2021, April 2). *Style in science communication.* 2022 LifeOmic, LLC. Retrieved December 13, 2021, from https://lifeomic.app.us.lifeology.io/viewer/lifeology/default/style-in-science-communication#d361025f704b

Corner, A., Webster, R., & Teriete, C. (2015, January). *Climate visuals: Seven principles for visual climate change communication*. Climate Outreach. https://climateoutreach.org/reports/climate-visuals-seven-principles-for-visual-climate-change-communication/

Cricket Media. (2017, February). Climate reality. *MUSE*. Retrieved December 11, 2021, from https://shop.cricketmedia.com/muse-magazine-february-2017-climate-reality

Diamond, J., Luke, J. J., & Uttal, D. H. (2009). *Practical evaluation guide: Tools for museums and other informal educational settings (American Association for State and Local History)*. AltaMira Press.

Dirnberger, J. M., McCullagh, S., & Howick, T. (2005). Writing & drawing in the Naturalist's journal. *The Science Teacher, 72*(1), 38.

Dutton, D. (2009). *The art instinct: Beauty, pleasure, and human evolution*. Bloomsbury Press.

ElShafie, S. J. (2018). Making science meaningful for broad audiences through stories. *Integrative and Comparative Biology, 58*(6), 1213–1223. https://doi.org/10.1093/icb/icy103

Entradas, M., Marcelino, J., Bauer, M. W., & Lewenstein, B. (2019). Public communication by climate scientists: What, with whom and why? *Climatic Change, 154*(1–2), 69–85. https://doi.org/10.1007/s10584-019-02414-9

Fairman Studios. (2020, December 30). *Jennifer Fairman publishes COVID-19 visualization article in JNSI*. Retrieved April 29, 2022, from https://www.fairmanstudios.com/project/covid-jnsi/

Falk, J. H., & Dierking, L. D. (2019, November 6). The 95 percent solution. *American Scientist*. Retrieved December 11, 2021, from https://www.americanscientist.org/article/the-95-percent-solution

Fischetti, M. (2008, September 5). Drowning New Orleans. *Scientific American*. Retrieved September 12, 2021, from https://www.scientificamerican.com/article/drowning-new-orleans-hurricane-prediction/

GNSI (n.d.). Visual science communication in action. https://www.gnsi.org/visual-science-communication-in-action

Grady, C. L., McIntosh, A. R., Rajah, M. N., & Craik, F. I. M. (1998). Neural correlates of the episodic encoding of pictures and words. *Proceedings of the National Academy of Sciences, 95*(5), 2703–2708. https://doi.org/10.1073/pnas.95.5.2703

Green, A. (2021, September 2). *Children's Institute 9: Children's books continue run of strong sales*. PublishersWeekly.Com. Retrieved September 9, 2021, from https://www.publishersweekly.com/pw/by-topic/childrens/childrens-industry-news/article/87280-children-s-institute-9-children-s-books-continue-run-of-strong-sales.html

Green Teacher. (2022, March 8). *Advertise with us*. Retrieved December 11, 2021, from https://greenteacher.com/advertise/

Guild of Natural Science Illustrators. (2019). *Education*. Retrieved April 29, 2022, from https://www.gnsi.org/education

Hahn, J. (2022, February 12). Blue bubbles helped "make the cause of climate change visible" say visualisers behind viral video. *Dezeen*. Retrieved December 12, 2021, from https://www.dezeen.com/2021/06/22/carbon-real-world-visuals-new-york-emissions-interview/

Haidt, J. (2012). *The righteous mind: Why good people are divided by politics and religion*. Pantheon.

Harrower, J., Parker, J., & Merson, M. (2018). Species loss: Exploring opportunities with art–science. *Integrative and Comparative Biology, 58*(1), 103–112. https://doi.org/10.1093/icb/icy016

Hassenzahl, D. M., Stephens, J. C., Weisel, G., & Gift, N. (2013). Art and climate change references. In B. C. Black (Ed.), *Climate change: An encyclopedia of science and history* (pp. 1–5). ABC-CLIO.

Hayhoe, K. (2017, October 9). Yeah, the weather has been weird. *Foreign Policy*. https://foreignpolicy.com/2017/05/31/everyone-believes-in-global-warming-they-just-dont-realize-it/

Hayhoe, K. (2021, September 12). no. Most politicians do understand it, even the ones who loudly deny it in public; it's their constituency that doesn't. [Twitter]. Twitter. https://twitter.com/KHayhoe/status/1437122870975598595?s=20&t=s4CFsmDTv7X3y13Kq3lBQw

Horst, M., & Davies, S. R. (2021). Science communication as culture: A framework for analysis. In *Routledge handbook of public communication of science and technology* (pp. 182–197). Routledge.

Houser, H. (2020). *Infowhelm: Environmental art and literature in an age of data*. Columbia University Press.

Ibrahim, A. M., Lillemoe, K. D., Klingensmith, M. E., & Dimick, J. B. (2017). Visual abstracts to disseminate research on social media. *Annals of Surgery, 266*(6), e46–e48. https://doi.org/10.1097/sla.0000000000002277

Inches, S. (2021). *Advocating for the environment: How to gather your power and take action*. North Atlantic Books.

Intergovernmental Panel on Climate Change. (2021). *AR6 climate change 2021: The physical science basis*. https://www.ipcc.ch/report/sixth-assessment-report-working-group-i/.

International Nature Journaling Week. (n.d.). Homepage. https://www.naturejournalingweek.com/

Iyengar, S., & Massey, D. S. (2018). Scientific communication in a post-truth society. *Proceedings of the National Academy of Sciences, 116*(16), 7656–7661. https://doi.org/10.1073/pnas.1805868115

King Abdullah University of Science and Technology. (2021, April 1). Adapting to climate change is business unusual. *KAUST Discovery, 11*. Retrieved December 13, 2021, from https://discovery.kaust.edu.sa/magazine/pdf/607feef298245f732c0796c5

King, H. M., Shubin, N. H., Coates, M. I., & Hale, M. E. (2011). Behavioral evidence for the evolution of walking and bounding before terrestriality in sarcopterygian fishes. *Proceedings of the National Academy of Sciences, 108*(52), 21146–21151. https://doi.org/10.1073/pnas.1118669109

Kipnis, A. (2018). Communication through playful systems: Presenting scientific worlds the way a game might do. *Integrative and Comparative Biology*. https://doi.org/10.1093/icb/icy087

Kiss the Ground. (n.d.). https://kisstheground.com

Koo, K., Aro, T., & Pierorazio, P. M. (2019). Impact of social media visual abstracts on research engagement and dissemination in urology. *Journal of Urology, 202*(5), 875–877. https://doi.org/10.1097/ju.0000000000000391

Landin, J. (2015, September 4). Rediscovering the forgotten benefits of drawing. Scientific American Blog Network. Retrieved September 12, 2021, from https://blogs.scientificamerican.com/symbiartic/rediscovering-the-forgotten-benefits-of-drawing/

Laws, J. M. (2017). *Wild wonder nature journaling conference*. John Muir Laws. https://johnmuirlaws.com/wildwonder/

Leiserowitz, A., Roser-Renouf, C., Marlon, J., & Maibach, E. (2021). Global warming's six Americas: A review and recommendations for climate change communication. *Current Opinion in Behavioral Sciences, 42*, 97–103. https://doi.org/10.1016/j.cobeha.2021.04.007

Lewandowsky, S. (2021). Climate change disinformation and how to combat it. *Annual Review of Public Health, 42*(1), 1–21. https://doi.org/10.1146/annurev-publhealth-090419-102409

LTER Network. (2020, March 27). *Schoolyard book series*. LTER. Retrieved April 29, 2022, from https://lternet.edu/schoolyard-book-series/

Mackay, I. P. M. (2021, September 4). *The Swiss cheese infographic that went viral*. Virology Down Under. Retrieved December 12, 2021, from https://virologydownunder.com/the-swiss-cheese-infographic-that-went-viral/

Markowitz, E., Hodge, C., & Harp, G. (2014). In C. St. John, M. Speiser, S. Marx, L. Zaval, R. Perkowitz, & Center for Research on Environmental Decisions and ecoAmerica (Eds.), *Connecting on climate: A guide to effective climate change communication*. Center for Research on Environmental Decisions and ecoAmerica. https://doi.org/10.7916/d8-pjjm-vb57

Marsh, O. (2020). Science, emotion and identity online: Constructing science and selves on 'I fucking love science'. In U. Felt & S. R. Davies (Eds.), *Exploring science communication: A science and technology studies approach* (pp. 171–189). Sage Publications. https://doi.org/10.4135/9781529721256.n9

McIntyre, L. (2021). *How to talk to a science denier: Conversations with flat earthers, climate deniers, and others who defy reason*. The MIT Press.

McMahon, R., Stauffacher, M., & Knutti, R. (2016). The scientific veneer of IPCC visuals. *Climatic Change, 138*(3–4), 369–381. https://doi.org/10.1007/s10584-016-1758-2

Merson, M., Allen, L. C., & Hristov, N. I. (2018). Science in the public eye: Leveraging partnerships—An introduction. *Integrative and Comparative Biology, 58*(1), 52–57. https://doi.org/10.1093/icb/icy034

Mesa Studios. (2021, January 15). *Microbiology*. Retrieved April 29, 2022, from https://www.mesaschumacher.com/microbiology/

National Intelligence Council. (2021). *Global trends 2040: A more contested world*. Retrieved December 12, 2021, from https://www.dni.gov/index.php/gt2040-home

National Research Council. (2009). *Learning science in informal environments: People, places, and pursuits*. National Academies Press.

National Science Foundation. (2021). *Future of Work | NSF National Science Foundation*. Retrieved December 11, 2021, from https://www.nsf.gov/eng/futureofwork.jsp

Norwegian Broadcasting Company. (2020). Our climate journey through Norway. In: NRK. Retrieved December 11, 2021 from https://www.nrk.no/chasing-climate-change-1.14859595

O'Neill, S. J., Boykoff, M., Niemeyer, S., & Day, S. A. (2013). On the use of imagery for climate change engagement. *Global Environmental Change, 23*(2), 413–421. https://doi.org/10.1016/j.gloenvcha.2012.11.006

Pelto, J. (n.d.). *Gallery*. Jill Pelto. Retrieved April 29, 2022, from https://www.jillpelto.com/gallery

Popovich, N., & Katz, J. (2021, July 21). See how wildfire smoke spread across America. *The New York Times*. Retrieved December 11, 2021, from https://www.nytimes.com/interactive/2021/07/21/climate/wildfire-smoke-map.html

Ramos, E., & Concepcion, B. P. (2020). Visual abstracts: Redesigning the landscape of research dissemination. *Seminars in Nephrology, 40*(3), 291–297. https://doi.org/10.1016/j.semnephrol.2020.04.008

Real World Visuals. (2012, October 19). *New York City's greenhouse gas emissions as one-ton spheres of carbon dioxide gas* [Video]. YouTube. https://www.youtube.com/watch?v=DtqSIplGXOA&ab_channel=RealWorldVisuals

Real World Visuals. (2017). *About*. Retrieved December 10, 2021, from https://www.realworldvisuals.com/about

Reason, J. T. (1990). *Human Error*. Cambridge University Press.

Remington, R. R., & Pontis, S. (2021). *Communicating knowledge visually: Will Burtin's scientific approach to information design*. RIT Press.

Roberts, S. (2020, December 8). The Swiss Cheese model of pandemic defense. *The New York Times*. Retrieved December 15, 2021, from https://www.nytimes.com/2020/12/05/health/coronavirus-swiss-cheese-infection-mackay.html

Romano Young, K. (2015). Log 200 October 8 2021.jpg | AntarcticLog Gallery. KarenRomanoYoung.Com. Retrieved December 15, 2021, from https://www.karenromanoyoung.com/antarcticlog-gallery?pgid=j9a88u0s-32f78ebf-b3c7-473a-80a5-af5a183877f6

Rosen, J. (2021, April 18). A climate change guide for kids. *The New York Times*. Retrieved December 11, 2021, from https://www.nytimes.com/interactive/2021/04/18/climate/climate-change-future-kids.html

Rudenko, A. (2018). Prehistoric body theater: Bringing paleontology narratives to global contemporary performance audiences. *Integrative and Comparative Biology*. https://doi.org/10.1093/icb/icy112

Sabraw, J. (2021). *FAQ*. John Sabraw. Retrieved December 14, 2021, from https://www.johnsabraw.com/faq

Say It with Science. (n.d.). *It's time to say it with science*. https://sayitwithscience.org

Schwabish, J. (Host). (2021, November 16). Steve Franconeri and Jen Christiansen a VisComm Workshop (No. 205) [Audio podcast episode]. In *PolicyViz podcast*. DataViz Community. https://policyviz.com/podcast/episode-205-steve-franconeri-and-jen-christiansen-a-viscomm-workshop/

SciArt Initiative. (n.d.). http://www.sciartinitiative.org

Statista. (2022, February 9). *New York Times Company: Digital news subscribers Q1 2014-Q4 2021*. Retrieved December 11, 2021, from https://www.statista.com/statistics/315041/new-york-times-company-digital-subscribers/

Støstad, M. N. (2020). NRK avslører: Slik blir klimaet i Oslo. In: NRK. Retrieved February 28, 2024 from https://www.nrk.no/klima/kommune/0301

STEM, ED Magazine. (2021). *About – STEM | ED Magazine*. Retrieved December 11, 2021, from https://www.stemedmagazine.com/about/

The Climate Explorer. (n.d.). https://crt-climate-explorer.nemac.org

The International Nature Journaling Week. (n.d.). https://www.naturejournalingweek.com

Thompson, C. (2016, June 27). The surprising history of the infographic. *Smithsonian Magazine*. Retrieved January 20, 2022, from https://www.smithsonianmag.com/history/surprising-history-infographic-180959563/

Thompson, A. (2018, December 3). Here's how much climate change could cost the U.S. Scientific American. Retrieved December 8, 2021, from https://www.scientificamerican.com/article/heres-how-much-climate-change-could-cost-the-u-s/

Thorp, H. H. (2021). It's not as easy as it looks. *Science, 374*(6575), 1537. https://doi.org/10.1126/science.abn7633

Tomb of Nebamun. (1350). [Painting]. British Museum, London, UK.

Upton, J. (2016, April 28). Climate change is 'devastating' The Great Barrier Reef. *Climate Central*. Retrieved April 29, 2022, from https://www.climatecentral.org/news/climate-change-devastating-great-barrier-reef%2D%2D20295

WashPostPR. (2020, August 6). The Washington Post's 2°C: Beyond the limit series recognized for outstanding explanatory reporting by the Society of Environmental Journalists. *The Washington Post*. Retrieved December 11, 2021, from https://www.washingtonpost.com/pr/2020/08/06/washington-posts-2c-beyond-limit-series-recognized-outstanding-explanatory-reporting-by-society-environmental-journalists/

Yale Program on Climate Change Communication. (n.d.). *What we do*. https://climatecommunication.yale.edu

Zaelzer, C. (2020). The value in science-art partnerships for science education and science communication. *Eneuro, 7*(4). https://doi.org/10.1523/eneuro.0238-20.2020

Zak, P. J. (2015, January). Why inspiring stories make us react: The neuroscience of narrative. In *Cerebrum: The Dana forum on brain science* (Vol. 2015). Dana Foundation.

Kalliopi Monoyios is driven by the conviction that science communicators operating in all spheres are a critical part of creating a scientifically literate public. After graduating from Princeton University with a degree in geology, she built her career as a science illustrator for the prominent paleontologist Neil Shubin at the University of Chicago. Her illustrations have appeared inside and on the covers of *Nature* and *Science* and in four popular science books, including *Your Inner Fish* by Neil Shubin. In 2011, she co-founded *Symbiartic*, a blog covering the intersection of science and art for *Scientific American*. From 2020 to 2022, she held the position of President of the Guild of Natural Science Illustrators, an international group of visual science communicators. She is currently developing new avenues of public engagement via her own art and curated exhibits that highlight the complexity of our relationship with plastic (www.kalliopimonoyios.com).

Kirsten Carlson is an illustrator, designer, photographer, and writer. She interprets topics relating to the ocean and sea life through the lens of science and art. Her focus is to develop creative ways to connect different audiences to nature. She strives to produce works that speak to audiences with diverse educational backgrounds—scientists, educators, children, and the public. She is a graduate of the University of California, Santa Cruz Science Communication Program, an alumni of the Artist-at-Sea Program with Schmidt Ocean Institute, and a grantee of the National Science Foundation Antarctic Artists and Writers Program. Her work can be seen at Fathom It Studios (www.kirstencarlson.net).

Taina Litwak has been a working illustrator since 1979 and a board-certified medical illustrator since 1994. She is currently the staff scientific illustrator with the Systematic Entomology Lab, US Department of Agriculture, at the Smithsonian Institution's Museum of Natural History. Her illustration work includes a wide spectrum of scientific subject matter. Hundreds of her illustrations have appeared in magazines and scientific journals, trade and textbooks, advertising campaigns, nature centers, and medical-legal exhibits. She works primarily in digital media. Taina is currently a board member of ASCRL, the American Society for Collective Rights Licensing. Her volunteer work with other professional organizations includes many years on the Board of Directors of the Guild of Natural Science Illustrators (GNSI), as president and as treasurer, and on the Board of the Vesalius Trust (www.litwakillustration.com).

Tania Marien is a podcast producer and educator. Her projects draw on her experiences working as an independent environmental education professional, first as the full-time editor, educator, and bookseller at ArtPlantae and now as a podcast producer, writer, researcher, and network builder. Tania connects independent environmental education professionals with new audiences to build partnerships and enhance environmental literacy in communities. She believes that independent professionals are overlooked and that their professionalism and contributions to lifelong learning need to be recognized. Documenting their work is important because it fills the knowledge gap about how people learn about science and the environment outside of the classroom. Tania has an Interdisciplinary Studies Master's degree in Biology & Student Learning and a professional certificate in Free-Choice Learning, which addresses the learning that occurs outside of the classroom throughout one's life. Tania is a contributor to The Carbon Almanac and The Carbon Almanac Podcast Network (www.talaterra.com).

Fiona Martin specializes in vector illustrations, infographics, and layout designs for scientific publications. She is passionate about marine, climate, and earth sciences, public health, and education. At the age of 2, her parents discovered she had a significant hearing loss. Despite being born almost deaf, she managed to adapt to a hearing world, attend public schools, and—out of necessity—learn the art of lip-reading. Fiona defied expectations, graduating as high school valedictorian, earning a BS in Marine Biology (summa cum laude), and obtaining a graduate certificate in Scientific Illustration. Fiona appreciates visual communication because it is like a universal language—transcending barriers and facilitating understanding of complex ideas "at a glance." She has 17 years of experience rendering scientific graphics, both from her studio Visualizing Science, and as a former Senior Illustration Editor at Annual Reviews. Major clients include NOAA's Climate Program Office, NPS, MBARI, Dr. Peter Vine, and numerous scientific authors (www.visualizingscience.com).

Music as a Vehicle for Climate Change Communication: The ClimateMusic Project

Catherine Emma Dixon, Laurie S. Goldman, Stephan Crawford, and Phoebe Camille Lease

ClimateMusic: Purpose and Goals

The ClimateMusic Project, a cross-disciplinary collaborative spanning the arts, science, technology, and public policy, aims to present music informed by climate science as a stimulus and space for listeners to process feelings of anxiety about the future, grief, and guilt about the past and the present and to transform negative feelings into positive actions. Within its compositions—including *Climate, Icarus in Flight, What if We?, Audyssey,* and *Voice of the Animals* (Fig. 1)—The ClimateMusic Project has highlighted two different scenarios: one where humanity continues "business as usual"—with insufficient further action to tackle climate change—and an alternative scenario where the world takes quick and significant action to reduce emissions in the first half of this century. The second path centers on the solutions aspect of climate change and can make people feel more confident in their ability to bring about change. In *What if We?* composed by Wendy Loomis, the "business as usual" scenario is sonified as "strident, loud and chunky" with a C minor chord structure; in contrast, the sonified mitigation scenario is played in major key rounds of B♭, A, and B, in a gospel-like sound. Loomis says it is "lightly syncopated [to represent] the possibility of change [...] a touch of hope is felt" (The ClimateMusic Project, 2019). The sonic difference between the two scenarios is significant and allows people to hear and feel the impact of our choices on our continued existence.

The ClimateMusic Project defines ClimateMusic as music that is directly informed by widely accepted, peer-reviewed climate science (ClimateMusic, 2022). It is fully rooted in the domain of the arts, but informed by reference data sets pertaining to climate science and reflected in the music composition. It has a didactic aim of communicating aspects of the science that underlie the urgency of the climate crisis. It has an artistic aim of captivating its intended audience with expressive

C. E. Dixon · L. S. Goldman (✉) · S. Crawford · P. C. Lease
The ClimateMusic Project, San Francisco, CA, USA
e-mail: info@climatemusic.org

© The Author(s) 2024
E. Coren, H. Wang (eds.), *Storytelling to Accelerate Climate Solutions*,
https://doi.org/10.1007/978-3-031-54790-4_12

Fig. 1 A visual slide from Wendy Loomis' piece, *What if We?*

music in different genres and therefore seeks out a diverse array of musicians in order to reach people with the types of music that resonate with them the most. This balance between didactic and artistic components is a defining characteristic of ClimateMusic and distinguishes it from purely data-driven works that sonically display climate information. The idea is to create compelling, relatable music that sparks new insight and conversations around the potential for individual and collective action to make a difference on climate change in the short amount of time that there is to tackle the issue.

The purpose of ClimateMusic is to deliver information about climate change in an emotionally engaging way. Its goal is to connect communities through music, providing inspiration, a space for reflection, and a supportive environment with opportunities for learning and action. The ClimateMusic Project leverages the universal appreciation of music to communicate scientific information, inviting audiences to engage with the crisis both analytically and emotionally. ClimateMusic utilizes the expectation-based underpinnings of musical expression to promote future-oriented thinking, encouraging listeners not only to imagine but to feel future scenarios, and inviting them to respond accordingly. In this way, listeners become emotionally invested in the continuation of the music and, correspondingly, in the future climate scenarios that shape the music. In delivering live performances of original music both in person and online through streaming platforms, The ClimateMusic Project unites people across the globe in experiencing new music informed by climate science, inviting audiences to learn and take action together through collaboration with science advisors and the provision of action pathways.

Information about the dangers of global warming is widely available and yet action toward mitigation remains insufficient. Results from public surveys show that, although the majority of Americans understand that global warming is happening and that it is mostly human-caused (Leiserowitz et al., 2021), far fewer are taking direct action to mitigate that risk (Carman et al., 2021). Moreover, public support for the immediate and transformational actions required is limited: According to a public survey by the Pew Research Center (Tyson et al., 2021), the majority of Americans oppose phasing out the production of new gasoline cars and trucks by 2035, and only a third of Americans support phasing out fossil fuels entirely.

Current methods of climate change communication are failing to motivate action appropriate to the scale and immediacy of the crisis. There is a growing awareness in the scientific community that quantitative information and objective analysis often fail to engage public interest and that new approaches are required to communicate in a way that is effective and meaningful (Downs, 2014; Marshall, 2014; Slovic & Slovic, 2015). This chapter reviews the evidence for music-based approaches to better communicate the risks posed by climate change in order to mobilize action and their application in The ClimateMusic Project. Since the efficacy of music-based approaches to climate change communication has not yet been studied, this chapter reviews the experimental evidence from more well-researched areas, such as music interventions for health, well-being, and learning. The authors discuss the methodologies used for communicating in ClimateMusic compositions and performances, and their outreach projects, partnerships, and impact. Finally, the authors examine future opportunities for music in climate communication, including potential funding sources and interdisciplinary projects.

Why Music?

Artistic mediums such as music have become an increasingly popular method of communicating science. Humans take in an extraordinary amount of information each day. Our neural mechanisms that allow us to process our environment evolved in the life-and-death conditions of our ancestors. For our brains, "information is meaningful [to us] insofar as it evokes emotion" (Martinez-Conde & Macknik, 2017, p. 8128). Humans have an intrinsic response to music, and its emotional impact can connect with audiences who do not regularly engage with scientific reports and information. Music is also a well-tested method of uniting disparate groups of people for a single cause (Reese, 2015). ClimateMusic gives complicated climate science an emotional boost to make the climate crisis—and the solutions needed to mitigate it—a more intimate, relatable, and pressing issue.

Historically, social movements have integrated their new values and ideals into popular culture in order to gain traction and attract new supporters. For example, music was a key tool in the push toward racial integration during the US Civil Rights Movement. Andrew Young, the former executive director of the Southern Christian Leadership Conference, explained: "[Political activists] knew how little chance they stood of gaining people's trust if they presented themselves as straight out organizers: people were too afraid to respond to that approach. So they organized gospel groups and hit the road." (Reese, 2015). Soul, gospel, and R&B groups such as The Staple Singers wrote songs that appealed to a mainstream audience while working in strategic messages of solidarity and racial justice for those in the know.

In the Vietnam War era of the 1960s and 1970s, explicit protest music hit the mainstream with anti-war songs like "Fortunate Son" by Creedence Clearwater Revival, "I-Feel-Like-I'm-Fixin'-To-Die Rag" by Navy veteran Country Joe MacDonald, and "I Ain't Marching Anymore" by Phil Ochs. While these songs raised awareness across everyday American households, they also served as a lifeline to soldiers on the battlefield. Listening via cassettes, underground radio stations, or around the campfire, protest music helped them to deal with the violence they had seen, and many veterans say it was an imperative part of their healing process back home. As author Michael Kramer notes, this music was "a sonic framework for thinking, feeling, discussing and dancing out the vexing problems of democratic togetherness and individual liberation" (Bradley, 2018).

Music and Emotional Engagement

Emotional engagement is essential to communicating meaningful information that can be used in judgment and decision-making (Slovic & Slovic, 2015, p. 28). Climate change is often communicated in a scientific manner that invites analytical processing, but fails to engage audiences emotionally. In his book *Don't Even Think About It: Why Our Brains Are Wired to Ignore Climate Change*, George Marshall (2014, p. 50) has argued that climate change communication fails to engage affective processing and as a result fails to motivate action:

> The divide between the rational brain and the emotional brain is embedded in the historical boundaries between science, the arts, and religion, and it is a particular risk for an issue that originates strongly in just one cultural domain—as climate change does with science—that finds it hard to engage our entire cognition. The view held by every specialist I spoke to is that we have still not found a way to effectively engage our emotional brains in climate change (Marshall, 2014, p. 50).

Music is often used to induce moods and to modulate and regulate emotional states. Music can provoke powerful emotions as well as physiological reactions in listeners including crying, thrills or chills, goosebumps (feeling as though one's hair is standing on end), and changes in hormone levels, heart rate, and other physiological parameters. Music is common to all known cultures and societies worldwide and has been since prehistory; thus, it is regularly used to support customs and behavioral contexts such as festivities and religious activities. There is evidence to suggest that psychological responses to music may be universal (Mehr et al., 2019) and that these responses to music are not significantly affected by age or formal training (Bigand et al., 2005; Hevner, 1935; Terwogt & Van Grinsven, 1991). As an accessible and enjoyable medium that can transcend cultural barriers and be understood by audiences of all ages, levels of training, and exposure, music may have the potential to enhance climate change communication by boosting audience engagement and emotional response.

Music and Narrative

Climate change communication typically delineates future scenarios in order to educate the public about the impact of current behaviors. George Marshall (2014, 63) has warned that, in the case of climate change, "The lack of a definite beginning, end, or deadline requires that we create our own timeline." Due to the complex and impending nature of the threat, climate change communication may benefit from a narrative approach to support information delivery and promote emotional engagement and future-oriented thinking. Narrative can help humans process complex information by personalizing the data and engaging the emotional response necessary to motivate action (Slovic & Slovic, 2015). Sociologist Julie Downs has recommended the use of narrative methods to communicate scientific information:

> Narrative can captivate the audience, driving anticipation for plot resolution, thus becoming a self-motivating vehicle for information delivery. This quality gives narrative considerable power to explain complex phenomena and causal processes, and to create and reinforce memory traces for better recall and application over time (Downs, 2014).

Narrative approaches are gaining traction in public health communication to motivate and support health-promoting behavioral change (Downs et al., 2004; Hinyard & Kreuter, 2007; McCall et al., 2019; Rieger et al., 2018). Use of storytelling and narrative approaches compared to statistical evidence alone has shown positive outcomes in studies on safer sex communication, patient education about medication, cancer screening and prevention, and vaccination rates (Donné et al., 2017; Downs et al., 2004; Hopfer, 2012; Larkey et al., 2009; Mazor et al., 2007) and has been shown to be especially effective in disadvantaged and minority populations (Larkey et al., 2009; Lee et al., 2016).

As a temporal art form, music relies on listener anticipation and the disruption and realization of expectations as a means for expression and as such is uniquely well-positioned to convey narrative (Huron, 2006). Music psychologist David Huron (2006, p. 361–366) theorizes that emotional responses to music act as "motivational amplifiers," inviting listeners to form expectations about the continuation of the music with the result that they feel emotionally invested in its continuation. Musical devices shape and manipulate listener expectations for expressive effects, such as harmonic tension and resolution, syncopation, cadential formulae, and large-scale tonal and motivic structures. Anticipation of musical pleasure in response to these expectations can stimulate dopaminergic activity, eliciting a pleasurable response and motivating the listener to want and pursue more of that musical experience (Ferreri et al., 2015).

Music and Learning

In the past, the concept of general intelligence (IQ) reigned popular as a theory that humans have a broad and measurable mental capacity that predicts cognitive ability. However, recent theories such as Gardner's Theory of Multiple Intelligences argue that humans actually possess multiple (eight or more) relatively autonomous entry points into a subject. Among these eight intelligences is music, providing a strong case for integrating sound into the learning process.

Music has been used as a tool to draw attention to significant data points. Data sonification helps scientists find abnormalities that may be harder to pick up on in spreadsheets of data. In 2018, Stanford's Experimental Physics Lab turned solar data from the Solar and Heliospheric Observatory into sound, allowing scientists to study different aspects of the sun like solar flares or coronal mass ejections that are not visible to the eye (Atkinson & Sosby, 2018). Scientists from MIT have sonified the COVID-19 spike protein by assigning each amino acid to a particular note on a musical scale and accounting for other characteristics with volume and duration changes. This sonic format, the researchers explained, helped to identify places in the protein to which antibodies or drugs could potentially bind (Venugopal, 2020). These sonification projects and others have been featured on music streaming platforms, podcasts, and popular science websites, demonstrating the potential of sound as an engaging and accessible science communication tool (Hermann et al., 2011).

There is evidence to suggest that music may enhance memory and facilitate information retention. Music may activate the dopaminergic mesolimbic system, with potentially beneficial effects on memory, attention, executive functioning, mood, and motivation (Sihvonen et al., 2017). Music is frequently employed as a memory-enhancing tool in early educational settings through nursery rhymes and songs. In a 2015 study, students from kindergarten to high school age were exposed to both educational music videos and non-music materials. The participants reported a higher enjoyment of the music-based information, suggesting that memories may be stronger if encoded in an emotional state (Crowther et al., 2016).

Many studies have also shown that music-based tools may enhance language learning. A study of 109 university students enrolled in a language course found that the use of Spanish-language music video facilitated engagement and narrative immersion, resulting in statistically higher levels of learning, thematic comprehension, and enjoyment (Benitez-Galbraith & Galbraith, 2019). A study with 22 participants found that adding musical information significantly improved language acquisition. The authors concluded that the learning process was supported by the motivational and structuring properties of music, and recommended simultaneous engagement of emotional and analytical processes to achieve optimum learning (Schön et al., 2008). A study on 22 young adults showed that background music improved episodic memory performance (Ferreri et al., 2015). Another study found that participant-selected music improved information retention, with a positive correlation between emotional arousal and enhanced recall, indicating the importance of audience preference to effective information delivery (Carr & Rickard, 2016).

Although there is no consensus on whether music directly improves information retention, the positive emotional connection could lead to more thoughtful reflection and boost engagement with the subject.

Music and Well-Being

A key challenge in climate communication is the overwhelmingly negative nature of the subject. As psychoanalyst Sally Weintrobe has suggested, "too much unacknowledged anxiety is one of the most important obstacles to our effective engagement with climate change" (2013, p. 9). Therapists have reported seeing more and more people with ecological grief, described as the anguish felt from experienced or anticipated ecological losses from climate change (Cunsolo & Ellis, 2018; Whitcomb, 2021). Music can be particularly effective in addressing feelings of ecological grief and anxiety because of its power to induce, modulate, and regulate emotional states. Scientists hypothesize that listening to sad music may evoke prosocial emotions that encourage sharing feelings with others (Eeerola et al., 2018). Indeed, music can even bring people to tears, but instead of being stressful, researchers believe this can induce physiological calming or catharsis (Mori & Iwanaga, 2017).

Art must provide a scaffolding for processing eco-grief and imagining a way forward outside of climate catastrophe. Although the effects of music interventions on climate-related psychiatric conditions have not yet been studied, evidence suggests that music may improve mental health outcomes for patients suffering from grief, depression, and anxiety. Music plays an important part in mourning and death rituals in many cultures and has been associated with positive memories of the dead (Viper et al., 2020). Evidence suggests that participants receiving music therapy show a greater decrease in grief symptoms compared to those receiving standard care (Iliya, 2015; Yun & Gallant, 2010). Meta-analysis of studies into the effects of music therapy on depression has shown that music therapy in conjunction with treatment as usual is more effective than treatment as usual alone for both clinician-rated depressive symptoms and patient-reported depressive symptoms and may decrease anxiety levels and improve functioning levels (Aalbers et al., 2017). There is also evidence to suggest that music-listening without the support of a music therapy professional may improve mental health outcomes. Meta-analysis has also shown moderate-quality evidence that listening to music may improve sleep quality in adults with insomnia and alleviate anxiety and depression in a variety of clinical contexts (Bradt & Dileo, 2014; Bradt et al., 2013; Jespersen et al., 2015).

Music as a Community Skill

As an experience that brings together large numbers of people with varied values and life experiences, live music performances provide ideal conditions for establishing social norms and supporting behavioral change. In this way, music functions not only as a performance art, but as a community skill that can bring people together to encourage socially positive behavior and improve social cohesion.

Evidence suggests that emotional engagement resulting from a musical stimulus can promote altruistic behavior. An experiment found that students were more likely to give aid after listening to music that they had selected to evoke a strong emotional response (Fukui & Toyoshima, 2014). Another experiment on students found that listening to soothing music promoted helpful behavior in comparison to stimulating music, aversive music, or no music at all (Fried & Berkowitz, 1979). A study on workplace behavior similarly found that listening to happy music significantly improved cooperative behavior among employees compared to unhappy music and no music at all (Kniffin et al., 2017).

Rhythmic synchronization through shared active or passive musical activity may be important in facilitating self-other merging and promoting social bonding. Functional MRI studies suggest that music-listening may activate a mirror-function mechanism, whereby the listener mimics internally not only the expression perceived in the music, but also the perceived action behind the musical stimulus, engaging a similar or equivalent motor network to that of the performer (Molnar-Szakacs & Overy, 2006). It is thought that this shared representation of sung or played musical experiences by performer and perceiver may facilitate emotional response, communication, empathy, and expression of intentions, without any active participation required on the part of the listener.

Music and Climate Action

There is a growing canon of art that reflects our existing ecological crisis; between the genre of dystopian climate fiction (or "cli-fi") and musical expressions from rock to hip-hop to classical, many artists are reacting to the consequences of climate change. But what is equally important yet vastly underused is the ability of pop culture to shift our society's values toward stewardship, regeneration, and that which author Favianna Rodriguez (2020, p. 121–127) has described as "building a cultural strategy."

Several artists have successfully used music to convey the urgency of climate change through emotional lyrics and instrumentation. The Hip-Hop Caucus, a non-profit that organizes young people toward action using pop culture, has created a new soundtrack to the climate movement with their "Think 100%" campaign: "Eschewing the gentle folk sounds of typical environmental anthems, we bring Hip Hop, R&B, and pop to bend the climate genre while feeding our souls with the power of music" (Think 100% Music, n.d.). Another young hip-hop artist, Xiuhtezcatl Martinez, is well known for his participation in the landmark Juliana v. United States of America case, which argued that the government had knowingly violated the plaintiffs' rights to life through the promotion of fossil fuel combustion (Our Children's Trust, 2021). His music, released with his sibling Tonantzin

Martinez under the name Earth Guardians, carries strong themes of earth steward-ship and indigenous rights to the land (Eyen, 2017).

There are several examples of analogous projects of science-driven climate music. In 2013, University of Minnesota student Daniel Crawford and geography professor Dr. Scott St. George composed a cello piece using 133 years of global temperature data. This piece, called *Song for Our Warming Planet*, highlights rising temperature through a rapidly increasing pitch. "We're trying to add another tool to that toolbox, to communicate these ideas to people who might get more out of this than maps, graphs or numbers," says Crawford (Yeo, 2013). The duo released another piece in 2015 called *Planetary Bands, Warming Planet*; building on the first composition, this newer piece uses different stringed instruments to represent tem-perature records in different planetary zones.

At Stanford University's Center for Computer Research in Music and Acoustics, Director and Composer Chris Chafe has created several environment-based data sonifications, including his project *Smog Music*, which uses air quality levels from global urban centers to create an electronic soundscape (Hart, 2010). He has also collaborated with graduate students at UC Berkeley to create an untitled piece that sonifies 1200 years of global temperatures and CO_2 concentration levels. Dr. Lauren Oakes and Dr. Nik Sawe wrote a conceptually similar composition using research on the decline of Alaskan cedar trees. Their 3-min piece, which sonified data on over 2000 conifer trees, opened avenues to a larger audience beyond academic jour-nal readers (Kahn, 2016).

Other science institutions are also starting to partner with arts-based groups. In 2018, an exhibit called "Sounding Climate" debuted at the National Center for Atmospheric Research in Boulder, Colorado (Gardiner, 2019). And in 2020, the World Bank partnered with seismologist and musician Dr. Lucy Jones on her initia-tive with Haitian artist Tafa Mi Soleil called "Creating Change Through Music: Understanding Risk" (Climate Music Initiative, 2020). Tafa Mi Soleil composed a song about hurricane preparedness, with lyrics such as "Prevention is better than cure...protect your families, protect your homes....my friends, watch out, streams may flood." She explains, "We Haitians have a very close relationship with music. Even if we are going through tough times, we'll be singing...I think the message we want to share spreads so much faster through music (Climate Music Initiative, 2020)."

Another noteworthy recent project is the collaboration of composer Jamie Perrera and filmmakers Leah Borromea and Katharine Round on *Climate Symphony*. The piece originated in a journalism competition to tell climate stories through a new medium, and they hope to alter the composition using localized data for live audi-ences around the world. The symphony is meant to be performed in a dome-shaped venue, where short film clips of the natural world are projected to play alongside the music. Borromeo says they hope the piece can also serve as a fact-checking method against climate change disinformation: "We want to create a formal record [that's] revealing. You're looking at it, and listening to it, and you find that [the music sounds] distorted. It's all distorted" (Simon-Lewis, 2017 para. 9). It is harder to deceive someone about the gravity of climate change when, through the medium of music, listeners can hear and feel it clearly for themselves.

As climate change becomes an increasingly pressing issue, people may turn to art for solace, calls to action, and even solutions. In the past, some musicians have turned to broad, anthemic "save the planet" tunes. While there is a time and place for this approach, it can be difficult for songs with vague messages to spur the type of behavioral change that is needed. In an episode of the podcast *Switched on Pop*, Nathan Sloane and Charlie Harding (2021) explain why a more detailed song about climate change that takes a stronger stance may resonate with listeners even more effectively:

> Successful music about the environment uses contemporary culture in the lyrics and instrumentation to grab people's attention and make them want to invest in a collective reimagining of the future. Artists can offer a reflection on who we are and what we desire. These songs are working on more of a natural, intimate level rather than a sermon (Sloane & Harding, 2021).

Both science-inspired and culture-inspired approaches can bring in new audiences in the fight against climate change. With these examples in mind, it is clear that the opportunities for science and art collaborations on climate content will continue to grow as more businesses, art organizations, and governments make addressing climate change a core issue.

ClimateMusic Methodologies: Creating ClimateMusic

ClimateMusic is created through an active collaboration between a composer and a small team of scientists and experts in public policy and engagement. The science advisory team includes renowned scientists who are leaders in their fields, affiliated with top academic institutions and key participants in the United Nations Intergovernmental Panel on Climate Change (IPCC) climate assessments. These distinguished scientists donate their time to the project because they see the value in using new and creative tools to communicate the science in a way that inspires; as science advisor Dr. Andrew Jones (2021) has explained, "we need more than just... scientific information in order to understand and act appropriately in response to something as complex as climate change [and to do so requires] our whole human selves, our imaginations, our emotions, our values, and our communities and culture." The frequent and active interaction between science advisory team members and individual composers during each project is the key ingredient in creating music that achieves both didactic and artistic aims.

Roles

The role of the ClimateMusic team, including the science team and public engagement experts, in relation to the composer is to (Fig. 2):

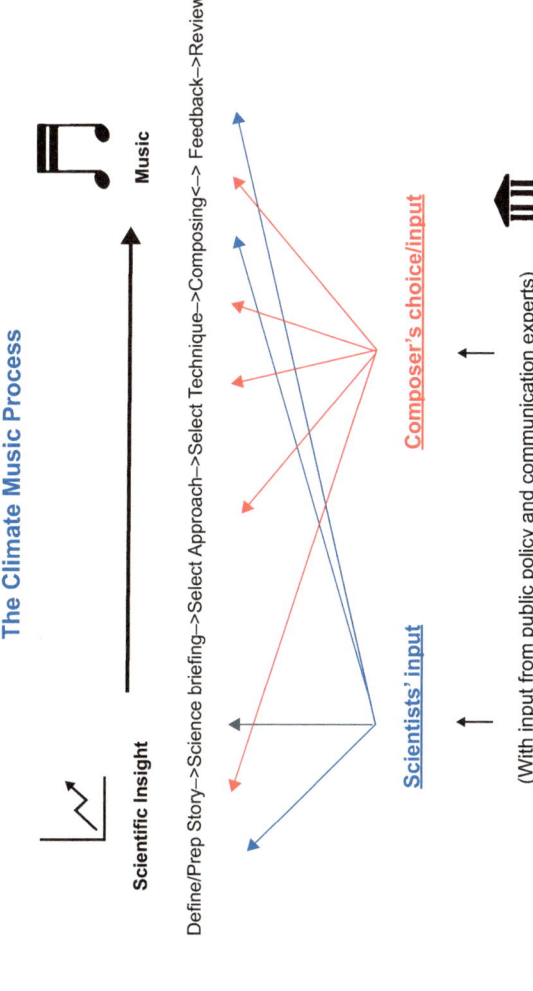

Fig. 2 An idealized illustration of the ClimateMusic process for creating new music. (Image source: The ClimateMusic Project)

- Co-define the story's (defined below) thematic focus and narrative
- Educate the composer on the science underlying the story
- Provide reference resources about the science
- Offer feedback throughout the process, and help calibrate the scientific references as integrated into each piece as necessary

The role of the composer is to

- Co-define the story's thematic focus and narrative
- Become familiar with the basic scientific insights reflected in the story (with the support of the science team)
- Refer to the reference materials and integrate the scientists' feedback at specific milestones during the compositional process

The expectation is that, within the scientific framework defined by each story, the composer has a maximum of creative freedom. This emphasis on the artist's unique expressive voice is what makes ClimateMusic familiar and approachable to audiences.

Composer and Scientist Selection Because ClimateMusic requires music to be composed within a set framework, it looks for artists who regard working under some constraints as an interesting creative challenge. Composer Erik Ian Walker has described his motivation to work with The ClimateMusic Project as a "desire [...] to bring the data to life as sound [...] bringing home the reality of the climate crisis in a way that nothing else could do" (Email message to Stephan Crawford on August 4, 2021). The composers do not require scientific aptitude, but an openness to learning about climate science is necessary. Experience with cross-disciplinary collaborations, e.g., in theater, dance, and film, is also important, since the process involves sharing ideas and working productively with people who may not have experience with music composition. Perhaps surprisingly, a strong interest in or active engagement on climate change—while certainly welcome—is not a prerequisite for the composers. An openness to working with the science team and absorbing its insights is sufficient and in practice has tended to leave each composer with a heightened awareness of the issue and the foundation to become an effective advocate. Walker has described his work on the project as "easily the most intriguing collaboration I've ever been involved with."

> I was already very aware of the dire situation we find ourselves in, but in working on the piece, I gained a deep respect for those in the sciences that work with this every day, year in and out. It is difficult to cope with it as a daily reality—as an artist, it was a challenge to not get overwhelmed at times. [...] I have often worked with the concept that music is but a veneer in front of what is chaos and noise, a chaos that is really just under the surface. Composing *Climate* was the ultimate opportunity to realize this lifelong thread that I have explored. It was a culminating piece to the central theme of my composing life (Email message to Stephan Crawford on August 4, 2021).

Scientists working on ClimateMusic collaborations do not need to have experience with music, though The ClimateMusic Project has recruited several who are themselves accomplished musicians. The chief qualifiers are relevant subject matter expertise and a willingness to participate in a unique creative cross-disciplinary collaboration. Ideally, they also possess the ability to communicate complex scientific concepts effectively to a lay audience. The scientists on The ClimateMusic Project team have tended to be self-selecting, i.e., they recognize that there is a critical need to communicate climate science in fresh ways that resonate with a broad public, and they are willing to devote time and energy toward this goal.

Stories

Stories in this context are not fictional accounts of climate change, but rather written descriptions of aspects of climate science that serve as starting points for new music. The story is the thread that weaves together the key scientific narrative being communicated and is buttressed by peer-reviewed data. Stories are varied in subject matter, which has so far included the physical science underlying climate change, human drivers of the crisis, how climate change affects sea level rise, energy solutions, and natural landscapes and biodiversity. Additional stories under development are energy access and ecosystem cycles. As varied as these stories are, they share three common denominators:

- They underline the urgency of the climate crisis.
- They address aspects of the climate crisis that are critical for the public to understand.
- They demonstrate the opportunities and potential for taking action to limit the severity of the climate crisis.

Each story includes a written description and accompanying data sets from relevant peer-reviewed studies, which are included in the story "package" as reference material. The composer can access the data in raw and graphical formats and also as sonifications. Both historical data sets and future scenarios created through powerful climate models are included.

Stories are created through a consultative process that includes subject matter experts in the sciences, public policy, and (usually) the collaborating artist. Once a story is produced, it can be used by any number of collaborating artists.

Story Preparation   The first step once a story has been defined is to provide the composer with a written summary and supporting materials. The science team completes this task, which includes identifying and accessing relevant peer-reviewed studies and data sets. In some cases, the data sets must be converted into a human-readable format and then further formatted to allow for an intuitive understanding for non-scientists, e.g., via visual and audio data displays. The ClimateMusic Project draws only from peer-reviewed sources and to date has relied mainly on studies associated with the Intergovernmental Panel on Climate Change (IPCC). This includes studies that explore possible future outcomes that are based on different mitigation and adaptation scenarios.

The Science Briefing Before the composer begins work, they receive a briefing on the science underlying the story. This is an opportunity for the composer to ask questions and for the science team to emphasize key insights and take-aways. The level of detail presented depends on the composer's level of interest and familiarity with scientific information. At a minimum, the briefing is intended to provide enough insight for the composer to be able to begin to formulate broad ideas for the composition.

Feedback and Calibration Multiple opportunities for the ClimateMusic team to provide feedback to the composer during the process of composing are at the core of each collaboration. These are typically scheduled as either in-person or virtual meetings, augmented by follow-on email communication. They provide an opportunity for the composer to ask questions and for the science team to help the composer interpret the scientific references at an appropriate scale in the music.

Documentation (Apollo) The creation of each ClimateMusic composition is documented according to a process template referred to internally as "Apollo," which serves three purposes. First, by templating certain functions, e.g., data identification, access, and preparation, Apollo introduces a standardized work process that is designed to reduce duplication of effort as each new composition is launched. Second, it documents precisely how the scientific story is reflected within the music, which is critical given The ClimateMusic Project's science-oriented and public-facing mission. Finally, it provides a detailed resource that allows team members who may not have worked on a given composition to speak about it authoritatively in public.

Compositional Approach A composer will take one of two compositional approaches

- *Constructive Approach*: In this case, the composer works with the science team from the start to discover ideas from the story and its supporting data that then develop into the composition.
- *Deconstructive Approach*: With this approach, the initial process of composition is completed independently of the science team. The composer then works with the science team to explore the musical consequences of "colliding" all or part of the score with the data.

In addition to being quite different creative pathways, the approach chosen also has workflow implications, since, in the case of a deconstructive approach, active collaboration would commence only after the composer has sketched out initial ideas for the music. These categories are not rigid, and in practice, a composition may include elements (e.g., movements) stemming from both approaches.

Techniques for Reflecting the Science Within the Music

To date, The ClimateMusic Project has applied three techniques for reflecting the science and data within the music: parameter mapping, audification, and information integrated into spoken word elements or lyrics. Others have written extensively on parameter mapping and audification (e.g., Hermann et al., 2011), and consequently, this section is limited to describing each technique as The ClimateMusic Project team has applied it, and specific limitations it has identified in each technique.

Parameter Mapping This technique relates data parameters to specific music characteristics or elements. These music analogs, e.g., characteristics such as tempo or pitch, track changes in the data over time as measured against a reference value. For example, in his string quartet *Icarus in Flight*, composer Richard Festinger relates global population data to the density of musical events. As the population numbers increase over time, the music begins to sound proportionally "crowded" and also gives the impression of increasing speed.

Parameter mapping is a versatile technique that leaves a composer considerable scope for the creative definition and application of mappings comparable to the roles assigned in a theatrical play. Obvious mappings might link data to musical characteristics such as pitch, tempo, or dynamics, but can also include anything that is countable on the score, e.g., the relative proportion of specific instrument types playing at any given point in time. One could imagine, for example, a mapping in which data on the number of invasive species established in a geographic area over time is represented through a proportional intrusion of a horn section into music for a string quartet.

Mapping design is important to determine the effectiveness of the music in conveying the scientific story (see Hermann et al., 2011, 385–388). Four design elements are noted here. First, the optimal number of mappings must be considered. Human ability to discern discrete elements in a soundscape is impressive, but it is not infinite (Hermann et al., 2011, 3 and 373). The maximum number of mappings used in ClimateMusic compositions has been four. This decision is not based on science, but rather has been based on an intuitive assessment of the number of mappings that will be easily discernible by most people and that are sufficient to convey key insights from each scientific story. This discernment is enhanced by the second critical design factor, which is how well each mapping analog can be distinguished from the others. Each musical feature should have a distinct timbre or other quality that is easy for an audience to identify. For example, in *Icarus in Flight*, Festinger selects analogs that are easily distinguishable and also work well together artistically to give the music a compelling tension and texture (Fig. 3).

For Festinger, "The challenge in composing *Icarus in Flight* was one of helping induce the listener to experience, through musico-dramatic means, the trajectory of three important human drivers of climate change over a span of 200 years." Festinger describes his compositional process below:

Icarus in Flight Mappings
Human drivers 1880-2080
1 year=8 seconds (27 minute total)

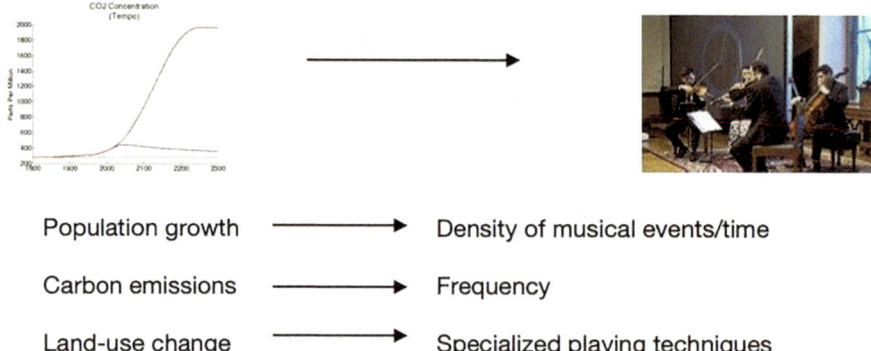

Population growth ⟶ Density of musical events/time

Carbon emissions ⟶ Frequency

Land-use change ⟶ Specialized playing techniques

Fig. 3 Graphic showing the mapping design of *Icarus in Flight*

The curves traced by the data sets are in general exponential in shape, and not conducive to musical processes, which typically have much more complex shapes, and wave-like characteristic[s], with successive peaks and troughs arranged to create a hierarchy of climactic moments designed for an overall integrated dramatic impact. For this reason, the strategy was adopted of mapping the data sets (population growth, carbon emissions and transformation of land use) to relatively non-specific musical elements (in this case, density, frequency range, and timbral distribution respectively), so as to govern the formal profile of the composition as a whole, without determining local musical details of melodic, harmonic or rhythmic materials (Email message to Stephan Crawford, August 20, 2021.)

Festinger's work presents a third critical design element, that of "polarity," or the direction of movement selected for each musical feature (see Hermann et al., 2011, 385). Carbon emissions are mapped to frequency range, in this case both rising and falling, beginning from the middle of the register, so that as the piece progresses, rising data values for carbon emissions trigger notes that are increasingly higher and lower in pitch. Mapping an increase in a data value to an increase in a musical feature's value seems to be intuitive. For example, mapping an increase in the value of carbon emissions data to an increase in tempo intuitively seems more effective than mapping this increase to a decrease in tempo. However, whether the polarity of a mapping is effective or not is highly context-dependent, and there are few absolutes. Festinger's bipolar mapping for carbon emissions results in very high and very low note frequencies by the end of the composition, representing carbon emissions in the late twenty-first century under a low mitigation scenario. Since music mainly occupies the mid-range frequencies, the extremes of register here create a powerful effect and communicate an urgent sense of alarm.

A key limitation—but also a strength—of this technique is the issue of how to scale the movement in the musical analogs in a way that accurately expresses the meaning in the data, not merely the individual data values themselves. For Festinger, this represented another core challenge to the compositional process:

> Tracking the evolution of the data sets to evolve in correct proportion over the duration of the composition requires detailed pre-compositional planning in which the total duration is broken into a succession of modules each representing roughly five years of historical (or projected) time, and each module assigned a density, frequency range, and timbral distribution commensurate with the corresponding range of data points. Thus, the large-scale profile of the piece will convey, through the data mapping, the overall progression of the three time series, while invention of melodic, harmonic and rhythmic materials, as well as the local flux and flow of high and low points in the music, is designed subjectively to evoke an emotive response in the listener appropriate to the ecological and societal implications of the data sets (Email message to Stephan Crawford, August 20, 2021).

The ClimateMusic Project's application of parameter mapping in new music development is tempered by active input from each project's consulting team of scientists. Especially with respect to communicating possible future scenarios, this introduces an element of informed subjectivity into the music: No one can say with precision what a four-degree warmer world would be like, but enough is known to understand that it would be very different from the world that humans have experienced until now. The scientist's role is to help the composer express an appropriate dynamism and magnitude of change in order to express the broader and deeper meaning in the raw numbers. Some may interpret this element of subjectivity as a limitation, but it is also a strength, because it allows a lay audience to perceive deeper meaning than would be possible, for example, by looking at a visual graph.

Audification This technique involves the "direct translation of data waveform into sound" (Kramer, 1994, cited in Hermann et al., p. 301). It can be thought of as the sonic equivalent of a visual graph. As such, The ClimateMusic Project has found it to be less versatile than parameter mapping in bringing the deeper meaning in the data to life for non-scientists. Nevertheless, it has applied this technique to good effect in one of its portfolio compositions. In *What If We...?*, composer Wendy Loomis embeds and contrasts two audifications of sea level rise data representing two possible future scenarios, a "business as usual" scenario in which little is done to rein in carbon emissions during this century and an aggressive mitigation scenario. The audifications are introduced consecutively and constitute an element of the bass line, upon which Loomis weaves harmony and melody. The audification of the "business as usual" scenario is at first only subtly audible, but over the course of about 90 s becomes a rhythmically driving element that is perceived by the listener as being increasingly alarming—indeed, threatening—as it overwhelms other elements in the music.

Lyrics and Spoken Word Creating a sung or spoken element that includes scientific content is another technique for communicating scientific insights in the music, especially when audification or parameter mapping is not feasible, or when it can augment these techniques. In the composition *What If We...?*, Loomis chose to integrate spoken fictional but scientifically defensible news headlines from the year 2045. The headlines were created by drawing from the conclusions of studies on the

Fig. 4 A 2017 performance of *Climate* by Erik Ian Walker at the San Francisco Performing Arts Centre with accompanying data visuals

future impact and effects of sea level rise and through direct input from The ClimateMusic Project's science advisors.

Visuals In addition, each ClimateMusic composition is paired with synchronized visuals, typically including data references such as animations. This visual element is intended to support the audience's understanding of how the music conveys the scientific story by providing visual cues. For example, *Climate*, by composer Erik Ian Walker, explores the effect of carbon emissions on near-earth atmospheric temperature and the earth's energy balance over 450 years, from 1800 to 2250. The accompanying visuals present a chronology of historical events in the United States and possible future city and landscapes, which serve to anchor the audience in time. Superimposed over these video images are data animations that track the movement of the data as referenced in the music in real time (Fig. 4). The visual element is intended to be supportive, not the main feature of the audience experience.

New Horizons

To date, The ClimateMusic Project has focused on refining its current approach to creating ClimateMusic. There remain, however, fertile areas of inquiry that could significantly enhance the realization of its mission.

Additional Sonification Techniques The field of sonification continues to develop, and there are additional techniques, such as model-based sonification (see Hermann et al., 2011, p. 399–425), that may yet find their way into The ClimateMusic Project's portfolio. Further innovation could yield entirely new approaches and techniques.

Interactive Music The introduction of interactive elements in the music could enhance the audience's ability to connect with and retain the insights in the scientific stories. During The ClimateMusic Project's 2018 collaboration with the San Francisco Conservatory of Music, composer Emily Pitt created an interactive soundscape entitled *2100*, in which the audience was able to influence how the composition played out via an app connected in real time that allowed the audience to make decisions that affected the representation of carbon emissions in the piece. For The ClimateMusic Project, collaboration with such institutions is an ideal environment in which to identify new approaches and innovations. The integration of interactive elements in ClimateMusic offers the prospect of strengthening personal insight gained through the music by each audience member, effectively converting the experience into one that is even more active.

The ClimateMusic Project has been working to introduce additional interactive elements to the performance program, including a series of action items designed to complement the musical program and to boost emotional engagement and willingness to act. In a 2021 concert, *Live! From Vienna and San Francisco: A Musical Call for Climate Action*, The ClimateMusic Project trialed the use of a word cloud generator following the performance of a ClimateMusic composition, in order to visually display audience responses across the globe in a single image. Later, in the program, the audience was encouraged to make an action pledge, dedicated to a loved one, and to invite several of their friends to do the same. These direct engagement activities will remain a key component of future engagements—both at live performances and virtual presentations.

Augmented/Virtual Reality In 2017, The ClimateMusic Project collaborated with a senior scientist from the Fuji Xerox Lab in Palo Alto, California, to explore the use of virtual reality (VR) to augment and enhance the ClimateMusic audience experience. The outcome was presented in public at a Project Showcase in San Francisco in 2017. Using an experimental platform called MUSE (Fig. 5), it allowed individual audience members to experience the sensation of being transported onto the imaginary control deck of a starship, complete with display screens, controls, and windows looking toward the earth from space.

From this deck, visitors could experience the music of The ClimateMusic Project's affiliated composition *Climate*, by composer Erik Ian Walker, while viewing visualizations of the climate data underlying the composition—CO_2 levels, near-earth atmospheric temperature, and the earth's energy balance. Visitors had the ability to choose among visualizations, including 3D animated graphs and even a dancer whose movements produced streaming particle trails in which color and density reflected data values. In VR, the visitor could also dance and see particles

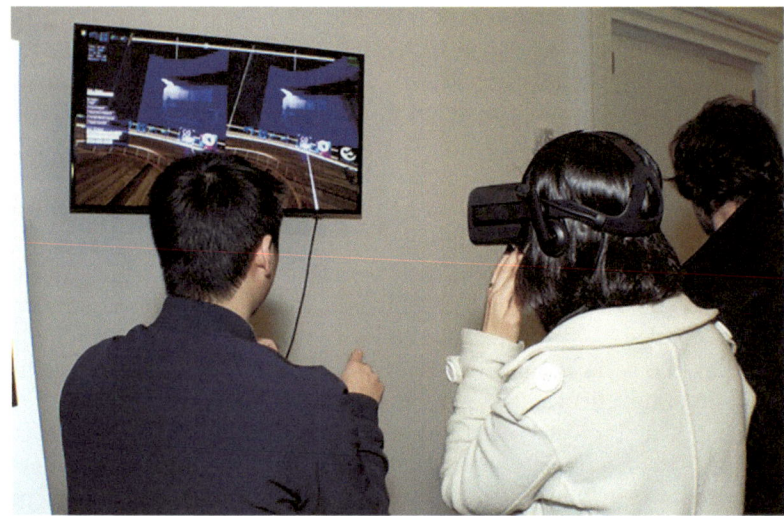

Fig. 5 MUSE demo at The ClimateMusic Project's Showcase, December 2017 San Francisco Performing Arts Center. (Photo credit: Tim Guydish)

streaming from their dancing. This promising exploration of new technological horizons was paused due to the still limited accessibility of the necessary technology infrastructure at public venues and also due to a strategic decision to focus on other competing priorities.

ClimateMusic in the Community

Citing a pivotal policy paper written by the Greater London Arts Association (GLAA) in the 1980s, Rod Brooks (1988, p.7) has described community arts as "an Arts activity [that is] defined by its method of work and aims, rather than by its art form," working within communities "in order to articulate, engage and address the needs, experience and aspirations of its communities". ClimateMusic encompasses a wide range of musical styles and does not describe a specific art form or genre, but rather a social purpose, to motivate climate action, and the methods employed to achieve that purpose as outlined above. As such, it is fundamentally a community art practice.

Community Engagement: Performing ClimateMusic

Community engagement is essential to the performance of ClimateMusic. Performances typically include three elements: an introduction to set the context, the performance itself, and a post-performance audience engagement segment. The

ClimateMusic Project invites audiences to engage with and ask questions of its science advisors and composers. Both live and virtual performances provide multiple opportunities for reinforcing and elaborating on the scientific story featured in each composition. The introduction, typically delivered by a scientist, is a brief (3 to 5-min) statement that sets the context for the concert by conveying the rationale for communicating scientific insights through music and how the science is reflected in the music to be performed. The post-concert audience engagement segment allows the audience to pose questions and to share new insights and impressions. This segment typically includes the participation of a scientist with expertise in the story conveyed, the composer, and one or more representatives of climate action organizations whenever possible.

Each performance segment plays a critical role in maximizing the potential for newfound audience insight and emotional energy to be converted into individual engagement and action. Engaging the audience in dialog after the music stops is a powerful way to help the audience process the experience. For example, during this interaction at a 2015 concert of Erik Ian Walker's *Climate*, one woman recounted how the experience of the music and visuals allowed her to connect the course of climate change—which had seemed an abstract concept to her—to her family's own history: Over the arc of 30 min, she was able to compare the "feel" of the music from the time she was born to that of the time her granddaughter might be born. The ClimateMusic Project personalizes the issue in this way to invite audiences to understand and reflect upon their personal stake in the issues raised, and their agency to affect the outcome, with the aim to motivate and provide opportunities for action.

The music stimulates emotional energy in audiences and a motivation to ask questions, learn, and act. At this writing, The ClimateMusic Project is building additional collateral resources around each composition to offer audiences easy access to further information about the underlying science and ways to engage. Because the science is complex and may involve concepts that require further explanation, the opportunity for audiences to explore the issues further helps to reinforce the message and allows audiences to learn more about the science and pathways to action. Follow-up discussion allows The ClimateMusic Project to delve into the details of each piece of music's creative process and methodology. Audiences may ask questions ranging from "How do we know how much CO_2 was in the atmosphere in the 1800s?" to "What is the tipping point for the planet?" The science advisors, all professional educators, provide insight into the science and share their perspectives on using music as a tool to communicate and to build community.

Partner Organizations

The ClimateMusic Project's objective is to ratchet up action so that those not yet active can learn more and begin to act, those somewhat active can learn what more they can do, and those highly involved are invited to amplify their activism by onboarding friends and family. In addition to collaboration with science advisors and composers, the Project invites partner organizations to discuss pathways for action. Engaging partner organizations provides audience members at different levels of knowledge and active engagement with a range of options to get involved, by learning more about their own carbon footprint and what they can do to reduce it (The Global Footprint Network); by helping people connect with communities already addressing climate change (Interfaith Power & Light); by working to draw down carbon through scientifically framed global projects (Cool Effect); by helping communities fund renewable energy systems to reduce their impact and save money (Re-Volv); and by working to make their city a showcase for urban planning (San Francisco Department of the Environment). Each partner also encourages civic engagement at the voting booth and with elected officials.

While The ClimateMusic Project has a few defined partners, the group encourages audience members to engage with any of the many other community organizations active on climate issues locally, nationally, and globally. The aim is to provide opportunities for audiences to channel the emotional energy, insight, and motivation gained from the musical experience into personal action that meets their interests and helps the planet.

Community Participation: Play for the Planet

Outside of the concerts that make up the core of the ClimateMusic program, The ClimateMusic Project has facilitated community music-making both in educational institutions and in the broader community.

As Lee Higgins has described, community music "challenge[s] us to dream of a politics of, and for, a musical future that is marked with active and meaningful participation" (174). This principle was central to ClimateMusic's 2018 series *Play for the Planet*, an official affiliate event of the Global Climate Action Summit (GCAS). GCAS, co-chaired by California governor Jerry Brown, Michael Bloomberg, Patricia Espinosa, Anand Mahindra, and Xie Zhenhua, took place in San Francisco in September 2018 and, as stated by the organizers, "brought together leaders and people from around the world to 'Take Ambition to the Next Level.'"

Play for the Planet was a departure from The ClimateMusic Project's standard performances. In order to demonstrate how performers can influence and drive action, *Play for the Planet* created an environment that offered musicians and artists a chance to share their non-scientific artistic expressions of climate change. Those expressions were powerful, for both audiences and performers, some of whom newly appreciated their ability to have an impact on a critical issue. Action at all levels is critical, and just like the official Summit delegates, the public and the artistic community has a vital role to play.

Play for the Planet featured more than twenty performers challenged to answer the question "What do you think the future will sound like, or what do you want it to sound like?" The performers, included Andrew Revkin, Strategic Adviser for

Environmental and Science Journalism at National Geographic Society; Rafael Jesús González, Berkeley Poet Laureate; COPUS, a San Francisco-based jazz and spoken word group; The Creative Liberation Network, an Oakland-based art, music, and educational organization; Nick Platoff, Associate Principal Trombonist for the San Francisco Symphony; and Guinevere Q, a slam rock and spoken word artist. In addition, artist and philosopher Carter Brooks provided an ice installation that slowly melted throughout the day—a concrete representation of humanity's current situation on earth. The ClimateMusic Project was able to give members of the public a taste of an original ClimateMusic composition, *Climate*, as composer Erik Ian Walker performed a short excerpt of this beautiful and haunting piece. Audience members were also invited to interact with world-renowned scientists including Dr. William Collins, Head of the Climate Readiness Institute at the Lawrence Berkeley National Lab (LBNL) and senior advisor to The ClimateMusic Project, and Alison Marklein, a post-doc researcher at LBNL. A few of The ClimateMusic Project's solutions partners—Cool Effect, RE-volv, and SF Environment—were available to discuss ways that the public can act on climate.

After the onset of the COVID-19 pandemic, The ClimateMusic Project relaunched the *Play for the Planet* initiative virtually with a global reach, both in September 2020, as an official event of Climate Week NYC, and in April 2021, for Earth Day. These virtual series saw artists from around the world respond to the question "What do you want the future to sound like?" and introduce their favorite environmental nonprofits or highlight The ClimateMusic Project's partner organizations in order to motivate action and engagement. Submissions were shared on The ClimateMusic Project website and social media channels including YouTube, providing hope and a sense of community and shared purpose during a particularly challenging and isolating time.

Collaboration with the San Francisco Conservatory of Music

The ClimateMusic Project has an ongoing collaboration with the San Francisco Conservatory of Music's Technology and Applied Composition program (Fig. 6). This joint effort provides a spring cohort of young composers with the opportunity to create their own science-guided music under the guidance of program's Professor Taurin Barrera and one or more scientist-musicians from The ClimateMusic Project. The student pieces have been performed for live audiences at venues around San Francisco including at the Conservatory's music halls and at the Exploratorium science museum. The ClimateMusic Project hopes to build upon the program developed with the Conservatory to facilitate the creation of ClimateMusic in classrooms. An aim for the future is to create a curriculum that reflects a multidisciplinary approach to science communication that educational institutions can adopt around the globe.

Fig. 6 Science briefing delivered by a scientist from The ClimateMusic Project via Zoom for student composers in the Technology and Applied Composition program at the San Francisco Conservatory of Music in 2023

The Future of ClimateMusic

Funding has been difficult to secure as an organization that does not fit squarely in the realm of art or science. While some of the funding barriers faced by The ClimateMusic Project are a result of needing more dedicated staff to seek out requisite opportunities, other obstacles are due to the nature of funding organizations and the grant process. Many traditional funding organizations have indicated that they require a certain level of previous funding and proof of funding past success. Because The ClimateMusic Project does not always align with a specific project category, it is often regarded by art funders as a science or educational organization and by science funders as an art organization, which can lead to a reluctance to fund a project that does not have a clear classification. Funders tend to be conservative and risk-averse when in fact what is needed is bold experimentation with new ways to reach people—especially those that integrate multiple disciplines, given that the climate crisis cannot be understood or solved from any single disciplinary silo. Current funding dynamics are inadequate to address the climate crisis.

Data from the National Endowment for the Arts (NEA) show that funding for not-for-profit art organizations has been largely dependent on earned income (40.7%), from sources such as ticket sales (National Endowment for the Arts, 2012). The ClimateMusic Project has aimed to keep ticket prices low to maximize audience numbers, and so it has not been possible to rely upon a performance revenue model. The NEA also reflects contributed income from all sources in the amount of 44.9%. That would include funds from corporations, public entities, individuals, and foundations. Conversations with non-for-profit art organizations in similar circumstances have indicated that this is a common problem with groups not affiliated with educational or scientific institutions. The broader community can help to incubate projects like The ClimateMusic Project in order to encourage a general understanding that these initiatives do help build knowledge and motivate necessary action on complex societal issues such as climate change. The Project presents a nascent opportunity for funders who have the foresight to incubate these multidisciplinary endeavors that break silos and reach audiences eager to learn and make a difference.

References

Aalbers, S., Fusar-Poli, L., Freeman, R. E., Spreen, M., Ket, J. C. F., Vink, A. C., Maratos, A., Crawford, M., Chen, X.-J., & Gold, C. (2017). Music therapy for depression. *Cochrane Database of Systematic Reviews, 11*. https://doi.org/10.1002/14651858.cd004517.pub3

Atkinson, K., & Sosby, M. (2018, July 23). *Sounds of the sun*. NASA. Retrieved January 1, 2022, from https://www.nasa.gov/feature/goddard/2018/sounds-of-the-sun/

Benitez-Galbraith, J., & Galbraith, C. S. (2019). Narrative engagement, enjoyment, learning and theme comprehension: Using an authentic music video in an introductory college language classroom. *RELC Journal, 52*(3), 397–411. https://doi.org/10.1177/0033688219874136

Bigand, E., Vieillard, S., Madurell, F., Marozeau, J., & Dacquet, A. (2005). Multidimensional scaling of emotional responses to music: The effect of musical expertise and of the duration of the excerpts. *Cognition & Emotion, 19*(8), 1113–1139. https://doi.org/10.1080/02699930500204250

Bradley, D. (2018, March 13). I served in Vietnam. Here's my soundtrack. *The New York Times*. Retrieved January 2, 2022, from https://www.nytimes.com/2018/03/13/opinion/vietnam-war-rock-music.html

Bradt, J., & Dileo, C. (2014). Music interventions for mechanically ventilated patients. *Cochrane Database of Systematic Reviews, 12*. https://doi.org/10.1002/14651858.cd006902.pub3

Bradt, J., Dileo, C., & Shim, M. (2013). Music interventions for preoperative anxiety. *Cochrane Database of Systematic Reviews, 6*. https://doi.org/10.1002/14651858.cd006908.pub2

Brooks, R. (1988). *Wanted! Community Artists*. Calouste Gulbenkian Foundation. London.

Carman, J., Lacroix, A., Leiserowitz, A., Maibach, E. W., Rosenthal, S. A., Kotcher, J. E., Neyens, L., Wang, X., Marlon, J., & Golberg, M. H. (2021). *Americans actions to limit and prepare for global warming, March 2021*. Yale University and George Mason University. Yale Program on Climate Change Communication.

Carr, S. M., & Rickard, N. S. (2016). The use of emotionally arousing music to enhance memory for subsequently presented images. *Psychology of Music, 44*(5), 1145–1157. https://doi.org/10.1177/0305735615613846

ClimateMusic. (2022). FAQs. https://climatemusic.org/faqs/

Climate Music Initiative. (2020). *Dr. Lucy Jones Center for Science and Society*. Accessed July 2021, from http://drlucyjonescenter.org/climate-music-initiative/

Crowther, G. J., McFadden, T., Fleming, J. S., & Davis, K. (2016). Leveraging the power of music to improve science education. *International Journal of Science Education, 38*(1), 73–95. https://doi.org/10.1080/09500693.2015.1126001

Cunsolo, A., & Ellis, N. R. (2018). Ecological grief as a mental health response to climate change-related loss. *Nature Climate Change, 8*, 275–281. https://doi.org/10.1038/s41558-018-0092-2

Donné, L., Jansen, C., & Hoeks, J. (2017). Uncovering factors influencing interpersonal health communication. *Global Qualitative Nursing Research.* https://doi.org/10.1177/2333393617711607

Downs, J. S. (2014). Prescriptive scientific narratives for communicating usable science. *Proceedings of the National Academy of Sciences of the United States of America, 111*(Suppl 4), 13627–13633. https://doi.org/10.1073/pnas.1317502111

Downs, J. S., Murray, P. J., Bruine de Bruin, W., Penrose, J., Palmgren, C., & Fischhoff, B. (2004). Interactive video behavioral intervention to reduce adolescent females' STD risk: A randomized controlled trial. *Social Science & Medicine (1982), 59*(8), 1561–1572. https://doi.org/10.1016/j.socscimed.2004.01.032

Eerola, T., Vuoskoski, J. K., Peltola, H. R., Putkinen, V., & Schäfer, K. (2018). An integrative review of the enjoyment of sadness associated with music. *Physics of Life Reviews, 25*, 100–121. https://doi.org/10.1016/j.plrev.2017.11.016

Eyen, L. (2017, November 29). *Xiuhtezcatl Martinez: The next generation - youth "artivist" and environmental advocate.* Santa Clara University. https://www.scu.edu/environmental-ethics/environmental-activists-heroes-and-martyrs/xiuhtezcatl-martinez.html

Ferreri, L., Bigand, E., & Bugaiska, A. (2015). The positive effect of music on source memory. *Musicae Scientiae, 19*(4), 402–411. https://doi.org/10.1177/1029864915604684

Fried, R., & Berkowitz, L. (1979). Music hath charms...and can influence helpfulness. *Journal of Applied Social Psychology, 9*(3), 199–208. https://doi.org/10.1111/j.1559-1816.1979.tb02706.x

Fukui, H., & Toyoshima, K. (2014). Chill-inducing music enhances altruism in humans. *Frontiers in Psychology, 5*, 1215. https://doi.org/10.3389/fpsyg.2014.01215

Gardiner, L. S. (2019, March 11). What does climate sound and look like? *The Plainspoken Scientist: American Geophysical Union.* Accessed July 2021, from https://blogs.agu.org/sciencecommunication/2019/03/11/what-does-climate-sound-and-look-like

Hart, H. (2010, October 7). Smog musicians turn pollution data into jagged melodies. *Wired.* Accessed July 2021, from https://www.wired.com/2010/10/smog-music/

Hermann, T., Hunt, A., & Neuhoff, J. G. (Eds.). (2011). *The sonification handbook.* Logos Publishing House.

Hevner, K. (1935). The affective character of the major and minor modes in music. *The American Journal of Psychology, 47*(1), 103–118.

Hinyard, L. J., & Kreuter, M. W. (2007). Using narrative communication as a tool for health behavior change: A conceptual, theoretical, and empirical overview. *Health Education & Behavior, 34*(5), 777–792. https://doi.org/10.1177/1090198106291963

Hopfer, S. (2012). Effects of a narrative HPV vaccination intervention aimed at reaching college women: A randomized controlled trial. *Prevention Science, 13*(2), 173–182. https://doi.org/10.1007/s11121-011-0254-1

Huron, D. (2006). *Sweet anticipation: Music and the psychology of expectation.* MIT Press.

Iliya, Y. A. (2015). Music therapy as grief therapy for adults with mental illness and complicated grief: A pilot study. *Death Studies, 39*(1–5), 173–184. https://doi.org/10.1080/07481187.2014.946623

Jespersen, K. V., Koenig, J., Jennum, P., & Vuust, P. (2015). Music for insomnia in adults. *The Cochrane Database of Systematic Reviews, 8*, CD010459. https://doi.org/10.1002/14651858.CD010459.pub2

Jones, A. (2021, April 22). *The sound of science: Artists and scientists discuss climate change.* Livestream by The Brooklyn Rail: Common Ground. https://brooklynrail.org/events/2021/04/22/the-sound-of-science-artists-and-scientists-discuss-climate-change/

Kahn, B. (2016, September 20). Tree loss is put to music [Audio]. *Scientific American.* https://www.scientificamerican.com/article/tree-loss-is-put-to-music-audio/

Kniffin, K. M., Yan, J., Wansink, B., & Schulze, W. D. (2017). The sound of cooperation: Musical influences on cooperative behavior. *Journal of Organizational Behavior, 38*(3), 372–390. https://doi.org/10.1002/job.2128

Kramer, G. (1994). An introduction to auditory display. In G. Kramer (Ed.), *Auditory display: Sonification, audification, and auditory interfaces*. Addison-Wesley.

Larkey, L. K., Lopez, A. M., Minnal, A., & Gonzalez, J. (2009). Storytelling for promoting colorectal cancer screening among underserved Latina women: A randomized pilot study. *Cancer Control: Journal of the Moffitt Cancer Center, 16*(1), 79–87. https://doi.org/10.1177/107327480901600112

Lee, H., Fawcett, J., & DeMarco, R. (2016). Storytelling/narrative theory to address health communication with minority populations. *Applied nursing research: ANR, 30*, 58–60. https://doi.org/10.1016/j.apnr.2015.09.004

Leiserowitz, A., Maibach, E. W., Rosenthal, S. A., Kotcher, J. E., Carman, J., Wang, X., Marlon, J. R., Lacroix, K., & Goldberg, M. A. (2021, March). *Climate change in the American Mind, March 2021*. Yale University and George Mason University. Yale Program on Climate Change Communication.

Marshall, G. (2014). *Don't even think about it: Why our brains are wired to ignore climate change*. Bloomsbury USA.

Martinez-Conde, S., & Macknik, S. L. (2017). Finding the plot in science storytelling in hopes of enhancing science communication. *Proceedings of the National Academy of Sciences, 114*(31), 8127–8129. https://doi.org/10.1073/pnas.1711790114

Mazor, K. M., Baril, J., Dugan, E., Spencer, F., Burgwinkle, P., & Gurwitz, J. H. (2007). Patient education about anticoagulant medication: Is narrative evidence or statistical evidence more effective? *Patient Education and Counseling, 69*(1–3), 145–157. https://doi.org/10.1016/j.pec.2007.08.010

McCall, B., Shallcross, L., Wilson, M., Fuller, C., & Hayward, A. (2019). Storytelling as a research tool and intervention around public health perceptions and behaviour: A protocol for a systematic narrative review. *BMJ Open, 2019*(9), e030597. https://doi.org/10.1136/bmjopen-2019-030597

Mehr, S. A., Singh, M., Knox, D., Ketter, D. M., Pickens-Jones, D., Atwood, S., Lucas, C., Jacoby, N., Egner, A. A., Hopkins, E. J., Howard, R. M., Hartshorne, J. K., Jennings, M. V., Simson, J., Bainbridge, C. M., Pinker, S., O'Donnell, T. J., Krasnow, M. M., & Glowacki, L. (2019). Universality and diversity in human song. *Science, 366*(6468), eaax0868. https://doi.org/10.1126/science.aax0868

Molnar-Szakacs, I., & Overy, K. (2006). Music and mirror neurons: From motion to 'e'motion. *Social Cognitive and Affective Neuroscience, 1*(3), 235–241. https://doi.org/10.1093/scan/nsl029

Mori, K., & Iwanage, M. (2017). Two types of peak emotional responses to music: The psychophysiology of chills and tears. *Scientific Reports, 7*, 46063. https://doi.org/10.1038/srep46063

N.A. Juliana v. United States. (2021). *Our Children's Trust*. Accessed December 2021 from https://www.ourchildrenstrust.org/juliana-v-us

National Endowment for the Arts (Ed.). (2012). *How the United States funds the arts* (3rd ed.). National Endowment for the Arts. https://www.arts.gov/sites/default/files/how-the-us-funds-the-arts.pdf

Reese, A. (2015, April 12). *How can music inspire social change? Facing history and ourselves*. Accessed July 2021, from https://facingtoday.facinghistory.org/how-can-music-inspire-social-change

Rieger, K. L., West, C. H., Kenny, A., Chooniedass, R., Demczuk, L., Mitchell, K. M., Chateau, J., & Scott, S. D. (2018). Digital storytelling as a method in health research: A systematic review protocol. *Systematic Reviews, 7*(1). https://doi.org/10.1186/s13643-018-0704-y

Rodriguez, F. (2020). Harnessing cultural power. In A. E. Johnson & K. K. Wilkinson (Eds.), *All we can save: Truth, courage, and solutions for the climate crisis* (pp. 121–127). Penguin Random House.

Schön, D., Boyer, M., Moreno, S., Besson, M., Peretz, I., & Kolinsky, R. (2008). Songs as an aid for language acquisition. *Cognition, 106*(2), 975–983. https://doi.org/10.1016/j.cognition.2007.03.005

Sihvonen, A. J., Särkämö, T., Leo, V., Tervaniemi, M., Altenmüller, E., & Soinila, S. (2017). Music-based interventions in neurological rehabilitation. *The Lancet. Neurology, 16*(8), 648–660. https://doi.org/10.1016/S1474-4422(17)30168-0

Simon-Lewis, A. (2017, June 19). Climate change data is being transformed into beautiful, haunting symphonies. *Wired*. Accessed July 2021, from https://www.wired.co.uk/article/climate-symphony-data-sonification

Sloane, N. & Harding, C. (2021, April 20). Pop on a Perilous Planet with Kyle Devine. *Switched on Pop, Vulture*. Podcast audio. https://switchedonpop.com/episodes/pop-on-a-perilous-planet-musicians-confront-the-climate-crisis

Slovic, S., & Slovic, P. (2015). *Numbers and nerves: Information, emotion, and meaning in a world of data* (1st ed.). Oregon State University Press.

Terwogt, M. M., & Van Grinsven, F. (1991). Musical expression of moodstates. *Psychology of Music, 19*(2), 99–109.

The ClimateMusic Project. Our Music. (2019). *The ClimateMusic Project*. Accessed July 2021, from https://climatemusic.org/our-music/

Think 100% Music. (n.d.) *Hip-Hop Caucus*. Accessed July 2021, from https://think100climate.com/music/

Tyson, A., Kennedy B., & Funk, C. (2021, May). *Gen Z, Millennials stand out for climate change activism, social media engagement with issue*. Pew Research Center. https://www.pewresearch.org/science/wp-content/uploads/sites/16/2021/05/PS_2021.05.26_climate-and-generations_REPORT.pdf

Venugopal, V. (2020, April 3). Scientists have turned the structure of the coronavirus into music. *Science Magazine*, Accessed July 2021, from https://www.sciencemag.org/news/2020/04/scientists-have-turned-structure-coronavirus-music

Viper, M., Thyrén, D., & Horwitz, E. B. (2020). Music as consolation-The importance of music at farewells and mourning. *Omega*. Advance online publication. https://doi.org/10.1177/0030222820942391

Weintrobe, S. (2013). *Engaging with climate change [electronic resource]: Psychoanalytic and interdisciplinary perspectives (New library of psychoanalysis (Unnumbered))*. Routledge.

Whitcomb, I. (2021, April 10). Therapists are reckoning with eco-anxiety. *Gizmodo*, republished in *Scientific American*. Accessed Jul 2021, from https://www.scientificamerican.com/article/therapists-are-reckoning-with-eco-anxiety/

Yeo, S. (2013). Climate change's greatest hit: Cellist writes song from NASA data. Accessed July 2021, from https://www.climatechangenews.com/2013/09/04/climate-changes-greatest-hit-cellist-writes-song-from-nasa-data/

Yun, S. H., & Gallant, W. (2010). Evidence-based clinical practice: The effectiveness of music-based intervention for women experiencing forgiveness/grief issues. *Journal of Evidence-Based Social Work, 7*(5), 361–376. https://doi.org/10.1080/15433710903323870

Catherine Emma Dixon is a Music graduate from the University of Oxford. Catherine has taken part in community music initiatives both as a researcher and as a performer, and has worked with organisations such as Sing Inside, a charity bringing singing workshops into prisons, Horatio's Garden, a charity promoting the wellbeing of people with spinal cord injuries, and The ClimateMusic Project, where she researched the evidence base for using music in climate change communication. During her time at university, Catherine campaigned for climate and sustainability policies with the Oxford Climate Society's "Decarbonise Oxford" campaign and participated in the Oxford School of Climate Change and the Oxford Programme in Sustainability Leadership at the Environmental Change Institute. Catherine now works as a Software Engineer in Cambridge and remains passionate about using her skills toward positive social change.

Laurie S. Goldman serves as Strategic Advisor to The ClimateMusic Project and has worked with the organization since 2016, including past service on the Executive Team. Laurie has held senior positions in academia, the private sector, and the US federal government. At UC Berkeley, Laurie managed campus-wide international strategy as Deputy Director of the Global Engagement Office and oversaw partnerships with leading international academic institutions. As part of her broader public policy remit at Levi Strauss, Laurie headed global trade policy and was appointed to the Obama Administration's Trade Advisory Committee. She also served on industry boards including the National Center for APEC, the European Branded Clothing Alliance, and the US Fashion Industry Association. Past positions include service as a trade negotiator for the US Department of Commerce and with the Public Affairs office at UC Berkeley. Laurie holds a master's degree from George Washington University and an undergraduate degree from Tufts University.

Stephan Crawford is an interdisciplinary artist based in San Francisco, California. He founded and co-leads The ClimateMusic Project, a cross-cutting collaborative that connects people to climate science and action through music. Since launching in 2015, ClimateMusic has reached thousands of people globally, partnering on programs with leading institutions such as the National Academy of Sciences and the World Bank. It has also garnered international media coverage, including profiles by *The New York Times and* the *BBC*. He left a distinguished parallel career in public service in 2017 to focus on ClimateMusic and his studio practice. Stephan holds an MS degree in environmental management with a focus on environmental science from the University of San Francisco and an MA in Law and Diplomacy from the Fletcher School of Law and Diplomacy. In 2024, The Explorers Club in New York elected Stephan into it's EC50 cohort of global change makers. For his creative art work, please visit https://www.sc2arts.com.

Phoebe Camille Lease served as a member of The ClimateMusic Project communications team from 2020 to 2021, where she researched the history of activism and climate change in popular music. She has previously conducted research on green finance for the nonprofit Future Coalition and interned with the National Oceanic and Atmospheric Association (NOAA). She now works for a leading clean technology public relations firm in Washington, DC. Phoebe graduated from Smith College with a BA in American Studies and concentration in climate change studies; her capstone project analyzed the advertisement methods and disinformation campaigns of major oil and gas companies.

Telling the Story of Climate Change through Food

Danielle L. Eiseman and Michael P. Hoffmann

Food is intended to provide us with the necessary nutrients and energy to survive, yet we all have a unique psychological, emotional, physical, cultural, and social relationship with food that contributes to our overall well-being (Eiseman, 2019). However, our relationship with food is at risk due to changes in the global climate. Scientists and researchers have conveyed their concern for climate change impacts and the risks it poses to global food security. The risks include increasing temperatures, drought, intensive rainfall events, and invasive pests or weeds all of which reduce crop yields and increase food prices (Goldenberg, 2014; Myers et al., 2017). Additionally, recent evidence has shown that due to a high concentration of carbon dioxide in the air, the nutritional value of staple crops such as rice and wheat is decreasing (Ebi & Ziska, 2018). Thus, climate change threatens our physical relationship with food.

On an emotional, social, and psychological level, individuals that rely on their morning cup of coffee, glass of wine with dinner, or a perfect vanilla bean ice cream cone in the summer may find these foods more difficult to acquire, either because they are too expensive or too difficult to produce. Tim Gore, the head of food policy and climate change for Oxfam, stated "The main way that most people will experience climate change is through the impact on food: the food they eat, the price they pay for it, and the availability and choice that they have" (Goldenberg, 2014).

The story of food and climate change is not how the climate change story has been told. Historically, communication about climate change has adopted the traditional deficit model of communication, whereby experts share scientific information with lay audiences, with the intention to process that information meaningfully. However, ongoing research on climate change communication shows that many Americans need help with understanding or accepting the causes of climate change, and even more fail to understand when and how it will impact them (Leiserowitz

D. L. Eiseman (✉) · M. P. Hoffmann
Cornell University, Ithaca, NY, USA
e-mail: dle58@cornell.edu; mph3@cornell.edu

© The Author(s) 2024
E. Coren, H. Wang (eds.), *Storytelling to Accelerate Climate Solutions*,
https://doi.org/10.1007/978-3-031-54790-4_13

et al., 2015). Thus, there is a critical need to engage people on a deeper level regarding the causes and risks associated with climate change in a way that connects with them beyond traditional means.

What Is Being Said About Food and Climate Change?

Given the relationship we have with food and the need for novel approaches to engage the public on climate change action, food seems like a logical choice for telling the climate change story. However, there is scant research on how the impacts of climate change on food affect people's attitudes and behaviors. Research has identified several ways to engage the public on climate change, for example, through values and norms, but progress remains hindered by it being politicized and that it is still disputed by many. It is possible, however, to overcome these impediments by increasing the dialog about climate change and making the topic part of everyday conversations rather than something generally not discussed. Research also suggests that simply conversing about it leads to deeper engagement. Fortunately, food looks to be an excellent – though understudied – communication tool. It is personally relevant to everyone, relatively non-partisan, and encountered multiple times per day. This creates a unique opportunity for us to examine the effects of how messages about climate change and food affect attitudes and behavior outcomes.

Despite a lack of empirical evidence on food as an engagement tool for communicating climate change, there are some examples in the media and the gray literature on food and climate change. One of the more prevalent bodies of work is a collaborative project called the EAT-Lancet Commission. The EAT-Lacent Commission is a small group of global scientists working to establish a framework for meeting global nutritional needs while also limiting the environmental impacts our diet has on the planet. The report published in 2019 generated a lot of attention and has been cited 3542 times. The EAT-Lacent Commission presents a framework for diets that optimize human and planetary health, each tailored to country-specific needs. The framework, however, has come under significant criticism for promoting an unaffordable diet for a majority of global populations and only considers health outcomes in its recommendations for meat consumption, not climate or environmental outcomes. Similar organizations have published works promoting low-carbon diets, for example, the Yale Program on Climate Change Communication published Climate Change and the American Diet in 2020. The report, consistent with the program's previous work, provides data from a nationally representative survey about public perceptions of eating a plant-based diet and access to food (Leiserowitz et al., 2020). Lastly, Bill Hawken's organization Project Drawdown (n.d.) discusses approaches to drawing down carbon emissions through individual and collective action. Food and diet are one of the methods of decreasing carbon emissions.

As discussions on food and climate change have begun to take precedence among the scientific community, mainstream media has also increased its attention on

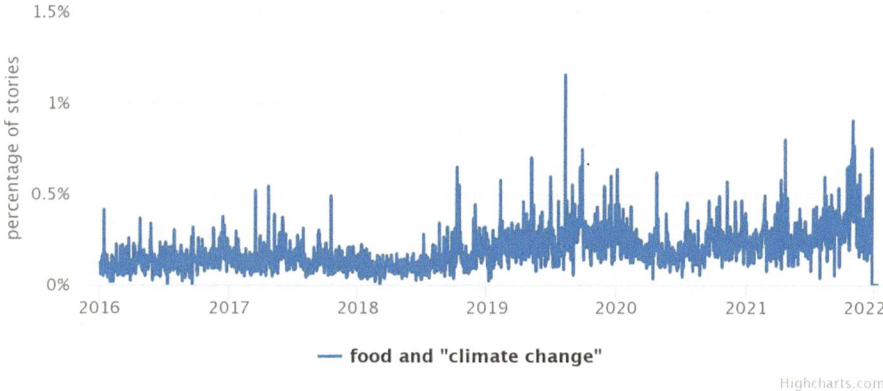

Fig. 1 Media Cloud Explorer Query on daily media stories covering food and climate change (created January 10, 2022)

climate change's impact on food. Using the Media Cloud (n.d.) explorer tool, an online database of media stories from 2015 to the present shows a small, but increasing trend of media coverage on climate change and food. In 2016, roughly 0.2% of all media stories in the United States mentioned the topic of food and climate change. Toward the end of 2021 and the start of 2022, the percentage of daily stories mentioning climate change and food is closer to 0.5% (Fig. 1).

Beyond media mentions of the topic of food and climate change, the home cook can explore an endless array of climate-friendly recipes. *The New York Times*, for example, has an entire online guide to climate-friendly cooking. Buzzfeed, the media outlet aimed toward younger audiences, provides low-carbon recipe ideas as well.

Small Strides in Effective Climate Change Communication

Issues about the environment encompass more than just presenting scientific evidence (of which there is an abundance). The examples in the previous section highlight in a small way how climate change has engendered a meek normative role in discussions about climate change. Communicating about climate change and the environment must examine more than just how people receive and interpret information. Hansen and Cox (2015) identify two key concerns the field of environmental communications should address: Environmental communication should "(a) improve our understanding of the cultures or locations in which such communication is produced and (b) strengthen the capacity of societies to deliberate and respond to conditions relevant to the well-being of both society and natural biological communities" (p. 15). Examining the history of environmental communication, however, demonstrates a failure to fully address these concerns.

A prime example is presented by Dunaway (2015), who highlights the ongoing pattern of depositing white bodies, especially children, as a universal symbol of vulnerability but masks the inequities people of color have endured when it comes to exposure to environmental hazards and pollutants. For example, the environmental justice movement's launch stemmed from community outrage in Warren County, North Carolina (Bullard, 2001). People in the community learned that companies were illegally dumping polychlorinated biphenyl (PCB), a chemical used as a coolant and in industrial processes, along the roads instead of designated landfill sites and began protesting the lack of community protection and governmental oversight. The protests led to a study by the US General Accounting Office, which looked at hazardous waste dumping sites and their correlation with the racial and economic status of communities near these landfills. The study revealed that three out of four commercial hazardous waste landfill sites among eight southern states were located in communities that were predominantly African American (Bullard, 2001).

This pattern further extends into the world of food and food access. Many areas that are economically disadvantaged and have a high concentration of minorities tend to have less access to fresh, healthy, and affordable food. These areas are widely known as "food deserts" (USDA, 2015) and are attributed to "supermarket redlining" (Eisenhauer, 2001). Supermarket redlining refers to the relocation of supermarkets away from urban areas to more profitable suburban areas (Eisehauer, 2001). Larger supermarket chains left urban and rural areas, areas that also happen to have greater levels of economically disadvantaged residents, for areas perceived to have fewer obstacles. Obstacles include higher property costs, labor costs, and utilities, as well as higher rates of crime and lower demand for perishable items (Zhang & Debarchana, 2016). Despite the abundance of food available in the United States and other wealthier countries, for millions of people, access to healthy, affordable food remains out of reach.

The main methods of climate change and environmental communication not only present environmental issues from one perspective but also further assume that we all experience environmental damage the same way, thereby obscuring environmental injustices experienced across populations, cultures, and societies. The failure to consider multiple voices in climate change and environmental communication is further compounded by the focus on individual responsibility in environmental campaigns. By focusing on individual actions, we ignore the actions of corporations and governments (Dunaway, 2015; Webb, 2012). Over the last 60 years, the emphasis on individual responsibility has been the dominant lens through which governments and organizations communicate about the environment and climate change, thereby failing to address strengthening capacity for societies as a whole.

Behavior Change Efforts Through Communication

With the emphasis on individual responsibility, large efforts were put forth to increase climate literacy and change the attitudes held by the public about environmentally friendly behaviors. The dominant approaches to tackling climate literacy and behavior change were based on norm theories (Cialdini, 2003; Schultz et al., 2007; White & Simpson, 2013), values-based models (Crompton, 2010; Stern, 2000; Thøgersen & Crompton, 2009), and economic incentives or nudges (Corner & Randall, 2011; Kristal & Whillans, 2020). Although there is no one-size-fits-all approach to promoting environmentally friendly behavior, there are significant barriers to overcome when trying to inform people about the impacts of climate change and encourage changes in behavior. Barriers to behavior change include social conformity, identity conflicts, moral and ethical concerns, habits, conflicting interpersonal goals, lack of perceived control or feelings of helplessness, and construal-level concerns.

Moser (2016) describes not just how environmental communication has worked, but more importantly, she highlights what needs to change moving forward, what has worked, and what has not. It is essential to understand that environmental and climate communication approaches should be context and audience-specific. But generally speaking, Moser concludes that the field of climate communication in practice needs to do more than educate people; it needs to empower people with appropriate actions to take. We also need to utilize more diverse and nontraditional communication channels, especially since mainstream media still presents climate change as a debate, and usually, it is a debate with Bill Nye. There is also a need to connect with broader social movements, given that the issue of climate change touches every aspect of human life. There is a need to embed climate change discourse in other forms of culture, such as literature, art, music, and theater. We need to move away from fear toward building hope and optimism to empower people to take action. And we must find ways to overcome some of the tensions between scientific practice and disseminating actionable policies.

How to Increase Support for Climate Change Action?

From the literature, three factors are helpful to consider for motivating greater engagement and action among the public with climate change (Sheppard, 2015). The first is to make the issue local. Making climate change more salient and immediate by pulling it into a community context that people care about and using the local landscape to express climate change issues and focus action are argued to promote more comprehensive action. The second factor is to make climate change visual. Harnessing the power of visual perception and imagery in producing concepts and realities of climate change and carbon, both explicit and compelling, helps people see climate change. Lastly, make climate change connected holistically with the big picture of climate change, integrating all aspects of climate change that interact with society and affected environments across scales. The challenge here is determining methods of communication that achieve or fulfill each of these goals. Food, although not widely applied within research, fits these three approaches.

A shift needs to happen, where climate change communication needs to adopt methods that cut through the psychological and social barriers to increased action. There is a need for novel approaches to make climate change and its impacts resonate more strongly with audiences, locally, visually, and holistically. Bolsen and Shapiro (2018) suggest that narratives on climate change need to be reframed so that evidence does not threaten or directly contradict an audience's existing opinions and worldviews. New narratives further need to convey the extent of scientific consensus on climate change causes and solutions while promoting efficacy to prevent further impacts of climate change. Stories need to encourage the accuracy of foundational climate science and new information to minimize the spread of disinformation and rely on credible spokespeople to share the story. Traditional methods of storytelling are one approach that could not only increase awareness of climate change impacts and action. Still, they could do so in a way that engenders local, visual, and holistic approaches.

When it comes to identifying a method for engaging in large-scale action on climate, two human activities both unify and connect us, storytelling and food. The remainder of this chapter will outline the connections between storytelling, food, and climate change as a method for raising awareness and spurring climate action. Both food and storytelling build relationships between people. Both stories and food evoke cross-cultural, political, social, and psychological boundaries, which build bonds and unite people.

Much of what is communicated about climate change reflects temperature, precipitation, sea-level rise, and storms. These impacts are rarely connected to an individual. Food, however, is something everyone can identify. There are memories, emotions, and images associated with food. Ask anyone to think about their favorite food, and it generates a complex response. Beyond providing a basic level of nutrition, food is connected more broadly to our overall health and well-being. Block et al. (2011; p. 6) define food well-being as "a positive psychological, physical, emotional, and social relationship with food at both the individual and societal levels." Applying food well-being as a framework extends our relationship with food across other disciplines, where food transcends a purely utilitarian function and engenders aspects of a community, kinship, comfort, and pleasure (Block et al., 2011).

Moreover, food carries symbolic meanings as we both consume and share it, physically and digitally, via social media (Kozinets et al., 2017; McDonnell, 2016; Wallendorf & Arnould, 1991). Food is simultaneously material and symbolic, connecting producers and consumers with various physical, social, and cultural meanings (McDonnell, 2016). Those who share food photos invite the public into another world, an aspirational world of travel and exotic flavors. Additionally, sharing a meal either physically or virtually promotes an inclusive space that fosters a sense of belonging and shared experience while also providing sustenance. The materiality of food overcomes the aspects of physical and psychological distance. The sharing of food breaks down opposing values and attitudes through inclusive experiences (Eiseman, 2019; McDonnell, 2016).

Lastly, within the international community, people have a right to access food, and governments should limit the barriers to individuals' sustenance access (Dumas, 2010). Food production is intrinsically linked to climate and environmental conditions. Therefore, failure to act on climate change inhibits individuals' right to food.

The right to food is a moral, ethical, and political imperative to ensure the global climate remains viable for food production. Ensuring equitable food production and distribution maintains the overall well-being and freedoms of a population.

Connecting Food to the Art of Storytelling

Just as food is a part of our everyday life, so too is storytelling. We tell stories every day and experience this phenomenon from a very young age (Alda, 2017). The story or narrative structure helps us understand the world around us and our place within it (Arnold, 2018). Examine the social and cultural components of storytelling to increase the human connection to what is happening globally and locally. Stories told as narratives are linear and share a sequence of events in terms of a beginning, middle, and end. Furthermore, narratives provide a structure such as suspense, tragedy, or comedy to convey a lesson (Shanahan et al., 1999).

Storytelling is one technique that has been shown to overcome challenges policymakers and practitioners face when trying to gain support for proactive resilience planning and climate action. The vivid imagery conveyed within a narrative taps into people's enduring memories, shared experiences, and sensations, while further informing listeners of how to negotiate similar situations (Goodchild et al., 2017; Shanahan et al., 1999). These sensations align with the emotive, social, psychological, and physical relationship we have with food. Thus, using food to tell the climate change story could be an effective way to scale up climate change action. Scientists often share information about climate change by providing facts and figures. However, people tend to assess technical knowledge within the whole of their worldviews and beliefs as opposed to rational thought (Tversky & Kahneman, 1981).

Additionally, Bednarek (2021) argues that the "divided brain" or the dual hemispheric structure of the brain results in many people's ability to carry on with everyday tasks while ignoring overwhelming information. The trauma theory of structural dissociation (Van der Hart et al., 2006) suggests, in times of stress, the left side of the brain can take over and allow an individual to continue in daily tasks. When the left side of the brain takes over, a person is able to stay positive and focused, while the right side of the brain interacts with the world in a contextual way. When the right side of the brain takes over, an individual can relate to the natural world and see life as more complex and interconnected (Bednarek, 2021). Thus, creating a narrative about climate change instead of facts and figures appeals to or activates multiple forms of information processing, resulting in a deeper understanding and holistic perspective of an issue. Given the challenges climate change poses (e.g., a wicked problem, existential problem, catastrophic problem), those who encourage action against climate change could benefit from a narrative approach. Stories can help because of the connections they forge between people and between people and ideas. Stories help to convey culture, history, and goals that unite communities. Stories help us learn from our mistakes; they turn failures into victories, into ways forward. Stories can create a culture of compassion, reminding us that we are all human.

Telling the Story of Climate Change Through Food

Telling a story is more than simply telling a story. Following the traditional narrative approach, a story should have a beginning, middle, and end. However, it should include the human element, some flair, or some emotion. Stories should be descriptive, so those listening can feel like they are walking in the storyteller's shoes. Ultimately, the audience should see the world through the storyteller's eyes. Although storytelling and eating are two things that we are all familiar with, it is helpful to have some parameters in producing stories, particularly about climate change and food. The following example provides an illustrative example of the impact storytelling has and how food and climate change can be integrated into a workshop with community groups to develop new and shared meanings. The outline was created in partnership with a local storytelling group, Trampoline. The guide below helps facilitate a public storytelling workshop for climate change engagement.

Using Food to Tell the Climate Change Story

A case has been made to use food to tell the climate change story, and *Our Changing Menu: Climate Change and the Foods We Love and Need* by Hoffmann et al. (2021) is one of the first major publications to advance this approach. The book is intended for a general audience along with its complementary website, Our Changing Menu (ourchangingmenu.com). The website's story is like the books but also includes a searchable database of potentially hundreds of food ingredients and how they are changing.

At the heart of the story is a menu that ventures from before-dinner drinks to desserts and describes the changes underway. Wakeup calls include the escalating risks to our food system but also increasing risks to chocolate, vanilla, and spuds; how flavors and aromas are changing; and how the loss of glacial meltwaters in Peru will affect blueberry availability to North America. The story also helps people to appreciate where their food comes from and the challenges it faces on the way to their table, understand the basics of climate change, and how the plants we depend on for life are being changed. This story also emphasizes solutions, such as what farmers, scientists, food businesses, and others are doing and, most importantly, what everyone can do. The number of stories to be told and the audiences are unlimited.

The story is not all doom and gloom, and there are no villains. It draws attention to climate change through food, offers solutions, and offers hope. We are all in this together, and through the common ground of the food, the goal is to catalyze a climate change social movement that brings about the transformations needed in every sector of society.

Climate Change and the Future of the Hamburger

The following is an example of how the climate change and food story can be told using the hamburger.

We tell a climate change and food story by visually portraying how climate change changes the menu using the culturally iconic hamburger, one of the world's most widely consumed meals. Using images and a narrated script, we describe the changes underway to this beloved meal. We believe this unique visual approach will affect people's emotions about how their foods are at increasing risk because of climate change and challenge all of us to take action to save our favorite foods and humanity as well.

Our story is part of a postgraduate research-led initiative from the Centre for Cultural Ecologies in Art, Design and Architecture at the University of Huddersfield Graduate School, Huddersfield, West Yorkshire, UK. We created digitally curated story about the hamburger which is one of many exhibitions at Climate Action and Visual Culture (Mateescu, 2021). The script below was complemented with images of the various components of a hamburger.

What's happening to your favorite foods? The answer might surprise you.

We see and hear about how climate change is causing glaciers to melt, more extreme weather, and heat waves, but what about closer to home? A changing climate doesn't just mean a change in our wardrobe and energy bills, it means a change in the conditions in which our food is grown. Water, soil, temperatures, and the composition of the air are all changing and affecting the food plants we depend on for life. This encompasses staples like wheat and rice to the spices that flavor the foods we love, need, and eat several times each day. It's all changing, in subtle and ominous ways. We'll tell this story by showing what's happening to a hamburger.

The tart in pickles.

Pickles are tart because they are soaked in vinegar to get "pickled." And vinegar is made by fermenting grains such as corn and barley. Globally, grains are under stress because of climate change, and the impacts will vary by region. For example, corn production in the USA is predicted to decline by 20% (Berlin et al., n.d.) and barley by 8–25% in the Mediterranean region (Cammarano et al., 2019) depending on the intensity of climate change.

Stressed but maybe sweeter tomatoes.

When it's too hot, tomato plants produce fewer fruit, and the tomatoes get sunburned and can be small in size. But when grown under higher levels of the greenhouse carbon dioxide expected in the future, they may be sweeter and have more vitamin C (Bisbis et al., 2018).

Onions may taste different.

Like many other vegetables, onions don't do well when it's too hot. But interestingly enough, higher temperatures can also make them more pungent (Coolong & Randle, 2003). If you like a strong onion, you might like this change.

Lettuce's ups and downs.

On the positive side, lettuce yields increase by over 40% when grown under higher levels of carbon dioxide, which is expected in the coming decades (Korres et al., 2016). In other good news, in regions where winters are already mild, lettuce might be grown over winter and provide a year-round supply as winters continue to warm (Bisbis et al., 2018). One challenge for lettuce, however, is if it gets too warm the lettuce plant sends up a flower stalk and develops seeds. This results in a strong and bitter taste, making the plant unmarketable. The tips of the leaves may also turn brown when it is too hot.

Cheese is changing too.

The thousands of kinds of cheese we enjoy are generally not yet being directly threatened by climate change, but cows are. When hot, they produce less milk (Wolfe et al., 2008). They also contribute to climate change by producing methane gas, but changing their diets can help reduce these emissions (Woodward et al., 2004).

Treating red meat as a delicacy.

Beef is essential to the lives of many people around the globe, but in rich countries, consumption of red meat often exceeds dietary recommendations. Also, consider that beef generates about 50 times more greenhouse gas emissions than wheat to produce an ounce of protein (Clark & Tilman, 2017). The plant-based meat substitutes are growing in popularity and also come with a lower greenhouse gas footprint (Peters, 2019). It's time that we consider reducing our consumption of red meat, and one approach is to consider it a delicacy and not a staple. This burger looks like a meaty hamburger, but it's meatless.

Higher nighttime temperatures cause stunted spuds.

In 2018, the British chip (almost like a French fry) was an inch shorter than normal because of climate change (Carrington, 2019). Increases in nighttime temperature alone could undermine global potato production (International Potato Center, 2013). Under a business-as-usual scenario, researchers estimate up to 95% of English and Welsh potato-growing land will become unsuitable for production by 2050 because of increasingly dry conditions (Daccache et al., 2012).

And then there is the bun.

Things are changing for the wheat in this burger bun. Some experts indicate that if we don't curb climate change, by the end of the century, much of the world's area used for wheat production will be simultaneously affected by severe water scarcity (Trnka et al., 2019). In the USA, impacts are expected to be modest with wheat grown over winter at higher latitudes actually benefiting (Crane-Droesch et al., 2019).

Sesame seeds too.

Africa is the source of 40% of the world's sesame, and climate change is making conditions there drier and hotter. Fortunately, the sesame plant is tolerant of these stressful conditions meaning that production could expand and become an important source of income for small-scale farmers in some regions (Dossa et al., 2017).

Ketchup.

Ketchup is typically made from processing tomatoes, most of which are grown in California where 11 million tons are produced each year. Processing tomatoes is expected to be in adequate supply until late in the century when production is predicted to drop by about 10% (Lee et al., 2011). The increasing threat of drought in California could create challenges well before the end of the century, however.

Does the burger have a future?

Yes! Scientists worldwide are working hard to develop more climate-hardy crops and farmers are changing how they grow crops to ensure that we have the wheat, potatoes, onions, and sesame seeds we love and need.

Find out what else is changing and how you can help at www.ourchanging-menu.com.

Conclusion

Scaling up rapid action for climate change requires various approaches to appeal to multiple audiences. Telling the story of climate change through food is one way to raise awareness about climate change and motivate people to take action to mitigate the future impacts of a changing climate. In this chapter, we outlined the current state of the environmental communication field. We introduced food and storytelling as a method for overcoming some of the existing barriers to broader engagement with climate change action. Lastly, we provided an illustrative example of how individuals can carry out a storytelling workshop and its effects on participants and how to apply these concepts to food and climate change. After all, everyone eats!

References

Alda, A. (2017). *If I understood you, would I have this look on my face? My adventures in the art and science of relating and communicating*. Random House.

Arnold, A. (2018). *Climate change and storytelling: Narratives and cultural meaning in environmental communication*. Springer.

Bednarek, S. (2021). Climate change, fragmentation and collective trauma: Bridging the divided stories we live by. *Journal of Social Work Practice, 35*(1), 5–17. https://doi.org/10.1080/02650533.2020.1821179

Berlin, J., Conant, E., De Seve, K., Nunez, C., Hartigan Shea, R., Stone, D., & Zuckerman, C. (n.d.). 5 ways climate change will affect you: Crop changes. *National Geographic*. http://www.nationalgeographic.com/climate-change/how-to-live-with-it/crops.html

Bisbis, M. B., Gruda, N., & Blanke, M. (2018). Potential impacts of climate change on vegetable production and product quality – A review. *Journal of Cleaner Production, 170*, 1602–1620. https://doi.org/10.1016/j.jclepro.2017.09.224

Block, L. G., Grier, S. A., Childers, T. L., Davis, B., Ebert, J. E. J., Kumanyika, S., Laczniak, R. N., Machin, J. E., Motley, C. M., Peracchio, L., Pettigrew, S., Scott, M., & Bieshaar, M. N. G. G. (2011). From nutrients to nurturance: A conceptual introduction to food well-being. *Journal of Public Policy and Marketing, 30*(1), 5–13. https://doi.org/10.1509/jppm.30.1.5

Bolsen, T., & Shapiro, M. A. (2018). The US news media, polarization on climate change, and pathways to effective communication. *Environmental Communication, 12*(2), 149–163. https://doi.org/10.1080/17524032.2017.1397039

Bullard, R. D. (2001). Environmental justice in the 21st century: Race still matters. *Phylon (1960-), 49*(3/4), 151–171. https://doi.org/10.2307/3132626

Cammarano, D., Ceccarelli, S., Grando, S., Romagosa, I., Benbelkacem, A., Akar, T., Al-Yassin, A., Pecchioni, N., Francia, E., & Ronga, D. (2019). The impact of climate change on barley yield in the Mediterranean basin. *European Journal of Agronomy, 106*, 1–11. https://doi.org/10.1016/j.eja.2019.03.002

Carrington, D. (2019, February 5). UK chips an inch shorter after summer heatwave – report. *The Guardian*. http://www.theguardian.com/environment/2019/feb/05/uk-chips-an-inch-shorter-after-summer-heatwave-report

Cialdini, R. B. (2003). Crafting normative messages to protect the environment. *Current Directions in Psychological Science, 12*(4), 105–109. https://doi.org/10.1111/1467-8721.01242

Clark, M., & Tilman, D. (2017). Comparative analysis of environmental impacts of agricultural production systems, agricultural input efficiency, and food choice. *Environmental Research Letters, 12*(6), 064016. https://doi.org/10.1088/1748-9326/aa6cd5

Coolong, T. W., & Randle, W. M. (2003). Temperature influences flavor intensity and quality in Granex 33' Onion. *Journal of the American Society for Horticultural Science, 128*(2), 176–181. https://doi.org/10.21273/JASHS.128.2.0176

Corner, A., & Randall, A. (2011). Selling climate change? The limitations of social marketing as a strategy for climate change public engagement. *Global Environmental Change, 21*(3), 1005–1014. https://doi.org/10.1016/j.gloenvcha.2011.05.002

Crane-Droesch, A., Marshall, E., Rosch, S., Riddle, A., Cooper, J., & Wallander, S. (2019). *Climate change and agricultural risk management into the 21st century (Economic Research Report No. 266; Economic Research Service, p. 63)*. United States Department of Agriculture.

Crompton, T. (2010). *Common cause: The case for working with our cultural values*.

Daccache, A., Keay, C., Jones, R. J. A., Weatherhead, E. K., Stalham, M. A., & Knox, J. W. (2012). Climate change and land suitability for potato production in England and Wales: Impacts and adaptation. *The Journal of Agricultural Science, 150*(2), 161–177. https://doi.org/10.1017/S0021859611000839

Dossa, K., Konteye, M., Niang, M., Doumbia, Y., & Cissé, N. (2017). Enhancing sesame production in West Africa's Sahel: A comprehensive insight into the cultivation of this untapped crop in Senegal and Mali. *Agriculture and Food Security, 6*(1), 68. https://doi.org/10.1186/s40066-017-0143-3

Dumas, G. F. (2010). Greener revolution: Using the right to food as political weapon against climate change. *New York University Journal of International Law and Politics, 43*(1), 107–158.

Dunaway, F. (2015). *Seeing green: The use and abuse of American environmental images*. University of Chicago Press. https://press.uchicago.edu/ucp/books/book/chicago/S/bo13666193.html

Ebi, K. L., & Ziska, L. H. (2018). Increases in atmospheric carbon dioxide: Anticipated negative effects on food quality. *PLoS Medicine, 15*(7), e1002600. https://doi.org/10.1371/journal.pmed.1002600

Eiseman, D. L. (2019). The coffee-drinking experience: Contributions to pleasure, wellbeing, and consumer engagement. In *Food and experiential marketing* (1st ed., p. 14). Routledge.

Eisenhauer, E. (2001). In poor health: Supermarket redlining and urban nutrition. *GeoJournal, 53*(2), 125–133.

Goldenberg, S. (2014). "Climate change 'already affecting food supply'–UN." The Guardian, https://www.theguardian.com/environment/2014/mar/31/climate-change-food-supply-un

Goodchild, B., Ambrose, A., & Maye-Banbury, A. (2017). Storytelling as oral history: Revealing the changing experience of home heating in England. *Energy Research and Social Science, 31*, 137–144. https://doi.org/10.1016/j.erss.2017.06.009

Hoffmann, M. P., Koplinka-Loehr, C., & Eiseman, D. L. (2021). *Our changing menu: Climate change and the foods we love and need*. Cornell University Press.

Hansen, A., & Cox, J. R. editors. (2015). The routledge handbook of environment and communication. Routledge.

International Potato Center. (2013, April 2). *Potato faces up to climate change challenges*. International Potato Center. https://cipotato.org/blog/potato-faces-up-to-climate-change-challenges/

Korres, N. E., Norsworthy, J. K., Tehranchian, P., Gitsopoulos, T. K., Loka, D. A., Oosterhuis, D. M., Gealy, D. R., Moss, S. R., Burgos, N. R., Miller, M. R., & Palhano, M. (2016). Cultivars to face climate change effects on crops and weeds: A review. *Agronomy for Sustainable Development, 36*(1), 12. https://doi.org/10.1007/s13593-016-0350-5

Kozinets, R., Patterson, A., & Ashman, R. (2017). Networks of desire: How technology increases our passion to consume. *Journal of Consumer Research, 43*.

Kristal, A. S., & Whillans, A. V. (2020). What we can learn from five naturalistic field experiments that failed to shift commuter behaviour. *Nature Human Behaviour, 4*(2), 169–176. https://doi.org/10.1038/s41562-019-0795-z

Lee, J., De Gryze, S., & Six, J. (2011). Effect of climate change on field crop production in California's Central Valley. *Climatic Change, 109*(1), 335–353. https://doi.org/10.1007/s10584-011-0305-4

Leiserowitz, A., Maibach, E., Roser-Renouf, C., Feinberg, G., & Rosenthal, S. (2015). *Climate change in the American mind: October, 2015* (Yale Program on Climate Change Communication). Yale University and George Mason University. http://environment.yale.edu/climate-communication-OFF/files/Climate-Change-American-Mind-October-2015.pdf

Leiserowitz, A., Ballew, M., Rosenthal S., & Semaan, J. (2020). *Climate change and the American diet.* Yale University and Earth Day Network. Yale Program on Climate Change Communication.

Mateescu, L. (2021). *Climate action and visual culture [Digital exhibition].* https://www.artsteps.com/view/60a5212af8f5287fae8f1c7c

McDonnell, E. M. (2016). Food porn: The conspicuous consumption of food in the age of digital reproduction. In P. Bradley (Ed.), *Food, media and contemporary culture* (pp. 239–265). Palgrave Macmillan. https://doi.org/10.1057/9781137463234_14

Media Cloud (n.d.). Homepage. https://www.mediacloud.org

Moser, S. C. (2016). Reflections on climate change communication research and practice in the second decade of the 21st century: What more is there to say? *Wiley Interdisciplinary Reviews: Climate Change, 7*(3), 345–369. https://doi.org/10.1002/wcc.403

Myers, S. S., et al. (2017). Climate change and global food systems: Potential impacts on food security and undernutrition. *Annual Review of Public Health, 38*(1), 259–277. https://doi.org/10.1146/annurev-publhealth-031816-044356

Peters, A. (2019, March 20). *Here's how the footprint of the plant-based impossible burger compares to beef.* Fast Company. https://www.fastcompany.com/90322572/heres-how-the-footprint-of-the-plant-based-impossible-burger-compares-to-beef

Project Drawdown (n.d.) Homepage. https://www.drawdown.org/

Schultz, P. W., Nolan, J. M., Cialdini, R. B., Goldstein, N. J., & Griskevicius, V. (2007). The constructive, destructive, and reconstructive power of social norms. *Psychological Science, 18*(5), 429–434. https://doi.org/10.1111/j.1467-9280.2007.01917.x

Shanahan, J., Pelstring, L., & McComas, K. (1999). Using narratives to think about environmental attitude and behavior: An exploratory study. *Society and Natural Resources, 12*(5), 405–419. https://doi.org/10.1080/089419299279506

Sheppard, S. R. J. (2015). Making climate change visible: A critical role for landscape professionals. *Landscape and Urban Planning, 142*, 95–105. https://doi.org/10.1016/j.landurbplan.2015.07.006

Stern, P. (2000). Toward a coherent theory of environmentally significant behavior. *Journal of Social Issues, 56*(3), 407–424.

Thøgersen, J., & Crompton, T. (2009). Simple and painless? The limitations of spillover in environmental campaigning. *Journal of Consumer Policy, 32*(2), 141–163. https://doi.org/10.1007/s10603-009-9101-1

Trnka, M., Feng, S., Semenov, M. A., Olesen, J. E., Kersebaum, K. C., Rötter, R. P., Semerádová, D., Klem, K., Huang, W., Ruiz-Ramos, M., Hlavinka, P., Meitner, J., Balek, J., Havlík, P., & Büntgen, U. (2019). Mitigation efforts will not fully alleviate the increase in water scarcity occurrence probability in wheat-producing areas. *Science. Advances, 5*(9), eaau2406. https://doi.org/10.1126/sciadv.aau2406

Tversky, A., & Kahneman, D. (1981). The framing of decisions and the psychology of choice. *Science, 211*(4481), 453–458.

U.S. Department of Agriculture. (2015). *Food deserts.* USDA. http://apps.ams.usda.gov/fooddeserts/fooddeserts.aspx. Last accessed 10 January 2022.

Van der Hart, O., Nijenhuis, E. R., & Steele, K. (2006). *The haunted self: Structural dissociation and the treatment of chronic traumatization.* WW Norton & Company.

Wallendorf, M., & Arnould, E. J. (1991). "We gather together": Consumption rituals of thanksgiving day. *Journal of Consumer Research, 18*(1), 13–31. https://doi.org/10.1086/209237

Webb, J. (2012). Climate change and society: The chimera of behaviour change technologies. *Sociology, 46.*

White, K., & Simpson, B. (2013). When do (and don't) normative appeals influence sustainable consumer behaviors? *Journal of Marketing, 77*(2), 78–95. https://doi.org/10.1509/jm.11.0278

Wolfe, D. W., Ziska, L., Petzoldt, C., Seaman, A., Chase, L., & Hayhoe, K. (2008). Projected change in climate thresholds in the Northeastern U.S.: Implications for crops, pests, livestock, and farmers. *Mitigation and Adaptation Strategies for Global Change, 13*(5–6), 555–575. https://doi.org/10.1007/s11027-007-9125-2

Woodward, S. L., Waghorn, G. C., & Laboyrie, P. G. (2004). Condensed tannins in birdsfoot tre-foil (Lotus corniculatus) reduce methane emissions from dairy cows. *Proceedings of the New Zealand Society of Animal Production, 64,* 6.

Zhang, M., & Debarchana, G. (2016). Spatial supermarket redlining and neighborhood vulner-ability: A case study of Hartford, Connecticut. *Transactions in GIS: TG, 20*(1), 79–100. https://doi.org/10.1111/tgis.12142

Danielle L. Eiseman is a climate change communication specialist, with a focus on pro-environmental behavior change. She is particularly interested in how different engagement tools, such as storytelling, food, and social media, can help accelerate public engagement with climate change policy and action. Specific interests include public engagement with climate change policy, audience segmentation, peer-to-peer learning, storytelling and stand-up comedy for public engage-ment, and the use of food experiences for telling the climate change story. She has a Ph.D. in Consumer Behavior, an M.S. in Carbon Management, an M.S. in Economics and Marketing, and a B.A. in Chemistry.

Michael P. Hoffmann dedicates all of his time to the grand challenge of climate change and often helps people understand and appreciate what is happening through food. He has published climate change articles in the popular press – *The Hill*, *Fortune*, and *USA Today* – and is the lead author of *Our Changing Menu: Climate Change and the Foods We Love and Need* (Cornell Press, 2021). Previous positions he has held at Cornell University include Executive Director of the Cornell Institute for Climate Smart Solutions, Director of the Cornell University Agricultural Experiment Station, Associate Dean of the College of Agriculture and Life Sciences, Associate Director of Cornell Cooperative Extension, and Director of the New York State Integrated Pest Management Program. He is a Professor Emeritus in the Department of Entomology. He received his B.S. degree from the University of Wisconsin, M.S. degree from the University of Arizona, and Ph.D. from the University of California, Davis.

Three Ways to Introduce More Stories of Climate Action into Climate Change News Reporting

Joe Whitwell

"So, the climate is changing, what can I do?"

We keep coming back to this question in panicked late-night Google searches and heated discussions with family, friends, and colleagues. As journalists, it is one of the questions readers ask the most. This is hardly a surprise when media leaders believe a key barrier to better climate change reporting is the depressive outlook of current journalism and the powerlessness it creates in audiences (Newman, 2022).

One way to help answer "what can I do?" is for journalists to share stories of people acting *right now*, all around the world. By giving these concrete and actionable examples, we can put the problem into a human context, discussing the real-world difficulties people are dealing with. By telling these stories, intelligent, articulated, and available solutions can spread. By telling these stories, readers can consider how to take action in their own lives. They can feel less alone, less isolated, and less doomed.

To be clear, journalists shouldn't stop reporting the science or the extreme weather events or the rest of the ways we report climate change. We have to keep telling this truth; it is the most important thing we do as journalists, but we can combine the climate *science* about cause and effect with stories of climate *action*. We can inform people that burning fossil fuels to power our homes contributes to climate change *and* tell the story of a man trying to install solar panels on his roof.

The truth is that the situation is getting worse, but people are acting to prevent it from becoming catastrophic. By telling the truth, we can acknowledge these changes and the feelings of despair many people struggle with. We can also inspire more human agency in positive change. Telling stories of climate action isn't new. There are examples included throughout this chapter, mostly from my employer, the BBC. Let's celebrate these stories. Then let's assess how we can gather more and include them more consistently in reporting.

J. Whitwell (✉)
BBC News Labs, Halifax, UK
e-mail: joe.whitwell@bbc.co.uk

E. Coren, H. Wang (eds.), *Storytelling to Accelerate Climate Solutions*,
https://doi.org/10.1007/978-3-031-54790-4_14

I propose three approaches. Firstly, we can tell these stories from across the newsroom and not just rely on the "science" or "climate" team. Secondly, we can engage with audiences to gather and share their own examples. Finally, we can take a systematic approach to incorporate stories of climate action at scale, by using technologies such as natural language templates that are reducing the cost of producing content while enhancing data-driven hyperlocal news reporting to promote human response and community engagement.

Climate Change Journalism

Journalism is one of the primary ways people learn about climate change, from its causes to its predicted impacts around the world. Over time, this has led to much of the scientific consensus around the topic being accepted by the public. To take my country, the UK, as an example, 67% of people support climate policies, believing that climate change is human-caused and a serious threat, and roughly half of them believe it is an *urgent* threat and *strongly* support climate policies (Leiserowitz et al., 2021). Given this acceptance of the situation, it isn't a surprise that readers are asking journalists what they can do to help.

Often, however, the response to "what can I do" looks like a list of climate commandments: "don't fly," "drive less," "drink less dairy," "eat less beef," and so on and so forth. To journalists the appeal is obvious. We can write a long list of suggestions for a broad audience, ostensibly addressing the question. However, for the reader, the recommendations are often impractical. Imagine it is 3 o'clock in the morning. You have just de-iced your car and you are driving through the fog to start a 10-h shift in an Amazon warehouse. How useful is the suggestion that you should take public transport to work?

Such lists also lack detail and appreciation for the trade-offs and choices involved. Don't fly? How do you propose I travel 700 miles for a family holiday? Should I take four trains? Should I drive? Should I stay at home? It is promising that newsrooms are engaging with the questions that audiences are asking about climate change, but the response needs to be more nuanced, more personalized, and more applicable.

What Do Stories of Climate Action Look Like?

Instead of prescriptive lists, we could answer the question "what should I do" differently. We could give illustrative examples of effective solutions that people are practicing *already*. We could tell stories of climate mitigation and adaptation, by both individuals *and* groups. We could report the truth, that people are trying their best, with limited time, limited options, and limited resources. In short, we can tell our readers stories of people like them.

Such storytelling can create communities of action cutting across language barriers and geographic boundaries. As one researcher put it, "Bangladesh has a lot to teach Germany about how to deal with floods" (Whitwell, 2021). From the examples below – some real, some illustrative – it's clear we all have a lot to learn from each other.

Climate Change Mitigation

So, there are no trains in your area to get you to work at 4 AM? Here are some stories of three people who carpool and share the cost of petrol. The three of them often argue about which radio station to listen to, but on Friday they treat themselves to a bacon sandwich with the money they save.

So, you are trying not to fly? Here is the story of a family who turned the challenge of not flying into a road trip and you know what? It was a *disaster*. Learn from their lessons! Pack car sickness tablets and treat yourselves to an overnight stay halfway. Here is another family who took a local holiday instead. Here is the story of a golfer who has to fly for work, but who is paying to carbon offset his flights (Carter, 2021). Want to rely less on fossil fuels to heat your home? Here is a man who saved the money to install solar panels on his roof, but who now is having trouble getting the permits from his local council. Here is the tenant who wants to insulate their rented house but is having trouble with their landlord. Do you have any suggestions to help?

These stories shouldn't be limited to the action of single actors but put into the context of larger workplaces. Want to make a difference through your employer? Here is a hospital procurement manager who got the rules changed so she can buy food from local farmers. The radishes were so peppery they made the patient's eyes water. Here is the software developer who reconfigured his company's cloud computing to use less electricity. Here is the waitress who encourages her customers to take away leftover food after their meal and who convinced her employer to introduce biodegradable takeaway boxes.

Beyond workplaces, we can tell stories of volunteers planting woodland (BBC News, 2021e) or restoring peatland (BBC News 2021b). We can tell stories of farmers using government funds to rewild their land (Bowman, 2022). We can show residents using infrared cameras to discover how to better insulate their homes, with help from local government (Bradley, 2022). We can show bike-share schemes in Rwanda (BBC News, 2021d). Crucially, these stories can make people realize that there are groups they can join, steps they can take, and authorities that can assist them. They are not alone in facing climate change.

Climate Change Adaptation

When audiences ask "what can I do?", they are often asking what they can do to prepare and adapt. They know climate change is happening, but it feels too global, too huge for many people to respond to. We have to be local when we describe climate impacts because the answer to "what can I do about climate change" differs

Fig. 1 Readers can enter their location to see how global warming is predicted to alter their local climate, originally published in Dale & Stylianou (2021)

depending on where you live. Articles that allow readers to enter their location to see how climate change manifests for them could be paired with case studies of climate adaptation (Dale & Stylianou, 2021; Fig.1).

Are you at risk of increased rainfall or flooding? Here is a floating school in Bangladesh (Beaudien, 2018). Here is a farmer reshaping a watercourse in his fields (Rebanks, 2020). Are you at risk from hotter weather? Here is a conservationist reintroducing beavers to restore rivers in the desert (Sherriff, 2021).

Introducing Stories of Climate Action into the Existing Mix of Climate Coverage

While climate action stories can do well on their own, they can also be embedded into other types of climate change coverage, such as scientific reports or predictions of the future. For example, a report of record-breaking temperatures can be accompanied by stories of people adapting to extreme heat around the world, like the story of slum residents in Ahmedabad painting their roofs white to reflect heat, after a housing association lent them the money (BBC News, 2021a), or the plans in Sydney (Lu, 2021) to follow a similar strategy.

Stories of adaptation are particularly powerful alongside climate change predictions. A report predicting increased monsoon rains and flooding can be accompanied by stories of first-time buyers considering flooding risk as they look for their first home or the psychologist who retrained to help people manage the trauma of losing their homes to flooding.

Three Ways to Unlock More Stories of Climate Action

Considering how to introduce more stories of climate action requires some introspection from the news industry, a look at the constraints we work under and the norms that guide our reporting.

Share Stories of Action Across the Newsroom

Most large newsrooms are typically split into different teams covering different topics such as education, business, and home affairs. While this setup allows journalists to develop expertise and monitor their domain closely, there are associated risks and downsides too. In particular, stories about complex issues like climate change can get siloed to small groups of journalists such as "science" or "climate" teams. Instead, they should be told across the newsroom, drawing on the domain expertise of different desks.

Stories of climate action are housing stories; how can we best insulate and power our properties (Dickins, 2021; Morton, 2021). They are investment and personal finance stories: Where should I put my pension investment and why are prices increasing (Timmins & Thomas, 2022)? They are political and legal stories: How can you exercise your rights or lobby your politicians? They are travel and transport stories, food stories (Pandey, 2021), sports stories (Stanton et al., 2021), community stories, education stories, and employment stories (Bearne, 2022). Sharing these stories can create curiosity and hope and even spark new ideas.

The pool of examples can be further broadened by directly connecting individual stories with the larger context not initially conceived as climate change stories. The entrepreneur working to get unwanted food to the hungry is also helping with climate change (Rose, 2022). People installing a smart meter to measure their electricity usage at home are saving money *and* reducing their emissions.

Ask Readers for Examples of Climate Action

One barrier to telling more stories of climate action is cost. These stories are perceived as expensive to collect by media leaders keeping an eye on travel expenses (Newman, 2022) and how newsroom staff are spending their time. Even if a particular story doesn't involve leaving the building there is an opportunity cost. Finding people, interviewing them, and putting their actions in a relevant context eat time that could be used for other types of reporting. As a journalist, should you spend half a day with a person talking about their allotment or instead turn a scientific report into an article? My hope is that we can do both, by asking readers to share their experiences with us as case studies. Not only does this reduce the cost of story collection, it is an effective way to broaden and diversify the stories we find and tell.

The BBC has its own user-generated content team and we've used platforms like Hearken to allow readers to ask questions or suggest stories for a number of years. For climate change specifically, it allows readers to also engage, comment, and question suggested solutions. Say there is a deluge of stories about people getting electric vehicles or heat pumps, audience questions and feedback will quickly reveal what is preventing these solutions being taken up by greater groups of people: cost (BBC News, 2021c).

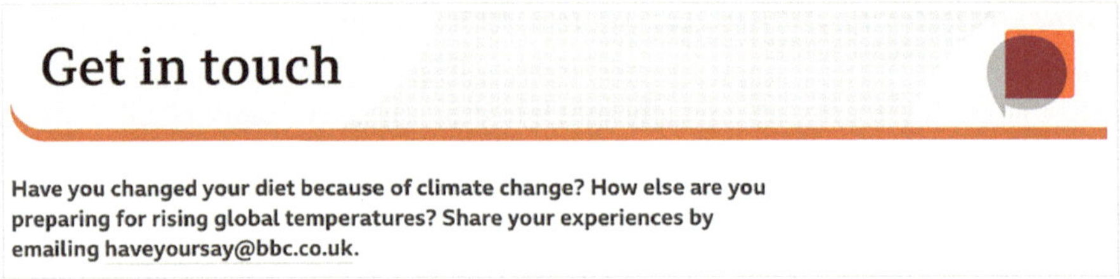

Get in touch

Have you changed your diet because of climate change? How else are you
preparing for rising global temperatures? Share your experiences by
emailing haveyoursay@bbc.co.uk.

Fig. 2 We've recently been asking readers for examples of climate action, as seen in Briggs (2022)

Recently, we've experimented with asking readers for their own stories of climate action (Briggs, 2022; Fig. 2). We took inspiration from prototypes like Climate Map (CHEP, 2022), which encourages people to submit images of them taking action on climate change. The response has been encouraging.

Tell Hyperlocal Stories Using Natural Language Templates

One barrier to better climate change reporting is that such a global problem can feel inaccessible or even irrelevant to readers, who want to know how the problem affects them where they live. By using Natural Language Generation technology (Leppänen et al., 2017) and natural language templates in particular, we have the opportunity to tell more locally specific stories of climate change, which we can pair with stories of climate action.

Let's take tree planting as an example. Imagine you wanted to write one news story about tree planting for every district in the UK. You start all revved up for the first story but begin to lose the will to live by the fifth story. It becomes clear that having an established template would make these stories much easier to produce and understand. In my team at BBC News Labs, we worked with journalists to write a template to tell such a story about government-funded tree planting. We took tree planting data, i.e., the number of trees planted and where, and slotted it into the template.

For example, the template for the headline might read:

[Number of trees] government-funded trees planted in [Name of place] in 8 years.

You then slot the data from the dataset into the template, giving a headline like "91,000 government-funded trees planted in Aylesbury Vale in eight years" (BBC News, 2019; Fig. 3).

You can use this slot-filling approach to generate text for a huge number of articles, turning data into text. You could do the same for stories about rainfall, or air pollution, or hundreds of stories related to climate change. With a good template and data from a reliable source, you can tell thousands of articles, all personalized to people depending on where they live (Molumby, 2019).

To each of these hyperlocal stories, you can pair case studies. For example, if the story is about harmful air pollution in East London, you might pair it with a case

18:29 2 Aug 2019

91,100 government-funded trees planted in Aylesbury Vale in eight years

There have been 91,100 government-funded trees planted in Aylesbury Vale between 2010 and 2018, Forestry Commission data shows.

This works out at 465 trees per 1,000 people.

Fig. 3 Sample new story generated from a natural language template, as seen in BBC News (2019)

study of people cycling instead of driving to work or making face masks. These examples of action don't necessarily have to be from East London, either. A bike scheme in Kigali, Rwanda (BBC News 2021d), could be just as inspiring, because it features people going through the same struggles, even if they are thousands of miles away.

The division of labor here is important; journalists can dedicate their time to finding and curating appropriate case studies while the template turns the data into comprehensible content. To return to our tree planting article in Aylesbury Vale, if the journalist already has a draft article, she could spend time finding a volunteer organization to plant those trees for inclusion in the story. By pairing the facts of the story, derived from data, to the human impact and human response, you can tell stories of action, at scale, to people about where they are.

Conclusion

There is an enormous number of stories we can share about climate change action. Each can add humanity and color to scientific reports, climate predictions, and other more traditional climate change stories. There are practical limitations, however; such storytelling is time-consuming, so if we want to make this the norm, introducing these stories into our reporting systematically, we need to find faster ways to collect and curate them.

Good progress has been made already. Platforms such as Hearken and prototypes like Climate Map show that readers are engaged in these topics, offering their own examples and reducing the time needed in the process. We know too that combining reader-submitted examples with template-generated news reports can deliver much more relevant, engaging, and inspiring climate news.

We don't have all the answers though. We may have more examples of climate action, but more work needs to be done to decide how best to present these stories. Should they be told with different tones or different media types for different readers? Only further experimentation will teach us. We should embrace this, continue getting readers' feedback, and share lessons learned across newsrooms.

References

BBC News. (2019, July 29). BBC Live: South East. *BBC News*. https://www.bbc.co.uk/news/live/uk-england-kent-49129645

BBC News. (2021a, October 7). Life at 50C: Keeping cool in India's heatwaves. *BBC News*. https://www.bbc.co.uk/news/av/world-asia-58820950

BBC News. (2021b, November 11). The young Cadishead conservationists fighting climate change. *BBC News*. https://www.bbc.co.uk/news/av/uk-england-manchester-59240608

BBC News. (2021c, November 11). COP26: How can an average family afford an electric car? And more questions. *BBC News*. https://www.bbc.co.uk/news/science-environment-58925049

BBC News. (2021d, November 18). Rwanda: Kigali's bike-sharing scheme to lower greenhouse emissions. *BBC News*. https://www.bbc.co.uk/news/av/world-africa-59318660

BBC News. (2021e, December 3). Woodland created in Hednesford by volunteers in two months. *BBC News*. https://www.bbc.co.uk/news/uk-england-stoke-staffordshire-59517701

Bearne, S. (2022, January 20). The people moving from high to low-carbon careers. *BBC News*. https://www.bbc.co.uk/news/business-60036245

Beaudien, J. (2018, September 12). 'Floating schools' make sure kids get to class when the water rises. *NPR*. https://www.npr.org/sections/goatsandsoda/2018/09/12/646378073/floating-schools-make-sure-kids-get-to-class-when-the-water-rises?t=1641998885500&t=1643147279482

Bowman, S. (2022, January 18) Sustainable farming: Mum and daughter farmers take part in pilot. *BBC News*. https://www.bbc.co.uk/news/uk-england-derbyshire-60025778

Bradley, R. (2022, January 17). Energy bills: Thermal imaging used to help with cost of heat loss. *BBC News*. https://www.bbc.co.uk/news/uk-england-somerset-60023918

Briggs, H. (2022, January 21). False banana: Is Ethiopia's enset 'wondercrop' for climate change? *BBC News*. https://www.bbc.co.uk/news/science-environment-60074407

Carter, I. (2021, November 16). Rory McIlroy pays thousands to offset carbon footprint of flying to golf events. *BBC News*. https://www.bbc.co.uk/sport/golf/59310338

CHEP. (2022). *Santa Cruz climate action stories*. Climate Health Equity Partnership (CHEP). https://climatemapsc.maps.arcgis.com/apps/Shortlist/index.html?appid=5cabaa68671c4b18b6fb8a6dd4a2d21c

Dale, B., & Stylianou, N. (2021, July 29). What will climate change look like near me? *BBC News*. https://www.bbc.co.uk/news/resources/idt-d6338d9f-8789-4bc2-b6d7-3691c0e7d138

Dickins, S. (2021, December 29). Climate change: Why do new homes not have solar panels? *BBC News*. https://www.bbc.co.uk/news/uk-wales-59668836

Leiserowitz, A., Carman, J., Buttermore, N., Wang, X., Rosenthal, S., Marlon, J., & Mulcahy, K. (2021). *International public opinion on climate change*. Yale Program on Climate Change Communication and Facebook Data for Good.

Leppänen, L., Munezero, M., Granroth-Wilding, M. & Toivonen, H., (2017, September). Data-driven news generation for automated journalism. In *Proceedings of the 10th International Conference on Natural Language Generation*. https://aclanthology.org/W17-3528/

Lu, D. (2021, November 20). Hot in the city: Can a ban on dark roofs cool Sydney? *The Guardian*. https://www.theguardian.com/australia-news/2021/nov/21/hot-in-the-city-can-a-ban-on-dark-roofs-cool-sydney

Molumby, C. (2019, November 28). Trees, machines and local journalism. *BBC News Labs*. https://bbcnewslabs.co.uk/news/2019/salco-trees/

Morton, B. (2021, December 18). Climate change: How can renters make their homes warmer and greener? *BBC News*. https://www.bbc.co.uk/news/uk-59223081

Newman, N. (2022). *Journalism, media, and technology trends and predictions 2022*. The Reuters Institute for the Study of Journalism, Oxford University.

Pandey, G. (2021, December 30). Why I switched to eating grandma's food. *BBC News*. https://www.bbc.co.uk/news/world-asia-india-59650408

Rebanks, J. [@herdyshepherd1] (2020, October 6). Twitter post commenting on reshaping a stream. https://twitter.com/herdyshepherd1/status/1313539154379190275

Rose, I. (2022, January 7). The woman working to get unwanted food to the hungry. *BBC News.* https://www.bbc.co.uk/news/business-59750333

Sherriff, L. (2021). The beavers returning to the desert. *BBC Future.* https://www.bbc.com/future/article/20210713-the-beavers-returning-to-the-desert

Stanton, J., Lockwood, D., & Gornall, K. (2021, November 12). Premier League: Should clubs stop flying to domestic matches for environmental reasons? *BBC Sport.* https://www.bbc.co.uk/sport/football/59213173

Timmins, B., & Thomas, D. (2022, January 20). Inflation: Seven reasons the cost of living is going up around the world. *BBC News.* https://www.bbc.co.uk/news/business-59982702

Whitwell, J. (2021, November 9). Climate change: What do scientists want from COP26 this week? *BBC News.* https://www.bbc.co.uk/news/science-environment-59212185

Joe Whitwell is a journalist with the BBC. He worked at BBC News Labs, where he collaborated with editors across the newsroom to reimagine and improve BBC Journalism. He enjoys speaking to academics about their research and how it can be applied in news reporting. He takes a particular interest in the psychological impact of climate change news coverage on readers. Whitwell studied Linguistics at the University of Cambridge, where he specialized in computational linguistics.

Community-Based Resilience: The Influence of Collective Efficacy and Positive Deviance on Climate Change-Related Mental Health

Maya Cosentino, Roni Gal-Oz, and Debra L. Safer

"Water that's far away can't put out the fire nearby," Jack says, translating from a Chinese expression. Jack is a Cool Block leader. A few months ago, he went door to door talking to neighbors, like Stuart.

"Hi Stuart! I'm Jack, and I live in 342"—he'd say, pointing to his house—"and I would like to invite you to my home to hear about a new program."

Stuart, drawn by the appeal of getting to know his neighbor, agreed. And he, along with other people living on the block, went to Jack's house the next week for a Cool Block meeting.

At the meeting, Jack explains Cool Block's goals to his neighbors: make the neighborhood safer, more connected, and, most of all, resilient against climate change. At the next meeting, the group discusses their dream block. For some, this means no cars so that children can play in the sectioned-off street. For others, it means exchanging homegrown fruits and vegetables. The new Cool Block members take turns hosting monthly meetings in their homes, while becoming experts on different parts of the Cool Block curriculum, such as water stewardship (e.g., reducing water use) or energy preservation (e.g., completely turning off appliances). The members exchange resources, advice, and general help to achieve a set of climate-related sustainability goals.

Jack explains, "Before Cool Block everybody would have been on their own in a disaster. They might have gotten to know each other in a hurry, but they didn't have the kind of social collaboration that Cool Block fosters. When—not if—a major bad thing happens, this is all going to be about neighbors helping neighbors. Cool Block feels like a family, a close-knit family."

Climate change presents an "unacceptably high and potentially catastrophic risk to human health" (Watts et al., 2015, p. 1). At the same time, global heating has been recognized as potentially "the greatest global health opportunity" of the twenty-first

M. Cosentino (✉)
Stanford University Affiliate, University Hospital of Child and Adolescent Psychiatry and Psychotherapy, University of Bern, Bern, Switzerland
e-mail: maya.cosentino@upd.ch

R. Gal-Oz
Barnard College, Columbia University, New York City, NY, USA
e-mail: rg3473@barnard.edu

D. L. Safer (✉)
Stanford University, Stanford, CA, USA
e-mail: dlsafer@stanford.edu

© The Author(s) 2024
E. Coren, H. Wang (eds.), *Storytelling to Accelerate Climate Solutions*,
https://doi.org/10.1007/978-3-031-54790-4_15

319

century (Watts et al., 2015, p. 1). In 2021, over two hundred medical journals called for emergency action to limit global temperature increases, restore biodiversity, and protect health (Atwoli et al., 2021).

This chapter aims to provide useful examples and tools to create climate resilience to decrease harmful climate change-related mental health impacts. First, we will review the effects of climate change on health (physical and mental), including the detrimental consequences on children's health. Second, we highlight the importance of social connection and resilience as mental health resources. The conceptual containers for this chapter are social identification (i.e., perceiving oneself as belonging to social groups; Hogg, 2018; Turner et al., 1994), collective efficacy (i.e., people's shared beliefs about their group's capability to accomplish collective tasks; Bandura, 1997), and positive deviance (i.e., a behavior-change approach that amplifies the successful actions of existing community members; Durá & Singhal, 2009; Singhal & Durá, 2017). The chapter will discuss the potential for local, intentional community building to serve as an effective strategy to improve resilience. The specific examples of Cool Block and the Transition Town movement (e.g., Eco Vista) will illustrate how communities can effectively support climate change adaptation, mitigation, and resilience strategies. Though mental health is inextricably related to larger public health mitigation strategies, this chapter will not address *how* to mitigate climate-related disasters that impact mental health. For example, while wildfires and forced migration clearly impact mental health, this chapter will focus on building collective efficacy and promoting mental health rather than strategies to reduce wildfires.

Impacts of Climate Change on Physical and Mental Health

As illustrated in Fig. 1, the health consequences of climate change are manifold. They range from the repercussions of food and water shortage to increased heat, allergy, vector-born, and air-pollution-related diseases, as well as to the devastating mental health effects of extreme weather, population displacement, and social conflict (Romanello et al., 2021; Watts et al., 2021;Watts et al., 2017). A sustainable energy transition is predicted to have major health benefits, with reductions in air pollution alone saving millions of lives yearly (World Health Organization, 2018).

Given the important connections between health, environmental destruction, and global heating, experts believe that "gains could be made by placing health at the very center of climate change and policy development" (Hamilton et al., 2021, p. e80). Growing Planetary Health and One Health initiatives recognize the complex interconnections and interdependencies of living things and their shared environments and strive to ensure a sustainable future by protecting and enhancing human, animal, and environmental health (Amuasi et al., 2020; Horton et al., 2014; MacNeill et al., 2021). These initiatives recognize the importance of "dismantling disciplinary and professional silos" and promoting forms of collaboration that enable the understanding and management of local and global health threats (Amuasi et al., 2020).

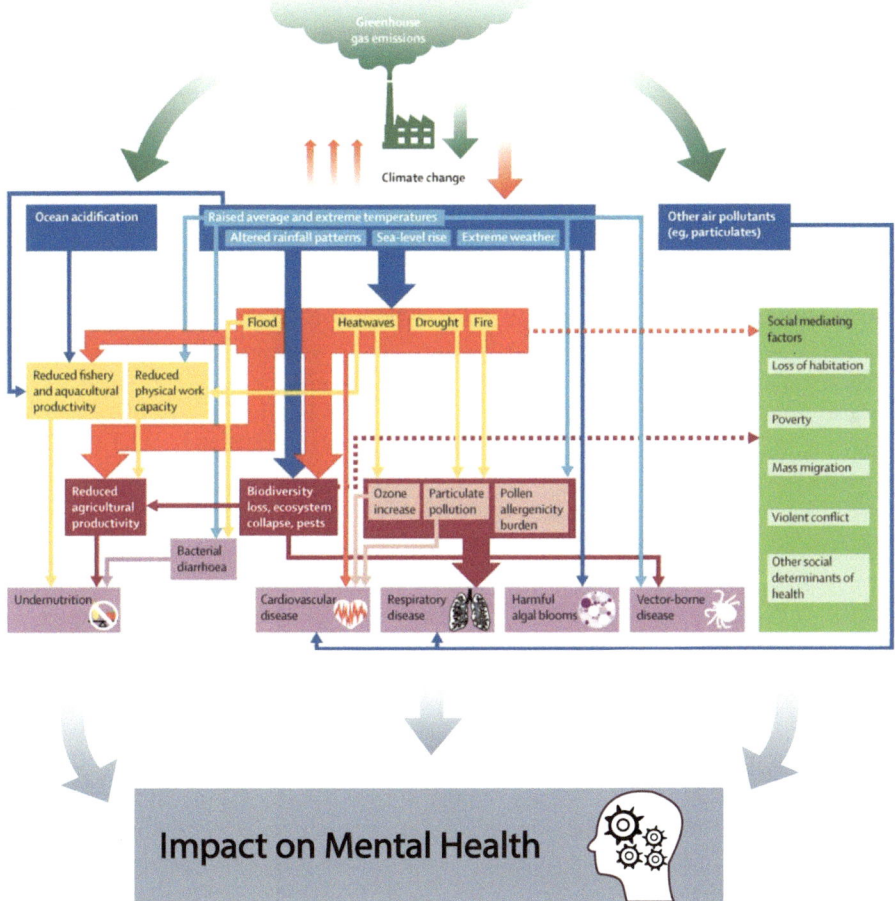

Fig. 1 The health impacts of climate change. (Adapted from Watts et al., 2017, copyright permission granted)

Negative Impacts of Climate Change on Mental Health

Mental disorders are estimated to be the second leading contributor to the burden of global disease, accounting for one-third of global disability and for 7–13% of disability-adjusted life years (Vigo et al., 2016). Climate change threatens the psychological health of individuals and communities (Cianconi et al., 2020; Lawrance et al., 2021; Obradovich et al., 2018; Ojala et al., 2021). For example, comprehensive reviews show connections between high temperatures and aggressive behavior (e.g., physical assaults, homicides, domestic violence) as well as suicide (Barlett et al., 2020; Burke M et al., 2018, Burke SEL, Sanson, & Van Hoorn, 2018; Miles-Novelo & Anderson, 2019; Palinkas & Wong, 2020; Rinderu et al., 2018; Ruderman & Cohn, 2021; Thompson et al., 2018). Air pollution and temperature variability are also associated with declining mental health (Xue et al., 2019). Additionally, extreme weather events such as hurricanes, wildfires, and drought increase the likelihood of depression, anxiety, post-traumatic stress disorder (PTSD), substance abuse, and suicidal thoughts (Cianconi et al., 2020; Palinkas & Wong, 2020).

The aforementioned mental health effects of extreme weather often strain social relationships which, in turn, negatively impact health on an immunological level (Segerstrom & Miller, 2004; Smith et al., 2020). More specifically, interpersonal conflict, low social support, and loneliness have been shown to increase pro-inflammatory immune molecule levels (i.e., cytokines; Kiecolt-Glaser et al., 2010; Smith et al., 2020) and the risk and development of both physical and psychological illness (e.g., anxiety and depression; Glaser & Kiecolt-Glaser, 2005; Kiecolt-Glaser et al., 2010). Insufficient social relationships and loneliness also present a mortality risk comparable to tobacco and alcohol abuse, and greater than physical inactivity, obesity, and air pollution (Holt-Lunstad et al., 2010). In other words, social relationships significantly influence our health, and, just like clean water and food, they should be recognized as a resource ("social capital") worth promoting and protecting (Almedom & Glandon, 2008; Umberson & Montez, 2010).

Another often under-recognized impact of climate change on psychological well-being involves the ongoing media coverage of catastrophic threats posed by extreme weather, which can lead to chronic anxiety and stress related to a heightened perception of risk. Climate change-related media coverage can result in sadness, guilt, pessimism, and the loss of belief in either or both individual and collective efficacy (USGCRP, 2016). These psychological symptoms are elements of an emotional state called *eco-anxiety* (Albrecht, 2011; Cunsolo & Ellis, 2018; Dodds, 2021; Panu, 2020), or the "chronic fear of environmental doom" (Clayton et al., 2017, p. 68). Media coverage or other forms of discouraging and/or inaccurate communication are more likely to evoke eco-anxiety than coverage that focuses on adaptive and preventive measures that individuals or groups can take (USGCRP, 2016).

Compounding the direct impacts of climate change are the related economic (e.g., job loss, displacement, homelessness) and social costs (e.g., isolation, reduced opportunities for exercise and other stress-relieving activities). These can also adversely influence mental health, resulting in a loss of a sense of identity and place as well as feelings of helplessness. The COVID-19 pandemic is a recent example of how mental health conditions, suicide attempts, drug overdoses, violence, abuse, and neglect can rise during times of increased stress and social confinement (Holland et al., 2021). While the mental health implications of climate change can affect everyone, these impacts tend to be greatest among individuals who are the most vulnerable and marginalized, such as those with pre-existing chronic health problems, senior citizens, women (especially if pregnant), the socioeconomically disadvantaged, and some ethnic minority groups (Hayes et al., 2018; Ingle & Mikulewicz, 2020; Kreslake et al., 2016; Palinkas & Wong, 2020; USGCRP, 2016).

Children are also particularly vulnerable to the mental health fallout of climate change (Burke M et al., 2018; Burke SEL et al., 2018; Uddin et al., 2021). Adverse childhood experiences (ACEs) include mental illness and violence within households, physical and/or emotional abuse, as well as neglect. The consequences of ACEs can create long-term problems in afflicted children, such as physical health problems, worsened academic performance, and risk of substance abuse (What Are ACEs?, 2018). ACEs compromise children's brain development, leading to poor decision-making and behavioral problems (Doppelt, 2019a, 2019b). Additionally, a majority of mass shooters were found to have faced ACEs at a young age (Peterson & Densley, 2019a, 2019b).

The mental health consequences of climate change in adult caregivers increase the likelihood of ACEs. In the face of persisting traumatic stresses, caregivers are more likely to develop PTSD, engage in self-protective behaviors such as avoidant withdrawal, self-medicate with substances, and/or engage in other maladaptive behaviors (Doppelt, 2019a, 2019b). In addition, climate change directly affects on children's mental health, with a recent global study reporting concerning levels of climate anxiety (Hickman et al., 2021). To address ACEs and toxic stress, creating a cross-sector network of care within communities is essential. The Network of Care Subcommittee, formed by the California Surgeon General, Dr. Nadine Burke Harris, created a guide for communities and health providers to reduce and treat ACEs, called the ACEs Aware Trauma-Informed Network of Care Roadmap (For Communities, n.d.). The guide outlines the importance of establishing a strong accountability system with collaborative group leaders, as well as identifying a set of shared goals across different leading organizations (ACEs Aware Trauma-Informed Network of Care Roadmap, 2021). Another strategy to protect children from ACEs brought on by stressors affecting caregivers involves building climate resilience.

Building Climate Resilience

The ability to recover from climate-related tragedies will be determined by "efforts that promote resilience" (Ingle & Mikulewicz, 2020, p. e128). Resilience is "the ability to maintain or regain health, despite experiencing adversity" (Herrman et al., 2011, p. 259). Climate resilience is defined by the Center for Energy and Climate Solutions (C2ES, 2015) as the ability to anticipate, prepare for, and respond to hazardous events, trends, or disturbances related to climate. Improving climate resilience involves assessing how climate change will create new, or alter current, climate-related risks, and taking steps to better cope with these risks.

A key component of climate resilience involves the access to resources and capacity to use these effectively (e.g., adaptive capacity). While individual adaptive capacity is important and can be strengthened via formal resources such as counseling and psychiatric care, many of the most effective adaptation measures require collective action (Adger et al., 2007).

Active engagement in problem solving (e.g., civic action) can result in a number of positive mental health consequences (Wikrama & Wikrama, 2010). Though active engagement in climate change actions does not guarantee reduced stress, it can offer a valuable buffer for managing stress and building emotional resilience—thereby lowering the likelihood of experiencing negative mental health impacts. Stress-resistant, resilient people have been found to exhibit a sense of group identity and perceived social support, thoughtful but active coping skills, and a perceived internal locus of control (i.e., sense of efficacy and determination; Agaibi & Wilson, 2005; Ozer et al., 2003; Schnurr et al., 2005).

The impact of natural disaster-related stress on depressive symptoms can be mitigated by social connections and perceptions of non-familial social support (Kaniasty & Norris, 1993). Supportive communication and social support are also associated with reduced reports of physical symptoms and depression as well as a greater sense of well-being (Albrecht & Goldsmith, 2003; Flannery, 1990). On the other hand, insufficient social support is a strong predictor of PTSD and has been associated with premature death (Flannery, 1990; Ozer et al., 2003). Hence, opportunities to connect with and support others is an important way to promote climate resilience.

Evolutionary psychology has shown that cooperation and mutual aid are as important for species survival (Ruskey et al., n.d.). Sustainability experts like Christian Berg emphasize that conflict and loss of social cohesion threaten our potential to overcome climate change (Berg, 2019). Overcoming global environmental challenges may therefore be partially dependent on solidarity rooted in social connection (Cosentino, 2021).

Social Identity and Cohesion, Collective Efficacy, and Positive Deviance

Social identity theory posits that an individual's sense of belonging to social groups is a critical part of one's self-concept (Hogg, 2018; Turner et al., 1994). By belonging to groups (e.g., family, community), aspects of a social environment are integrated into one's self-understanding. Group memberships and the social identity that one develops through these memberships are associated with enhanced health and well-being, as well as resilience against environmental stressors such as social isolation (Alcover et al., 2020; Fong et al., 2019; Jetten et al., 2017). These findings have led Jetten et al. (2012) to propose that the development of social identities constitutes a "social cure." In line with this proposal, community-based interventions that increase social identity have been found to mitigate the devastating health effects of loneliness (McNamara et al., 2021).

Social ties as well as community cohesion have been established as essential resources for health (Torres & Casey, 2017; Torres et al., 2018). More specifically, enhanced social cohesion (e.g., supportive social ties) is a protective factor of mental health outcomes (Hikichi et al., 2016) and can be healing post-disasters (Wikrama & Wikrama, 2010). Successful mental health and suicide prevention programs credit socialization as a predictor of program efficiency (Kahn et al., 2020). Taken together, social environments that build and maintain social ties improve mental health, community resilience, and adaptation.

According to Bandura (1997), human behavior is regulated by personal and social factors. These factors include personal and social identity (Turner et al., 1994), social modeling (i.e., learning from role models who demonstrate specific behaviors, attitudes, and emotional responses), self-efficacy (i.e., belief in one's ability to organize and follow through with the necessary actions to obtain desired results), as well as collective efficacy (i.e., people's shared beliefs about their group's capability to accomplish collective tasks; Bandura, 1997).

Higher degrees of collective efficacy are associated with a greater likelihood to coordinate and perform behaviors that increase a group's chance of succeeding despite initial setbacks or other barriers (Bandura, 2000; Goddard et al., 2004; Thaker et al., 2016). Collective efficacy also regulates how people or groups respond emotionally to challenging situations (Thaker et al., 2016). Finally, collective efficacy influences the decisions people or groups make in order to control their future (Bandura, 1997, 2000; Goddard, 2002).

Goddard and Salloum (2011) assert that collective efficacy beliefs may "foster decisions to gather health-related resources, eliminate environmental hazards to health, and promote communication among neighbors, each of which in turn could facilitate dissemination of health information, prevent disease, and increase the likelihood of treatment" (p. 11). High group collective efficacy establishes a strong normative influence of the group, affecting the persistence with which groups pursue goals (Goddard et al., 2004). Further, social norms are established through collective beliefs that encourage certain actions and discourage others (Goddard et al., 2004).

Given climate change's negative impact on psychological well-being, the influence of social identity on health, and the potential of collective efficacy to catalyze constructive, coordinated behaviors, it seems clear that climate mitigation and adaptation strategies should support community building. A promising approach to build collective efficacy within communities is *positive deviance*, a social change strategy that aims to identify what is already working in a community and amplify it (Pascale & Sternin, 2005; Durá & Singhal, 2009; Singhal & Durá, 2017; Sternin & Choo, 2000). Positive deviance posits that the wisdom to solve intractable social problems often lies within the community. Despite the odds, community outliers who deviate from conventional norms model desirable or positive behaviors. Positive deviance inverts conventional diffusion strategies via an "inside-out" process (Durá & Singhal, 2009; Singhal & Durá, 2017).

Collective Efficacy in Action: Two Examples of Positive Deviance

The following two organizations, Cool Block and Transition Town (e.g., Eco Vista), serve as models of positive deviance in how, at the community level, they showcase building collective efficacy for climate resilience.

Cool Block Program

Cool Block (coolblock.org) is a non-profit program established to shift behaviors within neighborhood blocks to reflect increased environmental awareness, with a focus on community building, crisis resilience, and carbon neutrality (Cool Block, n.d.). Cool Block provides individual blocks with specific actions to mitigate their carbon footprints. The program amplifies change by leveraging the behaviors of positive outliers or individuals already exemplifying desired behaviors. In other words, the program utilizes the power of social modeling (Bandura, 1997). Neighbors who are already proficient in a particular carbon-mitigation behavior serve as role models to motivate others on the block to follow suit. Behavior modeling is thus used to shift the culture of the local community. For example, if one neighbor has a composting bin and shares about its benefits (e.g., garden fertilizer, reduced garbage), their neighbors are more likely to use one as well. Similarly, the likelihood of someone purchasing solar panels or installing other climate change mitigation tools increases when a neighbor shares their personal experience. This phenomenon has been termed the familiarity principle or the mere-exposure effect (Zajonc, 2001). On a larger scale, each Cool Block acts as a positive deviant for surrounding neighborhoods. In this way, change snowballs from an individual's actions to a block's actions to, potentially, civic action at the city-wide or beyond level.

To create a block, the Cool Block organization first recruits and trains a "block leader," providing them with the resources (e.g., scripts) necessary to encourage neighbors to join an initial informational meeting. Cool Block recommends each block leader use a script when approaching a new neighbor (see Box A):

> **Box A**
> "Hi (neighbor's name if you know it). I'm (your name, if not already known), and I live (describe or point to where you live on the block or building, if not already known). I would like to invite you to my home to hear about a new program." (Cool Block, Team-Building, n.d.)

According to Sandra Slater, the Program Manager of Palo Alto's Cool Block branch, this script has been refined over years of trial and error to maximize its effectiveness, just as are all the scripts that the program recommends. The guide suggests that Cool Block leaders point to their own houses to emphasize the neighborliness of the proposed plan. Similarly, Cool Block suggests Cool Block leaders use specific terms in order to set a more welcoming and pleasant tone, such as "home" instead of "house."

According to Slater, the top reason people join Cool Block is to get to know their neighbors. The second is to foster a sense of safety, an integral element of mental health, through social connection. Slater explains, "We're tribal. The idea that we don't know someone that is sleeping thirty feet from our head is a completely alien thing to us as a tribe—as a species. We have to know who our neighbors are for our own safety."

Other motivations for joining the program include curiosity and fear of missing out. According to the manual for Cool Block leaders, "We human beings are innately curious. If we pass by someone's home every day, many of us have an impulse to peer inside. How do they live? What is their lifestyle like?" (Cool Block, n.d.).

By offering psychosocial support in a neighborly context, Cool Block is able to cater to people's desires for community, safety, along with other motivations, and use these to further a climate change and resilience agenda.

Cool Block also offers a script for use by Cool Block leaders when running the initial informational meeting (Cool Block, Information Meeting Guide, n.d.). In it, the following four questions are posed to guide potential participants toward building a collective vision for their block (see Box B):

> **Box B**
> - What do you most love about our block and why?
> - In the event of a natural disaster, how would our block ideally respond and function?
> - If our block were living more sustainably, what would it look like?
> - How would your life improve if you could rely on/get help from your neighbors?

The Cool Block manual explains that the purpose of these questions is "to inform, inspire, and empower" neighbors to participate (Cool Block, Information Meeting Guide, n.d.).

Climate resilience grows through the connections between households that, in turn, enhance collective efficacy. Strong community ties are established by working through the eight-topic, 6-month Cool Block curriculum (see Fig. 2), which teaches participants to better adapt to different climate emergencies (i.e., flooding, fires, hurricanes, and earthquakes). This curriculum is web-based with an interactive interface that enables participants to track their climate mitigation actions and carbon reduction progress.

The community ties neighbors develop when participating in Cool Block were shown to be helpful during the COVID-19 pandemic—a health and environment-related crisis. Despite the lack of resources experienced by many individuals and communities during the pandemic, Cool Block communities were able to quickly establish needed support for at-risk neighbors (e.g., the elderly and immunocompromised) by making use of their previously developed Cool Block infrastructures. In addition, social ties previously developed through the initial Cool Block process, facilitated opportunities for social connection despite the quarantine, which promoted mental health and well-being. For example, one block closed off their street,

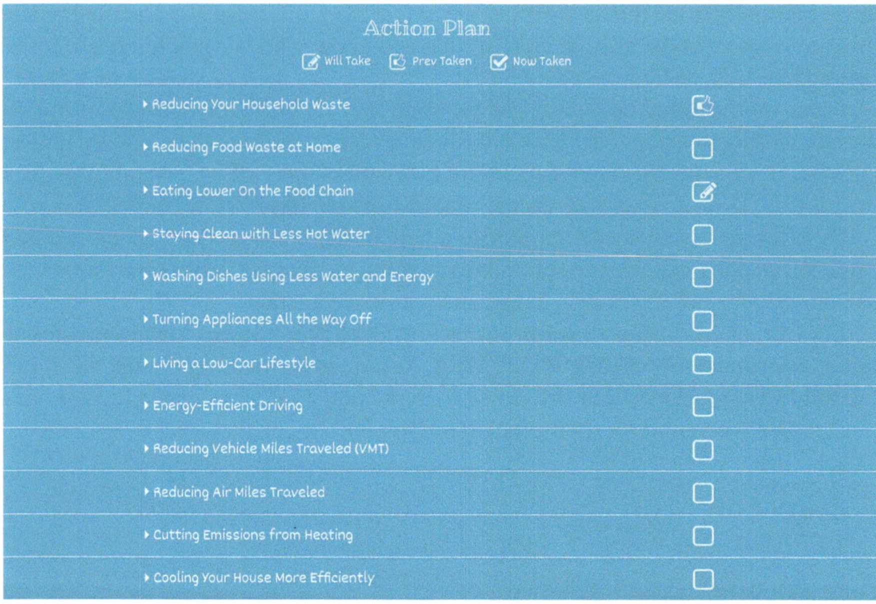

Fig. 2 Example of Cool Block's action plan curriculum. (Image used with permission)

played music, and encouraged children and their parents to gather in a physically distanced manner.

Cool Block has proven highly effective. Within the first pilot year, three large California cities—Los Angeles, San Francisco, and Palo Alto—recruited 45 blocks to participate, with participating households averaging 25 climate actions taken and a 32% household carbon reduction. Fifty-five percent of the people on the block who learned more about it agreed to participate (Lebeck, 2017).

The Palo Alto branch has grown so substantially that it has become municipally engaged. Cool Block is now officially endorsed by the city of Palo Alto, and the recruitment of Block Leaders has shifted to become a responsibility of the local government. This relationship between Cool Block and the city provides positive feedback and direct connections between citizens and government officials. For example, officials sometimes attend Cool Block meetings to provide information about government-sponsored initiatives like the "Home Efficiency Genie," and "Zero-Waste." (Home Efficiency Genie, 2021; Zero Waste, 2022). Engagement at the neighborhood level furthers the city's agenda to reduce its carbon footprint.

Despite its successes, the program also has limitations. Cool Block works most effectively when neighborhoods are organized on blocks. Slater describes a case where a woman went through the entire training program before revealing that her home was behind a drug store with only one neighbor. The lack of a tangible block made it very difficult for her to continue with the program. Cool Block has demonstrated its greatest success in affluent, elderly, and/or academic communities, where people are long-term homeowners and the city has the financial resources to hire

Cool Block staff to help organize. However, Cool Block is working to expand to different types of neighborhoods through its role in initiatives such as the Cool City Challenge (Cool City Challenge, n.d.). Cities are invited to compete for $1 million dollars by submitting plans to engage the whole community (e.g., businesses, government, technology, schools, and non-government organizations) in becoming carbon neutral.

Transition Town Movement

The Transition Town movement is a global network of grassroots community projects in which citizens join together to "reimagine and rebuild our world" in response to climate change (About the Movement, n.d.). Older than Cool Block, the Transition Town movement was founded in the United Kingdom in the mid-2000s and quickly spread to many other communities and countries. Transition Towns offers psychosocial support by helping individuals and communities process their fears, despair, and other emotional responses to climate change. Like Cool Block, the members of most Transition Towns are middle class and middle-aged. In addition, and also similar to Cool Block, members are given access to educational materials.

Recently, a group of university students living in Isla Vista, the 23,000-person community surrounding the University of California, Santa Barbara (UCSB), were searching for a climate change mitigation and resilience program for their university town (of which about 85% are students). After initially attempting to integrate Cool Block's curriculum, the group, joined by a few of their professors, quickly realized that Cool Block's structure, which was developed for long-term homeowners, was less suitable for the transient nature of their student community. With 96–99% of Isla Vista residents being renters (The Santa Barbara Independent, 2015; Lodise, 2019), the students chose to adopt Transition Town's structure, focusing on community projects. However, individual Cool Blocks exist within the Isla Vista community (Scully-Allison, 2020).

Eco Vista (ecovistacommunity.com), the name chosen for Isla Vista's student-led community initiative, provides members with the opportunity to participate in a variety of projects. "Food Forest," a collaboration with the Isla Vista Recreation and Parks District, enables Eco Vista members to plant fruit trees and other plants that are "edible, native, medicinal" and/or "[fix] nitrogen" (Eco Vista Community, n.d.; see Fig. 3). Another project, the "Radicle Zine," uses a magazine format to "spread [climate change] awareness through art" (Eco Vista Community, n.d.). The "Circular Economy" project collects donations of school supplies, household items, books, and more for a share shop to "foster a circular economy within Eco Vista" (Eco Vista Community, n.d.). Eco Vista also works to foster a sense of community through social events, such as shared meals.

Also among these projects is Eco Vista's "Green New Deal," which was created by 300 UCSB students, most of whom were taking sociology classes at the time (e.g., Earth in Crisis and Eco Vista: Creating Systemic Alternatives). This Green

Fig. 3 Two students planting medicinal plants as part of the "Food Forest" project. (Image used with permission)

New Deal proposes strategies to address renewable energy and resilience against climate emergencies with a strong focus on mental health.

Eco Vista's Green New Deal includes pathways to destigmatize mental health within the community. By making support more accessible, there are greater opportunities to address eco-anxiety. Specifically, the deal proposes a rideshare service for therapy appointments, additional mental health classes in the UCSB curriculum, an anonymous hotline for students, and general peer support (e.g., a group website; Eco Vista Community, n.d.). Jessica Parfrey, one of the founders of Eco Vista, explains that Eco Vista tackles eco-anxiety by allowing students the opportunity to directly impact climate change via the different Eco Vista projects. Helping people work together to directly impact their lives and futures fosters an increased sense of social identity and/or collective efficacy. As noted in the introduction, individuals with a greater sense of efficacy are more likely to take action, increasing their community's adaptive capacity (Thaker et al., 2016).

The ability to foster a sense of community that supports climate change adaptation, mitigation, and resilience strategies without the traditional structure of a neighborhood block is a notable strength of Eco Vista. The community is also an effective example of building a sense of social cohesion among individuals despite the typically transient aspects of student life. In addition, Eco Vista's focus includes the often under-recognized pertinent health impact of climate change on mental health (as discussed earlier in this chapter).

A significant limitation that applies to over half of Transition Towns is their need for more funding (Scully-Allison, 2020), which negatively impacts the funding of long-term sustainability projects. While Eco Vista shares this limitation with the Transition Town Movement, its interdependence with UCSB mitigates the necessity for independent resources to some degree. For example, while UCSB provides minimal financial support (<$2000/year), it offers college credit for students participating in the initiative (Scully-Allison, 2020). In this way, a recruitment pathway is created, facilitating a steady stream of students into the program without the need for additional financial resources. Eco Vista's relationship to UCSB, along with the adaptations it has made to serve unique community needs, act as positive outliers for the student-centric climate initiative, which could be scaled for use on other university campuses in the University of California system and/or nationwide in the United States.

Future Directions and Conclusion

According to Berry et al. (2018), "mental health could be a lead indicator for measuring progress on mitigating the human impacts of climate change" (p. 288). This chapter provides evidence for the many ways in which climate change negatively impacts mental health. Evidence is also given for the role of community in building mental health resilience in the face of these negative impacts. We presented Cool Block and the Transition Town movement (e.g., Eco Vista) as examples of positive deviance, by showing how they model practices (e.g., neighborhood meetings) and behaviors (e.g., initiating community projects such as gardening) that strengthen social cohesion and collective efficacy, and therefore, mental health. Social ties were also shown to increase the likelihood of disaster preparedness via social and material resource sharing (Torres & Casey, 2017).

Future directions include developing public health prevention strategies and designing community infrastructure to be resilient and health-promoting (Torres & Casey, 2017). For example, those displaced by climate change should be included under international refugee laws and provided pathways to integrate into new residences (Torres & Casey, 2017). In addition, effective climate communication (e.g., news, entertainment, and social media) can reduce eco-anxiety and channel awareness into active engagement in climate change adaptation and mitigation actions. The effect of community building and positive deviance stories can be amplified through public communications and also modeled via entertainment-education (Coren, 2024). Real-life examples (e.g., Cool Block, Eco Vista) and "reel" examples (e.g., transmedia entertainment shows like *East Los High* and *Main Kuch Bhi Kar Sakti Hoon*; Coren, 2024) can inspire, motivate, engage, and persuade (Singhal et al., 2013). Through working together and exhibiting compassion and solidarity, individuals can meet environmental changes and trauma with hope, a sense of belonging, and gratitude.

References

About the Movement? (n.d.). Transition Network. Retrieved on August 5, 2021 from https://transitionnetwork.org/about-the-movement/

ACEs Aware Trauma-Informed Network of Care Roadmap. (2021). Retrieved on August 5, 2021 from https://www.acesaware.org/wp-content/uploads/2021/06/Aces-Aware-Network-of-Care-Roadmap.pdf

Adger, W. N., Agrawala, S., Mirza, M. M. Q., Conde, C., O'Brien, K., Pulhin, J., Pulwarty, R., Smit, B., Takahashi, K., Enright, B., Fankhauser, S., Ford, J., Gigli, S., Jetté-Nantel, S., Klein, R. J. T., Pearce, T. D., Shreshtha, A., Shukla, P. R., Smith, J. B., et al. (2007). Assessment of adaptation practices, options, constraints and capacity. In *Climate change 2007: Impacts, adaptation and vulnerability. Contribution of working group II to the fourth assessment report of the Intergovernmental panel on climate change* (pp. 719–743). Cambridge University Press; University of East Anglia. https://ueaeprints.uea.ac.uk/id/eprint/25215/

Agaibi, C. E., & Wilson, J. P. (2005). Trauma, PTSD, and resilience: A review of the literature. *Trauma, Violence, & Abuse, 6*(3), 195–216. https://doi.org/10.1177/1524838005277438

Albrecht, G. (2011). Chronic environmental change: Emerging 'psychoterratic' syndromes. In I. Weissbecker (Ed.), *Climate change and human well-being: Global challenges and opportunities* (pp. 43–56). Springer. https://doi.org/10.1007/978-1-4419-9742-5_3

Albrecht, T. L., & Goldsmith, D. J. (2003). Social support, social networks, and health. In *Handbook of health communication* (pp. 263–284). Lawrence Erlbaum Associates Publishers.

Alcover, C.-M., Rodríguez, F., Pastor, Y., Thomas, H., Rey, M., & del Barrio, J. L. (2020). Group membership and social and personal identities as psychosocial coping resources to psychological consequences of the COVID-19 -confinement. *International Journal of Environmental Research and Public Health, 17*(20), 7413. https://doi.org/10.3390/ijerph17207413

Almedom, A., & Glandon, D. (2008). Social capital and mental health. In I. I. Kawachi, S. V. Subramanian, & D. Kim (Eds.), *Social capital and health* (p. 24). Springer. https://doi.org/10.1007/978-0-387-71311-3_9

Amuasi, J. H., Lucas, T., Horton, R., & Winkler, A. S. (2020). Reconnecting for our future: The Lancet One Health Commission. *The Lancet, 395*(10235), 1469–1471. https://doi.org/10.1016/S0140-6736(20)31027-8

Atwoli, L., Baqui, A. H., Benfield, T., Bosurgi, R., Godlee, F., Hancocks, S., Horton, R., Laybourn-Langton, L., Monteiro, C. A., Norman, I., Patrick, K., Praities, N., Olde Rikkert, M. G. M., Rubin, E. J., Sahni, P., Smith, R., Talley, N., Turale, S., & Vázquez, D. (2021). Call for emergency action to limit global temperature increases, restore biodiversity, and protect health. *New England Journal of Medicine, 385*(12), 1134–1137. https://doi.org/10.1056/NEJMe2113200

Bandura, A. (1997). Self-efficacy: The exercise of control. New York: Freeman.

Bandura, A. (2000). Exercise of Human Agency Through Collective Efficacy Current Directions in *Psychological Science 9(3)* 75–78. https://doi.org/10.1111/1467-8721.00064

Barlett, C. P., DeWitt, C. C., Madison, C. S., Heath, J. B., Maronna, B., & Kirkpatrick, S. M. (2020). Hot temperatures and even hotter tempers: Sociological mediators in the relationship between global climate change and homicide. *Psychology of Violence, 10*(1), 1–7. https://doi.org/10.1037/vio0000235

Berg, C. (2019). *Sustainable action: Overcoming the barriers (1st Edition)*. Routledge. https://doi.org/10.4324/9780429060786

Berry, H. L., Waite, T. D., Dear, K. B. G., Capon, A. G., & Murray, V. (2018). The case for systems thinking about climate change and mental health. *Nature Climate Change, 8*(4), 282–290. https://doi.org/10.1038/s41558-018-0102-4

Burke, M., González, F., Baylis, P., Heft-Neal, S., Baysan, C., Basu, S., & Hsiang, S. (2018). Higher temperatures increase suicide rates in the United States and Mexico. *Nature Climate Change, 8*(8), 723–729. https://doi.org/10.1038/s41558-018-0222-x

Burke, S. E. L., Sanson, A. V., & Van Hoorn, J. (2018). The psychological effects of climate change on children. *Current Psychiatry Reports, 20*(5). https://doi.org/10.1007/s11920-018-0896-9

Cianconi, P., Betrò, S., & Janiri, L. (2020). The impact of climate change on mental health: A systematic descriptive review. *Frontiers in Psychiatry, 11*, 74. https://doi.org/10.3389/fpsyt.2020.00074

Clayton, S., Manning, C., College, M., Krygsman, K., Speiser, M., Cunsolo, A., Derr, V., Doherty, T., Fery, P., Haase, E., Kotcher, J., Silka, L., & Tabola, J. (2017). *Mental health and our changing climate: Impacts, implications, and guidance* (p. 70). American Psychological Association, and ecoAmerica.

Cool Block. (n.d.). *Cool Block.* Retrieved January 9, 2022, from https://coolblock.org/

Cool City Challenge. (n.d.). Retrieved August 5, 2021, from https://coolcity.earth/

Coren, E. (2024). Rhythm and glue: An entertainment-education prototype for climate communication. In E. Coren & H. Wang (Eds.), *Storytelling to accelerate climate solutions.* Springer Nature.

Cosentino, M. (2021, August 4). *Die Verbundenheit von Körper, Geist und Natur Transform.* https://transform-magazin.de/verbundenheit/

Cunsolo, A., & Ellis, N. R. (2018). Ecological grief as a mental health response to climate change-related loss. *Nature Climate Change, 8*(4), 275–281. https://doi.org/10.1038/s41558-018-0092-2

Dodds, J. (2021). The psychology of climate anxiety. *Green Psychiatry Collection, 45*(4), 222–226. https://doi.org/10.1192/bjb.2021.18

Doppelt, B. (2019a, August 29). In my opinion: Climate change as ACE. *The Register-Guard.* https://www.registerguard.com/opinion/20190829/in-my-opinion-climate-change-as-ace

Doppelt, B. (2019b, September 26). In my opinion: Surviving climate trauma. *The Register-Guard.* https://www.registerguard.com/opinion/20190926/in-my-opinion-surviving-climate-trauma

Durá, L., & Singhal, A. (2009). Utilizing a positive deviance approach to reduce girls' trafficking in Indonesia: Asset-based communicative acts that make a difference. *Journal of Creative Communications, 4*(1), 1–17. https://doi.org/10.1177/097325861000400101

Eco Vista Community. (n.d.). *Eco Vista Community.* Retrieved January 9, 2022, from https://ecovistacommunity.com/

Flannery, R. B. (1990). Social support and psychological trauma: A methodological review. *Journal of Traumatic Stress, 3*(4), 593–611. https://doi.org/10.1002/jts.2490030409

Fong, P., Cruwys, T., Haslam, C., & Haslam, S. A. (2019). Neighbourhood identification and mental health: How social identification moderates the relationship between socioeconomic disadvantage and health. *Journal of Environmental Psychology, 61*, 101–114. https://doi.org/10.1016/j.jenvp.2018.12.006

For Communities. (n.d.). *ACEs Aware.* Retrieved August 5, 2021, from https://www.acesaware.org/provide-treatment-healing/for-communities/

Glaser, R., & Kiecolt-Glaser, J. K. (2005). Stress-induced immune dysfunction: Implications for health. *Nature Reviews Immunology, 5*(3), 243–251. https://doi.org/10.1038/nri1571

Goddard, R. (2002). A Theoretical and Empirical Analysis of the Measurement of Collective Efficacy: The Development of a Short Form. *Educational and Psychological Measurement, 62*(1), 97–110. https://doi.org/10.1177/0013164402062001007

Goddard, R. D., & Salloum, S. J. (2011). Collective efficacy beliefs, organizational excellence, and leadership. In G. M. Spreitzer & K. S. Cameron (Eds.), *The Oxford handbook of positive organizational scholarship* (pp. 642–650). Oxford University Press. https://doi.org/10.1093/oxfordhb/9780199734610.013.0048

Goddard, R. D., Hoy, W. K., & Hoy, A. W. (2004). Collective efficacy beliefs: Theoretical developments, empirical evidence, and future directions. *Educational Researcher, 33*(3), 3–13. https://doi.org/10.3102/0013189X033003003

Hamilton, I., Kennard, H., McGushin, A., Höglund-Isaksson, L., Kiesewetter, G., Lott, M., Milner, J., Purohit, P., Rafaj, P., Sharma, R., Springmann, M., Woodcock, J., & Watts, N. (2021). The public health implications of the Paris agreement: A modelling study. *The Lancet Planetary Health, 5*(2), e74–e83. https://doi.org/10.1016/S2542-5196(20)30249-7

Hayes, K., Blashki, G., Wiseman, J., Burke, S., & Reifels, L. (2018). Climate change and mental health: Risks, impacts and priority actions. *International Journal of Mental Health Systems, 12*(28). https://doi.org/10.1186/s13033-018-0210-6

Health benefits far outweigh the costs of meeting climate change goals. (2018, May 12). World Health Organization. https://www.who.int/news/item/05-12-2018-health-benefits-far-outweigh-the-costs-of-meeting-climate-change-goals

Herrman, H., Stewart, D. E., Diaz-Granados, N., Berger, E. L., Jackson, B., & Yuen, T. (2011). What is resilience? *The Canadian Journal of Psychiatry, 56*(5), 258–265. https://doi.org/10.1177/070674371105600504

Hickman, C., Marks, E., Pihkala, P., Clayton, S., Lewandowski, R. E., Mayall, E. E., Wray, B., Mellor, C., & van Susteren, L. (2021). Climate anxiety in children and young people and their beliefs about government responses to climate change: A global survey. *The Lancet Planetary Health, 5*(12), e863–e873. https://doi.org/10.1016/S2542-5196(21)00278-3

Hikichi, H., Aida, J., Tsuboya, T., Kondo, K., & Kawachi, I. (2016). Can community social cohesion prevent posttraumatic stress disorder in the aftermath of a disaster? A natural experiment from the 2011 Tohoku Earthquake and Tsunami. *American Journal of Epidemiology, 183*(10), 902–910. https://doi.org/10.1093/aje/kwv335

Hogg, M. A. (2018). Chapter 5 Social identity theory. In P. J. Burke (Ed.), *Contemporary social psychological theories* (pp. 112–138). Stanford University Press. https://www.degruyter.com/document/doi/10.1515/9781503605626-007/html

Holland, K. M., Jones, C., Vivolo-Kantor, A. M., Idaikkadar, N., Zwald, M., Hoots, B., Yard, E., D'Inverno, A., Swedo, E., Chen, M. S., Petrosky, E., Board, A., Martinez, P., Stone, D. M., Law, R., Coletta, M. A., Adjemian, J., Thomas, C., Puddy, R. W., et al. (2021). Trends in US Emergency Department visits for mental health, overdose, and violence outcomes before and during the COVID-19 pandemic. *JAMA Psychiatry, 78*(4), 372–379. https://doi.org/10.1001/jamapsychiatry.2020.4402

Holt-Lunstad, J., Smith, T. B., & Layton, J. B. (2010). Social relationships and mortality risk: A meta-analytic review. *PLoS Medicine, 7*(7), e1000316. https://doi.org/10.1371/journal.pmed.1000316

Home Efficiency Genie. (2021, October 22). https://www.cityofpaloalto.org/Departments/Utilities/Residential/Home-Efficiency-Genie

Horton, R., Beaglehole, R., Bonita, R., Raeburn, J., McKee, M., & Wall, S. (2014). From public to planetary health: A manifesto. *The Lancet, 383*(9920), 847. https://doi.org/10.1016/S0140-6736(14)60409-8

Information Meeting Guide | Cool Block. (n.d.). Retrieved August 5, 2021, from https://coolblock.org/information-meeting-guide

Ingle, H. E., & Mikulewicz, M. (2020). Mental health and climate change: Tackling invisible injustice. *The Lancet Planetary Health, 4*(4), e128–e130. https://doi.org/10.1016/S2542-5196(20)30081-4

Jetten, J., Haslam, C., & Haslam, S. A. (Eds.). (2012). *The social cure: Identity, health and well-being* (p. 390). Psychology Press.

Jetten, J., Haslam, S. A., Cruwys, T., Greenaway, K. H., Haslam, C., & Steffens, N. K. (2017). Advancing the social identity approach to health and well-being: Progressing the Social Cure research agenda. *European Journal of Social Psychology, 47*(7), 789–802. https://doi.org/10.1002/ejsp.2333

Kahn, J.-P., Cohen, R. F., Tubiana, A., Legrand, K., Wasserman, C., Carli, V., Apter, A., Balazs, J., Banzer, R., Baralla, F., Barzilai, S., Bobes, J., Brunner, R., Corcoran, P., Cosman, D., Guillemin, F., Haring, C., Kaess, M., Bitenc. U. M., et al. (2020). Influence of coping strategies on the efficacy of YAM (Youth Aware of Mental Health): A universal school-based suicide preventive program. *European Child & Adolescent Psychiatry, 29*, 1671–1681. https://doi.org/10.1007/s00787-020-01476-w

Kaniasty, K., & Norris, F. H. (1993). A test of the social support deterioration model in the context of natural disaster. - PsycNET. *Journal of Personality and Social Psychology, 64*(3), 395–408. https://doi.org/10.1037/0022-3514.64.3.395

Kiecolt-Glaser, J. K., Gouin, J.-P., & Hantsoo, L. (2010). Close relationships, inflammation, and health. *Neuroscience & Biobehavioral Reviews, 35*(1), 33–38. https://doi.org/10.1016/j.neubiorev.2009.09.003

Kreslake, J. M., Price, K. M., & Sarfaty, M. (2016). Developing effective communication materials on the health effects of climate change for vulnerable groups: A mixed methods study. *BMC Public Health, 16*(1). https://doi.org/10.1186/s12889-016-3546-3

Lawrance, E., Thompson, R., Fontana, G., & Jennings, N. (2021). The impact of climate change on mental health and emotional wellbeing: Current evidence and implications for policy and practice, *36*.

Lebeck, S. (2017, December 11). *Changing the game around climate action, one block at a time.* Meeting of the Minds. https://meetingoftheminds.org/changing-game-around-climate-action-one-block-time-24091

Lodise, C. (2019). *Isla Vista: A citizen's history* (2nd ed.). Independently Published.

MacNeill, A. J., McGain, F., & Sherman, J. D. (2021). Planetary health care: A framework for sustainable health systems. *The Lancet Planetary Health, 5*(2), e66–e68. https://doi.org/10.1016/S2542-5196(21)00005-X

McNamara, N., Stevenson, C., Costa, S., Bowe, M., Wakefield, J., Kellezi, B., Wilson, I., Halder, M., & Mair, E. (2021). Community identification, social support, and loneliness: The benefits of social identification for personal well-being. *British Journal of Social Psychology, 60*, 1379–1402.

Miles-Novelo, A., & Anderson, C. A. (2019). Climate change and psychology: Effects of rapid global warming on violence and aggression. *Current Climate Change Reports, 5*, 36–46. https://doi.org/10.1007/s40641-019-00121-2

Obradovich, N., Migliorini, R., Paulus, M. P., & Rahwan, I. (2018). Empirical evidence of mental health risks posed by climate change. *Proceedings of the National Academy of Sciences, 115*(43), 10953–10958. https://doi.org/10.1073/pnas.1801528115

Ojala, M., Cunsolo, A., Ogunbode, C. A., & Middleton, J. (2021). Anxiety, worry, and grief in a time of environmental and climate crisis: A narrative review. *Annual Review of Environment and Resources, 46*(1). https://doi.org/10.1146/annurev-environ-012220-022716

Outcomes of the U.N. Climate Change Conference in Paris. (2015). *21st Session of the Conference of the Parties to the United Nations Framework Convention on Climate Change (COP 21).* Center for Climate and Energy Solutions. https://www.c2es.org/content/cop-21-paris/

Ozer, E. J., Best, S. R., Lipsey, T. L., & Weiss, D. S. (2003). Predictors of posttraumatic stress disorder and symptoms in adults: A meta-analysis. *Psychological Bulletin, 129*(1), 52–73. American Psychological Association. https://doi.org/10.1037/0033-2909.129.1.52

Palinkas, L. A., & Wong, M. (2020). Global climate change and mental health. *Current Opinion in Psychology, 32*, 12–16. https://doi.org/10.1016/j.copsyc.2019.06.023

Panu, P. (2020). Anxiety and the ecological crisis: An analysis of eco-anxiety and climate anxiety. *Health and Sustainability, 12*(19), 7836. https://doi.org/10.3390/su12197836

Pascale, R. T., & Sternin, J. (2005). Your company's secret change agents. *Harvard Business Review, 83*(5), 72–81.

Peterson, J., & Densley, J. (2019a, February 14). School shooters show these signs of distress before they open fire. *Business Insider.* https://www.businessinsider.com/school-shooters-show-these-signs-of-distress-before-they-open-fire-2019-2

Peterson, J., & Densley, J. (2019b, August 4). Op-Ed: We have studied every mass shooting since 1966. Here's what we've learned about the shooters. *Los Angeles Times*. https://www.latimes.com/opinion/story/2019-08-04/el-paso-dayton-gilroy-mass-shooters-data

Rinderu, M. I., Bushman, B. J., & Van Lange, P. A. (2018). Climate, aggression, and violence (CLASH): A cultural-evolutionary approach. *Current Opinion in Psychology, 19*, 113–118. https://doi.org/10.1016/j.copsyc.2017.04.010

Romanello, M., McGushin, A., Napoli, C. D., Drummond, P., Hughes, N., Jamart, L., Kennard, H., Lampard, P., Rodriguez, B. S., Arnell, N., Ayeb-Karlsson, S., Belesova, K., Cai, W., Campbell-Lendrum, D., Capstick, S., Chambers, J., Chu, L., Ciampi, L., Dalin, C., et al. (2021). The 2021 report of the Lancet Countdown on health and climate change: Code red for a healthy future. *The Lancet, 398*(10311), 1619–1662. https://doi.org/10.1016/S0140-6736(21)01787-6

Ruderman, D., & Cohn, E. G. (2021). Predictive extrinsic factors in multiple victim shootings. *The Journal of Primary Prevention, 42*, 59–75. https://doi.org/10.1007/s10935-020-00602-3

Ruskey, A., Schundler, G., Gorton, I., Patrick, M., Lee, D., & Evans, E. (n.d.). *ClimateTree: A GIS "comparable communities" approach for sharing stories, building community resilience and reducing greenhouse gases.*

Schnurr, P. P., Lunney, C. A., & Sengupta, A. (2005). Risk factors for the development versus maintenance of posttraumatic stress disorder. *Journal of Traumatic Stress, 17*(2), 85–95. https://doi.org/10.1023/B:JOTS.0000022614.21794.f4

Scully-Allison, G. (2020). What role can student-led transition play in place-based sustainability? An investigation into Eco Vista using strategic niche management theory. [Unpublished master's thesis]. The University of Edinburgh

Segerstrom, S. C., & Miller, G. E. (2004). Psychological stress and the human immune system: A meta-analytic study of 30 years of inquiry. *Psychological Bulletin, 130*(4), 601–630. https://doi.org/10.1037/0033-2909.130.4.601

Singhal, A., & Durá, L. (2017). Positive deviance: A non-normative approach to health and risk messaging. In *Oxford research encyclopedia of communication*. https://doi.org/10.1093/acrefore/9780190228613.013.248

Singhal, A., Wang, H., & Rogers, E. M. (2013). The rising tide of entertainment-education in communication campaigns. In R. E. Rice & C. K. Atkin (Eds.), *Public communication campaigns* (4th ed., pp. 321–334). SAGE Publications.

Smith, K. J., Gavey, S., RIddell, N. E., Kontari, P., & Victor, C. (2020). The association between loneliness, social isolation and inflammation: A systematic review and meta-analysis. *Neuroscience & Biobehavioral Reviews, 112*, 519–541. https://doi.org/10.1016/j.neubiorev.2020.02.002

Staff, I. (2015, June 11). Isla Vista: Investor's Paradise. *The Santa Barbara Independent*. https://www.independent.com/2015/06/11/isla-vista-investors-paradise/

Sternin, J., & Choo, R. (2000). The power of positive deviancy. An effort to reduce malnutrition in Vietnam offers an important lesson about managing change. *Harvard Business Review, 78*(1), 14–15.

Team-Building/Topic 1 Meeting Guide | Cool Block. (n.d.). Retrieved August 5, 2021, from https://coolblock.org/team-building-topic-1-meeting-guide

Thaker, J., Maibach, E., Leiserowitz, A., Zhao, X., & Howe, P. (2016). The role of collective efficacy in climate change adaptation in India. *Weather, Climate, and Society, 8*(1), 21–34. https://doi.org/10.1175/WCAS-D-14-00037.1

Thompson, R., Hornigold, R., Page, L., & Waite, T. (2018). Associations between high ambient temperatures and heat waves with mental health outcomes: A systematic review. *Public Health, 161*, 171–191. https://doi.org/10.1016/j.puhe.2018.06.008

Torres, J. M., & Casey, J. A. (2017). The centrality of social ties to climate migration and mental health. *BMC Public Health, 17*(1), 1–10. https://doi.org/10.1186/s12889-017-4508-0

Torres, J. M., Epel, E. S., To, T. M., Lee, A., Aiello, A. E., & Haan, M. N. (2018). Cross-border ties, nativity, and inflammatory markers in a population-based prospective study of Latino adults. *Social Science & Medicine, 211*, 21–30. https://doi.org/10.1016/j.socscimed.2018.05.028

Turner, J. C., Oakes, P. J., Haslam, S. A., & McGarty, C. (1994). Self and collective: Cognition and social context. *Personality and Social Psychology Bulletin, 20*(5), 454–463. https://doi.org/10.1177/0146167294205002

Uddin, R., Philipsborn, R., Smith, D., Mutic, A., & Thompson, L. M. (2021). A global child health perspective on climate change, migration and human rights. *Current Problems in Pediatric and Adolescent Health Care, 51*(6). https://doi.org/10.1016/j.cppeds.2021.101029

Umberson, D., & Montez, J. K. (2010). Social relationships and health: A flashpoint for health policy. *Journal of Health and Social Behavior, 51*(Suppl), S54–S66. https://doi.org/10.1177/0022146510383501

USGCRP, (2016): The Impacts of Climate Change on Human Health in the United States: A Scientific Assessment. Crimmins, A., J. Balbus, J.L. Gamble, C.B. Beard, J.E. Bell, D. Dodgen, R.J. Eisen, N. Fann, M.D. Hawkins, S.C. Herring, L. Jantarasami, D.M. Mills, S. Saha, M.C. Sarofim, J. Trtanj, and L. Ziska, Eds. U.S. Global Change Research Program, Washington, DC, 312 pp. http://dx.doi.org/10.7930/J0R49NQX

Vigo, D., Thornicroft, G., & Atun, R. (2016). Estimating the true global burden of mental illness. *The Lancet Psychiatry, 3*(2), 171–178. https://doi.org/10.1016/S2215-0366(15)00505-2

Watts, N., Adger, W. N., Agnolucci, P., Blackstock, J., Byass, P., Wenija, C., Chaytor, S., Colbourn, T., Collins, M., Cooper, A., Cox, P. M., Depledge, J., Drummond, P., Ekins, P., Galaz, V., Grace, D., Graham, H., Grubb, M., Haines, A., et al. (2015). Health and climate change: Policy responses to protect public health. *The Lancet, 386*(10006). https://doi.org/10.1016/S0140-6736(15)60854-6

Watts, N., Adger, W. N., Ayeb-Karlsson, S., Bai, Y., Byass, P., Campbell-Lendrum, D., Colbourn, T., Cox, P., Davies, M., Depledge, M., Depoux, A., Dominguez-Salas, P., Drummond, P., Ekins, P., Flahault, A., Grace, D., Graham, H., Haines, A., Hamilton, I., et al. (2017). The Lancet Countdown: Tracking progress on health and climate change. *The Lancet, 389*(10074), 1151–1164. https://doi.org/10.1016/S0140-6736(16)32124-9

Watts, N., Amann, M., Arnell, N., Ayeb-Karlsson, S., Beagley, J., Belesova, K., Boykoff, M., Byass, P., Cai, W., Campbell-Lendrum, D., Capstick, S., Chambers, J., Coleman, S., Dalin, C., Daly, M., Dasandi, N., Dasgupta, S., Davies, M., Di Napoli, C., et al. (2021). The 2020 report of The Lancet Countdown on health and climate change: Responding to converging crises. *The Lancet, 397*(10269), 129–170. https://doi.org/10.1016/S0140-6736(20)32290-X

What are ACEs and Why Do They Matter? (2018). ACEs Aware Trauma-Informed Network of Care Roadmap. Retrieved on August 5, 2021 from https://www.acesaware.org/wp-content/uploads/2019/12/1-What-are-ACEs-and-Why-Do-They-Matter-English.pdf

Wickrama, K. A. S., & Wickrama, T. (2010). Perceived community participation in tsunami recovery efforts and the mental health of tsunami-affected mothers: Findings from a study in rural Sri Lanka. *International Journal of Social Psychiatry, 57*(5), 518–527. https://doi.org/10.1177/0020764010374426

Xue, T., Zhu, T., Zheng, Y., & Zhang, Q. (2019). Declines in mental health associated with air pollution and temperature variability in China. *Nature Communications, 10*. https://doi.org/10.1038/s41467-019-10196-y

Zajonc, R. B. (2001). Mere exposure: A gateway to the subliminal. *Current Directions in Psychological Science, 10*(6), 224–228. https://doi.org/10.1111/1467-8721.00154

Zero Waste. (2022, January 4). https://www.cityofpaloalto.org/Departments/Public-Works/Zero-Waste

Maya Cosentino obtained her medical degree at Witten/Herdecke University in Germany and her undergraduate degree in psychology and neuroscience at St. Mary's College of Maryland. She is currently completing her residency in child and adolescent psychiatry, as well as continuing her education as a postgraduate Global Health Policy student at the London School of Hygiene and Tropical Medicine (LSHTM). Maya is a member of Think Tank 30, the young think tank of the Club of Rome in Germany.

Roni Gal-Oz is a student at Barnard College, Columbia University, with a strong interest in psychology and environmental action. She is extremely grateful for the opportunity to help with this project and looks forward to further pursuing her writing and research career as a university student.

Debra L. Safer is an Associate Professor at Stanford in the Department of Psychiatry & Behavioral Sciences. She obtained her M.D. from U.C. San Francisco and completed her residency as well as a postdoctoral fellowship in eating disorder intervention research at Stanford University School of Medicine. In addition, she obtained a master's degree from U.C. Berkeley's School of Public Health focused on the intersection of fiction narratives and medicine. She has co-authored multiple peer-reviewed articles, books, and book chapters and presented her work both nationally and internationally. She recently broadened her clinical and research focus by addressing both the physical and mental health impacts of climate change, with a particular emphasis on the use of prosocial entertainment-education programs, based on the theoretical models of self-efficacy and social modeling developed by the renowned social psychologist and Stanford Emeritus Professor Albert Bandura.

Mapping Out Our Future: Using Geospatial Tools and Visual Aids to Achieve Climate Empowerment in the United States

Aviva Wolf-Jacobs, Nancy Glock-Grueneich, and Nathan Uchtmann

Climate empowerment is "[T]he capacity of peoples to see and achieve the changes necessary to reverse climate change, heal our planet and save all we still can, in ways that also lay down the conditions for a livable future..."

—United Nations Framework Convention on Climate Change

The Power of Maps and Stories

The disasters so long predicted for our *future* are now our *present* (Hansen, 2009; Myers et al., 2021). The climate catastrophes bursting upon us are felt around the world, their cost inescapable. Temperatures, multi-year droughts, famine, storms, sea level rise, and ever-greater numbers of people dying in violent conflict or compelled migration continue to set record upon record (Holland & Bruyère, 2014; Knutson et al., 2020; Mitchell et al., 2016; Samora-Arvela et al., 2017). As if the daily images of desperate people wading through waist-deep flooding in the aftermath of storms were not enough to bring home these dangers, a recent study found that for every degree of increase in average global temperature, there is an 25–30% increase in the proportion of "catastrophic" category 4 and 5 hurricanes (Holland & Bruyère, 2014). Monsoons are way out of season, too intense, or not showing up at all. Polar ice is melting under summer skies as hot as if in a temperate zone, even as in winter the temperate zones of Earth find polar air currents bringing deep, freezing storms lasting weeks way to the south of where they belong. Fires roar across whole continents, burning forests, animals, and vegetation and destroying millions of

A. Wolf-Jacobs (✉)
Spatial Sciences Institute, University of Southern California, Los Angeles, CA, USA
e-mail: wolfjaco@usc.edu

N. Glock-Grueneich
Within Reach Network for Strategic Citizen Action, Santa Cruz, CA, USA
e-mail: nglock@post.harvard.edu

N. Uchtmann
Natividad Medical Center, Salinas, CA, USA

© The Author(s) 2024
E. Coren, H. Wang (eds.), *Storytelling to Accelerate Climate Solutions*,
https://doi.org/10.1007/978-3-031-54790-4_16

acres, homes, and lives. Skies unbreathable and seashores cooking shore creatures by the billions, corals bleached, and the deeper oceans threatened by heat, acid, and eutrophication (Cheng et al., 2021; Koutsoyiannis, 2021).[1]

The news comes through each day, unrelentingly, of current and projected damage to our fragile planet from climate change,[2] and for many who let themselves listen, the news brings grief, erosion of hope, and loss of agency. At the same time, for many such news redoubles their determination, even outrage, and gives them even sharper focus on finding and scaling the many solutions we do in fact already have. Overall, there grows a miasma of distraction, overwhelm, grief, and despair or, perversely, continued outright denial, even as the climate chaos is worsening faster and more visibly than scientists had dared predict . And on every continent, "sensitivity" to this chaos defined as "the susceptibility of a system or population to harm from exposure to climate-related threats," grows rapidly, hitting hardest those least resourced to handle it (Cutter et al., 2009).[3]

Paths to Climate Empowerment

People need to act to use the climate solutions we already have, or are now finding. In order to act, we need to believe that the actions within our power to take matter. Ideally, we need to see that our actions are not just helpful, but even *necessary*, and that, taken together, the actions of us all, or enough of us, each doing our part, might yet prove *sufficient* to get us out of the mess we're now in. In this chapter, therefore, we focus on those uses of stories and mapping that help get us to climate empowerment. *We treat climate empowerment itself as a climate solution in need of acceleration* through the abundance of solution stories, coming from many directions, and mapped out concretely and at different levels of scale.

Using stories to achieve this goal is not a matter so much of more or different kinds of stories, it's rather that sharing the stories of solutions is a matter of spotlighting the agency of those making these solutions happen. Some stories turn on the—what they did, how they got there, where their courage came from, what help they required, and from whom, and what strategies worked, and how they kept going. We need to take that approach with enough different kinds of answers in enough diverse geographical, environmental, and demographic contexts, that

---

[1] Ocean temperatures, known to be critical indicators of trends in global climate systems, continue to hit record highs expanding still more our growing fears for what is yet to come.

[2] The term "climate change" as used in this chapter encompasses all effects traced by science to human-caused increases in greenhouse gases or their feedback loops but we also follow the common expansion of the term to include all anthropogenic threats to life on Earth, such as plastics, toxins, and soil and habitat destruction.

[3] Cutter et al. (2009) list access to technology, monetary resources, strength of infrastructure, and access to natural resources as factors in climate resiliency, i.e., a system's ability to prepare for, respond to, and recover from such threats.

hearers can gain not only new *direction* from the stories, with new ideas as to what can and should be done in their unique community, but new *courage* to take action as well.

If we're focusing not just on what needs to be done, but on what people find themselves *able* to do, then we have to ask, "What counts as a 'climate solution?'" It's a different question from "What can we do?"—with the often-implied rhetorical notion, "Nothing," to the question, "Of the things I can or could do, which if any of them could actually make a critical difference, i.e., could actually count as a 'climate solution?'" Such situations are on all sides now, of course, as we are now in what is really a *planetary crisis*, including though not limited to climate. It is in fact a systems' problem, the likes of which our human species has never seen before. Almost every single aspect of our lives is somehow implicated. But that also makes us *all part of the solution*—or potentially so—given *good answers*, the *power to act* upon these answers, *knowledge of* these answers, and *belief in* our power *to* act upon them, even take them to scale—given, that is, *climate empowerment.*

To Accelerate Climate Solutions, What Kinds of Stories Do We Tell, Map, and Show?

1. Answer the urgent question, *"What can we do?"* with *vivid stories of success.* Show *what* the *solutions* are. Show how they work and why they matter, the good they do, and how they came about.
2. Tell *where* current impacts and projected impacts of climate change are particularly high, along with *where* current actions are, with accurate, complete, and continuously updated *geospatial mapping tools* and *images* that pinpoint where the actual breakthroughs are happening so all can see for themselves what is happening on the ground.
3. Use these tools and connections to *help changemakers find each other and to amplify the voices of those most affected by the crisis*—to hang out, help out, learn from, share with, and work together, both locally *and* globally, in all the ways and with all the problems we face.
4. Where feasible, use the stories, maps, tools, and processes to *visualize how solutions connect and interact,* to better see the interdependencies and how real systems change, within and across all sectors, is actually emerging already, or is at least *possible.* Use maps to sketch what it would look like if the solutions we find were to be in place or to imagine what, in time, it would look like if they were still not in place. Use maps and geospatial modeling and analysis to see vividly before us the different futures to come someday from the plans and possibilities that we are weighing on this day.
5. Tell the stories of *what real people close to home are doing about the climate crisis*, often in surprising numbers, finding fulfillment and essential companions while making meaningful contributions to sustainability work. Use interactive

maps to provide opportunities and resources for *local collaboration and involvement.*

6. Show how those *newly or long-time aware* of the climate crisis and its vast threats and losses cope with the strains and the changes in their lives and make sense of our situation. Focus on the *growing millions of people not just at risk but already enduring the full impact of planetary breakdown,* dying or seeing all that they need and know gone forever. Tell stories that show how we can in fact plan globally, as well as locally, for all "4 R's" of resilience: *readiness, rescue, recovery,* and, where needed, *relocation;* how we share power in both *framing* and *making* the decisions that shape our futures; and how we share our hearts as *one* human family.

7. Tell of *the many, some half a world away, now meeting regularly on Zoom,* learning from each other of promising answers and finding potent leadership from ordinary people around the globe.

Stories to Accelerate Climate Empowerment

Climate empowerment as we define it at the beginning of this chapter suggests a process of discovery and development occurring over time. As envisioned by the UN, this process is to be optimized in each party to the Paris Agreement by the six elements of educational and participatory intervention that their Action for Climate Empowerment (ACE) program specifies (UNFCCC, 1992). In arriving at this conclusion, however, we've also seen the diverse ways to help people with different needs and situations. These approaches, summarized here, can be clustered as ones of *social support*; *comprehensive mapping*; and *effective action*. The starting point for all, and the first step toward climate empowerment, is awareness of the severity, the urgency, and the magnitude of the climate crisis.

Social Support "Circles" of support arise spontaneously and informally as people seek each other out, while more organized networks and methods are also coming into being (Hogan, 2021).[4] This social connection is essential as is the storytelling that elicits feelings and models actively sharing them, not suppressing or setting them aside. In this way, storytelling also builds communities of practice for solving localized challenges.

Comprehensive Mapping Letting go the burden of the whole and settling for just our own part can be made much easier if among the images and stories are some that *connect the dots,* helping make sense out of all that needs doing, and showing where and how the climate solutions needed for the other parts are indeed accounted for. What such communities require is a big picture: Working through smaller pieces

---

[4] See also social support efforts by climate psychiatrist Gayle Matson, Katherine Hayhoe, and Braiding Sweetgrass.

while being able to see the whole is helpful in understanding what it is overall that we are aiming for.

Effective Action Taking meaningful action, participating in real-world efforts, and seeing their successes mapped out can radically turn around our sense of the possible in a moment. It is this instant expansion of the possible, as well as the concrete direction stories offer, that puts storytelling at the heart of climate empowerment.

Stories show the magnitude of what is in fact possible. They can show the extraordinary breadth, complexity, and diversity of actions all over the world, at every level of scale, and social standing. They can be a curative to overly narrow or even dogmatic notions of what must be done and how, opening vast inventories of "adjacent possibilities" within and across every aspect of society. This diversity then is another of the reasons that stories, maps, and images are so pivotal. The climate crisis is so overwhelming not only because it is so big, but because it is so complex; it leaves nothing untouched. It seems beyond imagination how anything with so many facets and feedback loops could yield to even our biggest actions.

In describing climate empowerment, it is no accident then that the UN said it must be an *"all-society"* approach (UNFCCC, 1992). All systems, all sectors—connecting within and between them for the health of all—is a goal for how climate empowerment can manifest. In seeking to "mobilize all," it is *not* for all to become "climate activists" as such—walking off their jobs and into the streets. It is rather, or in addition, the way to the deeper transformations needed become clear, for us all to *embed climate action* into every aspect of every place, job, situation, network that we are part of anywhere. As one put it, *"don't just create green jobs, make all jobs green."* Reassess every policy, tool, rule, role, code, norm, behavior, process, structure, reward system, expectation, necessity, performance, recreation to see what changes to it—or preservations of it—are most vital in achieving near-term survival and foundational to a long-term livable future.

The Equalizing Power of Geospatial Tools

Tobler's First Law of Geography states that "everything is related to everything else, but near things are more related than distant things" (Tobler, 1970). This pattern can be seen reflected in residential demographics. Yee (1996) discusses the tendency of individuals belonging to the same racial or ethnic minority group to form "ethnic enclaves" in the form of hyper-segregated neighborhoods. While it is common around the world for people to choose to live in neighborhoods that are home to others who share similar racial and socio-economic attributes, there are also many "ethnic enclave" neighborhoods that exist as a direct result of oppressive power structures and discriminatory zoning laws (Rothstein, 2017). Robert Bullard, who is considered to be a founder of the Environmental Justice Movement, asserts that this form of neighborhood segregation, which he labels "residential apartheid," is a product of decisions made by "white slaveholders, merchants... and the white business elites, politicians, and workers in the periods since slavery" (Bullard, 1993).

Today, neighborhoods and city regions in the United States that are home to pre-dominantly Black and Brown people, and neighborhoods that are predominantly home to low-income people, are disproportionately impacted by extreme heat events, natural disasters, flooding, and a host of additional threats associated with climate change (Wilson et al., 2010; Hoffman et al., 2020). When discussing cli-mate empowerment and equity-oriented planning for climate change adaptation, it is necessary to name the histories of systematic oppression that have played into current disparities in vulnerability to the threats at hand. The US government has a long history of decision-making that has resulted in economic, social, and political violence toward groups of people not considered to be white or wealthy that dates all the way back to the inception of the country (Cooper et al., 2011; Krimmel, 2018; O'Connell, 2012).

In 1933, Franklin D. Roosevelt signed the Home Owners' Loan Act into law, establishing the Home Owners' Loan Corporation (HOLC) in hopes of rectifying mass foreclosures and economic struggle following The Great Depression (Greer, 2013). Although the intention behind the HOLC was to promote economic equity and growth, the HOLC's profit-driven design was rooted in returns rather than human well-being. This being the case, the HOLC constructed a series of "Residential Security Maps" of selected US cities, as can be seen in Fig. 1, breaking cities down by neighborhood and assigning each a grade based on perceived

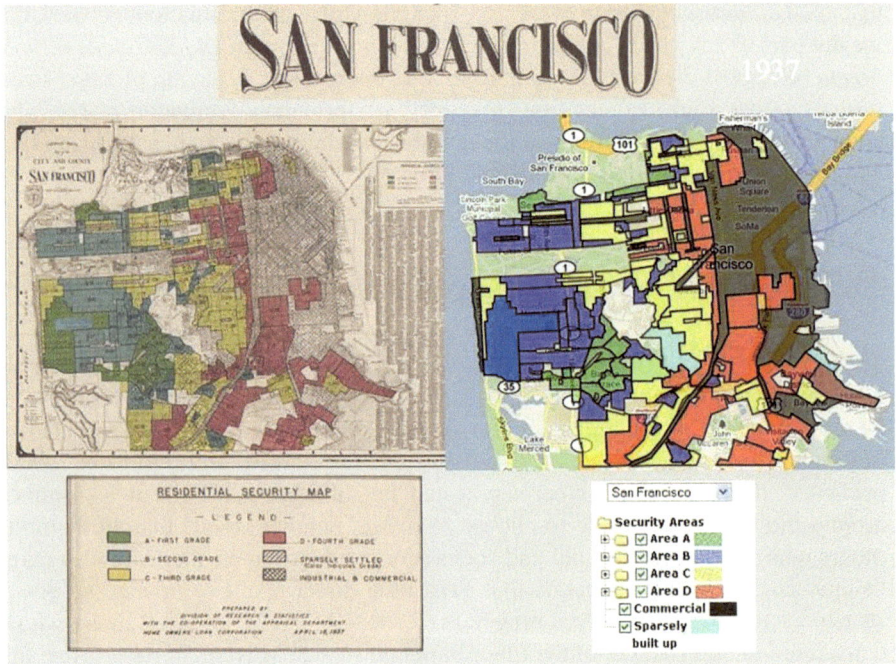

Fig. 1 Home Owners' Loan Corporation Residential Security map, sourced from the University of Maryland's T-Races project

investment risk for banks. Neighborhood grades ranged from A (shaded green on the maps) to D (shaded red on the maps). A-grades represent the lowest level of perceived neighborhood "blight," i.e., investment risk, and "redlined" D-grade represents the highest level of perceived risk (Nardone et al., 2021). This investment risk was determined largely by the presence of non-white people in a community, the presence of recent immigrants, and a range of additional population characteristics indicative of socio-economic status (Greer, 2013). The HOLC, as well as most private banking and real estate companies at the time, relied on these Residential Security Maps for residential, commercial, and industrial lending decisions (Jackson, 1980). Redlining practices have been shown to preclude heightened neighborhood segregation, limit generational wealth-building, and exacerbate long-term poverty cycles (Taylor, 2019). The Fair Housing Act of 1968 banned redlining, but many discriminatory lending and urban planning practices continued on in less explicit terms, leaving legacies of their own (Jackson, 1980; Taylor, 2019).

Neighborhoods that were home to discriminatory zoning and lending practices are now associated with present-day reduced green space, higher percentages of residential segregation, lower average income levels, and fewer economic and housing opportunities (Taylor, 2019). Beyond the disinvestment seen in formerly redlined neighborhoods, investment in environmentally detrimental development was simultaneously being encouraged by a number of federal initiatives, including programs incentivizing the construction of major roadways and buildings in low-income neighborhoods of cities (Wilson et al., 2020). Major highway and building development projects involve materials such as metal, concrete, brick, and other heat-retaining, impermeable materials that exacerbate heat (Namin et al., 2020; Pearcy, 2020). Vulnerability to the impacts of climate change, such as extreme heat, flooding, and natural disasters, is associated with the high proportions of impermeable land surface cover and reduced greenspace found today in formerly redlined neighborhoods (Hoffman et al., 2020; Nardone et al., 2021; Voelkel et al., 2018).

Government-sanctioned economic and land-use policies rooted in prejudice have not only been causal forces in creating place-based disparities in vulnerability to climate change, but their legacies have also taken the form of widespread hindrances to wealth-building that will likely prohibit many individuals from being able to move to a different city or neighborhood as climate conditions worsen (Hoffman et al., 2020; Namin et al., 2020; Gutschow et al., 2021).

It is critical to understand how discriminatory historic land-use policies have shaped modern disparities in climate change vulnerability and economic mobility if we are to move forward in a way that prioritizes repairing systematic harm in our fight for climate empowerment. Geospatial methods have been employed to facilitate the creation of unjust urban landscapes, as was the case with the HOLC's Residential Security maps, but these tools have also been pivotal in pinpointing environmental inequities and place-based differences in the built environment in order to center equity in current grassroots and governmental interventions. Geospatial tools may now be harnessed for equity-oriented urban planning around climate change, and to facilitate community-driven knowledge creation and communication around climate action.

How Even Good Intentions Sometimes Go Awry

Even the best of plans driving equity-centered climate change adaptation planning can have disastrous results if the interventions themselves are not developed and carried out in a manner that centers community empowerment. *Ecological gentrification* is a recurring problem here and a striking example of how poorly executed climate justice interventions can have harmful consequences for the very communities they were designed to benefit (Rice et al., 2020). In expanding green space in low-income neighborhoods with histories of systematic disinvestment and marginalization, and many Black and Brown or other residents of disenfranchised communities, planning agencies or entities can inadvertently cause quick and even extreme increases in local property values that drive out local low-income renters. In situations such as these, the ecological and esthetic benefits of the green infrastructure expansion can end up available only to the much better off, moving in, or able to stay behind and handle the higher rents or increased property tax costs. While not sufficient in completely preventing these free-market housing dynamics, climate empowerment approaches, including community-driven environmental planning processes that give a strong voice to all who are affected, go a long way to finding better solutions. No one knows a neighborhood's needs, wants, and realities better than the people who have made it their home for years, or even generations.

Stories can inspire, explain, and model the understandings and patterns of impact and justice that we need when addressing the climate crisis, and can effectively communicate their complexities. Maps do something else. Maps ground our stories like no other media—the when, where, who, and how—connections between peoples, places, and events; parts and wholes, ideas and resources, histories, scenarios, and choices. They situate narratives, memorialize grief, and give solid underpinnings to aspiration. Misused, they can sharpen division, conflict, and oppression—hardwiring human divisiveness. But put in service of life, of healing, transformation, and redesign, maps—like music and other images—can cut through barriers of language and shyness with shared seeing, pointing, even with fast action. By giving visualization and thus voice to the needs, concerns, aspirations, creativity, and mutual interdependence of all affected by high-stake decisions, maps go a long way toward making such decisions participatory and potentially wise, just, and inclusive. Accountability and justice are at the heart of the climate movement, and geospatial mapping is the storytelling tool of choice in supporting both.

How Will These Redesigns and High-Stake Decisions Be Made, and By Whom?

Community-driven collective action, along with far-reaching policy changes that emphasize regulation, culpability, and accountability for fossil fuel companies, is most needed. Mapping technologies offer a promising path toward shifting our collective mentality away from individual choices, helping us both contextualize current geographic disparities in vulnerability to climate change and recognize larger patterns of connectedness in climate empowerment narratives. Maps can help individuals visualize where they fit into their community's network of action and drive, and promote cohesion. Geospatial tools facilitate the connection of common threads of action, such as we saw with the Santa Cruz Climate Action mapping project. Carried out using participatory mapping to mobilize masses in a unified manner, such mapping initiatives have the potential not only to inform planning efforts, but also to bolster energy to push for tangible change and accountability that may reach all the way up to the corporations largely responsible for the climate crisis we now face. Unity is necessary if we are to tackle climate justice head-on, and unity can be both showcased and fostered through geospatial tools and participatory mapping initiatives.

History and Uses of Geospatial Tools

Dating back to the 1800s, maps have been used to paint a visual narrative of the distribution of benefits and burdens of both the built and natural environment in the United States. By the 1940s, scholars were already using detailed spatial and temporal data to explore trends and compile narratives around the built environment (DeBats and Gregory,, 2011). In the 1960s, computerized geographic information system (GIS) was established as a field of study and the Canada Geographic Information System (CGIS) was built (Goodchild, 2004). Since the inception of computational geographic technologies, geographic information science, or GIScience, has been a recognized subfield of information science with ever-evolving open-source and closed-source software systems for analyzing spatial patterns and illustrating place-based stories (Goodchild, 2004; Rey, 2009). Maps are accepted as intuitive, relatively universal platforms for sharing information and visualizing place-based narratives of how climate change is currently manifesting, and how it is projected to manifest.

As geospatial technologies have advanced, mapping tools have become increasingly useful for engaging members of the public with topics related to climate change, and for informing climate change preparedness urban planning initiatives. Earth observation data and machine learning techniques are now being used for hazard mapping and risk vulnerability mapping, as well as developing tools and platforms to model future risk (Albano & Sole, 2018). Geospatial technologies can be harnessed to (a) better understand relationships between built environment characteristics and climate change resiliency, (b) geographically target participatory climate change preparedness initiatives to regions most vulnerable to climate change, and (c) engage the public in climate action, centralize local expertise, and enhance community knowledge systems.

Responsible Data Sourcing and Usage

Input data are required to analyze, discern, and illustrate place-based patterns of past, current, and projected climate change impacts using geospatial tools. These data may be quantitative or qualitative in nature and can be demographic, environmental, or any combination of data classifications. For data to be represented or analyzed geospatially, the only requirement for input is that it have geographic information present, such as addresses or latitude and longitude coordinates. Compiling and analyzing place-based information, including precipitation levels, temperatures, flood damage, energy consumption, and many other variables directly related to climate change, is helpful for understanding and visualizing patterns of climate change. Outputs of place-based data compilation and visualization processes, whether in the form of maps, statistics, or interactive web-based stories, aid in communicating distributional patterns of climate-related challenges and may highlight positive and negative outliers in climate action and empowerment, all of which is helpful in strategically planning interventions for the climate crisis. Data may come from members of the public, trained scientists, or machines.

Demographic data are often used in geospatial projects to assess the distribution of social, economic, or health impacts of climate change in the United States. Data on race, income, education, age, and other factors that influence mobility, access to resources, and social vulnerability in the United States often come from the US Census Bureau (Voss, 2007).

Environmental data may be used to assess detectible environmental shifts and to communicate detected changes or distributions of manifestations of climate change. Data on climate and environmental characteristics such as precipitation, land and air temperatures, and land surface material can be found in a range of sources, many of them governmental. Depending on one's location and which specific variables happen to be of interest, environmental data can come from a variety of sources including the National Oceanic and Atmospheric Administration (NOAA), the National Aeronautics and Space Administration (NASA), the Environmental Protection Agency (EPA), and other state- and region-specific organizations and agencies. CalEnviroScreen, EnviroAtlas, and the Healthy Places Index offer organized and compiled place-based information on a wide range of demographic and environmental characteristics and should be used as a resource for climate mapping work. Table 1 below details a selection of established demographic and environmental data sources. Maps and spatial statistics can be used to illustrate gaps, or disparities in either susceptibility or impact depending on population characteristics such as income, race, or any of the other demographic variables included in the US census.

Vulnerability to climate change-related events, such as natural disasters, is often strongly associated with a population's age structure. Young children and senior citizens tend to be particularly vulnerable to harm in extreme climate events, along with the disabled populations and housing insecure (Cutter et al., 2009). Demographic data can be used in geospatial analyses to better understand the geographic distribution of very young or very old individuals, and assess risk factors related to climate

Table 1 Nonexhaustive list of data sources for environmental and demographic variables related to climate justice and climate empowerment in the context of the United States[a]

Variable of interest	Available data source(s)	Example
Average income	US Census Bureau 10-year survey, American Community Survey, Bureau of Labor Statistics	Mushore et al. (2018)
Race and ethnicity	US Census Bureau 10-year survey, American Community Survey	Azong and Kelso (2021)
Education	US Census Bureau 10-year survey, American Community Survey, USDA Economic Research Service	Muttarak and Lutz (2014)
Land surface temperature (LST)	USGS Landsat	Ozelkan et al. (2015)
Air temperature	NOAA's National Centers for Environmental Information, NOAA's climate.gov	Legates and Willmott (1990)
Ocean temperature	NOAA's National Centers for Environmental Information, NOAA's climate.gov	Selig et al. (2010)
Vegetation cover	Normalized difference vegetation index (NDVI), i-Tree software	Lee et al. (2017)
Sea level	NOAA, climate.gov, NASA Sea Level Change Portal	Li et al. (2009)
Precipitation level	NOAA's National Centers for Environmental Information	Ninyerola et al. (2000)
Ambient air pollution[b]	US EPA, AirNow.gov	Arheimer and Liden (2000)

[a]This table was created in 2021 as a baseline resource for data sourcing from external groups. Utilizing CBPR data collection methods and searching for additional, potentially more current data sources are strongly encouraged to optimize relevance and accuracy

[b]It is worth noting that air quality data are included in Table 1's climate empowerment variables of interest because of documented interaction between temperature increases and ambient air pollution levels in urban settings, as explored by Li et al. (2018)

change. Since 2010, the US Census Bureau's American Community Survey (ACS) has become the principal source for geographically specific population and economic information (Spielman & Folch, 2015). Population data at the census tract level are used frequently to tie demographic spatial data, e.g., a tract's average income, to data on climate vulnerability, e.g., percentage of surface area covered by tree canopies. Spielman and Folch (2015) argue that ACS in particular has a very high level of uncertainty, pointing out that ACS's estimate of the population size of children under 5 years old has a margin of error greater than the actual estimate in over 72% of US census tracts (Spielman & Folch, 2015). This reinforces the need for all data feeding into geospatial processes to be checked rigorously for uncertainty and validity.

When analyzing racial and socio-economic variables related to climate change vulnerability, it is important to note that the US Census Bureau's classifications of race have shifted dramatically as formerly non-white racial categories have assimilated into whiteness (Zuberi & Bonilla-Silva, 2008). "European ethnics," such as Irish populations, became classified as white as a result of mass international immigration at the start of the twentieth century (Zuberi & Bonilla-Silva, 2008). Temporal contextualization is essential when exploring how geospatial legacies of racial trauma relate to climate change vulnerability today. The racial "othering" that acted as a basis for economic and environmental discrimination has not lasted in a ubiquitous fashion across non-white racial categories, meaning that many Black and Brown communities and individuals who continue to identify as non-white bear the bulk of modern manifestations of the climate-related legacies in question.

Uses and Limitations of Geospatial Tools

Geospatial tools vary in platform design, computational and visualization capabilities, cost, and ease of use. Implications of the nature of these variations include how well the tools may potentially support marginalized populations needing greater access, on the one hand, and the power and flexibility critical, for example, in handling complex, large-scale, and quickly changing situations, such as climate disasters, on the other hand. Most geospatial tool systems allow users to integrate a geographic database with any other database, as long as the datasets have a common identifier for each item, such as a name, address, or unique identification number. This capability is very valuable for understanding the relationships between population or environmental characteristics, and quantifiable impacts of climate change (DeBats & Gregory, 2011). For example, many geospatial studies have combined housing datasets with demographic and environmental datasets to assess the distributional equity of extreme heat exposure (Hoffman et al., 2020; Wilson et al., 2020). This function can be applied to a wide range of global climate crisis challenges and can be used to work toward a safer and more just world through community-driven climate change adaptation planning.

One software company has monopolized the geospatial tool market: Environmental Systems Research Institute (Esri). Jack and Laura Dangermond founded Esri in 1969 and created their first commercially available GIS product, called ArcInfo, in 1981. Esri now offers a suite of geospatial ArcGIS tools, some of which are web-based, while others are only available on the downloadable desktop software that can be used offline. ArcGIS tools aid in organizing, analyzing, visualizing, and publicizing geospatial data. With a relatively straightforward and easy-to-navigate platform, ArcGIS is extremely popular with private companies, governmental agencies, and other nongovernmental or community-based organizations. Esri also offers a StoryMap tool, which allows users to create an interactive platform through which they can supplement the maps they have produced with text, audio, photo, and video content, thus enhancing the maps' storytelling utility.

Many geospatial analysis tools, such as Esri's ArcGIS, have been criticized for not being sufficiently accessible due to their subscription or download fees (Rey, 2009). Esri software products can be costly, but the company has tried to enhance product accessibility. By offering steep discounts to educational institutions, governmental agencies, nonprofit organizations, and other entities deemed worthy of subsidization, Esri has expanded its customer base and addressed many accessibility concerns (Rey, 2009). GIS technologies are now ubiquitous in nearly every field of research and continue to rapidly expand in their capabilities, as can be followed in a variety of GIS-focused journals, such as *Transactions in GIS* (Wilson, 2020b).

In response to criticism of closed-source geospatial tools not being inclusive enough, an influx of free and open-source geospatial tools was introduced to the circuit, and participatory approaches to mapping became popularized in both open- and closed-source mapping initiatives (Rey, 2009). Open source refers to any software for which the software and its source code are shared openly, so that anyone can recreate or modify the code (Steiniger & Hunter, 2012). Many geospatial experts argue that open-source geospatial software is inherently more accessible than its closed-source counterparts, because it is all public domain (Rey, 2009). Popular open-source geospatial tools include QGIS, GRASS, and FOSS GIS (Donnelly, 2010). InVEST software is one example of how open-source tools are being used to support climate change mitigation planning today. InVEST works by assessing benefits of green, or "natural" infrastructure by harnessing geospatial information related to the ecosystem and economic benefits of nature, and "prioritizing land use change and promoting inclusive planning" (Hamel et al., 2021).

Applications of both open- and closed-source GIScience technologies, especially participatory GIS, are being used in disaster risk reduction around the world (Cadag & Gaillard, 2012). Prudhomme et al. (2013) discuss the power of spatial modeling in predicting geographic vulnerability to drought. Global hotspots identified as being likely to experience an increase in drought frequency of more than 20% included regions in South America and Central and Western Europe (Prudhomme et al., 2013). A heightened drought frequency of this magnitude could be devastating for a region's agricultural systems and drinking water availability. In a different geospatial study, researchers apply spatial data infrastructures (SDI) to develop a method for quantifying vulnerability to climate change in the Arctic (Bernard et al., 2008).

Several climate change adaptation efforts using geospatial tools focus on identifying specific resources influencing preparedness, including facilities and social groups (Wood & Good, 2004. Wood and colleagues (Wood & Good, 2004) bring up the fact that a population or region's vulnerability to the effects of climate change is often compounded by other types of vulnerability, depending on the possible hazard or natural disaster. Geospatial tools allow for a better understanding of combined risks by integrating environmental and population features such as topography, population mobility, language abilities, weather, and population age structure. For instance, Wood and colleagues (Wood & Good, 2004) describe how coastal communities who are concerned about sea level rise use GIS to assess integrity of their built environment, which they break down into construction type, building code, and shoreline management planning. Coastal flood explorer tools are available all over the United States and the world, and offer viewers visualizations of what their local environments will look like, and which parcels are likely to be underwater, at a range of given sea level rise scenarios.[5]

Participatory Mapping: Tools, Processes, and Principles

The impacts of climate change on population health and well-being cluster geographically. For a wide range of stakeholders, such as members of the public, community and nongovernmental organizations, governmental agencies, and the private sector, geospatial tools have been useful in telling the stories of these clusters of environmental and public health crises and disasters related to climate change (Holland & Bruyère, 2014; Parente et al., 2016; Prudhomme et al., 2014).

Participatory mapping can stand alone as a complete process, or it can be used in tandem with a range of other qualitative or quantitative methods. (2016 Cochrane and Corbett (2020) define participatory mapping as the "creation of maps by non-expert individuals, groups and communities—often with the involvement of supporting organizations," such as community organizations, government agencies, nongovernmental organizations (NGOs), or universities. Participatory mapping projects are able to draw from large or small populations and provide opportunities to represent the experiences of individuals from a wide variety of backgrounds. Ground-truthing uses community-based participatory research (CBPR) methods, which are rooted in collaboration between academic institutions, individual community members, and community organizations (Sadd et al., 2014). Ground-truthing research strategies can be utilized for participatory mapping projects, as long as the community-collected data include a geographic identifier. Web-based participatory mapping is a common form in the United States and globally. Web-based mapping relies on online technology, emphasizes social and environmental justice, and is designed to enhance individual and community connectedness, agency, and power in decision-making processes and public narratives.

Corbett (2009) describes the different ways in which participatory mapping can be utilized, naming the following applications: providing a pathway for communities to articulate spatial knowledge to external agencies, allowing communities to record and archive local knowledge, assisting communities in land-use planning and resource management, supporting community advocacy for change, increasing capacity, connection, and cohesion within communities, and helping to address conflict related to resource management. These applications of participatory mapping all connect readily to challenges around climate change, as can be seen in the examples to come.

A study in the Kimberley region of Australia from 2013 to 2015 exemplifies how participatory GIS research can be included in mixed-methods CBPR. Aiming to

---

[5] See Adapting to Rising Tides (ART) Bay Area's interactive flood explorer map at https://explorer. adaptingtorisingtides.org/explorer

inform marine conservation planning efforts, researchers used a combination of semi-structured interviews, qualitative participatory mapping in the form of polygon marking exercises, and quantitative web-based participatory mapping exercises to assess which geographic polygonal regions' stakeholders see as most valuable, and to compile polygon-based coastal management preferences (Brown et al., 2017). These diverse expressions of community input, in the form of geographically specific community priority regions, helped planners make coastal conservation decisions that were best for community members, and helped build trust between community members and planning authorities.

Cadag and Gaillard (2012) illustrate how participatory community mapping can be used through showcasing a disaster risk reduction project in Masantol, Philippines. In this project, community members systematically identified regions considered particularly vulnerable to hazards through manual "stone" mapping, "sketch" mapping, and web-based geospatial mapping tools such as GIS (Cadag & Gaillard, 2012). By offering various options for participatory mapping involvement, community members were to decide which type of mapping felt most appropriate and most accessible. Hand-drawn community maps can easily be transferred to online geospatial systems to share broadly and analyze for whatever purposes community members see fit, and their physical manifestation within the community can aid planners and other community members in preparing for natural disasters. Drawing on local knowledge through participatory mapping makes for more effective and robust planning for climate change adaptation initiatives. After all, who knows the minute characteristics, culture, and vulnerabilities of a space better than the people who inhabit that land? On a similar note, participatory mapping tools used to highlight action happening on the ground locally have great potential to motivate individuals to join their community members, friends, and family members in mobilizing for change in neighborhoods, parks, and spaces that are visually familiar.

Participatory and Community-Based Mapping Projects

Participatory and community-based mapping projects have a distinct set of functions, best practices, and limitations. When carrying out volunteered geographic information (VGI) projects, informed consent is always mandatory. Publicizing geographic information, particularly in tandem with volunteer names, photographs, or other identifying media, poses issues of privacy that requires informed consent based on exactly what information will be made public and how and when the media will be shared. If participants change their minds at any point about consenting to share their information or media, the organizer must immediately comply with their content removal requests.

More broadly, there are recommended practices and limitations to be aware of when using geospatial tools in the realm of climate empowerment and justice. It is important, for example, to be especially clear in setting the geographic scales of spatial data when applying geospatial tools so as to avoid visual distortions.

Geographic scales of data refer to how spatial data are aggregated by region. Common scales include census tracts, cities, counties, and states. The choice of size and shape of areal aggregation schemes influence visual and statistical outcomes and have the potential to skew both study analyses and results (Räsänen et al., 2019). In other words, choosing to aggregate spatial data at one scale can result in a much different narrative than choosing to aggregate data on a different scale. This phenomenon is commonly known as the *modifiable areal unit problem* (MAUP) (Fotheringham & Wong, 1991).

Subject privacy is also of primary concern in geospatial storytelling and analysis. Unless specifically given permission to publish the names of project subjects or participants, de-identifying data points should be standard practice.

Experimenting with Mapping Tools

Climate Empowerment on the Central Coast of California

Here in Santa Cruz County, on the north end of the Monterey Bay, and with some hundreds of climate efforts and achievements among the six counties on this coast, we are in the earliest phases of thinking about how "climate empowerment" might best be realized in this bio-region. We have much in common on the Central Coast of California, our small cities strung out as they are along Hwy 1 on the tiny strip of land left between the mountains and the sea, with little and ever-lessening water and some of the most precious farming land anywhere in the nation or the world. We are passionate here about "climate," as its crisis is bearing down upon us, burning down our Redwoods and our homes, stealing our water, and bringing lethal heat to our fields and to the many individuals laboring in them. The question is: *How do we connect all this effort and focus it for greatest impact?*

Much of what we've discussed so far in this chapter has grown out of just that quest: *How might we in this bio-region best use mapping and stories to stop and reverse further climate damage, and lay the groundwork for a truly livable future?* We want to use both *storytelling and mapping* to help everyone here raise and answer the question, *"What can we do?"* in all the ways climate change does or will affect them. "Everything *is* local," as it's always real people in relationships with each other, near or far, who move the world. It is upon the empowerment of "ordinary" people, that in the end a new future must be forged.

In the County of Santa Cruz, with its three incorporated cities, Santa Cruz, Watsonville, and Scotts Valley, there are between one and two hundred organizations, maybe more, with direct or indirect connections to the issue of the climate crisis. Regionally, along the Central Coast, are several hundred more, so much of the informal groundwork for a substantial regional climate empowerment effort is already well established. Formally, we also have in place MBEP (Monterey Bay Economic Partnership), C3E (Central Coast Community Energy), an emerging

Central Coast Public Bank (an organization coming close to closure in their efforts to establish such a bank under California's new public banking law), and the informal and professional connections between City Staffs and elected officials from cities, counties, and dozens of district boards as well. These bodies appreciate of the value of working closely with citizens in achieving climate action goals, even where constrained in their ability to act upon it.

The City Santa Cruz is well into a participatory process of updating its 2010 to a 2030 climate *action* plan, with its 2-year-old climate *adaptation* plan; *Watsonville* has a 2030 Climate Action & Adaptation Plan and *San Luis Obispo*, midpoint on the Central Coast and regarded by many as the leader in such efforts in this area, is well into implementing its own 2020 Climate Action Plan. There is considerable overlap in the values, agendas, and challenges among these.

Regionally, we also have three universities and eight community colleges and, of course, hundreds of citizen and student leaders, faculty, climate scientists, experienced campaign strategists, citizen journalists, community media, and associations of small business, tourism, farmers, fishers, and faith communities.

"The RE-AMP Network is a collaboration of 175 nonprofits and foundations across eight Midwestern US states ... that used a systems map developed by [participants with the help of] Scott Spann to identify [4] leverage points to ... [reduce GHGs by] ... 80% by 2050...[These were to simultaneously]...(1) stop new coal plants from being built, (2) shut down existing coal plants, (3) increase renewable power, and (4) increase energy efficiency (Spann, S., 2012). As a result of their collaborative efforts, the RE-AMP network has helped retire over 150 coal plants, five states have implemented rigorous renewable standards, six states have adopted or approved energy efficiency resource standards, and the Network has regranted over $25 million to support strategic climate action in the Midwest." *Key to this success was that from the outset representatives from every part of the overlapping systems* that together made the Midwest the most polluting part of the United States, from coal-coal users, coal companies, solar and alternative energy, and local and government college—*all came together.* Each helped spell out how they understood the whole system to work, while coming to trust each other across the usual barriers and appreciate the constraints and opportunities of each other's situations. These dynamics elicited exceptional collaboration and creativity (Fig. 2).

The challenge, then, and the opportunity, is to further strengthen this network involving potentially hundreds of knowledgeable and committed climate-concerned citizens and officials and make the most out of these efforts. Our efforts so far have been particularly influenced by two models, the *RE-AMP Network* and the *US ACE Framework.*

The accomplishments of the RE-AMP program, ongoing, continue to be among the most highly regarded in the United States and among the climate "success stories" most needing to be told. It offers a model of effective strategy for systems change. Part of this strategy, the idea of bringing all parts of a system together to identify leverage points and achieve change has been picked up, branded as "Collective Action" and promoted by among others the SSIR. But in that process, a critical piece may have been dropped, one fundamental to climate empowerment as

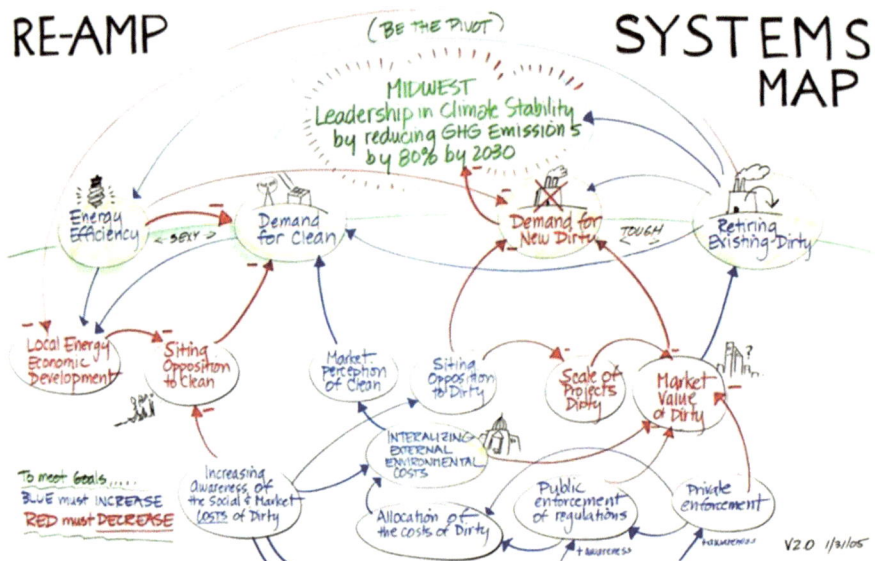

Fig. 2 A systems map of the RE-AMP network

we see it, their distinctive and powerful "map and stories" technique. If we can manage the resources available, it is this element we would seek to emulate, making it fundamental to a systems approach to climate empowerment on this coast, and as a pilot of US ACE, if we can manage the resources it could take.

The UN ACE commitment, though the first to use the term, does not define "climate empowerment" on its website, but rather says this about it:

> Action for Climate Empowerment (ACE) is a term adopted by the United Nations Framework Convention on Climate Change (UNFCCC) to denote work under Article 6 of the Convention (1992) and Article 12 of the Paris Agreement. The overarching goal of ACE is to empower all members of society to engage in climate action, through education, training, public awareness, public participation, public access to information, and international cooperation on these issues.

In 2020, absent federal leadership, an independent group of US citizens, convened a Talanoa dialogue with diverse communities nationwide to develop an ACE National Strategic Planning Framework for the United States. In March 2021, the Biden government adopted the Framework as US Policy, appointing Frank Niepold as the domestic "focal point" to implement it. Climate Heath and Equity Partnership (CHEP) team members offered to help develop an "ACE Pilot'" here on the Central Coast. Local work described in this chapter is in part an early stage in such an effort.

In 2021, CHEP, the authors, and others in the network began to explore what it would look like for the Central Coast to pilot US ACE elements on a regional level and share what they learned. From that, effort so far has come the components listed here and the experiments, research, and reflections shared in this chapter.

Exploring Participatory Processes with Geospatial Tools and Maps

The Santa Cruz Climate Action Photo Contest

Among the first CHEP experiments focused directly on mobilizing "ordinary" people through storytelling, was the Santa Cruz StoryMap Project, which took place in the spring of 2021. Technically, the project was a *mixed-media participatory mapping exercise* using a local climate action volunteer geographic information (VGI) process. Socially, as an experiment in *self-story-driven climate empowerment*, it was set up as a photo contest with scholarship money as prizes for student entries, and "reputation" rewards for other participants.

The StoryMap project was a collaborative effort between the Climate and Health Equity Partners (CHEP), a group of activists, educators, health professionals, and scholars and the Santa Cruz Climate Action Network (SCCAN), a group with similar goals also based in the northern part of the County of Santa Cruz. Young community members were invited to submit photos they had taken relating to a climate action they had been involved in that fit into one or more of the following categories: education or outreach, environmental justice, sustainability and waste reduction, energy, transportation, civic engagement, or food. CHEP worked with SCCAN to design the survey as a contest entry form and make sure the geographic coordinates were included.

Participatory mapping has the potential to incorporate different forms of media sourced from community members, such as videos, photos, and audio, and organize the media into maps or other types of place-based outputs. Such outputs, including ArcGIS StoryMaps, create compelling and personalized narratives of climate action on the ground, or adverse climate change impacts at the local level. We wanted to see how these potentials would play out on the Coast with a small-scale model.

Thorough outreach is an important step in optimizing contribution accessibility and including participants from a wide range of backgrounds. The survey, conducted through Esri's Survey123 platform, was distributed to all local community organization contacts available in the group, as well as a long list of administrators and teachers in the Santa Cruz and Watsonville educational districts.

The photo contest received 32 survey submissions from individuals, all willing to be included in a public mapping project. CHEP exported the survey responses, which included website URL links for the photo submissions, to a CSV file. Spreadsheets formatted as CSVs, including addresses, coordinates, or other digestible geographic information, can be directly uploaded to ArcGIS Online. In order to de-identify the photos and data points featured on the map outputs, CHEP replaced the names of submitters with random numbers, keeping the original spreadsheet with contest submission data including both names and de-identified numbers in a private and safe place. After exporting and cleaning the CSV file to include only geographic coordinates, climate action description information, and climate action photo URLs, CHEP uploaded the CSV file into Esri's ArcGIS Online application to map the submissions. After creating maps showing the photo submissions with their locations, CHEP made a StoryMap, as a cohesive narrative on the SCCAN website.

We hoped the contest would lure in a few people beyond "the usuals," targeting engagement from younger participants, and seeking novel takes on what counts as "climate action," and it did. We also wanted to test out the potential impact on climate empowerment, across the many distinct constituencies ultimately to be reached, if people could, for example, see in their own "hometown," the diverse range of climate actions already occurring there. Looking ahead, we hoped that such mapping tools and visualizations located near people's homes, and in different places and by numbers of people, organizations, or individuals they knew or would come to know, and perhaps reach out to, as a result of similar mapping projects in the future that might be done on a larger scale.

Connecting Climate Action and Health Communities

There also exist opportunities to use geospatial tools in the future to inventory and map all of the well over a hundred climate actions and organizations to be found in the Santa Cruz area, with the hope of a more regional approach extending down the coast. Figure 3 illustrates another promising use of geospatial tools by helping to visualize how two networks, each vital to achieving climate solutions, can be so geographically mixed and yet so socially separate.

As a foundational step toward both connecting and integrating climate and health at the community level, mapping accessible health, climate, and environmental groups, services, and locations holds great potential. Figure 3 displays the early stages of a sample map for Santa Cruz County stakeholders and prospective CHEP partners who share an interest in this type of climate and health initiative, aimed at ensuring that health facilities, health professionals, and climate advocates can readily partner and comprehensively advance community resiliency, health promotion, and robust prevention.

Similar to other counties, Santa Cruz offers a wide variety of health facilities and services to serve the public's medical needs. Examples of health facilities that have a significant impact and play a vital role in climate and health are hospitals and county health departments. Santa Cruz County has three main hospitals and a Health Services Agency that includes primary care clinics, environmental health, behavioral health, and public health. These facilities form the health-focused data points in Fig. 3. As a foundational step toward both connecting and integrating climate and health at the community level, maps displaying accessible health, climate, and environmental groups, services, and locations hold a lot of potential.

Hospitals are often top regional employers and pillars of local economies. They are also often some of the top emitters of greenhouse gases, which both drive climate change and cause severe health impacts. Consequently, it is vital that hospital leaders and planners seriously consider and reliably implement climate-informed priorities. This is particularly the case in California, where climate change and

Santa Cruz Climate Actions 20210908 Health

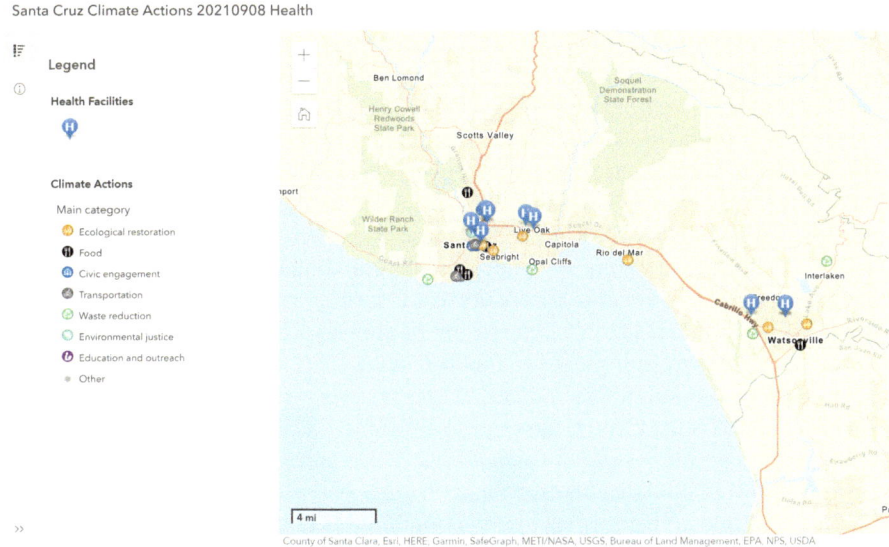

Fig. 3 A regional map showing where healthcare facilities and climate actions connect and coordinate currently disparate processes

equity are embedded in public health planning by the Climate Change and Health Equity Section (CCHES).

Given the goal of developing dynamic maps with responsive, action-oriented data points, for anyone who is interested in developing Climate and Health County- or Community-level maps, it will likely be most useful to prioritize inclusion of the health facilities and systems that are both early adopters and reliable champions for climate, health, and equity priorities. When patients, providers, families, and community members all understand these connections and apply lessons about how human health and climate are fundamentally interdependent, we can collaboratively achieve a vision of durable social and ecological wellness.

With these overarching goals setting the larger context, this chapter has focused primarily on the near term and the close to home. *Couple climate success stories with the climate problems when telling of them.* Thus, the first step toward climate empowerment are stories told with this intention in mind:

In considering the full range of possible impacts storytellers may seek to have different kinds of stories in different kinds of situations, we hxave noticed a pattern among possible response, their *internal* sense of *agency*, and transformations of the *external* conditions within or through which the concerns can be effectively acted upon. This is to say that while climate empowerment is centrally a matter of the growth and evolution of our own knowledge, compassion, and courage, it is also a matter of an *increasingly potent democratization* of the means by which we make and act upon collective decisions.

It is this external dimension that we focused on in the second section of this chapter, *Realizing Climate Justice*, The Equalizing Power of Geospatial Tools, the dimension of *equity*. Participatory planning, budgeting, and problem-solving are the growing edge of climate empowerment, globally, assuring that all affected by decisions have substantial say in them—and the tools and processes that will make the most of that say. We focused on one of the most promising, the appropriate use of geospatial mapping tools as these are particularly valuable in helping all peoples to see for themselves precisely the challenges that we are up against in the climate crisis, what our choices are, and the effects of these choices on all the different populations affected. Well-used and to these ends such tools, skills, and processes help assure that the *external* systemic or structural dimensions of climate empowerment do themselves become a core focus of the climate movement, especially for those who must never again be closed out of the very climate decisions where their own interests and even lives are so much at stake.

In the third section of this chapter, "*Experimenting with Mapping Tools*, Climate Empowerment on the Central Coast of California," we also focus on the two dimensions of climate empowerment, *efficacy* and *community*. Recognizing how much any move toward climate empowerment on this Coast will depend on stories, we are inviting multiple conversations across communities that care deeply about climate change, climate justice, community resilience, and a number of related questions and causes. Listening, imagining, and mapping the many "climate success stories" already occurring helps to bolster anever-growing inventory of resources for the public to work with.

And, in all this, finally, we come full circle, a third way "through" the dilemma, *living into it:* acceptance without resignation; determination. Mapping tools that turn hard data into human stories, helping make just-in-time decisions. Stories at the edge of good planning, daring innovations, and mutual aid—these can bring us into alignment with the natural world, each other, our near-term challenges, and our long-term potentials.

Acknowledgments We thank Daniel Caraco and John P. Wilson for insightful discussions and feedback.

References

Albano, R., & Sole, A. (2018). Geospatial methods and tools for natural risk management and communications. *ISPRS International Journal of Geo-Information, 7*(12), 470. https://doi.org/10.3390/ijgi7120470

Arheimer, B., & Liden, R. (2000). Nitrogen and phosphorus concentrations from agricultural catchments—Influence of spatial and temporal variables. *Journal of Hydrology, 227*(1–4), 140–159.

Azong, M. N., & Kelso, C. J. (2021). Gender, ethnicity and vulnerability to climate change: The case of matrilineal and patrilineal societies in Bamenda Highlands Region, Cameroon. *Global Environmental Change, 67*, 102241. https://doi.org/10.1016/j.gloenvcha.2021.102241

Bernard, L., & Ostländer, N. (2007). Assessing climate change vulnerability in the arctic using geographic information services in spatial data infrastructures. *Climatic Change, 87*(1–2), 263–281. https://doi.org/10.1007/s10584-007-9346-0

Bishop, M. P., James, L. A., Shroder, J. F., Jr., & Walsh, S. J. (2012). Geospatial technologies and digital geomorphological mapping: Concepts, issues and research. *Geomorphology, 137*(1), 5–26.

Brown, G., Kangas, K., Juutinen, A., & Tolvanen, A. (2017). Identifying environmental and natural resource management conflict potential using participatory mapping. *Society & Natural Resources, 30*(12), 1458–1475.

Bullard, R. D. (1993). The legacy of American apartheid and environmental racism. *John's Journal of Legal Commentary, 9*, 445.

Cadag, J. R. D., & Gaillard, J. C. (2012). Integrating knowledge and actions in disaster risk reduction: The contribution of participatory mapping. *Area, 44*(1), 100–109.

Cheng, L., Abraham, J., Trenberth, K. E., Fasullo, J., Boyer, T., Locarnini, R., Zhang, B., Yu, F., Wan, L., Chen, X., Song, X., Liu, Y., Mann, M. E., Reseghetti, F., Simoncelli, S., Gouretski, V., Chen, G., Mishonov, A., Reagan, J., & Zhu, J. (2021). Upper ocean temperatures hit record high in 2020. *Advances in Atmospheric Sciences, 38*(4), 523–530. https://doi.org/10.1007/s00376-021-0447-x

Cochrane, L., & Corbett, J. (2020). Participatory mapping. In J. Servaes (Ed.), *Handbook of communication for development and social change* (1st ed., pp. 705–713). Springer. https://doi.org/10.1007/978-981-10-7035-8_6-1

Cooper, A. D. (2011). From slavery to genocide: The fallacy of debt in reparations discourse Allan D. Cooper. *Journal of Black Studies, 43*(2), 107–126. https://doi.org/10.1177/0021934711410879

Corbett, J. (2009). Good practices in participatory mapping: A review prepared for the International Fund for Agricultural Development (IFAD).

Corbett, J., Cochrane, L., & Gill, M. (2016). Powering up: Revisiting participatory GIS and empowerment. The Cartographic Journal, 53(4), 335–340.

Cutter, S. L., Emrich, C. T., Webb, J. J., & Morath, D. (2009). Social vulnerability to climate variability hazards: A review of the literature. *Final Report to Oxfam America, 5*, 1–44.

DeBats, D. A., & Gregory, I. N. (2011). Introduction to historical GIS and the study of urban history. *Social Science History, 35*(4), 455–463. https://doi.org/10.1017/s0145553200011639

Donnelly, F. P. (2010). Evaluating open source GIS for libraries. *Library Hi Tech, 28*(1), 131–151. https://doi.org/10.1108/07378831011026742

Fotheringham, A. S., & Wong, D. W. S. (1991). The modifiable areal unit problem in multivariate statistical analysis. *Environment and Planning, A23*, 1025–1044.

Goodchild, M. F. (2004). GIScience, geography, form, and process. *Annals of the Association of American Geographers, 94*(4), 709–714.

Greer, J. (2013). The home owners' loan corporation and the development of the residential security maps. *Journal of Urban History, 39*(2), 275–296.

Gutschow, B., Gray, B., Ragavan, M. I., Sheffield, P. E., Philipsborn, R. P., & Jee, S. H. (2021). The intersection of pediatrics, climate change, and structural racism: Ensuring health equity through climate justice. *Current Problems in Pediatric and Adolescent Health Care, 51*(6), 101028.

Hamel, P., Guerry, A. D., Polasky, S., Han, B., Douglass, J. A., Hamann, M., Janke, B., Kuiper, J. J., Levrel, H., Liu, H., Lonsdorf, E., McDonald, R. I., Nootenboom, C., Ouyang, Z., Remme, R. P., Sharp, R. P., Tardieu, L., Viguié, V., Xu, D., et al. (2021). Mapping the benefits of nature in cities with the InVEST software. *Npj Urban Sustainability, 1*(1). https://doi.org/10.1038/s42949-021-00027-9

Hansen, J. (2009). *Storms of my grandchildren*. Bloomsbury Press.

Hoffman, J. S., Shandas, V., & Pendleton, N. (2020). The effects of historical housing policies on resident exposure to intra-urban heat: A study of 108 US urban areas. *Climate, 8*(1), 12.

Hogan, C. (2021). *Force of nature*. https://www.forceofnature.xyz/

Holland, G., & Bruyère, C. L. (2014). Recent intense hurricane response to global climate change. *Climate Dynamics, 42*(3), 617–627.

Jackson, K. T. (1980). Race, ethnicity, and real estate appraisal: The home owners loan corporation and the federal housing administration. *Journal of Urban History, 6*(4), 419–452.

Knutson, T., Camargo, S. J., Chan, J. C., Emanuel, K., Ho, C. H., Kossin, J., et al. (2020). Tropical cyclones and climate change assessment: Part II: Projected response to anthropogenic warming. *Bulletin of the American Meteorological Society, 101*(3), E303–E322.

Koutsoyiannis, D. (2021). Rethinking climate, climate change, and their relationship with water. *Water, 13*(6), 849.

Krimmel, J. (2018). *Persistence of prejudice: Estimating the long-term effects of redlining. SocArXiv.* https://doi.org/10.31235/osf.io/jdmq9

Lee, S. J., Longcore, T., Rich, C., & Wilson, J. P. (2017). Increased home size and hardscape decreases urban forest cover in Los Angeles County's single-family residential neighborhoods. *Urban Forestry & Urban Greening, 24*, 222–235.

Legates, D. R., & Willmott, C. J. (1990). Mean seasonal and spatial variability in global surface air temperature. *Theoretical and Applied Climatology, 41*(1), 11–21.

Li, X., Rowley, R. J., Kostelnick, J. C., Braaten, D., Meisel, J., & Hulbutta, K. (2009). GIS analysis of global impacts from sea level rise. *Photogrammetric Engineering & Remote Sensing, 75*(7), 807–818.

Li, H., Meier, F., Lee, X., Chakraborty, T., Liu, J., Schaap, M., & Sodoudi, S. (2018). Interaction between urban heat Island and urban pollution Island during summer in Berlin. *Science of the Total Environment, 636*, 818–828.

Mitchell, D., Heaviside, C., Vardoulakis, S., Huntingford, C., Masato, G., Guillod, B. P., Frumhoff, P., Bowery, A., Wallom, D., & Allen, M. (2016). Attributing human mortality during extreme heat waves to anthropogenic climate change. *Environmental Research Letters, 11*(7), 074006. https://doi.org/10.1088/1748-9326/11/7/074006

Mushore, T. D., Mutanga, O., Odindi, J., & Dube, T. (2018). Determining extreme heat vulnerability of Harare Metropolitan City using multispectral remote sensing and socio-economic data. *Journal of Spatial Science, 63*(1), 173–191.

Muttarak, R., & Lutz, W. (2014). Is education a key to reducing vulnerability to natural disasters and hence unavoidable climate change? *Ecology and Society, 19*(1).

Myers, K. F., Doran, P. T., Cook, J., Kotcher, J. E., & Myers, T. A. (2021). Consensus revisited: Quantifying scientific agreement on climate change and climate expertise among earth scientists 10 years later. *Environmental Research Letters, 16*(10), 104030.

Namin, S., Xu, W., Zhou, Y., & Beyer, K. (2020). The legacy of the home owners' loan corporation and the political ecology of urban trees and air pollution in the United States. *Social Science & Medicine, 246*, 112758.

Nardone, A., Rudolph, K. E., Morello-Frosch, R., & Casey, J. A. (2021). Redlines and greenspace: The relationship between historical redlining and 2010 greenspace across the United States. *Environmental Health Perspectives, 129*(1), 017006.

Ninyerola, M., Pons, X., & Roure, J. M. (2000). A methodological approach of climatological modelling of air temperature and precipitation through GIS techniques. *International Journal of Climatology: A Journal of the Royal Meteorological Society, 20*(14), 1823–1841.

O'Connell, H. A. (2012). The impact of slavery on racial inequality in poverty in the contemporary US South. *Social Forces, 90*(3), 713–734.

Ozelkan, E., Bagis, S., Ozelkan, E. C., Ustundag, B. B., Yucel, M., & Ormeci, C. (2015). Spatial interpolation of climatic variables using land surface temperature and modified inverse distance weighting. *International Journal of Remote Sensing, 36*(4), 1000–1025.

Parente, J., Pereira, M. G., & Tonini, M. (2016). Space-time clustering analysis of wildfires: The influence of dataset characteristics, fire prevention policy decisions, weather and climate. *Science of the Total Environment, 559*, 151–165.

Pearcy, M. (2020). "The Most insidious legacy"—Teaching about redlining and the impact of racial residential segregation. *The Geography Teacher, 17*(2), 44–55.

Prudhomme, C., Giuntoli, I., Robinson, E. L., Clark, D. B., Arnell, N. W., Dankers, R., et al. (2014). Hydrological droughts in the 21st century, hotspots and uncertainties from a global multimodel ensemble experiment. *Proceedings of the National Academy of Sciences, 111*(9), 3262–3267.

Räsänen, A., Heikkinen, K., Piila, N., & Juhola, S. (2019). Zoning and weighting in urban heat Island vulnerability and risk mapping in Helsinki, Finland. *Regional Environmental Change, 19*(5), 1481–1493.

Rey, S. J. (2009). Show me the code: Spatial analysis and open source. *Journal of Geographical Systems, 11*(2), 191–207.

Rice, J. L., Cohen, D. A., Long, J., & Jurjevich, J. R. (2020). Contradictions of the climate-friendly city: New perspectives on eco-gentrification and housing justice. *International Journal of Urban and Regional Research, 44*(1), 145–165.

Rothstein, R. (2017). *The color of law: A forgotten history of how our government segregated America*. Liveright Publishing.

Sadd, J., Morello-Frosch, R., Pastor, M., Matsuoka, M., Prichard, M., & Carter, V. (2014). The truth, the whole truth, and nothing but the ground-truth: Methods to advance environmental justice and researcher–community partnerships. *Health Education & Behavior, 41*(3), 281–290.

Samora-Arvela, A. F., Ferrão, J., Ferreira, J., Panagopoulos, T., & Vaz, E. (2017). Green infrastructure, climate change and spatial planning. *Journal of Spatial and Organizational Dynamics, 5*(3), 176–188.

Selig, E. R., Casey, K. S., & Bruno, J. F. (2010). New insights into global patterns of ocean temperature anomalies: Implications for coral reef health and management. *Global Ecology and Biogeography, 19*(3), 397–411.

Spielman, S. E., & Folch, D. C. (2015). Reducing uncertainty in the American community survey through data-driven regionalization. *PLoS One, 10*(2), e0115626. https://doi.org/10.1371/journal.pone.0115626

Steiniger, S., & Hunter, A. J. (2012). Free and open source GIS software for building a spatial data infrastructure. In *Geospatial free and open source software in the 21st century* (pp. 247–261). Springer.

Taylor, K. Y. (2019). *Race for profit: How banks and the real estate industry undermined black homeownership*. UNC Press Books.

Tobler, W. R. (1970). A computer movie simulating urban growth in the Detroit region. *Economic Geography, 46*(sup1), 234–240.

United Nations Framework Convention on Climate Change (UNFCCC). (1992). Report FCCC/INFORMAL/84

Voelkel, J., Hellman, D., Sakuma, R., & Shandas, V. (2018). Assessing vulnerability to urban heat: A study of disproportionate heat exposure and access to refuge by socio-demographic status in Portland, Oregon. *International Journal of Environmental Research and Public Health, 15*(4), 640.

Voss, P. R. (2007). Demography as a spatial social science. *Population Research and Policy Review, 26*(5), 457–476.

Wilson, B. (2020a). Urban heat management and the legacy of redlining. *Journal of the American Planning Association, 86*(4), 443–457.

Wilson, J. P. (2020b). GIScience research at the 2020 Esri user conference. *Transactions in GIS, 24*(3), 553–555.

Wilson, S. M., Richard, R., Joseph, L., & Williams, E. (2010). Climate change, environmental justice, and vulnerability: An exploratory spatial analysis. *Environmental Justice, 3*(1), 13–19.

Wood, N. J., & Good, J. W. (2004). Vulnerability of port and harbor communities to earthquake and tsunami hazards: The use of GIS in community hazard planning. *Coastal Management, 32*(3), 243–269.

Yee, S. O. (1996). Ethnic enclaves as teaching and learning sites. *The Social Studies, 87*(1), 13–17.

Zuberi, T., & Bonilla-Silva, E. (Eds.). (2008). *White logic, white methods: Racism and methodology*. Rowman & Littlefield Publishers.

Aviva Wolf-Jacobs is a Ph.D. student at the University of Southern California in Los Angeles. Aviva's primary research interests include climate justice, environmental epidemiology, and geographic information science (GIS) tool utilization to promote racial and economic justice. Having grown up in Northern California, Aviva has long been appreciative of the natural world and its influence on human health and well-being. Aviva is enthusiastic about the ways in which maps provide accessible narratives of justice and injustice. Highlighting and sharing place-based stories of environmental inequalities, innovation, and people power may help pave the way toward much-needed resource redistributions and exciting community-driven climate action.

Nancy Glock-Grueneich created the local TV series *The Future We Need and How to Get It,* founded *Within Reach*, a support network for climate solution strategists, and is currently developing the climate action strategy for the Pacific Yearly Meeting of the Religious Society of Friends (i.e., Quakers of California, Nevada, Hawaii, Mexico, Central America, and Guam). She authored the Curriculum Standards Handbook for the then 108 colleges of the California Community College system, approving new programs, overseeing faculty development, and initiating establishment of online learning for the system. She co-edited the International Journal of Public Participation, co-developed the Antioch University Program on Systems Transformation and Civic Development, and both taught courses and trained faculty in Shandong Youth University of Political Science in Jinan, China. Her doctorate in Philosophy and Education is from the Harvard Graduate School of Education.

Nathan Uchtmann is an internal medicine and pediatric physician and licensed New York attorney who currently practices in California. He is a Voluntary Clinical Professor for UCSF Pediatrics and completed the UCSF HEAL Global Health Fellowship. Nathan did his residency training and medical school at the University of Illinois, where he also received his law degree and a master's degree in Natural Resources and Environmental Sciences. He works to bridge unjust divides and cultivate ethical stewardship through a wide range of local and global projects that use complementary frameworks such as Planetary Health, Climate and Health, and Child Health to link human health with environmental sustainability.

Exploring Climate Science in the Metaverse: Interactive Storytelling in Immersive Environments for Deep Learning and Public Engagement

Stacey Spiegel and Hua Wang

In our contemporary society, digital media dominates our lives and it is well understood that technology and media are the primary tools to influence culture and facilitate social change. It was back in the early 1980s with the arrival of the Atari and Amiga consumer computers that the beginnings of what would become known as the metaverse started to emerge. While the term Metaverse itself was first coined by science-fiction fantasy writer Neal Stephenson in his 1992 novel *Snow Crash*, early computers in the 1990s had already pioneered solutions for users to interact in computer-generated virtual worlds. Early examples of social interaction in these virtual worlds were often found in arcades where people played computer games together on terminals. With the ubiquity of the Internet, gamers quickly found solutions for creating parallel lives in a digital paradigm. These connected virtual worlds have allowed millions of people around the world to play in simulations and story worlds together with others while physically remaining at home. Technologies have been evolving rapidly from the early Nintendo consoles and PlayStations to Oculus head-mounted displays for virtual reality (VR) and smartphone apps for augmented reality (AR), to the latest in mixed reality (MR/XR) with the Microsoft HoloLens.

Each of these technologies has excellent examples of how storytelling can be incorporated into the user experience as a powerful tool for enhanced entertainment and engagement and, in more recent decades, for informal education and social change such as the serious games movement that began in the late 1990s (Ritterfeld et al., 2009). America's Army (2002) is an early example. While this game was designed as a popular first-person shooter by the United States Army, it also had real-world content built in. At first, it was designed solely to engage the users with

S. Spiegel (✉)
Parallel World Labs Inc., Georgian Bluffs, ON, Canada
e-mail: stacey@staceyspiegel.com

H. Wang
University at Buffalo, The State University of New York, Buffalo, NY, USA
e-mail: hwang23@buffalo.edu

the hope of recruiting them and offset the need for subscriptions, but later it became a tool for the actual training of new recruits (Shen et al., 2009). A well-known example of a game designed for changing social perspectives is Darfur Is Dying (2006). This was a Web-based flash game designed to raise awareness of the genocide taking place in Darfur with the aim to elicit empathy and motivate players to help stop the crisis (Peng et al., 2010). Another example is BREAKAWAY (2010) that uses character role modeling in a story-driven scenario-based soccer game to address the complex issue of violence against women and girls. The game employs a branching narrative structure that facilities self-reflection as it highlights the behaviors that lead to gender inequity and bullying (Wang et al., 2019). Games for Change (2022) announced a national challenge, supported by NOAA, on student game development about climate change. There is an increasing number of commercial games addressing the climate crisis, and what they all have in common is that the players play on their own discrete personal devices. They are separated by geography and can only participate in the Metaverse virtually without any social interactions in a shared physical space.

Also in the 1990s, the same decade when the term metaverse was coined, the I-MMERSiON (Immersion Studios Inc.) was launched. Inspired by the cinematic innovations of IMAX, our dream was to expand audience engagement and understanding of the natural world by combining state-of-the-art digital technologies with large-scale and real-time social interaction all in one physical space. The idea was to create an immersive cinematic experience where audiences could participate in a self-directed exploration along with many others simultaneously through various narrative elements and gamification mechanisms. While loosely based on the analog version of "choose your own adventures" books developed in the 1920s, our ambition was to design a solution for large audiences to be emotionally immersed in a story world and to affect the learning outcomes of that narrative experience through mediated and nonmediated social interactions in real time.

The earliest example of this approach was in the year 2000, called the "Immersion Theater." And for 2 years, it was installed in the Museum of Natural History at the Smithsonian Institute. The experience we created and later called "Immersion Reality" utilized a large-format 220° wrap-around immersive screen combined with an interactive touchscreen terminal at every seat, allowing each audience member to explore the subject, gain new insights, and, by either competing or voting, influence the outcome of the story. This combination of game and movie that we called a "govie" at I-MMERSiON and as such seemed logical that this format would be an ideal platform for the entertainment industry.

As one considers the growing enthusiasm for social media and digital games, the obvious big difference an Immersion Theater offered was the physical social engagement of the audience. The Immersion Theater offered compelling immersive cinematics along with shared physical contact and dialog as the players interacted with each other to collectively influence the outcome of the story. The opportunity to connect in the physical setting while exploring the Metaverse was a unique and attractive feature for museums worldwide interested in developing compelling digital experiences. By sharing several exemplary projects in this chapter, we hope to demonstrate how combining interactive storytelling in immersive environments with simulation, games, telepresence, social learning, and govies can provide mass audiences with a deeply meaningful and transformative learning experience about the natural world and help inspire climate action.

Interactive storytelling is a technique that experience designers use to create a narrative world and weave through original content to share the information with users in an engaging manner such as real scientific problems, instantaneous feedback, and aggregating data through emerging technologies (Hoguet, 2014). By default, the individual users will not have the same experience in the same sequence and they will have to make their own choices along the way as they explore new content to solve a problem or overcome a challenge with other users. Immersive environments are computer-generated spaces that allow the users to have a sensory experience of physical immersion with the awareness of one's self and often the presence of others. Three popular types of immersive environments are massively multiplayer online role-playing games (MMORPGs) such as *World of Warcraft*, multiuser simulated virtual worlds such as *Second Life*, and cave automatic virtual environments (CAVEs) with projections on large-size surrounding screens such as *Ars Electronica* (Lui, 2014). The CAVEs may be less known among these three types but they have been around for decades and have become more accessible to the public through recent exhibits such as the *Van Gogh: Immersive Experience* and *Immersive Klimt*. The advantage of CAVEs lies in the immersive experience taking place in a shared physical space that allows the users to engage in meaningful dialogs about the content in real time, which is critical for deep learning.

Simulation: *Storm Over Stellwagen*

The growing popularity of museums around the world meant they were constantly searching for meaningful ways to better communicate contemporary science with the public. As the content they wished to share evolved to a higher degree of complexity, the demand for new ways to entertain and educate the public became critical. This led to the first series of Immersion Cinema experiments in social interaction. We were looking for evidence that interactivity used as a control element within a storyline would provide a social dialog that engaged the audience more deeply in the meaning of the content.

The first large-scale test of this approach was a collaboration between the US National Oceanic and Atmospheric Administration (NOAA) and the New England Aquarium in Boston, who engaged I-MMERSiON to produce a show titled Storm Over Stellwagen (1999). The pressing issues these groups wanted to highlight were the impact of climate change and human behavior in a particularly sensitive marine sanctuary near Boston called Stellwagen Bank. The approach was to create an interactive and immersive experience that combined a fictional storyline with a scientifically sound marine environment, a host of accurately represented marine species, and a unique simulation engine. The storyline set the stage for the audience to determine by interacting on their terminals the combined impact of environmental, climatic, and human intervention on Stellwagen Bank. The audience was able to choose from a host of factors (which were expressed interactively), and the culmination of these factors was fed to the simulation engine for evaluation. The result was a set of parameters and each set matched one of five potential future scenarios. Once the potential future scenario was identified by the software, the cinema system called up a sophisticated pre-rendered and animated sequence of what Stellwagen Bank would look like in that projected future.

After several years of running this show, the New England Aquarium found that audiences were often emotionally involved in the impact they had caused and returned to test out alternative solutions. Students responded particularly well to the association they experienced between cause and effect, recognizing the complexity each future scenario entailed. The embedded games that the audience interacted with were all text-based with simple simulation parameter controls, rather than the real-time 3D graphical games we would later come to employ. One notable finding was the level of dialog among the audience members as they experienced the show. This was often a very noisy experience with people engaging those around them in a discussion to find alternative answers and debate about potential strategies to approach the simulation.

Simulation turned out to provide an effective emotional connection and audience engagement with the story. A key characteristic of the way simulation was used in this show was how the results were shown to the audience. They were presented on two levels: The individual choices were seen both privately (on their personal touch-screen terminal), and the accumulation of everyone's input was seen publicly (on the large immersive screen). These public-private interactions became a common characteristic in many productions that followed, providing a specific result that encouraged enthusiastic dialog both during and after the experience. In addition, we decided to recognize the highest achievers through a scoring system, something we carried forward in all our future productions as an additional means of fostering a high level of participation. Scores were shown throughout the experience to the individual participants with increased interaction and content discovery, and top performers were highlighted to the entire audience at the end of the experience. Tying it back to the concept of the metaverse, in this scenario-based experience, the choices the audience made could actually predict the future in reality, with a one-on-one direct connection from each of the five scenarios to each of the five outcomes.

Immersive Game: *Sharks: Predator/Prey*

An example pointing to the challenge of changing contemporary thinking was an immersive game called Sharks: Predator/Prey (2001). As marine science evolved, it had become evident how important it was to communicate to the general public that the relationship between species, which in earlier times was called the food chain, was now more accurately understood as the food web. How can we show that

Fig. 1 *Sharks: Predator/Prey* Immersion Cinema

changes in the life cycle of one species can influence the links in the web in minor or catastrophic ways? And that the diversity of the food web itself can affect its own ability to withstand any changes in water temperature or the impact of other animal populations? The solution we came up with was *Sharks: Predator/Prey*. This was a fully 3D real-time multiplayer game that could be played by a hundred or more people simultaneously in the Immersion Theater (Fig. 1). The experience began with players starting as the smallest creature in the ocean, a type of phytoplankton, and maturing through the game to try and become a great white shark. Each level of success resulted in players taking the role of more complex creatures in the ecosystem and trying to meet increasingly complex goals for survival. Players had to discover who was a predator and who was prey in order to survive among an ocean of creatures all controlled by the audience members. The game was set in a very competitive environment against the backdrop of a large-format immersion screen showing high-resolution animations of who is eating who – in graphic detail. The players were introduced to the action through an opening narrative and then thrust into the immersive game and allowed to unearth the key relationships more deeply through their gameplay. As one might expect, this particular game was far more popular with audience members who were avid gameplayers and more of a challenge to those who needed time to absorb game mechanics instead of focusing on the game content. *Sharks: Predator/Prey* represented a good example of a multiplayer experience designed for experienced gamers with readily available motivations and skills. Digital games have been powerful tools for entertainment and in more recent years evolved to include serious games purposefully designed for learning, development, and change (Ritterfeld et al., 2009; Spiegel, 2006; Spiegel & Hoinkes, 2009). *Sharks: Predator/Prey* was our first project to push for a real-time game concept in an immersive environment. At that time, most museum visitors could only endure about 30 minutes of cinematic presentations due to the intensity of the visual effects. But what we found from this project was that the game-based experience significantly extended the time people spent in the Immersion Theater with dramatically improved user engagement. In this case of the metaverse, players were learning about the interdependency of species in the food web and the immersive experience helped them to better connect to the real-world conditions and consequences of what happens in the marine environment and their relationship with these events.

Telepresence: *Ring Road*

As we continued to explore new technological solutions to enhance large-scale social interactions in interactive storytelling and immersive environments, we had an opportunity to experiment with telepresence – the use of VR for remote control of machinery to participate in a distance event – in Ring Road (2001). In this project, we collaborated with Dr. Robert Ballard, the scientist best known for discovering the *Titanic*, and connected the underwater habitats of Monterey Bay in California to the visitors to Mystic Aquarium in Connecticut. A remotely operated vehicle (ROV) equipped with a live high-definition camera was tethered in the water in Monterey Bay and connected via Internet2 (an advanced technology for streaming video at that time) to visitors across the United States, sitting in the Mystic Aquarium in a dome structure configured as an Immersion Cinema. In this design, the high-definition video captured by the ROV was projected on the large-format immersive screen in the dome through telepresence using a live stream instead of prescripted and pre-rendered content. At the same time, visitors could use their individual touchscreen terminals to interact with a matching but simulated undersea environment (Fig. 2). All of these events were happening simultaneously and not driven by a preprogrammed narrative. This parallel world experience was dynamic and allowed instructors to take the visitors on a personal investigation of the animal and plant species in the virtual habitat in relation to what was observed in their natural habitat in reality. Instructors also had the option to activate simple trivia games as part of the experience and provide the winning participants with control of the ROV in Monterey Bay (Fig. 2). If actual underwater conditions were poor on a particular day, alternative video recordings of the live environment could be called up on the Immersion Cinema screen as a comparison for the audience. The live environment inputs and controls were seen as very compelling and made the content experience real to the participants in a way that virtual environments and gameplay could not achieve (Fig. 2). Metaverse is about leveraging technologies in the virtual world to enhance the experience and impact in the real world. In this case, we used the interactive and immersive experience to create conditions for the participants to better understand what was happening in real time by observing the changing conditions in Monterey Bay.

Fig. 2 Screenshots from the *Ring Road* Project Video (left: A student in Mystic Aquarium uses the controller to maneuver the ROV on the dry land of Monterey Bay; right: The live stream of videos projected on the immersive screens in the dome while visitors explore the content at their individual touchscreen terminals and Spiegel explains how immersive environments coupled with real-time content feeds and meaningful group dialogs can help visitors understand complex scientific phenomena through self-directed and socially engaging learning experiences)

Telepresence + Social Learning: *Exploration: Sea Lions*

Another collaboration we had with Bob Ballard that combined both live and virtual immersive experiences was Exploration: Sea Lions (2003) as part of the JASON Project, Ballard's initiative to promote scientific exploration among school-aged groups across America (Cohen, 2020). We offered a significant expansion to Ballard's proposition of using only telepresence as an educational tool by creating enhanced features for social learning through Immersion Cinemas and Immersive Classrooms. The show ran at Mote Marine Laboratory and Aquarium in Florida, the Mystic Aquarium and Institute for Exploration in Connecticut, and in The Lamphere Schools in Michigan with over 2000 students (mostly 9–14 years old) in a 2-week period. A real-life scientist Kathleen presented to the students with an urgent problem that the California sea lion pups on San Miguel Island were facing through a live broadcast and allowed them to connect with each other as well as experts to figure out what the major cause was. These pups were dying at a higher rate that year than in previous years. Students were invited to help the scientists prove one of the three competing hypotheses: predators, pollution, or climate (Fig. 3).

Students across all locations had live stream videos of the sea lions on San Miguel Island projected on a large-format Immersion Cinema screen while they had a touchscreen terminal with an interactive interface that allowed them to conduct research on their own, with other students in the same physical space, discuss with students virtually across all locations, and the possibility to ask the expert (Fig. 3). Their discoveries got transferred to their field notes, and when the time was up, students reviewed their notes and send the facts that they believed to support or reject their hypothesis. Participants could then reflect upon their hypotheses after viewing each of the alternatives and supporting presentations to make a final assessment of the cause of the problem. A final presentation on each hypothesis was made

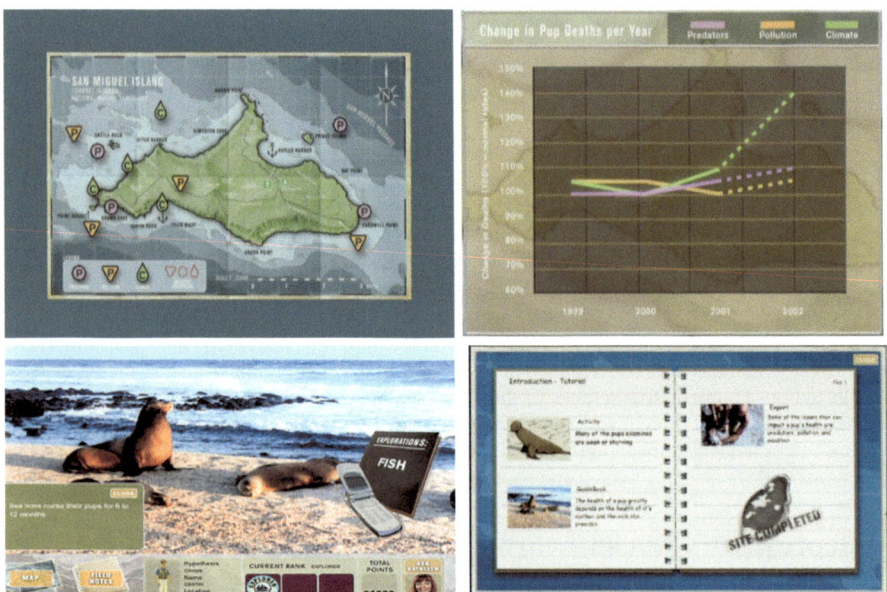

Fig. 3 Screenshots from the *Exploration: Sea Lions* Project Video (top left: map of San Miguel Island showing different locations where sea lion pups were dying for different reasons; top right: an example of data visualization the host used to explain the correct hypothesis: climate; bottom left: touchscreen terminal interface with facts about the sea lion pups on San Miguel Island and other resources for scientific exploration to solve the problem; bottom right: personalized field notes based on each student's interaction with various features in the program to facilitate self-reflection and group discussions)

virtually to the entire audience, and the correct hypothesis was revealed and explained by the scientist host Kathleen (Fig. 3).

In the end, the telepresence experience with enhanced social learning features for such a scientific exploration was found to be a powerful means for more significant concern and active discovery among the students. One of the unique features introduced in this project was the use of avatars to represent the students (note that this was the same year when *Second Life* was released). This introduction of a virtual representation of the students themselves added to the value of play and personal identity in the pursuit of a scientific inquiry. As students mastered their chosen area of inquiry in this program, they could directly offer assistance to other students through their avatars in advancing their collective understanding of the problem.

Exploration: Sea Lions is similar to *Ring Road* in that they both connected live on-site experiences and virtual investigations. However, they had very different results. *Ring Road* tried to weave the live into the virtual directly and found itself challenged by constant context-switching to establish a strong enough presence for the participants, thereby reducing the motivation and engagement level and ultimately the learning outcomes. *Exploration: Sea Lions* put the live and virtual experiences next to one another but left each one to remain in their own formats and was by far the more successful approach. The live broadcast helped to anchor the interactive elements of the experience and still left room for personal exploration and discovery. Students and teachers both rated highly of the *Exploration: Sea Lions* experience, noting that it encouraged dialog, was of strong value to weaker students, and motivated them to learn more about the topic (Ritterfeld et al., 2004).

Govie: *Dolphin Bay*

The most intriguing and complex narrative highlighting the human impact on the marine environment that we created is called *Dolphin* Bay (2005). It represents the state-of-the-art approach to storytelling and digital technologies in blending narrative in movies and interactivity in gameplay, the type of immersive experiences we called govies. *Dolphin Bay* was a compelling action-adventure production that illustrated the human impact upon dolphin habitat that had a seamless integration of gameplay with the narrative throughout. This "govie" was played for many years exclusively at the Immersion Theater of the Mote Marine Laboratory and Aquarium in Sarasota, Florida. This highly evolved storyline interweaves a series of complex games that through personal interaction transform the participant into the role of a marine biologist trying to save dolphins within a fictitious bay. The production utilized movie directors, screenwriters, and Hollywood actors combined in a sophisticated game development. This experience blurred the lines between cinema and games to an extent that a player would not be able to identify either as the clear driving factor. Intense emotion and action fostered engagement with the storytelling and pushed audience dialog on environmental issues to occur during gameplay portions of the experience. The intertwining of narrative and gameplay can serve a key entertainment/educational role in engaging emotional participation as evident in this production due to the highest quality of each aspect.

Real Change Is Possible: *Sparking Reaction*

Sparking Reaction (2001) is an excellent example of how a self-directed, interactive, and immersive experience enabled the public to influence a major corporation to get out of the business of nuclear energy production. As part of a visitor center that we built for the British Nuclear Fuels Ltd. (BNFL) headquarter in Sellafield, Cumbria, *Sparking Reaction* was intentionally designed as an open public forum on energy where visitors were empowered to directly join the debate about nuclear energy vs. alternative solutions through a unique experience. Anchored through simulator engines in an Immersion Cinema, visitors from all walks of life were challenged to build nuclear reactors, generate the fuel, power the grid, and then dispose of the waste (Fig. 4). The outcomes resulted from their individual choices led to a collective discovery that nuclear energy production left behind a major toxic

Fig. 4 A young visitor learning about nuclear energy production at *Sparking Reaction*

waste problem that was extremely difficult to resolve, as we know is a fact today. The visitors were able to explore multiple alternative and renewable energy solutions such as solar and wind and explore the same power demands and challenges for production. We worked closely with BNFL and the content experts from the London Science Museum to ensure that the interactive narrative experience we created was grounded in the scientific evidence to date. So when the visitors came to the Immersion Theater, they would be put at the center of this real-world energy debate and use the simulated experiences to explore and better understand the pros and cons of each option, weighing the risks and benefits of generating nuclear energy, and voice their personal opinions. In fact, BNFL was at a crossroad in defining its business priorities and directions for future development. With most of its business in nuclear energy generation at that time, BNFL was contemplating a new role in fuel reclamation. *Sparking Reaction* served as a tool for public engagement in gauging the role of energy production from the perspective of the nuclear industry. This experience, combined with a series of related interactive exhibitions in the venue, provided BNFL with direct inputs from 500,000 members of the public (both adults and children) as to what corporate roles and responsibilities the company should have regarding the nuclear industry. Based on this and other public feedback, BNFL ultimately got out of the business of nuclear energy generation and moved its primary focus to the reprocessing of spent nuclear fuels.

Conclusion

Raising awareness and engaging the public in meaningful dialogs are critical for social change. Digital media can help transform scientific information into an aesthetic and compelling experience. When combined with interactivity in an immersive environment, the narrative experience can be even more engaging as users explore the content in a self-directed way. Early public immersive experiences can be traced back to the nineteenth century when large-scale 360° circular panoramic paintings allowed the public to appreciate religious tales, historical events, and urban landscapes through realistic representations known as cycloramas in America) and panoramas in Europe (Fletcher's Mutiny Cyclorama, n.d.). These immersive settings allowed for social interactions and meaningful discussions, but the narrative content was not at all interactive. Fast forward to more recent years, the success of cinematic environments such as IMAX prepared the public for the *Van Gogh Immersive Experience* and *Immersive Klimt*, both of which have become international attractions drawing huge audiences. Again, they missed out on the opportunity for interactive storytelling and meaningful engagement for deeper learning. There is a constant demand for innovation in digital media to create new and unique experiences in the metaverse: 360 videos, AR, VR, MR/XR, and most recently AI-generated experiences, with each technology appearing more seductive than the one before. However, most of these commercially popularized and personalized digital technologies can be physically, socially, and emotionally isolating, separating people from the reality – especially the meaningful dialogs that can take place in a shared space – leaving individual users feeling disembodied, dislocated, and disoriented and not knowing where they are in the real world.

Climate change is urgent and complex. Urgent stories need to be told yet complex stories are hard to tell. What we have learned in the past 20+ years is that technological solutions exist and will continue to improve for interactive storytelling in immersive environments, with powerful capabilities such as simulation, gamification, telepresence, and cinematic presentation shown in the examples in this chapter. Equally important is the seamless integration of science-based content, dramatic narratives, interactive mechanisms, and meaningful dialog in a shared physical, social, and emotional experience through self-directed exploration and collective discovery. When people can get up close and personal with the challenges and then are put at the front and center to contemplate the moral dilemmas and seemingly intractable problems, people would be more compelled to step up and respond in action, with faith and reverence. These examples from around 20 years ago are from a different era, as we were experimenting with various cutting-edge technologies in our creative solutions while working with different museums, aquariums, and research institutions. But 20 years later, we see some of these creative solutions becoming accessible and popular public attractions. And we firmly believe that they provide exciting opportunities for climate storytelling and empowerment.

For example, Arcadia Earth (n.d.) uses video projects, AR, VR, and large-scale installations by more than a dozen environmental artists to provide a multisensory immersive experience to awaken public conscience about the ecological crisis and inspire climate action. "Entertainment, coupled with enlightenment, is the purpose of this pop-up exhibition," and "Together, they evoke the landscapes, marine depth and life-forms that global warming threatens" (Graeber, 2019). ARTECHOUSE (n.d.) is another public venue for metaverse experiences with nature, from traversing through the cherry blossom in Washington D.C. in PIXELBLOOM to an imagined future in 100 years RENEWAL 2121. In the meantime, researchers have been working with indigenous communities to use technologies like InstaOneX 360 cameras to create immersive videos and share their stories about culture and the environment they live in (Westervelt et al., 2022) and the scientific community has been collaborating with technology experts to develop online tools such as the En-ROADS simulator (n.d.) from Climate Interactive and the MIT Sloan Sustainability Initiative.

The new wave of AI-generated media has arrived with many cautionary tales and unprecedented ethical concerns (Mets, 2023). As with any new technological innovation, it will serve as a double-edged sword and will require dedicated support, effective guidelines, and thoughtful collaborations to serve the public good. Despite the steep challenges we face, one can only envision what possibilities there are when the tech innovators, digital artists, climate scientists, and professional storytellers all come together to help engage the citizens in the world and the policymakers to better understand what humanity is facing and what can be done right here and right now.

References

American's army [Digital game]. (2002). U.S. Army. https://www.americasarmy.com

Arcadia Earth. (n.d.). *About us*. https://www.arcadia-earth.com/about

ARTECHOUSE. (n.d.). *What's on*. https://www.artechouse.com/current/

Dolphin Bay. (2005). MOTE Marine Laboratory.

BREAKAWAY [Digital game]. (2010). https://breakawaygame.champlain.edu

Cohen, C. (Writer, Producer). (2020). *Bob Ballard: An explorer's life* [film]. National Geographic.

Darfur is dying [Digital game]. (2006). https://www.darfurisdying.com

En-ROADS simulator. (n.d.). https://en-roads.climateinteractive.org/scenario.html?v=22.3.0

Exploration: Sea Lions. (2003). Mystic Aquarium/The JASON Foundation.

Fletcher's Mutiny Cyclorama. (n.d.). *History of cyclorama*. https://www.norfolkcyclorama.com/history-of-cycloramas/

Games for Change. (2022). *Climate change student challenge*. https://gamesforchange.org/studentchallenge/nyc/climate-change/

Graeber, L. (2019, October 23). At 'Arcadia Earth,' dazzle illuminates danger. *The New York Times*. https://www.nytimes.com/2019/10/23/arts/design/arcadia-exhibition-climate-change.html

Hoguet, B. (2014). *What is interactive storytelling?* https://benhoguet.medium.com/what-is-interactive-storytelling-46bfdd2a8780

Lui, M. (2014). Immersive environments. In R. Gunstone (Ed.), *Encyclopedia of science education*. Springer. https://doi.org/10.1007/978-94-007-6165-0_39-1

Mets, C. (2023, May 1). 'The Godfather of A.I.' leaves Google and warns the danger ahead. *The New York Times*. https://www.nytimes.com/2023/05/01/technology/ai-google-chatbot-engineer-quits-hinton.html

Peng, W., Lee, M., & Heeter, C. (2010). The effects of a serious game on role-taking and willingness to help. *Journal of Communication, 60*, 723–742. https://doi.org/10.1111/j.1460-2466.2010.01511

Ring Road. (2001). Mystic Aquarium/The JASON Foundation.

Ritterfeld, U., Weber, R., Fernandes, S., & Vorderer, P. (2004). Think science! Entertainment education in interactive theaters. *Computers in Entertainment, 2*(1), 11. https://doi.org/10.1145/973801.973819

Ritterfeld, U., Cody, M. J., & Vorderer, P. (2009). *Serious games: Mechanisms and effects*. Routledge.

Sharks: Predator/Prey. (2001). Immersion Studios, Inc.

Shen, C., Wang, H., & Ritterfeld, U. (2009). Serious games and seriously fun games: Can they be one and the same? In U. Ritterfeld, M. J. Cody, & P. Vorderer (Eds.), *Serious games: Mechanisms and effects* (pp. 48–62). Routledge.

Sparking Reaction. (2001). London Science Museum/British Nuclear Fuels Ltd.

Spiegel, S. (2006). Foreword. In P. Vorderer & J. Bryant (Eds.), *Playing video games: Motives, responses, and consequences* (pp. ix–x). Routledge.

Spiegel, S., & Hoinkes, R. (2009). Immersive serious games for large scale multiplayer dialogue and cocreation. In U. Ritterfeld, M. J. Cody, & P. Vorderer (Eds.), *Serious games: Mechanisms and effects* (pp. 469–485). Routledge.

Storm Over Stellwagen. (1999). NOAA.

Wang, H., Wu, Y., Choi, J. H., & DeMarle, A. (2019). Players as transitional characters: How youth can "breakaway" from gender-based violence. *Well Played, 8*(1), 27–40.

Westervelt, M., Castillo, J., & Pant, S. (2022, December). *Indigenous visions: Immersive 360 video storytelling for social change*. SBCC Summit. For more information about this project: https://initialeyes.org

Stacey Spiegel is an artist, entrepreneur, and creative director. He was a fellow at MIT's Center for Advanced Visual Studies and the winner of the New Media Visionary Award in Canada. As the Co-Founder and CEO of I-MMERSiON (Immersion Studios Inc., 1997–2006) and PWLabs (Parallel World Labs Inc. since 2007), he has been leading world talents to develop innovative media experiences to promote culture, education, and health by connecting technology with humanity and nature in compelling and transformative ways. His teams have designed and developed over 50 projects around the world for prestigious clientele such as the Smithsonian, the Science Museum London, the Government of Norway, the Government of Canada, the International Olympic Committee, La Citè des Sciences et de l'Industrie Paris, Kunsthalle Mannheim, among others. At the core of Spiegel's creative visions is the combination of immersive environments and interactive storytelling for large-scale social interaction and public engagement. By leveraging emerging technologies, such mediated experiences can enable profound transformation through self-directed exploration and socially engaging dialogs for deep learning among diverse audiences.

Hua Wang is a communication scientist who is passionate about using innovative strategies to design, implement, and assess entertainment programs that leverage powerful storytelling, emerging technologies, and communication networks for health promotion, behavior change, and social justice. All of her areas of interest are labeled with an oxymoron: entertainment education, serious games, positive deviance, and liberating structures. After earning her Ph.D. from the University of Southern California's Annenberg School for Communication and Journalism, she joined the faculty and is now a Professor of Communication at the University at Buffalo, The State University of New York. She has contributed to interdisciplinary projects funded by the U.S. Department of Health and Human Services, the Canadian Social Sciences and Humanities Research Council, the National Institutes of Health, the National Science Foundation, the Robert Wood Johnson Foundation, and the Annenberg Foundation among others. Her research has appeared in high-impact journals such as the *American Journal of Public Health*, *Journal of Medical Internet Research*, *Communication Research*, and *Computers in Human Behavior* and received prestigious awards from the American Public Health Association and the International Communication Association.

Bird's Eye View: Engaging Youth in Storying a Survivability Future through Performance and Interspecies Friendship

Beth Osnes, Chelsea Hackett, Molly T. McDermott, and Rebecca Jo Safran

Under a bridge near Discovery Drive, tucked behind the Environmental Studies building at the University of Colorado (CU), there is a population of Barn Swallows (*Hirundo rustica*) who swoop from their nests catching insects in their bills to feed their chicks. The creek beneath the bridge runs as slowly as time on a hot summer day. Bikes zip past on the path that winds beside the creek. Wild grass along the water's edge sways slowly in the breeze. A Red-Winged Blackbird (*Agelaius phoeniceus*) darts towards a Great Blue Heron (*Ardea herodias*) who stands in the water, likely too close to the Red-Winged Blackbird's nest. In this scene during the summer of 2021, eleven human high school students (*Homo sapiens*) – all either female-identifying or gender nonbinary – leisurely observed the birds as they flew and noted their behavior in relation to each other and their environment. They filled their journals with scientific fieldnotes, modeled for them by the evolutionary biologists. Besides those notes, they wrote poems and sketched the birds. They danced in the field adjacent to the creek, led by the applied theater artists. They meditated together on what they might learn from this population of Barn Swallows and this merging of often disparate ways of knowing: art and science. Day after day, these young people returned to see the birds, and over time, they began to slow down and immerse themselves in the more-than-human world through a direct relationship with one group of the Barn Swallows. They began to see these birds as friends.

Bird's Eye View is an art-science, informal learning project for youth engagement in storying our way beyond the climate crisis. The project is situated within the larger *Side by Side* initiative at CU and was designed to lead a group of Boulder youth in observing and embodying birds through puppetry and costumes within multiple outdoor settings. After a few weeks spent with the Barn Swallows under the bridge, we – the artists, scientists, and educators who led this immersive

B. Osnes (✉) · C. Hackett · M. T. McDermott · R. J. Safran
University of Colorado, Boulder, CO, USA
e-mail: beth.osnes@colorado.edu; Chelsea.hackett@gmail.com;
Molly.Mcdermott@colorado.edu; rebecca.safran@colorado.edu

© The Author(s) 2024   379
E. Coren, H. Wang (eds.), *Storytelling to Accelerate Climate Solutions*,
https://doi.org/10.1007/978-3-031-54790-4_18

program – expanded the scope of the program beyond the bridge. We held the last four weeks of our 6-week program at various locations in Boulder's Open Space Mountain Parks (OSMP), the protected greenspace that makes Boulder a popular destination for diverse species (including humans). As the program leaders, we invited the young people to branch out beyond Barn Swallows to seek out a connection with a local bird of their choosing. We challenged them to take art-science field notes on their observations and research on this species, which came in the form of sketches, poetry, and traditional observational and natural history field notes. Drawings from these field notes informed the young people as they constructed and painted articulated wings made of recycled cardboard of the local birds they chose – crows, magpies, hawks, and beyond. These human-sized wings allowed the students to embody the movements of the bird species they had chosen and further deepened their understanding and partnership with the more-than-human world. The youth then constructed headpieces inspired by what they had learned about their birds from various sustainably sourced materials, such as thrift store hats, papier mâché with newspaper, and discarded foam.

The written field notes and scientific research the youth conducted on these bird species became raw materials for written scripts, which highlighted the interspecies connection the youth co-formed with the local bird they chose. Embodying these birds through performance in Boulder's OSMP marked our assertion of unity with these birds and the natural world. Two professional filmmakers created videos of *Art Hikes* that include each youth moving in the bird costumes within the bird's environment, multiple types of large-scale bird puppets, and the spoken scripts by the youth. The videos layer each of these elements into an artistic whole – five-minute videos for each of the three OSMP locations – that we posted on the OSMP website and on placards with QR codes posted at the three trailheads where each video was shot. An artistically edited video compiling all of these videos from each location has been shown by several film festivals, including the Colorado Environmental Film Festival.

Throughout this process, we engaged as a group in conversation and action for equity and environmental concerns to invite and support young people in coauthoring an equitable, survivable, and thrive-able future for all life and the ecosystems upon which life depends. We offer these brief definitions for each of these goals: *Equitable* means each person has the opportunity to fulfill their needs, without barriers based on identity or ability. *Survivable* means all beings have access to what they need. *Thrive-able* means all beings, humans and beyond, have an environment conducive to the pursuit of happiness and meaning. The ritual of these meetings provided space for personal exploration, open and safe discussion about belonging in the natural world, and a place to talk and dream.

Introduction

The *Side by Side* initiative described above aims to story a new way out of our climate crisis by utilizing an "art-science" approach to foster a feeling of belonging as a part of the natural world for youth and adult leaders. For too long, we have been telling a story of difference that keeps us disconnected and reinforces false hierarchies: humans above other species, adults above youth, and certain types of knowledge above others. We strive with this project to disrupt these hierarchies to cocreate an equitable, survivable, and thrive-able future for all forms of life on this planet and the ecosystems upon which life depends. If we need a new story, we likely need new storytellers. This project attempts to make deep shifts in perception toward an honorable relationship between humanity and the natural world through a story of interspecies friendship that centers the voices of youth and values multiple pathways of knowing: the head, the heart, and the soul.

In this chapter, we share moments and highlights of our time spent alongside youth and birds in *Side by Side* during the summer of 2021. We connect this experience to the broader questions we face in informal learning through art-science as a field:

- How can we create the rapid and lasting shifts needed to ensure our collective futures through creative climate engagement?
- How can we disrupt current ways of being and hierarchical models through interdisciplinary approaches?
- How can we story a new future?

We offer our core findings as guideposts for our future work and for the collective efforts of all who work in informal learning spaces. Our findings center on the importance of fostering a sense of belonging, using narratives to create interspecies friendship and solidarity, and supporting youth to participate in climate action through positive frameworks, while providing opportunities to process climate-related grief.

The Nest

Side by Side is an ongoing art-science initiative created and facilitated by two coauthors of this chapter: Dr. Beth Osnes (Professor of Theatre and Environmental Studies) and Dr. Rebecca (Becca) Safran (Professor of Evolutionary Biology), cofounders of the CU Inside the Greenhouse, an interdisciplinary project for creative climate communication (https://insidethegreenhouse.org). We (Beth and Becca) are drawn to the power of what feminist scientist Donna Haraway (2016, p. 4) calls "unexpected collaborations and combinations" in which we "become-with each other or not at all." As interdisciplinary artists and researchers, we have witnessed the power of dismantling academic hierarchies and blending both qualitative and quantitative approaches to ensure that learning experiences and climate communication speak to all parts of our collective humanity. Furthering this work, we created *Side by Side* as an ongoing pursuit to dismantle other forms of hierarchy that have limited our ability to work our way out of the climate crisis. Two core focuses of this work are patriarchal structures that reinforce notions of detachment and neutrality as necessary elements of knowledge creation and the humanist notion of separation from nature. *Side by Side* benefits from Becca's decades-long career researching Barn Swallows and Beth's equally long career utilizing theater and performance-based methods for community building, education, and climate communication.

Through *Side by Side*, Beth and Becca activate the theories of primarily feminist ecologists and Indigenous scholars as they seek a re-evolution of how we conceptualize our relationship with each other and the Earth beyond harmful colonizing and commodifying ways. In *Decolonizing Relationships with Nature*, philosopher and ecofeminist Val Plumwood (2003) focuses on characteristics of both colonial and anthropocentric approaches to the *Other*, namely a strong focus on dualism, exaggerating differences and denying commonality. One aspect is *hyper-separation* or *radical exclusion*, marking "out the Other for separate and inferior treatment" (Plumwood, 2003, p. 51). A natural extended group of partners in this work are the Barn Swallows and local birds of Boulder, Colorado. We have centered this project around humanity's relationship with Barn Swallows because, with this species, many humans have a shared story of expansion and home. Barn Swallows have followed humans in their migration across the planet and currently build their nests almost exclusively in human-made structures (Safran & Levin, 2019). Settler communities – some with their colonizing legacies of domination that require healing and reconciliation – can benefit from guidance toward mindful action from this interspecies friendship in coauthoring a survivable future for all. In this new story, we join together in the spirit of *mutualism*, a biological term for a mutually beneficial relationship among members of different species (Biologydictionary.net, 2017). As we seek to find an equitable way for life to survive on this planet, we look to this relationship for inspiration.

Extending interspecies friendship to birds is a gesture toward dismantling species hierarchies to establish a "side-by-side" relationship. Towards this aim, through our collaboration with birds, we offer our protective care to birds whose lives have been lost and whose lives we are losing (Rosenberg et al., 2019) because of the impacts of climate change (Langham et al., 2015) and who have historically been put forth to ensure human survivability in the pursuit of fossil fuels, including the caged canaries taken into the coal mines where their death would warn miners of the presence of lethal gases. This painful fact highlights that birds have long been considered an excellent indicator species (Rosenberg et al., 2019), such that studying the presence and behavior of birds can inform us about changing ecosystems, which directly affects our shared, interspecies ability to continue "becoming."

Another core goal of *Side by Side* is to ensure that all voices and ideas are given platforms to be heard so that we might benefit from all, especially those who are historically and still routinely marginalized and silenced. Becca and Beth have both spent much of their careers embracing and promoting diversity. A goal of *Side by*

Side is to increase access to STEM learning for young women, nonbinary students, and members from minoritized groups. Beth has worked directly on supporting young women's voices through SPEAK (SPEAK, 2022), a nonprofit she co-founded with Dr. Chelsea Hackett, who holds a PhD in Educational Theatre from New York University and serves as the Executive Director of SPEAK. Chelsea was brought in as a core educational and voice consultant for *Side by Side* during the summer of 2021.

Notably, Beth and Chelsea are applied performance artists and scholars. Several distinct features characterize applied performance: "Performers" are mostly members of the community rather than professional artists; creators utilize nontraditional performance spaces; positive social change generated through the performance experience is a core objective of the work; and the artistic process is valued as highly as any final performance (Prendergast & Saxton, 2009). Essential to this work is a focus on equity and social justice. These values are mirrored in Becca's anti-racist, gender-aware pedagogical and scholarly work, which is expressed through an article she coauthored with the students in her laboratory entitled *Belonging in STEM: An Interactive, Iterative Approach to Create and Maintain a Diverse Learning Community*. This project marks a coming together of these approaches for a unique art-science experiment that broadens belonging beyond just STEM and the arts. This project seeks to inspire a feeling of great wide belonging as part of the natural world and as part of a community taking on the response-ability to preserve and protect it for those who participate in its making and those witnessing creative outputs.

Pilot

Side by Side is a long-term initiative that has, and will continue to have, many iterations. The focus of this chapter is on the pilot of the *Side by Side* initiative during the summer of 2021 – *Bird's Eye View* – and the many people and birds who made our shared learnings possible, as well as the initial work done on this project in the summer of 2020.

Summer of 2020

Our first launching of *Side by Side* occurred in the summer of 2020 with seven female-identifying youth who had been active with SPEAK's Young Women's Voices for Climate group (Osnes et al., 2022). During this time, we began with observations at the Discovery Drive research site and made scientific notations and drawings of barn swallows. We also explored other species as a group, such as Turkey Vultures (*Cathartes aura*), together and created puppets that we performed in various outdoor locations. There is a short video made of this work that was shown at two film festivals, the Eugene Environmental Film Festival and the Hague

Global Cinema Festival (Inside the Greenhouse, 2020). During the summer of 2020, Beth and Becca worked with Molly T. McDermott, a PhD candidate in Becca's group, to create a pilot curriculum that integrated a prescribed set of art-science observations and explorations, which was utilized during the 2021 summer.

Summer of 2021

For work in her Barn Swallow research lab, Becca was awarded large-scale funding from the National Science Foundation. Through this funding, we were able to secure supplemental funding for the summer of 2021 – the Research Assistantships for High School Students (RAHSS) – to foster interest in the pursuit of studies in the Biological Sciences, particularly those who are women and members of minoritized groups. We offered these RAHSS-paid positions to the seven young women who had participated in 2020 first. We then informally recruited four more students who were either female-identifying or gender nonbinary students based on interest in the project. Our final group of 11 youths consisted of Eliza Anderson, Lola D'Onofrio, Olympia Kristl, Ting Lester, Uli Miller, Lerato Osnes, Sofia Wendell, Gwen Harker, Finella Guy, Leela Stoede, and Noah Barbosa. Numerous CU undergraduate and graduate students from both Ecology and Evolutionary Biology and Theatre and Performance Studies joined the project as interns and served as near-peer mentors to the high school-aged participants. Their names and contributions are listed in Appendix 1.

Community Partner, Boulder City Open Space Mountain Parks

Our primary community partner in this ongoing work is Juanita Echeverri, the director of Outreach and Engagement with Boulder City's OSMP. She did much more than help choose and reserve the space for our sessions and filming; she helped us conceive of our work within and as a part of the Open Space Mountain Parks *Artist Hike* series. Since the mission of Open Space is to preserve and protect these natural spaces for all species, this was a perfect match. She was also helpful in ensuring equitable access and relevance of our project for all members of our community by assisting with Spanish subtitles for our films to make them accessible for Spanish-speaking community members.

At the start of our 2021 pilot at the Discovery Drive bridge, the youth participants were led to a nearby tree where 11 netting-wrapped gifts were hung with ribbon, left mysteriously by the birds, we told them. Inside each was a note, "Let's dip our paint brushes into a test tube to coauthor a future that's looking up" along with an actual test tube and a paintbrush. Each youth was handed their blank field journal and was asked to fill their test tube with water from the creek to watercolor a personalized cover for their journal. Our journey began with this literal coming together of tools for an art-science approach, and this playful invitation to see ourselves as participants in a narrative-approach to climate action. This activity was designed to intentionally invite participants into the act of storytelling through immersion in their setting and grounding in a specific art-science approach. We wanted the youth to author a future that's looking up – by fostering a sense of belonging and friendship in the more-than-human world; we hoped to counter the doom and gloom narratives about climate change with a creative approach that invites hope and thus action. From May 29 to July 1, 2021, we met three to four days a week from 9:00 AM to 1:00 PM. Every session began with a physical and vocal warm-up including running about a quarter of a mile together, in equal parts running sideways facing one direction, sideways facing the other direction, backward, and forward. We did this to remind ourselves daily that we are capable of a wide range of ways for running and moving about our environment. Participants were asked to pay attention to what they noticed when they diversified their perspective and how they traveled. Through this activity, we sought to embody the notion put forth by Queer Ecology that all species, including *Homo sapiens*, are capable of a much wider range of behavior than usually credited by the status quo (Bagemihl, 1999). Part of the widening of our own behavior is to unlearn the humanist delusion of being separate and above the rest of the planet and instead be "in service of world-making," recognizing "nonhuman animals as agential others swirling in, and of, this shared earthly substrate" (Lloro-Bidart & Banschbach, 2019). This concept was woven into our daily rituals.

After moving through space in a new way, we explored meditation. Lying on the grass, Beth asked all to close their eyes and breathe with the Earth, exhaling what the plants need and inhaling what we need, grounding ourselves in our breath and its rhythm of reciprocity between ourselves and the natural world. Then, facing the grass on our hands and knees, we were invited to travel back beyond 350 million years to our human ancestors before they lost their boney tail for better upright movement. We were asked to imagine the remnants of an embryonic bony tail buried in our lower backs – the coccyx or tailbone. We were asked to move that tail while tucking our head and tail down between our legs, then upward with arched backs, then both head and tail to the left, and both to the right. Then, we were asked to achieve a synchronized circling of our heads and tails up, left, under, right, and then in the opposite direction. Next, all were asked to move around quadrupedally on our hands and feet around the space. Slowly rising to stand on two feet again, we were invited to imagine having tail feathers attached to our tails and to notice how we, like birds, are bipedal when moving about on the land. Imagining our arms as wings, we were asked to move about the space as a bird accomplishing the following tasks: Eat a worm from the grass, drink from a puddle, intimidate a predator, and attract a mate. This physical and vocal exploration not only produced laughter but also prepared us for artistic expression alongside the science we were exploring.

For the first six sessions, we focused on Barn Swallows using an art-science field guide developed by Molly with Becca and Beth. These lessons focused on honing observation skills through all of our senses, then applying those skills to observing patterns of movement, behaviors, and interactions. Each scientific concept addressed in the curriculum had a corresponding performance-based exploration. What follows is content from a session entitled *Coming to Our Senses*, in which we learned that animals perceive the world in distinct ways through specialized sensory organs.

> Often, but not always, animals have organs that sense light (sight), vibration/sound (hearing), chemicals (taste and smell), and texture (touch). There is a lot of variation in how sensitive each organism is to each type of sensory input. This allows them to see things we can't – For example, many bird species have plumage which appears black to us, but actually reflects UV light, making them appear colorful to other birds (Tedore & Nilsson, 2019). Humans are primarily a visual animal: We rely on our eyes to tell us about the world. What are we good at noticing, and what might we miss as a result?

To artistically explore this material, we asked the group to spread out around the site, with most of them on the bridge looking down to minimize stress for the birds. Once settled, we asked them to close their eyes for 5 min and observe the site with only their ears. Once done, we asked them to draw and/or write what they observed in their journals. Seated in a circle, we announced we would create a soundscape of what we heard. One at a time, going in alphabetical order by first names, each person added one sound they heard. Sounds included the approaching and the retreating sound of a truck crossing the bridge, the caw of a bird, the gurgling of the stream, and beyond. Once everyone had joined in, we allowed it to continue so we could all experience this recreation of the sounds of our site.

Once we stopped, we reflected together on what it was like to rely on sound for observation and what was noticed by creating a soundscape together. Many were surprised by how much the sounds could vividly represent the site. These explorations in our local context were an enactment of a form of activism known as *grounded normativity* – practices that are "inherently informed by an intimate relationship to place" expressed through "other people and nonhuman life forms in a profoundly nonauthoritarian, nondominating, nonexploitative manner" (Coulthard & Simpson, 2016, p. 254). Our extended time at the Discovery Drive bridge ecosystem and the various locations in OSMP fostered a familiar relationship with these places. At first glance, our Discovery Drive site seemed a most ordinary concrete bridge over which cars drove and under which bikes passed. Over extended and numerous visits, the complex ecosystem beneath was revealed to us as we became more immersed and familiar with our surroundings. This was deepened by activities such as the soundscape, which helped us to isolate aspects of the space and see (and hear it) through new refined senses.

Although there was a basic plan for the 6 weeks, extensive theoretical research, and practical preparations, there was also a feeling that we were building the bridge before us during each session with each step across. This was experienced generally as exhilarating and appropriate rather than as stressful. Since we were doing this as informal learning and were not bound to any specific student learning outcomes, we tried to stay nimble and responsive to what emerged from the overall experience and from any one of the participants. Beth created the design for each session – drawing from the guidebook created by Molly for early sessions – and oversaw the arc of the overall experience with regard to an enriching process along the way for the participants, opportunities for the CU students to facilitate activities, and for completion of needed benchmarks to achieve what we were setting out to collaboratively create by the end of our time together. After Beth had created the activity focusing on the tail, she asked Avani Fachon, CU Ecology undergraduate student intern, if she would be willing to create a similar exploration on the similarities between human arms and bird wings. That next week, Avani led a magical visualization during which we traced the evolutionary split between arms and wings as she helped us feel the anatomical differences and vast similarities between the two (see Appendix 2). Ben Stasny, CU Theatre graduate student intern, created a meditation for the group that began with each of us as a seed buried beneath the cold soil and then slowly led us through sprouting and growing fully toward the summer sun.

The sessions were designed and prepared, but we left room for being responsive to the environment and living organisms – human and beyond human – that arrived. During one session at Discover Drive, the sky above the site came to life as we witnessed a team of small black common grackles (*Quiscalus quiscula*) pester a much larger Red-Tailed Hawk (*Buteo jamaicensis*) who had dared to fly too close to their nests. Later on in the program, we grew familiar with and affectionate for each of our three OSMP locations, especially the one on Flagstaff Mountain, and the life that called it home. One day while discussing the construction of the wings, we halted our conversation, remained still and quiet, and watched with awe and delight as a fox approached.

Since we never knew exactly when the science students from Becca's lab would finish their morning fieldwork, we never knew when or if they would be able to join in our sessions, though they all often expressed their desire to be involved and present. Our sessions naturally evolved as being permeable with a welcoming quality that easily absorbed those who could join in and made time for them to share what they were doing in their field research. When there, they would be expected to jump right in and participate in the performance-based and the science-based activities. This was new ground for many of them, and the team, including Becca, reported feeling challenged because scientists are trained to be objective and as separate as possible from the subjects of their study. It became extremely refreshing and wonderful to witness the science team participate with a positive attitude, an open mind, and curiosity. The CU undergraduate intern and member of the Colorado Ballet Academy, Grant Gonzalez, was with us for most of our sessions and often had an update on what he had witnessed in his fieldwork in regard to the birds, such as announcing the hatching of chicks. Heather Kelley, CU Theatre graduate student intern, led us in an interactive session on augury, the ancient art of divining guidance from birds' movement or their sounds. This was based on research she had done on the use of augury in Ancient Greek drama and the horrid fates that befell those who did not listen to the birds – a useful and relevant lesson for our contemporary society and this project.

A high point for many of us was the times we sat on the grass after experiencing one of the experiential lessons on Barn Swallows, and Becca was there to respond in real time to questions from participants. This was a conversation truly led by the students' curiosity, which seemed to vividly hold their attention, evident by their sustained visual focus on whomever was speaking, their participation in the conversation, and their lack of side talk with each other. It was just as interesting when Becca *didn't* know the answer to one of their questions as when she did, since her explanation of why that was not known or had not yet been researched was just as interesting and revealing about the nature and limitations of science. In the field notes written about June 8 by one of the high school students, it was shared that:

> When Becca arrived nearly everyone seemed to direct their attention fully to her stories of bird research and the findings that have come of it. Many in the group raised great questions, and even those who didn't were audibly gasping and otherwise reacting to Becca's descriptions in ways that seemed genuinely invested in what she was saying.

After these times, many of us reflected on a sort of magic circle that seemed to form around us, as if we were of one focus with a shared curiosity that led us on an improvised path of inquiry.

Indigenous biologist Robin Wall Kimmerer (2013) describes humanity's broken relationship with the living world as a state of isolation and disconnection referred to as *species loneliness* – a deep, unnamed sadness stemming from estrangement from the rest of Creation, from the loss of relationship. Actively involving adolescents while they are still relatively young is important regarding climate-related issues since research reveals that pessimism about addressing climate change increases with age, particularly from early to late adolescence (Ojala, 2012). With increased knowledge of climate challenges and biodiversity loss, youth mental health and well-being can be impacted. Researchers, such as CU Professor Emerita of Environmental Design, Louise Chawla, offer some key ways adults can help youth work through these feelings and maintain hope to actively address environmental problems constructively, such as sharing and listening to each other, spending time outdoors connecting with nature, building a community of shared concerns, valuing young people as partners in addressing environmental challenges, and actively authoring and enacting solutions (Chawla, 2020). These moments with Becca showed the impact that Chawla's suggestions can have. The youth were engaged not only in the facts that Becca shared, but the emotions beneath her stories. Moments spent together in a circle in nature asking extemporaneous questions reframed the youth as researchers doing fieldwork. We were all together as partners, and we all wanted to find answers.

Once we finished with the Barn Swallow curriculum, we began to meet at various OSMP locations and our focus shifted to supporting each student in choosing a local bird to observe, explore, and embody with whom they felt a personal affinity. Choosing a bird was easy for some and difficult for others. Becca offered to create an online survey of personal traits to match people with a local bird with similar traits, and several of the participants filled out the survey and chose the bird she recommended. This survey was designed so that Becca could help each participant feel connected to their chosen bird either through their love of a shared habitat, diet, or color. Early on we met at Pella Crossing in Boulder County Open Space, a location that features several irrigation ponds separated by dams. Due to the abundance of water and islands in the ponds, it hosts a vast abundance of bird species, including

Great Blue Herons (*Ardea herodias*), American White Pelicans (*Pelecanus erythro-rhynchos*), American Coots (*Fulica americana*), Red-Tailed Hawks (*Buteo jamaicensis*), Ospreys (*Pandion haliaetus*), and beyond. After we had settled in and done our warmups, we asked everyone to hydrate, leave their cell phones and field journals behind, and find their own location anywhere in the expansive park. We challenged them to find a place with no one else in their field of vision, get comfortable, and do nothing but dissolve into belonging within the natural world. We agreed upon a vocal call that would call them back to our meeting place at the end of 45 min. Afterward, Noah said that it wasn't until after sitting for 20 min in one space that a sudden rustling in the reeds revealed an American White Pelican (*Pelecanus erythrorhynchos*) only a few feet away that had likely been there the entire time they had been there. Others noted losing track of time in the way they usually experienced it. This queering of time, away from productive use with expected and measurable outcomes, was intended to release them into a different sort of observation free from the necessity for a resulting outcome (Halberstam, 2005). This location and the activity combined were designed to offer a possible affinity to arise unbidden between them and any of the local birds present that day. In her field notes on this session, Gwen Harker wrote:

> As a group, we immersed ourselves in the prompt of 'Do nothing but dissolve into belonging'. As I walked by my peers, I noticed their body language was focused on the watery ecosystem and their attention to the natural world. From this, I concluded that everyone was very immersed in belonging to the environment and connecting with our friends the birds. A highlight of this experience for me was the connection to one bird. In simply existing and coming to terms with belonging and existing, one bird species kept coming back to my locations, and I felt that I was called to learn more about the species. The exploration of the water ecosystem helped me to find my bird to study (Harker, 2021).

Gwen had been chosen by and had chosen the Violet-Green Swallow (*Tachycineta thalassina*).

Each student was encouraged to use that same art-science approach that we had used to explore the Barn Swallows while exploring a local bird species of their choosing. We did the following work together described below to create three videos that serve as *Art Hikes* for three different OSMP locations. These videos are designed to share the gift of our exploration this summer into connection, a feeling of wholeness, and a sense of belonging as a part of the natural world. We considered the following questions to guide the creation of our Art Hike videos – What happens if connection, wholeness, and belonging occur deeply in our human species? What choices would be widely made? What societal structures would change to accommodate that belief? What actions would be inspired? How might this contribute towards a world that is more equitable, survivable, and thrive-able for all life and the ecosystems upon which all life depends?

We requested that each high school participant create the following on their own outside of our sessions together: three pages of art-science field notes and a headpiece for the bird costume made of repurposed materials. During our sessions together, each student relied on their research to inform creating a costume of articulated wings of the bird, by first cutting out 16 scrap corrugated cardboard feathers for each wing from a template, then painting these in the colors of the bird, attaching the feathers to the reinforced underside of a long-sleeved shirt purchased from a thrift store, and carefully securing the feathers to each other with string on each side of each wing. The shirts were also painted to convey the bird. This process took three sessions – a total of 12 hours – to complete and required careful observation of the bird's coloring and patterns. Subsequent sessions were spent with Grant giving coaching to each student while wearing the wings on movement specific to each species. During that time, others worked on writing their script which was to integrate: (1) some scientific aspect of the bird chosen by each student, (2) each person's point of connection, and (3) anything else that rises unbidden from each person's heart. These individual pieces were to be about 40 seconds when read aloud or roughly 70–100 words each. The youth described the birds with language along a continuum of art-science: Some used more poetic language, some more straightforward and scientific. We recorded each person speaking their script to be used in the videos as a voice-over to images of them moving costumed within Open Space embodying the bird of their choosing.

As we created costumes and puppets based on various local birds and sought to embody their movements, we were guided by the notion put forth by Jack Halberstam (2011) in *The Queer Art of Failure* against "gross and crude forms of anthropocentrism (p. 33)," where the human "projects all of his or her [or their] uninspired and unexamined conceptions about life and living onto animals, who may actually foster far more creative or at least more surprising modes of living and sharing spaces (p. 34)." We based this embodiment on our art-science observation, leaning into the notion expressed in the book *Becoming Animal,* "The simple act of perception is experienced as an interchange between oneself and that which one perceives—as a meeting, a participation, a communion between beings" (Abram, 2011, p. 268). We proceeded with the intention of feeling *embodied empathy*:

> …a concept which describes feelings/seeing/thinking bodies that undo and redo each other, reciprocally though not symmetrically, as partial perspectives that attune themselves to each other. Therefore, empathy is not experiencing with one's body what the other experiences, but rather creating the possibilities of an embodied communication (Despret, 2013, p. 51).

Throughout this process, we remained in conversation about the nature of the relationship between each student and the bird they chose. We considered not referring to each bird as "my bird" to remember each bird's autonomy and self-agency, especially in light of humanity's history and continued practice of owning animals. Also, we did not want to deny the implication of "my bird" as "my friend" in the manner of friendship being claimed and celebrated. We explored the nature of empathy in the sense of understanding and being sensitive to, even vicariously experiencing the feelings or possible thoughts or perspectives of another, yet we also held the risk of our imaginative projections on each bird as turning each into an object infused only with our projections. To guide this balance, we send out a collective ode of gratitude for Alexis Pauline Gumbs' (2020) book *Undrowned: Black Feminist Lessons from Marine Mammals,* part of which we all read that offered guidance on how to approach interspecies observation, friendship, and love. Her voice celebrates a kind of generosity of scholarship toward interspecies friendship that supports love. She writes:

> My hope, my grand poetic intervention here is to move from identification, also known as that process through which we say what is what, like which dolphin is that over there and what are its properties, to *identification*, that process through which we expand our empathy and the boundaries of who we are become more fluid, because we *identify with* the experience of someone different, maybe someone of a whole different so-called species.

We aimed to carry the spirit of her underwater exploration for connection to the skies above.

Holding the dignity and sovereignty of each bird in our consciousness guided us in this process of offering interspecies friendship that was being extended without assumptions or expectations of how it might be received. Notably, each time a student was observing the bird of their choosing, it was nearly certain that that bird was also watching them and noting their behavior as either a threat or an admiring benign presence. We supplied high-quality binoculars for students to check out to avoid getting too close and causing stress to the birds. The scripts created by each student for the *Art Hike* videos reflect this sensitivity and dance adeptly along the edges of these distinctions. In one of our warm-up activities, everyone stood across from a partner making eye contact that would be held throughout the entire activity. The person wearing the darker shirt was asked to go first and move slowly at a consistent pace facing their partner while the other person mirrored their movements. Without stopping the movement, the other person was asked to be the leader and the other to follow; then again without stopping the movement, partner groups were asked to continue moving with both leading together, leaning into co-becoming with the movement. Partners were encouraged to continue moving and investing in seeing if a co-led movement is possible. Several groups afterward reported achieving moments that felt like authentic co-becoming.

We filmed all day on June 30 and the morning of July 1, 2021, at each of the three locations. The *Art Hike* videos made through this art-science process are available for viewing (Inside the Greenhouse, 2021). Each student wore the wings and the headpiece based on their observations of the bird of their choosing and moved extemporaneously as their bird within a chosen location at one of the three OSMP trails. We all collaboratively decided where to place each person for the filming based on the deep familiarity we had developed with each of the sites and based on each student's knowledge of where that bird might most likely dwell within this space. Grant was present with each student to guide and encourage their movements based on their exploration of each bird. Milo Lewon, a CU undergraduate student in Music, was present with his accordion to provide musical accompaniment and, when being filmed within the frame, was wearing a Barn Swallow papier mâché headpiece. Milo's accordion music was also recorded as the soundtrack along with each student's recorded script. Many of the CU students were also present for the filming days to help manipulate some of the larger three-person puppets and to attend to other sundry details. After seeing a mature Prairie Rattlesnake (*Crotalus viridis*) at the head of the South Mesa Trail, we included a rattlesnake check along the edges of each trail before using it for movement. We tried to stay on the trails rather than in the grass to model good stewardship with the plant life in Open Space, which challenged our movement during filming somewhat, but lessened our risk of snakes. Since the dissemination of the films through our partners has not yet occurred, we cannot yet comment on their reception. For those of us who have seen the first drafts, we feel ridiculously pleased and satisfied. Each of the three five-minute videos for each different location prompts a profound, unexpected, and joyful shift in how humans relate to the natural world and other species. They rejuvenate a new way of interacting.

Doom and gloom narratives that emphasize problems, costs, and adverse impacts are the most dominant narratives about climate change (Hinkel et al., 2020). In our work, we were guided by climate change communication researchers who agree that narratives motivate climate action more than climate information (Chapman et al., 2017; Fløttum & Gjerstad, 2016; Hulme, 2009; Moser & Ekstrom, 2010). In fact, many researchers believe that doom and gloom narratives are counterproductive for climate action as fear may demotivate climate action (Chapman et al., 2017; Hinkel et al., 2020). Narratives can contribute to people's agency for climate action through the creation of transformative narratives that tell a positive and engaging story, articulate an aspirational vision, and provide solutions for attaining this vision, rather than articulating problems to avoid (Hinkel et al., 2020). Research by Beth and Max Boykoff, CU Professor of Environmental Studies and cofounder of Inside the Greenhouse, shows that involvement in "good-natured" arts-based performance supports youth in processing negative emotions regarding climate change, feeds hope, and sustains engagement (Osnes et al., 2019). The transformative narratives created through *Bird's Eye View* highlight the stories of friendship between Boulder youth and a wide variety of local birds enacted by the youth within the City of Boulder's OSMPs. The design for our final *Side by Side* performance, *Bird's Eye View*, intentionally included ways for supporting youth well-being, which in turn feeds into a sense of belonging and, importantly, hope. Only with an intact sense of hope – a belief that your actions can make a difference – does climate action make sense, and only with a feeling of well-being is that action sustainable.

Lessons Learned and Reflections

Art can accentuate and foster a connection and feeling of belonging within the natural world. At Sawhill Ponds at the very end of our first filming day, we all stood for a "viewing this natural world" meditation – just using our sense of sight to relax into seeing all that was around us – while Milo improvised a song on his accordion in response to his feeling of place. The experience was extraordinary – five minutes of being music-assisted into seeing this place with deeper sight – an enhanced feeling of connection carried by the live music and coming through Milo in this shared co-presence. One of the participants, Lerato, wrote:

> During our last meeting, when we had our meditation accompanied by Milo's accordion playing, I felt a deep sense of belonging while listening and just being, absorbing and exploring the natural world by looking at the sky, moving with the wind, and feeling the gravel on the ground. It was almost like the music brought me to a place where I really felt like I could be in that environment and feel a connection to nature through the music (Osnes, 2021)

There is something powerful about arts-assisted connections to the natural world that can augment, deepen, and increase what connections might have otherwise happened on their own. Perhaps that is not always true or even true for everyone, but it could be a sweet spot that is reached by our summer's offering when the urgency for unification is so strong, and our species is so disconnected from the natural world.

During many of our sessions, we spoke of "unlearning" what we have been taught of domineering ways of learning and viewing our relationship with the natural world. What Lola D'Onofrio wrote offers another way of considering this notion.

> I have found that belonging in nature is not a process, not a struggle. we have othered ourselves so very much that we think to come into belonging with the natural world is a journey, that it's work, that it's unlearning of so many things. but all it is, really, is embracing that which we already are. through the explorations we have done this summer, i realized the real work, the real struggle, is distancing yourself from nature. to belong is our natural state. when i listen, meditate, jump in rivers, hike mountains, it is not work to attune myself to their natural rhythms and beauties. it is work to get into my car, realign my brain in a 'human' way that is selfish and materialistic. understanding the birds is not hard. it is just a reversion to my natural, wondrous, childlike mindset of understanding, love and joy for all things. belonging is something we all have bursting from us. It is only held back by a wall our society & culture has told us to build. that wall breaks down every time we go outside, take a deep breath. now i think all we need to learn is not to rebuild it (D'Onofrio, 2021).

From a place of belonging and inspired by our interspecies friendship, we continuously brought forward our hopeful narratives into climate action. During our time together this summer, we invited participants to opt into several public enactments of climate. When many from our group costumed as all sorts of birds arrived at the 350.org march in support of a state-level climate bill, the already-gathered activist crowd seemed to be visually arrested by our arrival, took it in, and then exploded into applause. We added a visual spectacle to the march that was exceptional and joyful. When the Plastic Pollution Act was awaiting our governor's signature to be made into law, we created a sign with the name of the bill up top and then a message saying, "Governor Polis, Please sign the bill!" with a drawing of a duck's bill signed by each of them and sent to his office. When one member of our group's sister planned an LGBT Pride event, we created a sign in support reading "Science says, being queer is natural!" The modeling of how to be in consistent and active partnership with local efforts and challenges seems the most useful in this effort. It seems to support their process of "becoming" actors in creative and public civic action. Our youth are learning a nimble responsiveness in which they are ready to show up. They are cultivating a community in which to discuss various perspectives on an issue. They are gaining the courage to be visible and speak in the public sphere, as one of them, Leela, was interviewed at the march for the local radio station. They are engaged in enacting climate through imagination and performative acts that are publicly shared through multiple venues and media. They are learning how to shepherd these imagined solutions from acts of public expression into public policy. They are participating in contacting city, state, and national governmental representatives to encourage and thank them for introducing new policies. They are being facilitated in tracking these bills through the political process via government websites. They are growing in understanding of how to advocate for these policies all along the way until they become law. They are enacting climate through performance and policy in a way that makes literal the double meaning of the word *enact*: (1) to act out as play or to perform and (2) to make into policy or law.

This experience supported them in authoring an account or insight into their experience of an interspecies friendship – first as a group with Barn Swallows and then individually with a local bird of their choosing. The art-science approach toward this friendship was more formally structured for the first portion focused on Barn Swallows, and then the process was opened up to be more self-led for the second portion when they chose, observed, and researched a local bird of their own choosing. A hope built into this entire experience design was that an increased feeling of belonging would be cultivated. Digging deeper, there are three aspects to this feeling of belonging that are all intertwined and that gain sustenance from the same experiential root system: (1) a feeling of belonging as part of the natural world, (2) a feeling of belonging as part of a local community engaged in action for climate solutions, and (3) a feeling of belonging within STEM and the artistic community focusing on climate solutions. By focusing on female-identifying, gender nonbinary youth and members of minorized groups, we hope to contribute under-represented perspectives for coauthoring an equitable, survivable, and thrive-able way forward for all life and the ecosystems upon which all life depends. We trust that the empowered presence of female-identifying, gender nonbinary, and members of minoritized groups youth in STEM and in the arts will diversify the approaches, perspectives, and solutions made possible. We look forward to learning what will be experienced through the finished videos of the Art Hikes that we created. We look forward to codesigning future iterations of this project that invite and support young people in enacting a climate for a future that is looking up.

What Is Needed Now

This program offers the coming generation a narrative of hope that if we can see through the eyes of another species, we can not only solve our greatest challenge, climate change, but we can also find our own place of belonging and unity within the natural world. This focus on Barn Swallows is serving as a portal to our larger relationship with the natural world. At this moment in history, to claim our responseability to the planetary ecological crisis we all face, we look to our relationship with Barn Swallows and local birds in our open spaces and through their eyes for an expanded interspecies perspective. High school student Sofie Wendell described this summer's experience:

As I move to the beautiful music of the accordion and sense the sunlight warming my skin, I feel at peace. As I open my eyes and watch the organic movements of my closest friends and dearest mentors, I feel belonging. Here, protected by tall pine trees, surrounded by strong mountains, and among such beautiful individuals, I feel as though I belong. Here, my ideas hold weight, they matter. Here, I am not judged or overlooked. Here, I am given the opportunity to connect not only with nature but with new friends and unique ideas. Together, we are building the bridge to an equitable, survivable, and thrive-able future through science, art, and love. I know I belong because this feels like home.

Appendices

Appendix 1: Contributions by CU Students to Bird's Eye View

The two CU students who were able to attend most consistently and contributed the most to the sessions were Avani Fachon, an undergraduate student in Ecology and Evolutionary Biology who completed an honors project with Becca as her advisor based on an art-science approach to Barn Swallows, a website entitled *Rituals of This Good Earth* (http://ritualsofthisgoodearth.com/) and Ben Stasny, a graduate student in Theatre and Performance Studies who first volunteered in the summer of 2020 and is focused on the study of theater and climate change. Our side-by-side balance was perfectly represented by our two PhD students who participated in some sessions each – our "art Heather" (Heather Kelley, a PhD student in Theatre and Performance Studies) and our "science Heather" (Heather Kenny, a PhD student in Ecology and Evolutionary Biology). A community college student and a dancer with the Colorado Ballet Academy, Grant Gonzalez, joined our project through an internship sponsored by CU's Cooperative Institute for Research in Environmental Science. He contributed enormously by coaching participants in the movement and embodiment of the various bird species. Recent and current CU Ecology and Evolutionary Biology undergraduate students included Sage Madden, Aleea Pardue, and Marina Ayala. To create the filmed versions of the Art Hikes, we worked with Jonah Sublette and Sara Herrin, who created our 2020 video. Herrin was a student in Beth's CU Creative Climate Communication several years back and is currently a film producer focusing on climate change stories.

Appendix 2: Excerpt from Evolutionary Relationship Between Arms and Wings

Assembled by Avani Fachon, National Science Foundation-funded Intern with *Side by Side* and CU Undergraduate student in Ecology and Evolutionary Biology.

This is an abbreviated version of the script used to guide participants through a moving meditation in which they explored their own connection with birds through the common anatomy of human arms and bird wings. Ask participants to stand far enough apart that they can raise and move their arms. Invite participants to close their eyes, and read the following script slowly, leaving pauses for them to explore through movement each prompt you provide.

> I invite participants to imagine this-- long ago, birds and humans shared a common ancestor-- a primitive reptile-like, vertebrate creature. This creature split into two groups-- one group went on to become dinosaurs, birds, crocodiles, and lizards, while the other group split to become mammals (Lungmus et al.). Mammals started developing diversity in their forelimbs much before dinosaurs even existed (Lungmus et al.). After some time, however, dinosaurs began to develop wing-like structures, and started flapping and fluttering. Sixty million years later, they became the modern bird that we see today. From this common ancestor, both birds and humans developed their own unique forearms.

> Because we evolved from a common ancestor of birds, we have the exact same types of bones in our forearms as birds do-- they're just arranged a bit differently. Birds have the same elbow, wrist, hand, finger, and thumb bones as we do. Now let's change our human arms into bird wings. The palm of your hand is lengthening, and in doing so your thumb is growing closer to your wrist joint. Your four other fingers are morphing into three less mobile ones. You begin to realize that your elbow in wrist joints are connected, so that as you extend your elbow your wrist automatically extends as well. As you contract your wing, your wrist automatically closes down as well. Take a few moments to explore this new relationship between your muscles and bones.

References

Abram, D. (2011). *Becoming animal: An earthly cosmology*. Vintage.

Bagemihl, B. (1999). *Biological exuberance: Animal homosexuality and natural diversity*. Macmillan.

Biologydictionary.net. (2017). Mutualism. In L. Li (Ed.), *Biology dictionary*. https://biologydictionary.net/mutualism/

Chapman, D. A., Lickel, B., & Markowitz, E. M. (2017). Reassessing emotion in climate change communication. *Nature Climate Change, 7*(12), 850–852. https://doi.org/10.1038/s41558-017-0021-9

Chawla, L. (2020, November 16). Connecting to nature is good for kids—But they may need help in coping with a planet in peril. *The Conversation*.

Coulthard, G., & Simpson, L. B. (2016). Grounded normativity/place-based solidarity. *American Quarterly, 68*(2), 249–255. https://doi.org/10.1353/aq.2016.0038

D'Onofrio, L. (2021). *Field otes*. Unpublished Manuscript.

Despret, V. (2013). Responding bodies and partial affinities in human–animal worlds. *Theory, Culture & Society, 30*(7–8), 51–76. https://doi.org/10.1177/0263276413496852

Fløttum, K., & Gjerstad, Y. (2016). Narratives in climate change discourse. *WIREs Climate Change, 8*(1). https://doi.org/10.1002/wcc.429

Gumbs, A. P., & Brown, A. M. (2020). *Undrowned: Black feminist lessons from marine mammals (emergent strategy)*. AK Press.

Halberstam, J. (2005). *In a queer time and place: Transgender bodies, subcultural lives*. NYU Press.

Halberstam, J. (2011). *The queer art of failure*. Duke University Press Books.

Haraway, D. J. (2016). *Staying with the trouble: Making Kin in the Chthulucene*. Duke University Press Books.

Harker, G. (2021). *Field notes.* Unpublished Manuscript.

Hinkel, J., Mangalagiu, D., Bisaro, A., & Tàbara, J. D. (2020). Transformative narratives for climate action. *Climatic Change, 160*(4), 495–506. https://doi.org/10.1007/s10584-020-02761-y

Hulme, M. (2009). *Why we disagree about climate change: Understanding controversy, inaction and opportunity* (4th ed.). Cambridge University Press.

Inside the Greenhouse. (2020). *Side by side.* https://vimeo.com/473564810.

Inside the Greenhouse. (2021). Bird's eye view. vimeo.com/user/24633009/folder/5246141.

Kimmerer, R. W. (2013). *Braided sweetgrass.* Milkweed Editions.

Langham, G., Schuetz, J., Soykan, C., Wilsey, C., Auer, T., LeBaron, G., Sanchez, C., & Distler, T. (2015). *Audubon's birds and climate change report: A primer for practitioners.* National Audubon Society, Version 1.3.

Lloro-Bidart, T., & Banschbach, V. S. (2019). *Animals in environmental education: Interdisciplinary approaches to curriculum and pedagogy* (1st ed.). Palgrave Macmillan.

Moser, S. C., & Ekstrom, J. A. (2010). A framework to diagnose barriers to climate change adaptation. *Proceedings of the National Academy of Sciences, 107*(51), 22026–22031. https://doi.org/10.1073/pnas.1007887107

Ojala, M. (2012). Regulating worry, promoting Hope: How do children, adolescents, and young adults cope with climate change? *International Journal of Environmental and Science Education, 7*(4), 537–561.

Osnes, L. (2021) *Field notes.* Unpublished manuscript.

Osnes, B., Boykoff, M., & Chandler, P. (2019). Good-natured comedy to enrich climate communication. *Comedy Studies, 10*(2), 224–236. https://doi.org/10.1080/2040610x.2019.1623513

Osnes, B., Hackett, C., Fahmy, S., & Nixon, L. (2022). Young Women's voices for climate. In L. S. Brenner, C. Ceraso, & E. D. Cruz (Eds.), *Applied theatre and youth: Education, engagement, activism* (pp. 223–231). Routledge.

Plumwood, V. (2003). Environmental culture: The ecological crisis of reason. *Environmental Values, 12*(4), 535–537.

Prendergast, M., & Saxton, J. (2009). *Applied theatre international case studies and challenges for practice.* Intellect.

Rosenberg, K. V., Dokter, A. M., Blancher, P. J., Sauer, J. R., Smith, A. C., Smith, P. A., Stanton, J. C., Panjabi, A., Helft, L., Parr, M., & Marra, P. P. (2019). Decline of the North American avifauna. *Science, 366*(6461), 120–124. https://doi.org/10.1126/science.aaw1313

Safran, R. J., & Levin, L. L. (2019). The sexual and social behavior of the barn swallow. In J. C. Choe (Ed.), *Encyclopedia of animal behavior* (Vol. 3, 2nd ed., pp. 173–180). Elsevier, Academic Press.

SPEAK. (2022). https://speak.world

Tedore, C., & Nilsson, D. E. (2019). Avian UV vision enhances leaf surface contrasts in forest environments. *Nature Communications, 10*(1), 1–12.

Beth Osnes, Ph.D., is a Professor of Theatre and Environmental Studies at the University of Colorado. She is the codirector of Inside the Greenhouse (www.insidethegreenhouse.net) for creative climate communication and cofounder of SPEAK (http://speak.world) for women's vocal empowerment. Her books include *Theatre for Women's Participation in Sustainable Development* and *Performance for Resilience: Engaging Youth on Energy* and *Climate through Music, Movement, and Theatre.* She is a theater and performance studies artist/scholar who is active in applied performance and creative climate communication. She engages in performance to coauthor and actualize an equitable, survivable, and thrive-able future for all life and the ecosystems upon which all life relies. This applied approach to performance is explicitly for positive social change and is characterized by process-oriented work on an issue identified by the community. She is featured in the award-winning documentary *Mother: Caring for 7 Billion* (www.motherthefilm.com) and lives in Boulder, Colorado.

Chelsea Hackett, Ph.D., is the cofounder and Executive Director of SPEAK, a nonprofit that focuses on Vocal Empowerment for women and girls. In addition, she is an interdisciplinary artist, researcher, and adjunct professor at the University of Colorado at Boulder. She completed her doctorate in Educational Theatre at New York University. In addition, she served as a professional development coach and teaching artist in New York City, working with educators to integrate the storytelling elements of theater into classrooms to aid with social-emotional learning, literacy, community building, and more. She now brings her experience with educational theater to the *Bird's Eye View* project, working with her collaborators toward a shared focus on addressing youth climate grief and increasing a sense of belonging. She is passionate about using the power of the performing arts to create collective meaning and engage in critical conversations.

Molly T. McDermott is a Ph.D. Candidate at the University of Colorado Boulder in Ecology and Evolutionary Biology. She studies interactions between environmental conditions and sexual selection in migratory songbirds. For her Ph.D. research, she tracks individual barn swallows – familiar farm residents across much of North America – to measure how nutrition and stress affect plumage traits and breeding performance. Previously, she studied how climate-driven vegetation change in the Arctic may impact insect communities and the songbirds that depend on them for food. In addition to her biological interests, she is a musician interested in communicating scientific information via creative disciplines.

Rebecca Jo Safran, Ph.D., is a Professor of Ecology and Evolutionary Biology at the University of Colorado where her research group focuses on questions related to the evolution of new species using ecological, genomics, and behavioral studies. Safran earned her doctorate at Cornell University followed by a postdoctoral fellowship at Princeton University. Along with Beth Osnes, Max Boykoff, and Phaedra Pezzullo, she is the founding codirector of Inside the Greenhouse whose mission is to create new narratives of hope and inspiration about solving the climate change crisis. Safran's teaching explores the communication of climate change through science communication and filmmaking. Safran is passionate about interdisciplinary collaboration, art-science connections, issues related to belonging in STEM, natural history, and providing learning opportunities for students at all stages of their educational journeys, all of which are integrated into the *Bird's Eye View* project.

Instructional Strategies for Climate Education in the Classroom: Storytelling about Our Place in the Earth System

Jessica R. Bean

Before you dig into this chapter, I want you to think about a place that you care about in the world. It could be anywhere. It might be your living room. The tree outside your house. The park where your dog plays. A lake where you fish or camp. Now think about how and why you feel connected to this place. Is it where you live? You find comfort? You are excited to explore? What stories could you tell about this place?

Now think about how you want this place to look in the future. How and why might it change? What do you hope it will look and feel like?

If you are reading this book, this chapter, in particular, you are looking for practices to leverage the power of storytelling for climate education in the classroom. It turns out that constructing coherent stories about what we care about and how we understand the world around us is aligned with what we know about how people learn, and research-based pedagogical practices that are responsive to the Next Generation Science Standards (NGSS Lead States, 2013). Answering the questions above can always provide you with a place to start planning for instruction, and can ground your students as they learn about our changing world.

Whether or not we acknowledge it every time we talk about climate change, the consequences of a warming world will have implications for all the places we care about, including where and how we live, and the resources that determine our quality of life such as air, water, and food. Some of your students already know this because they have lived through heat waves, wildfires, or other extreme weather events, or know someone who has. Because of this reality, there are four major shifts we must make in climate education:

1. Create learning sequences ("storylines") in which students actively seek to coherently connect their learning to climate change.

J. R. Bean (✉)
University of California Museum of Paleontology, University of California,
Berkeley, CA, USA
e-mail: jrbean@berkeley.edu

© The Author(s) 2024

E. Coren, H. Wang (eds.), *Storytelling to Accelerate Climate Solutions*,
https://doi.org/10.1007/978-3-031-54790-4_19

2. Interweave climate stories throughout curricula; do not teach climate as an isolated topic because everything is connected to the climate crisis.
3. Connect students' lived experiences and the places they care about to their learning about climate change.
4. Provide opportunities to envision and plan for a resilient future in your community.

I work with K-16 educators to make these instructional shifts and integrate climate education into science, math, social science, and language arts instruction. Exploring the connectedness of our world and grounding our responses to the climate crisis in the places we care about are essential to engaging in meaningful science learning and action (Bowman & Morrison, 2021). As a teacher said to me recently, students need to "see that these topics are relevant; bring the work back to their community, their city block."

Research has shown that understanding the Earth as an interconnected system and contextualizing instruction about a place in the world makes us more likely to be concerned about the climate crisis (Ballew et al., 2019; Zeidler & Newton, 2017). Again, many of us know this from our own experiences, yet somehow, do not consistently provide opportunities for this kind of sensemaking and application of knowledge in the classroom because we feel we have to "cover" certain material, lack of time, or other reasons (Petersen et al., 2020). But creating an equitable and inclusive classroom means leveraging student experiences and prior knowledge (Windschitl et al., 2018), and it is now more urgent than ever that we take small steps to embed connections to climate change into what we teach every day. I am not saying it is easy because most of us have never experienced this kind of learning, nor been trained in providing this kind of instruction. I personally had discrete moments in my own education where opportunities for this kind of learning could have been attempted, but ultimately fell short in various ways. I'll tell you a story about a couple of my own learning experiences, which we will reflect on throughout the chapter.

Reflections on Meaningful Learning

In seventh grade English class, I was asked to write a first-person account of a memorable experience. I wrote about a hike in Lassen Volcanic National Park in northern California from the meadow around King's Creek into the hydrothermal area called Bumpass Hell, named for a settler who lost a leg after falling into one of the many bubbling mud pots. I wrote about the meadows, trees, flowers, and rocks I observed, and how I felt exploring, interacting with, and being a part of this place. I described the topography as we viewed distant lakes that I hoped to swim in on my next trip. Thinking about it now, there were so many opportunities to connect my observations to learning about ecology and geology, but there was no integration with the science curriculum or any other subject, for that matter. I haven't read the essay in at least 15 years and can't reread it now (for reasons I will explain), but I saved, and still remember doing this assignment, because it was about a place that I love. I returned to Lassen Volcanic National Park year after year, into adulthood, most recently in 2019. I suspect that if I wrote a similar story today, the scientific content would deepen, but I would still focus on what I value and how I perceived my place within, not separate from, the biological and physical system I was exploring.

One of the few assignments I recall from high school was for my biology course. We were told to explore a local environment and present to the class what we learned about that ecosystem. I chose to document the accelerating development of hundreds of homes in the hills above my neighborhood in Santa Rosa, California. This area had originally been slated for a recreational park. The new roads tore through the rolling hills of oak woodland, where we would regularly see deer, wild turkeys, skunks, and opossums. I took pictures of the construction for my report and remember thinking I was glad I would be leaving for college in a few years so I wouldn't have to watch as the trees on the ridgeline of our little valley disappeared. Again, in hindsight, this was an assignment ripe with the opportunity to discuss the consequences of and practices for living at wildland–urban interfaces, yet we were never encouraged to investigate these larger issues related to our city's changing landscape.

Fast forward 20 years, and the significance of what I documented in these assignments is apparent. The work, handwritten and saved in a box in my childhood room, was lost in the Tubbs Fire in Sonoma County in October 2017, along with most of the evidence of my youth and four generations of family photos, books, and other artifacts. This wind-driven fire was sparked by downed power lines in dry and low humidity conditions, and significantly fueled by those homes built in the hills above my neighborhood (Kramer et al., 2019; Watkins, et al., 2017). And, as I write this, those meadows and trees in Lassen Volcanic National Park are burning in the Dixie Fire of 2021 which has now swelled to become the second-largest fire in California history (Ilati & Moriarty, 2021). The combined effects of extreme drought in a warming world along with human ignition from failed electrical infrastructure, humans inhabiting wildland–urban interfaces, a century of fire suppression, and seasonal winds made these fires more explosive and destructive. Many of us are living with the consequences of these fires—losing homes and places of refuge and living with hazardous air quality. I am experiencing ecological grief (something I will touch on in the Introducing Phenomena section below) and learning about the policies and infrastructure that are needed to build safer communities, make decisions about how and where my family and I will live, and how we reduce the chances of even more catastrophic consequences of climate change.

Why share this story? First, it can serve as an example we will return to reflect on how to intentionally interweave climate change throughout instruction. The Earth's climate and ecosystems are one interrelated systems, so let's frame and shift instruction to reflect this reality. Second, constructing scientific understandings and narratives that are contextualized in the places we value—where we live, work, and play, and associated emotional connections—are important for meaningful climate change instruction (Hufnagel, 2015; Peel et al., 2017). These assignments, although lacking explicit connections to climate change or human impacts, gave me opportunities to think about my place in the world and what I value. Climate change education needs to engage learners' hearts and minds in the here and now, not just in far-off places or in the future. My reflections on this story and how we can enhance instructional experiences are highlighted in boxes throughout this chapter.

Now think about your own educational experiences. What made learning memorable or meaningful? How could you connect that learning to climate change and local environments?

As you read, think about how you can shift instruction to engage students in constructing explanations and narratives about climate change. This chapter will give you an introduction to some of the practices and tools for:

- Supporting systems thinking and understanding the Earth as an interconnected system (that includes humans) to design unit storylines
- Contextualizing learning using focal, locally relevant phenomena
- Modeling to represent students' understanding of how the world works and how they individually and collectively can influence the Earth system

These instructional practices can be supported by a suite of tools from the Understanding Global Change (UGC) Project at the University of California Museum of Paleontology developed in collaboration with classroom educators (Bean, 2020; Bean et al., 2020).

Systems Thinking and Climate Connections

As noted above, the causes of and solutions to climate change (as well as other health and environmental issues) are multifaceted and complex, and threaten the resources we need to survive and thrive. To be able to explain and solve current and future problems, tomorrow's scientists, engineers, and informed communities must understand the multidimensional causes of climate change and have the skills to synthesize interdisciplinary knowledge. One such skill set is *systems thinking*, which is defined as the "ability to recognize, describe, model and to explain complex aspects of reality as systems" (Riess & Mischo, 2010, p. 707). The value of systems thinking to tackle complex problems has been recognized for decades (Sweeney & Sterman, 2007; Plate & Monroe, 2014), and "systems and system models" are identified as one of the Cross-Cutting Concepts that are common across all STEM disciplines in the Next Generation Science Standards (NGSS Lead States, 2013). Exploration of systems supports the development of a "holistic" perspective, in which understanding the dynamics of the "system as a whole" is emphasized, and phenomena are explained as "emerging from the dynamic interactions between components across different levels of organization" (Verhoeff et al., 2018, p. 5). People who use systems thinking are more likely to understand the role of human activities in causing climate change and that the consequences are cause for concern (Ballew et al., 2019; Roychoudhury et al., 2017). Research indicates, however, that classroom teachers and college-educated adults have poor systems thinking skills (Sweeney & Sterman, 2000, 2007). For these reasons, educators need to provide students with opportunities to develop an understanding of the dynamic and inter-connected nature of the Earth as a system, and why it changes through time. Systems thinking has been applied across science disciplines to engage learners in construct-ing coherent understandings of complex phenomena (Hmelo-Silver et al., 2017; Verhoeff et al., 2008, 2018; York et al., 2019).

This coherency of understanding from a systems perspective has been described by the classroom teachers I work with as "storytelling" to plan instructional sequences, and the word "storylines" is used to describe coherent instructional units aligned with the Next Generation Science Standards (NGSS Lead States, 2013; Next Generation Science Storylines, 2018). This makes sense because engaging storytelling and systems thinking both require defining causal relationships and temporal sequences of events that will allow learners to construct explanations of phenomena and processes (Young & Monroe, 1996; Roychoudhury et al., 2017). This approach is very different from the way science is often taught, i.e., as discon-nected, discrete bits of information that are not applied or actively used by students to explain the world around them (NRC, 2012). Instruction needs to be designed such that students explore various parts of the Earth system in relation to each other to explain how and why the climate and environments are changing (Roychoudhury et al., 2017; Shepardson et al., 2012). While the NGSS and various curricula in response to these standards support instructional coherence through carefully sequenced content and activities (Achieve, 2016; NGSS Lead States, 2013), curri-cula should also ensure that students are "seeking" to construct their own coherent explanations (Sikorski & Hammer, 2017). For this to happen, we need to welcome student questions and ideas and support them in connecting learning about various parts of the Earth system.

Understanding Global Change Framework

To support planning and instruction from an Earth systems perspective, the UGC Project developed a Framework of three categories of components that interact in the Earth system (Fig. 1): (1) *Measurable changes* at the center of the diagram are the changes that we can monitor in the Earth system over time; (2) *How the Earth system works*, which are ongoing processes in the middle ring of the diagram that shape the Earth through time; and (3) *Causes of global change* in the outer ring of the diagram, which are the ultimate human and nonhuman forcings that change the rate and magnitude of system processes, resulting in measurable changes in the Earth system. Key global change processes and phenomena (e.g., precipitation, atmospheric circulation, erosion, biodiversity) are visually represented as icons. Importantly, human activities and needs are integrated throughout this Framework, not separated from the rest of the Earth system. This is because we, as part of the

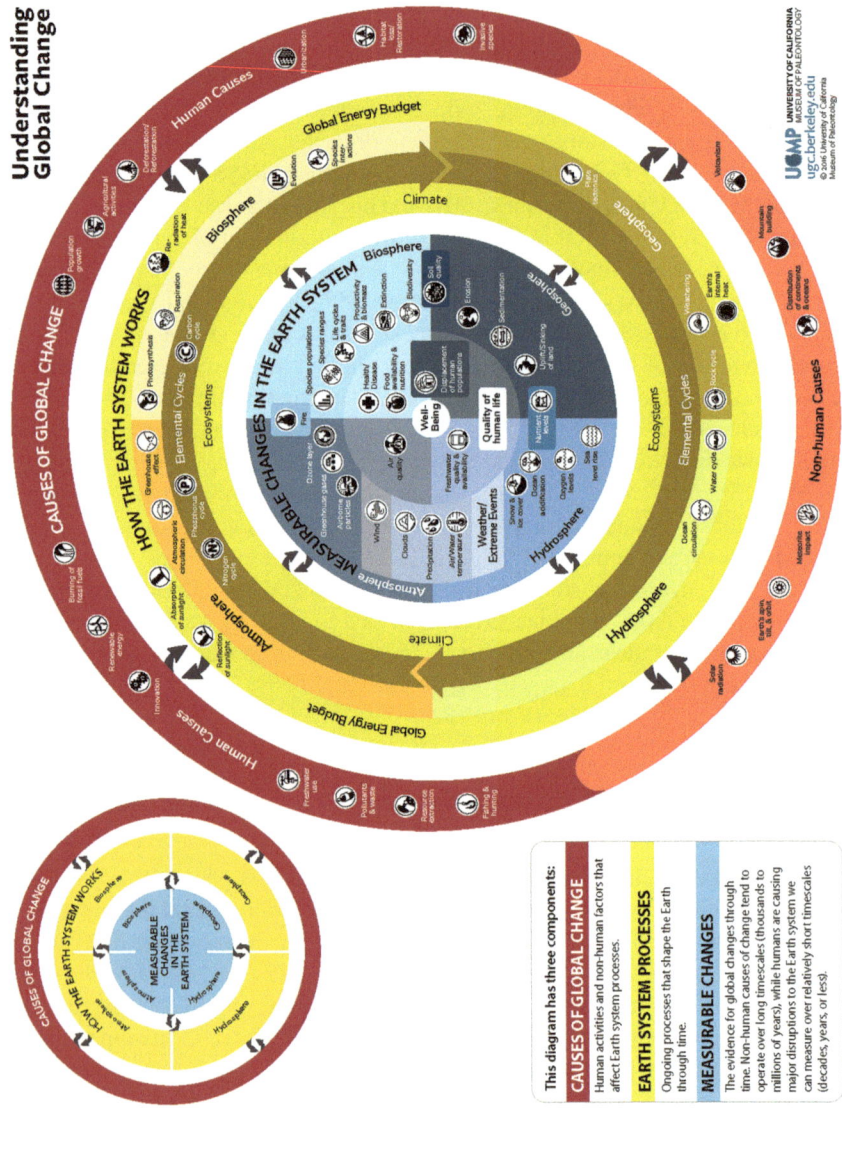

Fig. 1 The Understanding Global Change Framework

Earth system, cause changes and experience the results of changing climate and ecosystems.

I want to emphasize that this UGC Framework is not something to introduce to K-12 learners all at once, but can be used for planning instruction among educators and shared with young learners over time as a series of topics for exploration. The "Simple" Framework with the spheres (i.e., atmosphere, hydrosphere, biosphere, and geosphere) in the upper left corner of Fig. 1 has been used by classroom teachers as an organizational scheme for subsets of the concepts, allowing them to build up to the complexity of the full diagram over time. For more about how to use the Framework with students, please visit the Planning for Instruction page on the UGC website (Bean, 2020). For a more comprehensive introduction to the Framework appropriate for high school and undergraduate students and educators, view the slide show called Understanding Global Change 101 (Bean et al., 2020).

Regardless of what grade level you teach, you can find topics on this Framework that are part of your science standards. Elementary school standards—including energy, the carbon and water cycles, organismal growth and life cycles, and ecosystems—lay the essential foundation for understanding the climate system. In middle and high school, almost all science standards can be connected back to the causes and consequences of, and solutions to, climate change. Standards that do not explicitly include the phrases "climate change," "environmental change," or "human impacts," still provide countless opportunities to apply a systems perspective and explore various connections. Just asking ourselves as we plan, and then asking students, "How will this (biological, physical, or chemical) process respond to climate change?" opens the door for students to share ideas and explore how and why climate change is relevant and connected to what they are learning. Table 1 shows some more concrete examples that students could explore using various data sets and activities. Additional examples of climate connections are on the Understanding Global Change Website content pages (Bean & Marshall, 2020).

Take a minute to look over Fig. 1 and identify the topics you teach. The UGC-NGSS Crosswalk spreadsheet on the UGC Planning for Instruction page (Bean, 2020) can also help you explore the K-12 NGSS standards relevant to each topic/

Table 1 Example climate connections to foundational science topics

Science topic	Climate connection examples
Photosynthesis	Examine seasonal fluctuation of CO_2 in the Keeling Curve related to plant growth in the Northern Hemisphere (as levels continue to rise overall due to human activities). Determine how evapotranspiration rates affect plants and soils. Interpret maps that show how agricultural practices in your region have and will change in a warming world.
Freshwater ecosystems (lakes, streams, etc.)	Explore local historical and current data about stream flow, temperature, and eutrophication and discuss how these conditions will change in a warming climate.
Energy and chemical reactions	Determine what energy sources are used in your community. Explain how fossil fuels are a source of energy and the alternatives that can be used.

icon in the UGC Framework. You can determine which grade level standards you plan to address, and then explore how you would coherently "connect the dots" from these topics to climate change.

Finding the right data sets to make climate connections in your curriculum takes time, but remember your students are also a resource that can help guide your planning. Provide opportunities for them to ask questions reflecting how they might seek to coherently connect climate change to their personal experiences and learning across the curriculum. Climate change is a theme we should come back to again and again throughout instruction so that students can use an Earth systems perspective to figure out how seemingly disparate topics are unequivocally interrelated.

Connecting My Story to Climate Change and the Earth System

If we come back to the examples discussed from my own education above, we can identify many key concepts and pathways to connect these assignments to climate change using the UGC Framework. Specifically, if we reflect on the essay I wrote about Lassen Volcanic National Park, there were opportunities to connect my observations to explore seasonal patterns of plant and animal growth (productivity and biomass) and biodiversity (measurable change icons in the center). The biology of this place is related to the Earth system processes (middle ring icons) that shape the land (rock, water, carbon, nitrogen, phosphorus cycles), which influence the soil quality and nutrient levels (measurable change icons). These are all processes that are affected by climate change. In that essay, I remember writing a lot about the water in King's Creek, streams running through the meadows, and the lakes. Understanding water use and availability, precipitation patterns, how our water cycle is changing, and in turn patterns of precipitation, snow, and ice cover are all fundamental to understanding the impacts of the climate crisis and how we will need to respond now and in the future. In this essay, I also reflected on my place in this relatively remote ecosystem, which could be more deeply explored in relationship to human impacts and land management practices (habitat loss/restoration and deforestation/reforestation, causes of change icons in the outer ring). Again, these were not connections my English or science teachers made, but my essay would have provided ample material to connect my personal experience to these topics.

Although not everyone would write about a hike in a national park if given an essay prompt about a personal experience, a similarly open-ended assignment in the twenty-first century can be framed with climate connections in mind. For example, to connect students' stories to climate change, ask them to pay attention to the presence (or absence) of water in their narratives, or to note if the place they are writing about has changed over time. Students could also be asked to envision this place in the future and think about how they hope it will look and feel (see the reflective questions at the beginning of this

(continued)

chapter). Again, climate change touches all corners of the Earth, whatever our students write about, so let's support our students to connect the dots between their lived experiences and the consequences of, and solutions to, a changing world.

Similarly, the project about the new homes constructed in the hills above my neighborhood could be connected to so many climate-related topics: urbanization (in this case severe suburban sprawl), burning of fossil fuels (because while the city allowed the construction of hundreds of million-dollar homes, we lacked investment in public transit and everyone drove EVERYWHERE), habitat loss, deforestation, pollution, and freshwater use, as well as habitat restoration, reforestation, and renewable energy, topics related to the future I hoped for and envisioned. And, of course, the connection I could not foresee as a 15-year-old was how these housing projects at wildland–urban interfaces made our community extremely vulnerable to devastating wildfires (Mann et al., 2014). There are many interdisciplinary connections that can be explored about climate, land use, and ecosystem changes through time.

To reiterate, standards do not have to include the phrase "climate change" to be connected to learning about the climate crisis and our personal stories. We can work to identify and interweave climate connections throughout curricula, and we can also frame learning around climate-related issues in our communities. In the next section, we will explore how to frame learning around exploring climate issues and solutions.

Anchoring Coherent Instructional Storylines with Climate Phenomena

When you examine the UGC Framework, you will likely recognize topics that you teach that are in your standards, or that you know are connected to content in your curriculum. Based on the standards alone, we could likely map out a unit of study about various topics in the Earth system that are related to climate change appropriate for your grade level. We could imagine a unit that starts with understanding the carbon cycle, and then addresses human emissions, and the greenhouse effect, and then maybe explores some of the effects of climate change, such as sea level rise or wildfires, and then maybe engages students in discussing possible solutions, but without diving in too deeply to any one climate change topic. This could be similar to the structure of most units of study that I experienced in high school or college in the early 2000s. I might have learned some new (horrifying) facts, but I'm not sure I would have known what to *do* with this information or understand how *my life* is connected to the climate crisis. Let us reframe how we teach to support students in constructing explanations about how and why the climate changes and connecting learning back to their own experiences.

Suppose we want students to figure out how to *use and apply* scientific concepts and practices, as well as leverage their prior knowledge and lived experiences. In that case, we need to anchor their learning around observable and measurable climate phenomena. An appropriate *anchoring phenomenon* is a complex real-world event or process that requires students to synthesize various science concepts to formulate an explanation (NGSS, 2016; Reiser et al., 2017; Windschitl et al., 2012). Because anchoring phenomena are complex and multifaceted, students should investigate and revise explanations of the phenomenon over weeks of instruction. Anchoring phenomena should be distinguished from *investigative phenomena*, which are often smaller in scope and can be explored in one or maybe two lessons. In other words, you can have a series of investigative phenomena in a unit that support the exploration of a unit anchoring phenomenon. A compelling anchoring phenomenon should also be relevant to students' lives and allow them to explore the phenomenon by engaging in the practices of science, such as analyzing and interpreting data, and arguing from evidence, as outlined in the NGSS (NGSS, 2016; NGSS Lead States, 2013).

An anchoring phenomenon about climate change should challenge your students to *co-construct explanations and stories about specific places and times*, meaning that learning can be highly contextualized in places that are familiar to your students. That could involve a place as accessible as your school, a local park, your watershed, or a larger region such as your county or state. That is not to say that students should not learn about far-off places, including the shrinking polar bear populations in the Arctic and the burning of the Amazon rainforest over the last 30 years (these are important investigative phenomena!). However, having opportunities to apply and connect what they learn about places around the world to the phenomena in their local environments can be more effective for the students to build a systems perspective with enhanced personal relevance. Additionally, learning about regional phenomena should be connected to global, large-scale processes, such as the greenhouse effect, ocean circulation, or resource extraction and distribution (Lehtonen et al., 2019). Once you have settled on an appropriate phenomenon based on the guiding criteria explained below, you can then decide on the appropriate learning resources and activities that can be used to help students construct their understanding of the phenomenon. (Note: Based on my experiences working with professional learning communities of teachers, selecting a phenomenon and finding relevant data can be the *most* challenging part of the planning process. Be patient with yourself and try to discuss your phenomenon with colleagues or find local experts to help you if you feel stuck.)

To identify anchoring phenomena about climate change, you can start by thinking about the measurable changes in the Earth system at the center of the UGC Framework because phenomena should be observable, either through direct or indirect observation using instruments and technology for monitoring these changes. These are changes that can be observed and measured over time (the time scale will

Table 2 Criteria checklist for identifying anchoring phenomena

Description of the anchoring phenomenon:	
Understanding Global Change Measurable Change(s):	
Does an explanation of this phenomenon require connecting ideas from all three categories of the UGC Framework (Causes of Global Change, How the Earth System Works, and Measurable Changes in the Earth System)?	Yes/No, Explain:
Is this phenomenon an observable event that happens over time? (*Phenomena could occur over a short or long time period, and can be experienced by direct observations or second hand through images, video, and/or datasets.*)	Yes/No, Explain:
Does this phenomenon happen in a particular place? (*Phenomena can occur over small areas or large geographic regions, and should be events and changes that are context-rich.*)	Yes/No, Explain:
Does this phenomenon have the potential to be explored through a variety of engaging resources (e.g., observations, pictures, videos, datasets) and investigations (classroom and outdoor experiences)?	Yes/No, Explain:
Does this phenomenon have the potential to motivate and sustain students' interest and purpose for learning?	Yes/No, Explain:
Does this phenomenon connect to prior student classroom or out-of-school-time experiences?	Yes/No, Explain:

vary depending on the topic and context for the unit of study you design) and could be compared in different locations. The measurable changes in the UGC Framework are by no means a comprehensive list of potential phenomena, but provide a place to start thinking about which aspects of climate change you want students to investigate.

At the very center of the UGC Framework are the measurable changes that most directly affect human life—air and water quality, food availability and nutrition, health/disease, and where we can live (or where we are displaced from due to changing conditions). Ensuring that anchoring phenomena connect back to the quality of human life (hint, most of them do) can focus, motivate, and sustain students' interest in learning as they have opportunities to apply their knowledge to construct explanations and explore solutions to issues that they face in their communities (e.g., Taylor et al., 2019). Additionally, to explain the consequences of and solutions to issues posed by these measurable changes requires that students consider the ultimate causes of change as well as the Earth system processes that are altered with climate change. Table 2 provides a checklist of criteria for identifying phenomena for climate change units. Some of these criteria are adapted from NGSS (2016) and Ambitious Science Teaching resources (Windschitl et al., 2014, 2018).

For example, we could start with the measurable changes of fire and sea level rise for two new instructional units. We could then contextualize these phenomena and connect these topics to students' lives by exploring locally relevant data sets. When working with school districts in Maryland, we focused on the phenomenon of sea level rise by investigating how the frequency of sunny day coastal flooding events in Annapolis and the western shore have changed over the last 100 years (see graphs and photos from Boesch et al., 2018; Marder, 2020). In San Diego, California, where wildfires repeatedly threaten communities, we can compare housing development and land management practices in neighborhoods that have been resilient in wildfires to the practices used in neighborhoods that were destroyed. We can then determine how these practices can inform how and where we build homes in our own community (Sommer, 2019). Both sea level rise and wildfires, of course, have very direct impacts on the quality of human life, both threaten to displace human populations. Additionally, sea level rise reduces freshwater availability and soil quality, and wildfires affect air quality and respiratory health, and contaminate water sources and systems.

Introducing Phenomena

Once you have decided on an anchoring phenomenon, you can determine how students will be introduced to the phenomenon. Students should not be "told about" the phenomenon, but presented with images, videos, data sets, or a combination of these that represents this measurable change in the Earth System. On a somewhat personal but relevant note, as someone dealing with trauma from wildfires and ecological grief, I recommend being sensitive to the experiences of the students in your classroom when selecting the images, videos, and data sets that will be used to introduce the anchoring phenomenon. For example, I know that some of the students in communities where I work in northern California would be sensitive to images of homes burning, as I am after the loss of my family home. For this reason, I would not use these images but focus on starting a unit with data in the form of graphs, maps, and information that might be *useful for understanding and responding to wildfires in the future*. While an image of a burning house, or a flooded neighborhood after a hurricane may catch some students' attention, it may not best serve other instructional purposes. Research indicates that learners can have very strong emotions in response to climate change topics (Lombardi & Sinatra, 2013). This

---

Returning to My Story: Understanding Our Changing Communities
Recall that for my high school biology course, I was asked to document a local ecosystem and report back to the class about what I observed. This assignment was probably for a unit about ecology and ecosystems (I honestly cannot remember), and everyone gave presentations about different ecosystems in our county with varying depth and focus. I really do not remember much of the specifics, but one student reported on oak trees that were dying of sudden oak death (Cobb et al., 2020), but not really about the trees in the larger context of the ecosystem. I imagine that most of my classmates wondered "Why does this matter?". I likely recall this presentation because I've always loved oak trees, as demonstrated by the acorns I collected, and my copies of Ansel Adams oak tree photos.

Looking back on that experience now, I think about how much more engaging that (presumably ecology and ecosystems?) unit could have been if the

---

(continued)

learning had been anchored in an exploration of our changing landscape and the effects of climate change in Santa Rosa. To launch this unit, we could have made observations from historic and current photographs of familiar places in and around the city that would elicit our thinking about land use (urbanization, deforestation, agriculture) and Earth system processes (water, carbon cycle, etc.). Additionally, we could have explored local data about biodiversity, temperature, and precipitation patterns, or how our local waterways have been altered by dams and agricultural needs over time. Framing the unit in this way would have grounded our learning in opportunities for us to *use* this information to tell the story of where we live and how it has and will change. This exploration could have happened over the course of an entire unit (or even two or three units, there are so many system connections to explore!). The ecosystem in which we lived should not have been a side project as we learned some concepts about ecosystem structure and species interactions. I'm guessing that's what we were learning about? Again, I can't remember the details because the learning was not grounded in examples in the real world, or *my* world.

A unit about local ecosystems could have introduced me to important historical events, like the Hanly Fire of 1964, which burned in almost the exact footprint of the Tubbs Fire that destroyed my family home (Van Niekerken, 2017), or the land management practices used by various native peoples to reduce the risk of destructive fires (Flores & Russell, 2020; Long & Lake, 2018; Marks-Block & Tripp, 2021). We could have explored ecosystems in other parts of the world and then applied that learning to thinking more deeply about our own environment. And there still would have been a place for a project like the one I completed about the construction in the hills above my neighborhood, but I envision students working in groups to share their information in service of *answering* an investigative question about a local issue. For example, "How should we manage water use in our city given current drought conditions and future climate projections?" If this had been the question to guide my original project, I would have reported back about the sources of water that would be needed for the new homes, and the landscaping choices that could be made in response to limited water supplies (there was *a lot* of grass in those yards). This allows students to tell the story of what they have observed, and hopefully, the resilient community they will help to shape in the future.

may happen even if learning is framed in service of understanding how we can build a more promising future. It is best to acknowledge the emotional responses that are felt in your classroom about climate phenomena (Hufnagel, 2015, 2017).

Once you have identified the anchoring phenomenon, you can formulate *unit-driving questions* to help focus and engage students in their learning. For example, using the phenomena of sea level rise and wildfires described above: "How and why have sunny day flooding events become more common in the last 30 years?" Or

"Why are some communities more resilient in wildfire-affected areas than others?" I would also add the question, "How can we make our communities more resilient in response to these changes (flooding or wildfires)?" The driving question(s) should not be answerable with a yes/no response and should require students to connect ideas throughout the unit to explain the focal phenomenon. Driving questions that are sufficiently complicated often include the words HOW and WHY. For this reason, the unit will need to include a series of learning opportunities for students to construct an explanation of the phenomenon. In other words, all learning experiences in the unit should serve a purpose and contribute in some way to students' ability to answer the driving question(s). Student responses to the driving question can be revised and revisited as the unit progresses.

Final thoughts on anchoring phenomena:

- As I mentioned before, selecting a phenomenon can be a very challenging part of the planning process for embedding systems thinking and local connections in your curriculum. It takes time to identify a compelling phenomenon, and do the research to find engaging photos, videos, and data sets. Be patient, try things out, and revise as necessary. Planning for this kind of instruction is a dynamic, iterative process, just like the nature of science itself!

- I also want to remind you that climate change is complex, therefore, you will not always know the answer to student questions, *and that is OK!* Let me repeat, you (and nobody else, Ph.D. climate scientists included) cannot have *all the answers* to *all the questions* we face about this existential threat facing humanity and ecosystems across the entire globe. Your phenomenon might even leave the door open for new questions and ideas from students that were not originally envisioned as part of your unit. That is also OK! Meaningful learning comes from students seeking their own coherence, and it's better if they collaborate with you in researching topics of interest and constructing an understanding of the phenomenon. We need to start feeling more comfortable teaching without knowing all the answers, and finding time to explore climate issues with more depth and purpose.

- Climate change is a multifaceted, difficult problem that forces us to think about how we live our lives, thus asking us to pay attention to our own individual and societal ethics and values. For these reasons, climate change is often referred to as a socio-scientific problem (Peel et al., 2017; Sadler et al., 2004). All of us need food, clean air and water, healthcare, and a safe place to live. But the way we currently live is unsustainable and will lead to more human suffering and ecological destruction unless scientific, economic, and political decisions are made to stop our use of fossil fuels, draw down greenhouse gas levels in our atmosphere, and find ways to adapt to the changes that are now unstoppable. Your climate stories will not always have happy endings. But we desperately need our students to be armed with a scientific understanding of climate change and Earth systems that empower them to fight for the societal changes that could allow us to sustain human populations and ecosystems around the globe for generations to come.

- Finally, make sure your students have ways of sharing their thinking and explanations of phenomena. Learners will have misconceptions about climate change (Shepardson et al., 2017). Therefore, it is important that students feel safe to show what they know, and what they do not know and want to learn. Modeling activities, discussed in the next section, can serve as assessment tools for understanding student thinking.

Storytelling with Modeling

When we tell a story about something that has happened in our lives, to family, friends, or colleagues, we do not always communicate coherently and clearly. Maybe we forgot to include information that the recipient needs to put the pieces together to understand why something happened. Maybe we needed to provide more context for the story about the people or the places involved, or we gave superfluous information that was not relevant to the story, or we recounted something out of sequence that confused. Any number of things can happen. But through questioning and responding to the person receiving the information, hopefully, a more complete story can be provided. If you have ever repeated the same story a few times or edited a manuscript, usually the second or third attempt is more coherent and streamlined than the first. For this reason, students should have opportunities throughout instruction to use what they know and construct and refine explanations of the world around them. The scientific practice of modeling allows us to externalize our stories and explanations, and visualize systems connections (Passmore et al., 2017; Windschitl et al., 2008).

The National Research Council (2012) *Framework for K-12 Science Education* defines models as "concrete 'pictures' and/or physical scale models (e.g., a toy car)" in lower grades "to more abstract representations of relevant relationships in later grades, such as a diagram representing forces on a particular object in a system" (p. 58). Furthermore, modeling can and should serve a purpose in developing ideas, by inspiring new questions, explanations, and predictions, and is not simply an end product to represent scientific concepts (Gouvea & Passmore, 2017). Evidence suggests that the construction of models can support sensemaking and the development of a holistic systems perspective as learners tackle complex problems or explain phenomena (Gilissen et al., 2020; Hmelo-Silver et al., 2017). While the use of modeling is not always tied explicitly to systems thinking, modeling phenomena can support instructional coherence and the development of systems thinking skills (Svoboda & Passmore, 2013). Riess and Mischo (2010) defined systems thinking as "the ability to recognize, describe, *model* and to explain complex aspects of reality as systems" (p. 707). Three of the four competency dimensions used by Schuler et al. (2018) to measure systems thinking skills in students and student teachers explicitly include the learners' ability to construct, evaluate, and *use models* to solve problems or make predictions.

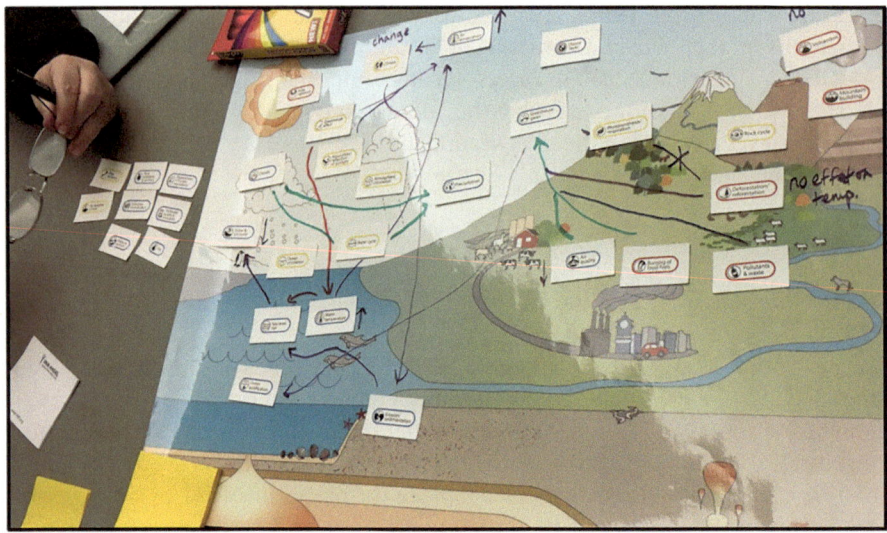

Fig. 2 An example Earth System Model constructed with the Understanding Global Change Earth Scene and icons

The UGC processes and phenomena icons (e.g., precipitation, atmospheric circulation, burning of fossil fuels, biodiversity) can be used to construct models about climate change either using physical materials (cards, paper, pens) or an online modeling tool (Bean & Nielsen, 2018). Modeling can be used both for planning instruction to visualize conceptual links within a unit and as student formative and summative assessments within units of study. Learners can use the UGC Earth Scene as a background for drawing arrows and providing written explanations of the connections among the components at various scales (Fig. 2). Alternatively, students could use a diagram or photograph of their local environment as the background for models, depending on the location and scale of the processes that students want to represent. Additional smaller-scale models of specific processes, such as the greenhouse effect or the combustion of fossil fuels, could also be constructed to help students visualize chemical and physical mechanisms in the system. Iterative modeling allows students to visualize the climate system and determine what they know or want to know about the phenomenon under investigation, and helps the instructor gauge the coherency of students thinking as they construct stories and explanations.

Ideally, students should work collaboratively and share ideas, especially early on in a unit of study, and modeling should be a low-stake activity. As the unit progresses, students could alternate between individual work, group model revisions, and whole class discussions to critique and refine explanations expressed in the model. Models should be revised over time as students explore the cascading effects that a single cause or measurable change can have on other parts of the system and the well-being of their local communities. For example, in a unit about sea level rise, students have constructed increasingly more complex models as they expand

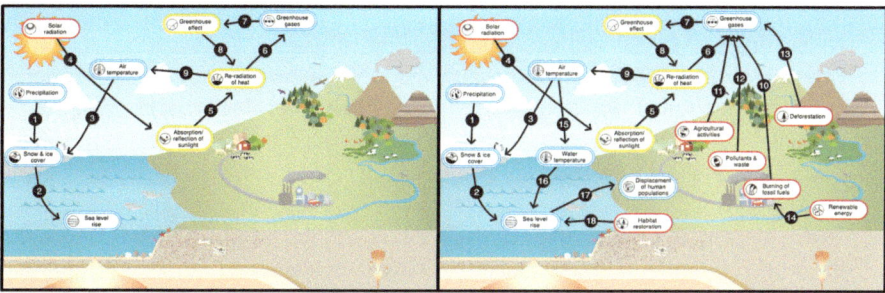

Fig. 3 Models of sea level rise that explain the mechanisms of land ice melt and global warming (left). A more complex model (right) includes the ultimate causes of global temperature increases (emissions from the burning of fossil fuels, agricultural activities, etc.), mitigation (renewable energy, reducing emissions), and adaptation strategies (coastal habitat restoration). These models were constructed in the UGC online interactive (Bean and Nielsen, 2018), a tool that also allows for annotation of the Earth system connections (not shown)

their understanding of the causes and consequences of, and solutions to sea level rise in coastal communities (Fig. 3).

As you plan for instruction and students create models, it will become apparent that all the UGC Framework icons can be interconnected to construct one, complicated (and difficult to interpret) model of the Earth's system. For this reason, if students are allowed to select which icons to include in their models, they are likely to include icons and ideas that are irrelevant to the focal phenomenon, inaccurate, or incomplete. To help guide learners through revising and refining their models, they can use the seven system characteristics to help define their models: boundaries, components, interactions, inputs and outputs, feedback, dynamics, and hierarchy (Gilissen et al., 2020). Exploration of these system's characteristics during model construction can help to structure group and class discussions and clarify student thinking about the part of the climate system under investigation. These models then become resources for students when they complete written assignments or presentations about the phenomenon.

Finally, modeling is a mechanism for envisioning the world we want in the future, and the stories we *want* to tell generations to come. Models can help us explain the changes we need to make individually and collectively to solve the climate and environmental problems we have caused. For example, we can examine the various methods for reducing atmospheric greenhouse gases presented by Project Drawdown (Hawken, 2017), connect them to our system models, and think about how these solutions should play out at local and regional scales. Models can focus learning and thinking on what we value, and guide us toward the actions, both large and small, that will help keep communities around the world safe and sustain the resources on which we depend.

Returning to My Examples: Modeling My Stories

I often wondered about the purpose of the assignments I was given in school. I must admit that I was always curious, which is probably what made me ultimately complete my schoolwork. I was worried I would miss something important. But I often asked, "How could I *use* this information now or in the future?" Modeling climate phenomena in your community answers these questions. It provides an opportunity for learners to recognize how connected the world is, and how much our well-being depends on the state of the Earth system around us.

I wish my classmates and I had been asked to model our stories and explorations of local ecosystems. I doubt that any of us would have predicted how rapidly our community would be changed by wildfires, but our models would have been records of what we knew then and what we valued. We could have shared and collaborated to connect and compare elements of our experiences and investigations instead of working individually and then being evaluated for a grade. We could have collectively thought about how our community is changing, and how we want it to be in the future. These are the stories that will help us understand and respond to the climate crisis.

Final Thoughts

The climate crisis is here, and we need to make instructional shifts now that will prepare the next generation for current and future challenges. The tools and practices presented here support learning about climate change by:

- Leveraging *anchoring phenomena* to *contextualize* instruction in local environments. This makes climate change relevant and immediate and brings purpose and meaning to learning experiences.
- Connecting science learning to climate change *across the curriculum* to understand how the places we live and the resources we need across the globe are affected by climate change.
- Making student thinking visible through *modeling* the Earth system, thus creating space for students to share ideas and construct knowledge.
- Allowing learners to envision a *resilient* future for themselves and their communities.

Don't work on these instructional shifts alone. Find colleagues and networks, such as the Climate Literacy and Energy Awareness Network (CLEAN, 2021), with which you can share your struggles and achievements in teaching climate change.

These practices and tools do not have to be implemented all at once (this is not an all or nothing situation). We should examine and revise curricula from an Earth systems perspective and seek opportunities that allow students to explore their place in the Earth system. Instruction must explore the multifaceted solutions and support learners to envision a future that sustains human communities and ecosystems. Using these techniques, we can explicitly connect learning about the climate crisis to what we value and provide students with opportunities to construct stories and explanations about climate phenomena relevant to their current and future well-being.

Acknowledgments I would like to acknowledge the University of California Museum of Paleontology team, C.R. Marshall, L. White, T. Roque, and H. Chin, and many educators and colleagues, including K. Carlson, J. Levine, J. Kiehl, I. Aiello, A. Oshry, J. Taylor, V. Brunsing, H. Howett, D. Jackson, S. Machado, and J. Lee, who contributed to the design and development of the Understanding Global Change Project. I also want to thank B. Mitchell, J. Totino, and M. Simani for useful feedback and discussions about the project. The Understanding Global Change is supported by funding from the Gordon and Betty Moore Foundation (#3416), the Novim Foundation, and the California Science Project awarded to C.R. Marshall and J.R. Bean.

References

Achieve. (2016). *EQuIP rubric for lessons & units: Science (version 3.0)*. http://www.nextgen-science.org/sites/default/files/EQuIPRubricforSciencev3.pdf

Ballew, M. T., Goldberg, M. H., Rosenthal, S. A., Gustafson, A., & Leiserowitz, A. (2019). Systems thinking as a pathway to global warming beliefs and attitudes through an ecological worldview. *Proceedings of the National Academy of Sciences, 116*(17), 8214–8219.

Bean, J. R. (2020). Planning for instruction. *Understanding Global Change*. https://ugc.berkeley.edu

Bean, J. R., & Marshall, C. R. (2020). Understanding global change 101. *Understanding Global Change*. https://ugc.berkeley.edu/what-is-global-change/

Bean, J. R. & Nielsen, M. (2018). Understanding global change online interactive.. https://www.biointeractive.org/classroom-resources/understanding-global-change

Bean, J. R., Marshall, C. R., & Chin, H. (2020). Understanding global change 101. *Understanding Global Change*. https://ugc.berkeley.edu/what-is-global-change/understanding-global-change-101/

Boesch, D. F., Boicourt, W. C., Cullather, R. I., Ezer, T., Galloway, G. E., Jr., Johnson, Z. P., Kilbourne, K. H., Kirwan, M. L., Kopp, R. E., Land, S., Li, M., Nardin, W., Sommerfield, C. K., & Sweet, W. V. (2018). *Sea-level rise: Projections for Maryland 2018* (27 pp). University of Maryland Center for Environmental Science.

Bowman, T., & Morrison, D. (2021). *Empowering climate action in the United States (Volume 7) (Resetting our future, 7)*. Changemakers Books.

Climate Literacy and Energy Awareness Network (CLEAN). (2021). *CLEAN: Committed to climate and energy education*. cleanet.org

Cobb, R. C., Haas, S. E., Kruskamp, N., Dillon, W. W., Swiecki, T. J., Rizzo, D. M., Frankel, S. J., & Meentemeyer, R. K. (2020). The magnitude of regional-scale tree mortality caused by the invasive pathogen *phytophthora ramorum. Earth's Future, 8*. https://doi.org/10.1029/2020EF001500

Flores, D., & Russell, G. (2020). Integrating tribes and culture Into public land management [Chapter 5.5]. In R. K. Dumroese & W. K. Moser (Eds.), *Northeastern California plateaus bioregion science synthesis. Gen. Tech. Rep. RMRS-GTR-409* (pp. 177–185). *US Department of Agriculture, Forest Service, Rocky Mountain Research Station*, 409, 177–185.

Gilissen, M. G. R., Knippels, M. C. P. J., & van Joolingen, W. R. (2020). Bringing systems thinking into the classroom. International Journal of Science Education, 0(0), 1–28. https://doi.org/1
0.1080/09500693.2020.1755741

Gouvea, J., & Passmore, C. (2017). Models of' versus 'models for. Science & Education, 26(1–2),
49–63. https://doi.org/10.1007/s11191-017-9884-4

Hawken, P. (2017). Drawdown: The most comprehensive plan ever proposed to reverse global
warming. Penguin Books.

Hmelo-Silver, C. E., Jordan, R., Eberbach, C., & Sinha, S. (2017). Systems learning with a conceptual representation: A quasi-experimental study. Instructional Science, 45(1), 53–72. https://
doi.org/10.1007/s11251-016-9392-y

Hufnagel, E. (2015). Preservice elementary teachers' emotional connections and disconnec-
tions to climate change in a science course. Journal of Research in Science Teaching, 52,
1324–1296.

Hufnagel, E. (2017). Attending to emotional expressions about climate change. In
D. P. Shepardson, A. Roychoudhury, & A. S. Hirsch (Eds.), Teaching and learning about
climate change: A framework for educators (pp. 69–84). Essay, Routledge, an imprint of the
Taylor & Francis Group.

Ilati, M., & Moriarty, D. (2021, September 17). Anatomy of a wildfire: How the Dixie Fire became
the largest blaze of a devastating summer. The Washington Post. https://www.washingtonpost.
com/climate-environment/interactive/2021/dixie-fire/

Kramer, H. A., Mockrin, M. H., Alexandre, P. M., & Radeloff, V. C. (2019). High wildfire damage in interface communities in California. International Journal of Wildland Fire, 28(9), 641.
https://doi.org/10.1071/wf18108

Lehtonen, A., Salonen, A. O., & Cantell, H. (2019). Climate change education: A new approach
for a world of wicked problems. In J. W. Cook (Ed.), Sustainability, human well-being, and the
future of education. Essay, Palgrave Macmillan. https://doi.org/10.1007/978-3-319-78580-6_11

Lombardi, D., & Sinatra, G. (2013). Emotions about teaching about human-induced climate
change. International Journal of Science Education, 35, 167–191.

Long, J. W., & Lake, F. K. (2018). Escaping social-ecological traps through tribal stewardship on
national forest lands in the Pacific Northwest, United States of America. Ecology and Society,
23(2), 1–14.

Mann, M. L., Berck, P., Moritz, M. A., Batllori, E., Baldwin, J. G., Gately, C. K., & Cameron,
D. R. (2014). Modeling residential development in California from 2000 to 2050: Integrating
wildfire risk, wildland and agricultural encroachment. Land Use Policy, 41, 438–452.

Marder, J. (2020). NASA sea level change, observations from space: Beating back the tides. https://
sealevel.nasa.gov/news/203/beating-back-the-tides/

Marks-Block, T., & Tripp, W. (2021). Facilitating prescribed fire in Northern California through
indigenous governance and interagency partnerships. Fire, 4(3), 37. https://doi.org/10.3390/
fire4030037

National Research Council. (2012). A framework for K-12 science education: Practices, crosscut-
ting concepts, and core ideas. The National Academies Press.

Next Generation Science Storylines. (2018). www.nextgenstorylines.org/

NGSS. (2016). Phenomena. https://www.nextgenscience.org/resources/phenomena

NGSS Lead States. (2013). Next generation science standards: For states, by states. The National
Academies Press.

Passmore, C., Schwarz, C. V., & Mankowski, J. (2017). Developing and using models. In
C. V. Schwarz, C. Passmore, & B. J. Reiser (Eds.), Helping students make sense of the world
using the next generation science and engineering practices (pp. 109–134). NSTA Press.

Peel, A., Sadler, T. D., Kinslow, A. T., Zangori, L., & Friedrichsen, P. (2017). Climate change as an
issue for socio-scientific issues teaching and learning. In D. P. Shepardson, A. Roychoudhury,
& A. S. Hirsch (Eds.), Teaching and learning about climate change: A framework for educators
(pp. 69–84). Routledge.

Petersen, C. I., Baepler, P., Beitz, A., Ching, P., Gorman, K. S., Neudauer, C. L., Rozaitis, W.,
Walker, J. D., & Wingert, D. (2020). The tyranny of content: "Content coverage" as a barrier to

evidence-based teaching approaches and ways to overcome it. *CBE—Life Sciences Education, 19*(2), ar17. https://doi.org/10.1187/cbe.19-04-0079

Plate, R. R., & Monroe, M. (2014). A structure for assessing systems thinking. *The Creative Learning Exchange, 23*(1), 1–12.

Reiser, B. J., Novak, M., & McGill, T. A. W. (2017). *Coherence from the students' perspective: Why the vision of the framework for K-12 science requires more than simply "combining" three dimension of science learning.* Paper presented at the Board on Science Education Workshop "Instructional Materials for the Next Generation Science Standards. http://sites.nationalacademies.org/cs/groups/dbassesite/documents/webpage/dbasse_180270.pdf

Riess, W., & Mischo, C. (2010). Promoting systems thinking through biology lessons. *International Journal for Science Education, 32*, 705–725.

Roychoudhury, A., Shepardson, D., Hirsch, A., Niyogi, D., Mehta, J., & Top, S. (2017). The need to introduce system thinking in teaching climate change. *Science Educator, 25*(2), 73–81.

Sadler, T. D., Chambers, W. F., & Zeidler, D. L. (2004). Student conceptualizations of the nature of science in response to a socioscientific issue. *International Journal of Science Education, 26*(4), 387–409.

Schuler, S., Fanta, D., Rosenkraenzer, F., & Riess, W. (2018). Systems thinking within the scope of education for sustainable development (ESD) – a heuristic competence model as a basis for (science) teacher education. *Journal of Geography in Higher Education, 42*(2), 192–204. https://doi.org/10.1080/03098265.2017.1339264

Shepardson, D. P., Niyogi, D., Roychoudhury, A., & Hirsch, A. (2012). Conceptualizing climate change in the context of a climate system: Implications for climate and environmental education. *Environmental Education Research, 18*(3), 323–352. https://doi.org/10.1080/1350462 2.2011.622839

Shepardson, D. P., Roychoudhury, A., Hirsch, A., & Top, S. M. (2017). Student's conception of a climate system: Implications for teaching and learning. In D. P. Shepardson, A. Roychoudhury, & A. S. Hirsch (Eds.), *Teaching and learning about climate change: A framework for educators* (pp. 69–84). Routledge.

Sikorski, T.-R., & Hammer, D. (2017). Looking for coherence in science curriculum. *Science Education, 101*(6), 929–943. https://doi.org/10.1002/sce.21299

Sommer, L. (2019). *This California neighborhood was built to survive a wildfire. And it worked KQED.* https://www.kqed.org/science/1941685/this-california-neighborhood-was-built-to-survive-a-wildfire-and-it-worked

Svoboda, J., & Passmore, C. (2013). The strategies of Modeling in biology education. *Science and Education, 22*(1), 119–142. https://doi.org/10.1007/s11191-011-9425-5

Sweeney, L. B., & Sterman, J. D. (2000). Bathtub dynamics: Initial results of a systems thinking inventory. *System Dynamics Review, 16*(4), 249–286. https://doi.org/10.1002/sdr.198

Sweeney, L. B., & Sterman, J. D. (2007). Thinking about systems: Student and teacher conceptions of natural and social systems. *System Dynamics Review, 23*(2–3), 285–311. https://doi.org/10.1002/sdr.366

Taylor, J., Thomas, L., Penuel, W., & Sullivan, S. (2019). Food fight! *The Science Teacher, 87*(1), 42–49.

Van Niekerken, B. (2017). *Wine country fire of 1964: Eerie similarities to this week's tragedy.* San Francisco Chronicle.

Verhoeff, R. P., Waarlo, A. J., & Boersma, K. T. (2008). Systems modelling and the development of coherent understanding of cell biology. *International Journal of Science Education, 30*(4), 543–568. https://doi.org/10.1080/09500690701237780

Verhoeff, R. P., Knippels, M.-C. P. J., Gilissen, M. G. R., & Boersma, K. T. (2018). The theoretical nature of systems thinking. Perspectives on systems thinking in biology education. *Frontiers in Education, 3*(June), 1–11. https://doi.org/10.3389/feduc.2018.00040

Watkins, D., Griggs, T., Lee, J.C., Park, H., Singhvi, A., Wallace, T., & Ward, J. (2017, October 21). How California's most destructive wildfire spread, hour by hour. *The New York Times.* https://www.nytimes.com/interactive/2017/10/21/us/california-fire-damage-map.html

Windschitl, M., Thompson, J., & Braaten, M. (2008). Beyond the scientific method: Model-based inquiry as a new paradigm of preference for school science investigations. *Science Education, 92*, 941–967.

Windschitl, M., Thompson, J. & Braaten, M. (2014). *Planning for engagement.* http://ambitious-scienceteaching.org/wp-content/uploads/2014/08/Primer-Plannning-for-Engagement.pdf

Windschitl, M., Thompson, J., & Braaten, M. (2018). *Ambitious science teaching* (Illustrated ed.). Harvard Education Press.

Windschitl, M., Thompson, J., Braaten, M., & Stroupe, D. (2012). Proposing a core set of instructional practices and tools for teachers of science. *Science Education, 96*(5), 878–903. https://doi.org/10.1002/sce.21027

York, S., Lavi, R., Dori, Y. J., & Orgill, M. K. (2019). Applications of systems shinking in STEM education. *Journal of Chemical Education, 96*(12), 2742–2751. https://doi.org/10.1021/acs.jchemed.9b00261

Young, R. D., & Monroe, M. C. (1996). Some fundamentals of engaging stories. *Environmental Education Research, 2*(2), 171–187. https://doi.org/10.1080/1350462960020204

Zeidler, D. L., & Newton, M. H. (2017). Using a socioscientific issues framework for climate change education: An ecojustice approach. In *Teaching and learning about climate change* (pp. 56–65). Routledge.

Jessica R. Bean is the leader of the Understanding Global Change Project at the Museum of Paleontology at University of California, Berkeley. She designs tools for learning about the Earth as a dynamic, interconnected system, and partners with K-12 educators to develop and implement new resources for learning about the nature and process of science and climate change. Jessica studies the effects of environmental change on marine invertebrates along the California coast, uses stable isotope analyses to reconstruct past nearshore oceanographic conditions, and has taught college and graduate-level biology and Earth science courses. Previously, she ran an NSF Graduate STEM Fellows in the K-12 Education Program at the University of California, Davis, where she received her Ph.D. in Earth and Planetary Sciences.

What We Need Now to Accelerate Climate Solutions through Storytelling

Emily Coren and Hua Wang

We are in a planetary race, and the climate crisis deserves the use of all tools at our disposal to achieve the recommended mitigation and adaptation goals. Effective communication strategies are necessary to accelerate climate solutions at the required speed, scale, and scope, and they can be designed and implemented based on decades of research in behavior science. The future is ours to choose (Figueres & Rivett-Carnac, 2021; Hawken, 2021). It is not too late to change the climate stories we tell ourselves and each other; difficult is not the same as impossible and nothing is inevitable (Solnit & Lutunatabua, 2023).

Time Is Now and the Future Lies in Us

The window for mitigating global warming is closing and closing fast before its impact becomes irreversible. However, we do have the capacity to avert this crisis in time. The urgency to scale climate solutions and avoid an unprecedented catastrophe for humanity is imminent. The Intergovernmental Panel on Climate Change published its latest reports (IPCC, 2022a, 2023a) with the press release emphasizing messages like, "Pace and scale of climate action are insufficient to tackle climate change" and "The evidence is clear: The time for action is now" (IPCC, 2022b, 2023b).

Although the climate crisis may seem insurmountable and often triggers anxiety and fear (Wang et al., 2023), many individuals and groups have come together, especially in recent years, to collect, evaluate, and share evidence-based climate

E. Coren (✉)
Stanford University, Stanford, CA, USA
e-mail: emilycoren@gmail.com

H. Wang (✉)
University at Buffalo, The State University of New York, Buffalo, NY, USA
e-mail: hwang23@buffalo.edu

421

solutions and organize collaborations and movements to advocate for climate empowerment and action (Figueres & Rivett-Carnac, 2021; Hawken, 2021; Solnit & Lutunatabua, 2023). For example, nongovernmental organizations such as Project Drawdown (n.d.) provide resources for achieving climate goals. These resources include detailed listings on source reduction for carbon dioxide emissions (e.g., transitions to renewable energy and increasing electrification) and carbon sinks (e.g., regenerative agricultural practices and habitat restoration). All these resources are periodically updated and provide excellent recommendations for climate actions.

Climate solutions are no longer just for governments, scientists, and activists; they must include and engage everyone. Leading an inclusive regeneration movement, the creator of *Drawdown* and environmentalist Paul Hawken (2021) urges us that, "Tens of thousands of organizations, teachers, companies, architects, farmers, Indigenous cultures, and native leaders know what to do and are active in implementation. The current growth of the climate movement is magnificent, but it remains a small fraction of the world. Hundreds of millions of people need to realize that they have agency, that they can take action, and that collectively it is possible to prevent runaway global warming" (p. 10).

Now the question is how?! It is well known in the medical community that, on average, it takes 17 years for scientific solutions to be translated into clinical practice (Green et al., 2009; Morris et al., 2011). Yet, accelerated solutions can be developed, validated, and scaled at an unprecedented rate in a polarized world as we have witnessed through the COVID-19 global pandemic in the last few years (Collins & Stoffels, 2020; Corey et al., 2020; Suran, 2022). In addition, public responses to changing guidelines such as COVID masking, indoor air filtration, and vaccination measures provide insights into how different the outcomes can be between public communication strategies that were thoughtfully designed and coordinated versus those that were not (Sachs et al., 2022).

Current climate change communication sometimes can be confusing, hard to relate, and difficult to act on or to scale up (Scott, 2023; Stoknes, 2017). To improve public engagement, we need messages that are "simple and clear, repeated often, by a variety of trusted and caring messengers," and to promote climate action, we need to make the recommended behaviors "easy, fun, and popular" (Maibach et al., 2023, p. 54). The future of humanity and all life on Earth lies within us but we cannot afford to waste any more time in response to the climate change crisis. We need to admit that climate change communication that creates enabling conditions and facilitates concerted and transformative actions is just as important, if not more, as finding effective climate solutions. Information is helpful but information alone is not sufficient to foster real change. Storytelling—purposefully designed, implemented, and integrated into a larger and funded public communication system to help coordinate multilevel and interdisciplinary efforts—is what we need now to accelerate climate solutions at the necessary speed, scale, and scope to foster real change.

Storytelling as a Climate Change Communication and Action Strategy

Storytelling is the practice of human communication with information shared in a narrative format (Fisher, 1987). As humans, we are wired as storytellers and crave stories (Abbott, 2002; Brown, 2015; Gottschall, 2012). From cave paintings and oral histories to Sundance documentaries and Netflix series, storytelling has been ingrained in our everyday experiences and shaping our cultures and societies since the dawn of humanity (Abbott, 2002; Brown, 2015; Gottschall, 2012; Wang & Coren, 2024). So, what makes a story? Although definitions and terminologies may vary across disciplines and over time, storytellers and narrative researchers generally agree that key elements and a path or arc should be in place. For example, Kreuter et al. (2007) defined narrative as "a representation of connected events and characters that has an identifiable structure, is bounded in space and time, and contains implicit and explicit messages about the topic being addressed" (p. 222).

What are the advantages of storytelling over expository or didactic non-narrative communication? Research in neuroscience, social sciences, and humanities has shown that information presented in a narrative format is easier to attract and sustain attention and creates shortcuts to facilitate comprehension and recall because it is more accessible and engaging (Gottschall, 2012; Polkinghorne, 1988; Schneider-Mayerson et al., 2023). Powerful stories can change people's knowledge, attitudes, practices, and social norms (Bálint & Bilandzic, 2017; Bilandzic & Kinnebrock, 2009; Green et al., 2002; Murphy et al., 2015; Riley et al., 2022; Singhal & Rogers, 1999; Singhal et al., 2013). We rely on storytelling to organize information, generate meaning, and make sense of ourselves in the world and how we are related to each other (Gottschall, 2012; Polkinghorne, 1988). More importantly, for the purpose of this book, stories can evoke empathy and compassion, foster cooperation and collaboration, and drive human agency and action (Bietti et al., 2019; Bruner, 1990; Keen, 2006; Schneider-Mayerson et al., 2023; Smith et al., 2017). Storytelling can serve as an effective strategy for science communication (e.g., Dahlstrom, 2014; Martinez-Conde & Macknik, 2017) and climate action (e.g., McComas & Shanahan, 1999; Morris et al., 2019).

Climate change communication research to date has largely focused on either endogenous (e.g., demographic characteristics, individual knowledge, and cultural beliefs) or exogenous factors (e.g., geographic locations, extreme weather experiences, and news media framing) (Jones & Peterson, 2018). While they have provided important understandings of individual responses to and public opinions about climate change, their investigations follow the rational and linear knowledge-attitude-practice (KAP) model with considerable empirical studies stopping at behavioral intention or likelihood of policy support as outcome variables. Yet, more scholars are pointing out now that we should move beyond the information deficit model and overcome the historically documented attitude-behavior gaps by prioritizing factors that facilitate and accelerate individual and collective empowerment and action (De Meyer et al., 2021; Harris, 2020; Hendersson & Wamsler, 2020; Jones & Peterson, 2018).

The idea of using storytelling in climate change communication is not new. Influential figures like Thomas Berry started advocating for conscious climate storytelling in the 1970s and the 1980s (Berry, 1988, 2003). "It's all a question of story. We are in trouble just now because we do not have a good story. We are in between stories" (Berry, 1988, p. 123). In more recent years, we see the efforts of applying the narrative policy framework, helping scientists become better storytellers, as well as individuals and groups working together to claim authorship of their climate stories (e.g., Climate Beacon Newsroom, 2022; Climate Central, 2021; Climate Solutions Cohort, 2022; De Meyer et al., 2021; Harris, 2020; Hendersson & Wamsler, 2020; Jones & Peterson, 2018). The capacity of narrative engagement through deep emotions, perspective-taking, and imagination is undeniably powerful for reducing counterarguing and complimentary to rational communication approaches (Bilandzic, 2023; Green et al., 2019). While we acknowledge storytelling as an effective tool for persuasion when used consciously for public good, we want to emphasize that, by communicating universally accepted truths through narratives, storytelling is an indispensable way to bring people together and build communities of practice (e.g., Coren & Tiwathia, 2023; El Amiri et al., 2020; De Meyer et al., 2021; Harris, 2020; Maibach et al., 2021; Sarfaty et al., 2022; Scott, 2023; Stoknes, 2017). And we need to bring people with diverse backgrounds together now and connect them in meaningful ways that enable, foster, and catalyze climate solutions. Some have even advocated incorporating storytelling in IPCC reports for better public understanding and engagement (Bloomfield & Manktelow, 2021).

Power of Storytelling in Popular Entertainment for Social and Behavioral Change

In a recent review, Maibach et al. (2023) thoughtfully articulated the rationale, evidence, and strategies for harnessing the power of communication and behavior science to promote climate empowerment and action. They used exemplary initiatives such as the Climate Matters program to demonstrate guiding heuristics. They recommended: Keeping climate change communication messages simple and clear, having them designed by an interdisciplinary science communication team, presenting them with diverse and trustworthy messengers across different platforms several dozen times, and making climate actions easy, fun, and popular to overcome personal, social, and situational barriers (Maibach et al., 2023). The practices, experiences, and insights in our book are complimentary to this review, and other similar efforts not only with a focus on narrative communication for social and behavioral change but also on specific ways stories can be created to help speed up and scale up change, especially through popular entertainment.

Decades of research (e.g., Green et al., 2002, 2019; Frank & Falzone, 2021; Riley et al., 2022; Singhal et al., 2004, 2013; Singhal & Rogers, 1999; Storey & Sood, 2013) have shown that:

- Messages embedded in popular entertainment can quickly reach mass audiences and attract significant public attention.
- These messages are stickier than conventional campaigns because of underlying mechanisms such as narrative transportation, character identification, and para-social interaction that can enhance cognitive, emotional, and social engagement during and after the exposure.
- Many narrative genres and communication platforms naturally allow the space for repeated messaging and extended exposure for longer-term and sustained engagement.
- The more thoughtfully designed and implemented narrative entertainment experiences are, especially when behavioral modeling is incorporated and connected to matching resources in the real world, the more effective they can be in facilitating prosocial behaviors, which can range from starting conversations about taboo topics to collective actions that lead to policy change.

In the introductory chapter, Wang & Coren, (2024) reviewed the evolution and exemplary projects in entertainment-education: a social and behavioral change communication strategy that is built on over half a century of theory, practice, and research to purposefully leverage the power of storytelling in popular entertainment—with deliberate and collaborative efforts through creative content production, implementation across communication platforms, as well as program monitoring and evaluation—to tackle seemingly intractable problems in the real world and create enabling conditions for desirable and sustainable change locally, regionally, nationally, and globally (Wang & Singhal, 2021).

Here, we want to point out the importance of alignment of six key elements of the entertainment-education strategy (Fig. 1). (1) *Interdisciplinary partnerships*: The nature of complexity and urgency of seemingly intractable issues requires expertise in many different domains. No one can single-handedly solve a global crisis like the COVID-19 pandemic or climate change. Forging complimentary yet interdisciplinary partnerships is a prerequisite for creating any meaningful initiative. (2) *Social objectives*: Having clear and measurable objectives for social impact will guide the project team's priorities in planning, implementation, and assessment. It is nice to raise awareness and shift attitudes, but without enabling the actual behaviors, the change will only stay in our heads while lives are slipping away in the real world. (3) *Narrative contents*: The art and science of storytelling will continue to evolve along with centuries-old traditions and emerging technologies and trends in contemporary society. Translating social objectives and scientific recommendations into thought-provoking, emotionally engrossing, and action-oriented change-making narratives will lie in the quality collaboration between professional storytellers and content experts. (4) *Communication platforms*: There are a myriad of ways to reach and engage the public while tailoring the experiences for different audience segments these days. Meeting them where they already are and

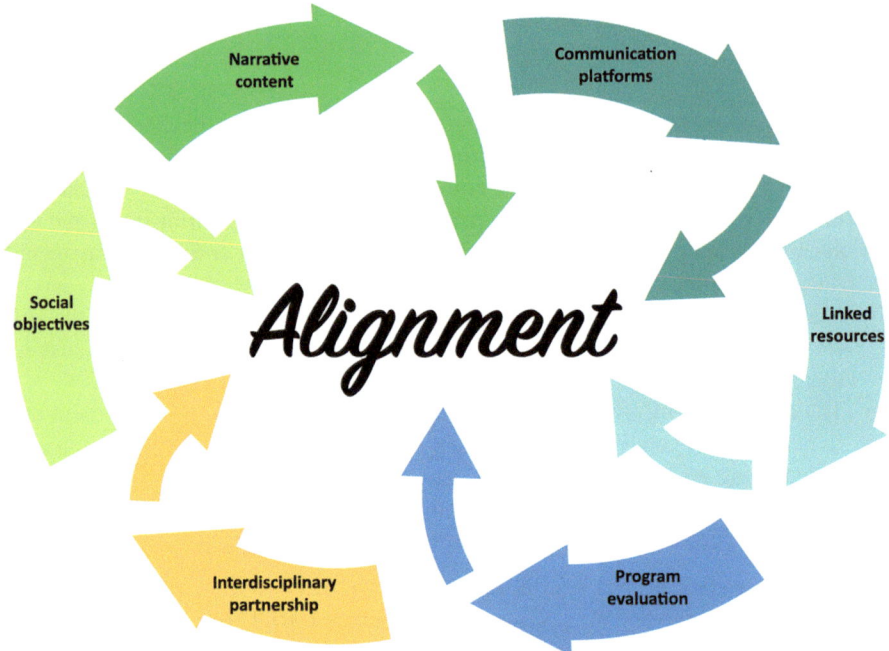

Fig. 1 Alignment circle of key elements in entertainment-education strategy

communicating with them in compelling ways over time are imperative to achieving the desired outcomes. (5) *Linked resources*: Providing linkage to existing or newly created resources is an indispensable step for facilitating real actions from information-seeking and interpersonal discussions to community outreach and policy advocacy. This gets people closer to behavior change right when they are inspired and motivated to contribute to the cause. (6) *Program evaluations*: Rigorous research before (formative evaluation), during (process evaluation), and after (summative evaluation) a social and behavioral change communication initiative can inform message design, monitor project progress, and document intended and unintended outcomes and larger social impact. This element sets entertainment-education apart from many popular entertainment projects with good intentions but fails to facilitate meaningful change.

This approach that leverages the power of storytelling in popular entertainment has become a critical social and behavioral change communication strategy (Wang & Singhal, 2021). Creative professionals in the entertainment industry are increasingly interested in addressing health and social issues of our time, and more people are making deliberate efforts to work with subject matter experts to create compelling and life-changing narrative experiences (Frank & Falzone, 2021; Rippberger, 2022; Singhal et al., 2013; Skoll Center for Social Impact Entertainment, 2019; SIE Society, n.d.; Wang & Singhal, 2021). While social impact entertainment and entertainment-education have been expanding and thriving in the United States and around the world to promote healthy behaviors and human rights in many domains, there is still a "glaring absence" of climate change according to a recent review of 37,453 scripted TV episodes and films from 2016 to 2020 by the USC Norman Lear Center (Giaccardi et al., 2022). This is a tremendous, missed opportunity to accelerate climate solutions, especially when the field has established evidence on the low costs such as Population Media Center's (n.d.) estimate of $2.54USD per person to start family planning, high returns of investment such as the 24:1 ratio in Hutchinson et al. (2022) report on youth development by Shujaaz, and significant societal impacts such as the reduced Mexico's national population growth by millions as Sabido recounted in the Sundance documentary about entertainment-education (Friedman, 2013). The chapters in this book fill in the gap of climate storytelling in popular entertainment by providing timely and rich knowledge, grounded in theory and praxis across disciplines and geographies.

Narrative Strategies Across Disciplines and Geographies

This book, *Storytelling to Accelerate Climate Solutions*, brings together scholars and practitioners from different fields to share knowledge, experience, and insight about how stories can be purposefully designed and effectively told to engage, enable, and empower various populations in climate communication and action (Fig. 2). As summarized in the introductory chapter (Wang & Coren, 2024), we present a wide range of current strategies and exemplary applications of climate storytelling, in terms of professional practices (e.g., popular media, journalism, literature, performing arts, and education), narrative genres (e.g., drama, comedy, and fiction), media platforms (e.g., television, radio, mobile, and immersive virtual environments), communication modalities (e.g., text, visual, audio, and multisensory), and geographic locations across five continents (e.g., US, UK, India, Nigeria, and Peru). This is our first step to leverage the work that has been done, build a multidisciplinary community of practice, facilitate coordinated efforts to promote agency, and advocate for increased support.

The first part of this volume focused on evidence- and practice-based narrative strategies of entertainment-education and social impact entertainment. Sood et al. (2024) provided us with a comprehensive review in a 20-year span of social and behavioral change communication initiatives tackling the climate crisis from the Global South to the Global North. Brown (2024) showcased PCI Media's "My Community Methodology" in collaboration with environmental sustainability initiatives in West Africa by engaging local actors through a participatory approach to narrative development to optimize program impacts. Bish (2024) offered the Population Media Center's perspective on connecting the climate crisis with global population growth and the role of gender equity for a more just and sustainable living on Earth. Garg et al. (2024) shared BBC Media Actions' recent collaboration in Indonesia that uses storytelling across media platforms for youth engagement on traditional and social media. Falzone and the PVI team (2024) told

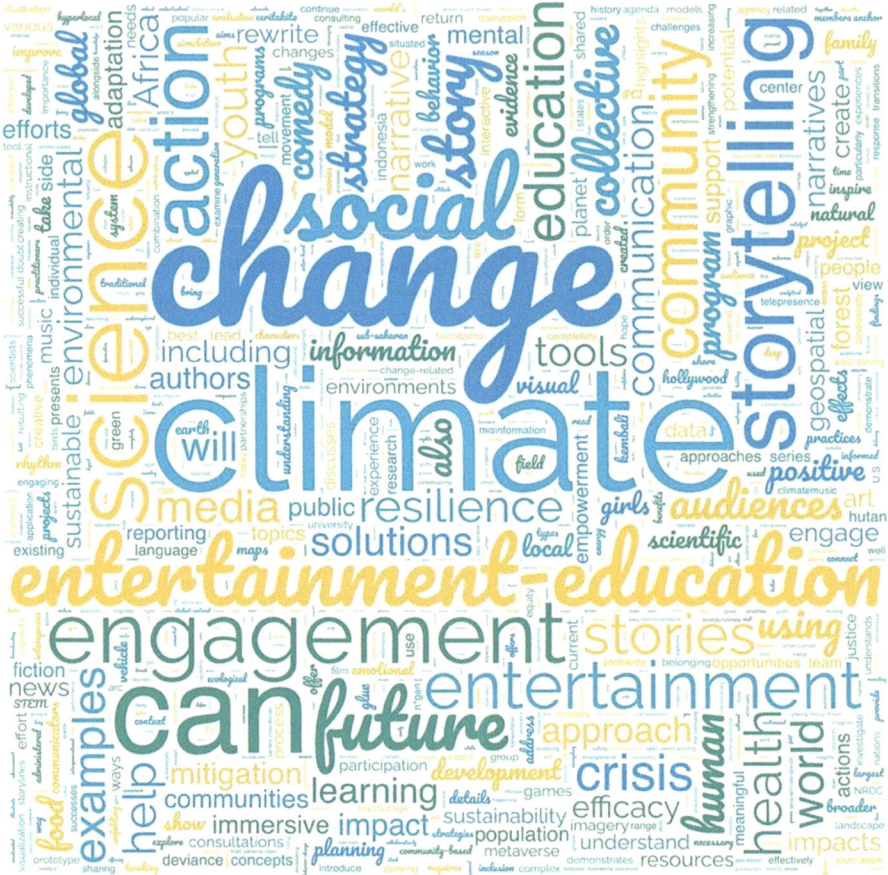

Fig. 2 Word cloud of the titles, keywords, and abstracts Chapters 2–19

us the tales of now their Peabody-nominated children's television series NGen in Sub-Saharan Africa, inspiring holistic learnings of science to promote climate education and human agency. Then, we transitioned to explore how entertainment-education and social impact entertainment strategies have been used and can potentially be improved in the Global North, represented by the United States. A comedy-drama prototype for climate communication tailored toward young adult Americans is proposed with detailed character development and story arcs based on theory and research (Coren, 2024). A newly established Rewrite the Future program at the Natural Resources Defense Council (NRDC) has been helping Hollywood better incorporate climate change themes into their entertainment programming to promote effective solutions (Hinderfeld et al., 2024). Despite the seriousness of the climate crisis, comedians are having a ball firing their secret weapons to wake up their audiences and get them involved in climate action (Gurney & N'Diaye, 2024).

In the second part of this volume, we have included many other creative storytelling strategies for climate change communication. Climate fiction (or cli-fi) is a literary genre that uses fictional characters and storylines to inspire green actions (Baden & Brown, 2024). Visual illustrations created by professionals have a long history in science communication and should be increasingly supported to produce quality imagery to effectively convey messages about climate change (Monoyios et al., 2024). Music, with powerful melodies and lyrics, can evoke deep human emotions to care about planet Earth, as shown in The ClimateMusic Project (Dixon et al., 2024). The evolution of the food we consume can also tell compelling stories about the impact of global warming (Eiseman & Hoffman, 2024). Practices in journalism can benefit from cross-disciplinary teamwork, audience engagement, and artificial intelligence-driven technologies to generate hyperlocal news reporting that promotes actionable solutions (Whitwell, 2024). Understanding community-based resilience and learning from positively deviant initiatives can help us boost collective efficacy and reduce mental health challenges related to climate change (Cosentino et al., 2024). Maps and geospatial software applications can be powerful tools for climate storytelling and empowerment (Wolf-Jacobs et al., 2024). Interactive storytelling in immersive environments can create conditions for self-directed deep learning and large-scale public engagement in climate science communication (Spiegel & Wang, 2024). Inviting youth to observe and embody birds through puppetry, costumes, and movement in outdoor settings can better integrate the themes of equity and inclusion into environmental preservation (Osnes et al., 2024). In addition, instructional storylines developed by the Understanding Global Change Project use coherent sequences of lessons to help students connect their lived experiences with scientific concepts of systematic change (Bean, 2024).

Actionable Next Steps for Climate Storytelling

So far, in both the introductory chapter (Wang & Coren, 2024) and this concluding chapter, we have discussed the role of storytelling in climate change communication and the value and power of leveraging narrative experiences to promote climate action. Building on the collective wisdom of all contributors in this book as well as individuals and organizations that have inspired us, we leave you with a tiered model of actionable next steps. The three tiers represent what could be possible with varying levels of funding and institutional support (Fig. 3).

Tier 1: Grassroots Organizing

At the very minimum, assuming little to no investment in a unified communications effort, we provide guidance for existing programs on how to best support their current communications teams in crafting climate action stories that boost human

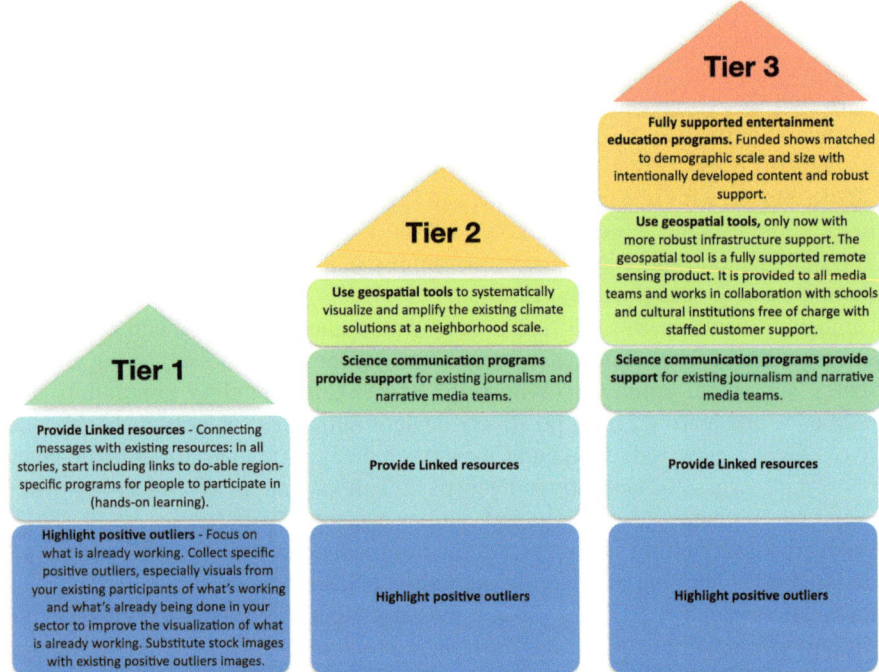

Fig. 3 Tiered model of storytelling for climate change communication and action

agency. We recommend highlighting *positive outliers* and providing *linked resources*. First, start by identifying positive outliers by focusing on what is already working and shine a spotlight on those behaviors inspiring others to amplify it through their own actions. Positive outliers are behavioral traits that support people, individually or collectively, to achieve significantly better outcomes than their peers despite facing similar challenges (De Meyer et al., 2021; Wang et al., 2023). For example, they can be early adopters of electrification transitions in buildings and public transport. They can also be youth groups that have built a movement and advocated for policy change. These positive outliers are not limited to certain professions or social roles. What sets them apart is that their actions are highly effective but may be invisible to others. For example, a medical facility may be transitioning to all-renewable energy that can substantially reduce emissions, but members of its local community or the public may never hear about it. Therefore, we recommend intentionally seeking positive outliers within your domain of practices, especially increasing the images of existing participants that showcase what is working and already being done so you can feature their climate solutions in your story, linguistically, visually, and culturally. Stories about positive outliers are empowering, can boost self and collection efficacy, and foster actions especially when linked to enabling resources. Therefore, the second part of this recommendation is to provide linked resources to existing programs. For example, climate journalists can start including links to action-oriented and region-specific programs for people to obtain hands-on experience in their own communities. As part of our effort to advocate for positive and action-oriented framing in climate storytelling (De Meyer et al., 2021; Wang et al., 2023), we want to make a special note here that stock images do not represent realities. There is nuance to authentic visual representations that cannot be captured by stock photography. Public health messaging has developed an overreliance on stock images, but they impair our ability to relate at a personal level. This opacity is a consequence of systematic underfunding for communications efforts. These are technically resolvable issues with a little additional time and energy. Visual representations intersect with equity issues, and they can be either a tool for improving health equity or, when inadequately funded, inadvertently reinforcing health disparities (Charani et al., 2023; Chichester et al., 2023). Images that are specific to the real communities and exemplary members are necessary for people to see themselves and their peers participating in climate actions and stewardship.

Tier 2: Added Geospatial and Professional Support

One level up, given some investment in coordinated climate change communication efforts, we recommend programs establishing dedicated science communication teams. These teams can include content experts who can provide scientific knowledge and behavior solutions that need to be promoted, social scientists who can understand and analyze people's information processing and behavioral patterns, and science communicators who can craft the messages and facilitate high-quality production of narrative experiences (Maibach et al., 2023). In addition, the geospatial tools previously discussed (Wolf-Jacobs et al., 2024) can be constructed into a base layer for generating an up-to-the-moment visual database of what climate action looks like regionally. This reference-material database can enable a range of linked resources, with topics sortable under subheadings specific to climate mitigation and adaptation solutions. We can also build in program supports for media teams similar in function to existing programs like the Climate Matters program for weathercasters connecting media programs to these real-time reference material feeds. The Climate Matters program already provides data analysis, localized graphics, science explainers, and other reporting resources (Maibach et al., 2023). Media partners could use these high-quality materials to tell science-based stories about how to participate in existing climate change actions. They can provide graphics support showing how the existing efforts interlink while highlighting needed areas of growth. These tools can be subspecialized for niches in communities of practice. For example, food storytelling and music teams would have different needs than journalists, and programs can be developed to meet the needs of each. The combination of a geospatial storytelling database and structured support for communities of practice such as environmental journalists and screenwriters provides reference materials and linked resources incorporated into existing media production pathways.

Tier 3: Fully Supported Communications Interventions

Ultimately with favorable policies and sufficient funding, we craft geospatial tools that are fully supported remote sensing products. They are provided to all media teams and in collaboration with schools and cultural institutions free of charge with staffed customer service support. They serve as a timing belt, of a sort, to provide coordination and reference materials to media teams. More importantly, we now can also employ fully supported entertainment-education programs that can be planned to match the needs of different audience segments and implemented at scale. This includes all media and distribution formats. You can work backward from what changes that you need to create and then plan the intervention to match your community's needs and social objectives. If funding limitations weren't an issue, we can map out a custom strategy for each region that is representative of the population groups and climate action priorities. Popular entertainment programs can be constructed in a similar pattern to the Rhythm and Glue prototype (Coren, 2024) launched for each demographically representative population and connected to actionable, regional implementation programs. Intentionally developed narrative contents can be created from scratch and reinforced through transmedia platforms (Singhal et al., 2013; Wang & Singhal, 2016; Wang et al., 2019) and crosslinked with media programs that are using the geospatial tool for reference materials but are not entertainment-education programs themselves. This tier, unlike the others, includes full research support for formative, process, and summative program evaluations that can shape the programming, monitor audience engagement, and assess the social impact it has. These formal program evaluations can more systematically guide the design of messaging embedded in the entertainment narrative experience and associated media products. In a fully developed entertainment-education plan, there are components that would improve the systemization of the whole process such as including data-powered positive deviance (Albanna et al., 2022) for identifying positive outlier behaviors at scale and the use of digital markers (Lutkenhaus et al., 2022) for tracking meaningful audience engagement and prosocial behaviors. Creating entertainment-education programs would empower us to meet our full potential for climate action.

Effective climate solutions must integrate technical solutions from both natural and social sciences with communication strategies to foster public engagement and prosocial behaviors. Media practitioners and creative professionals can create stories that reflect the changes already underway and help us connect with the people and resources in our own communities to engage fully in the self-preservation of climate change mitigation and adaptation. Streamlined funding can support this work and the people who do it. Investments made in developing a coordinated mechanism with equitable salaries for professional science communicators and research in behavioral science will allow us to craft and implement more effective interventions. In addition, most importantly, these investments will help us reach our climate mitigation and adaptation goals faster. Being slow on these tasks means incalculable loss and damage to human lives. We acknowledge that the climate crisis is complex but emphasize that there are ways to improve the communication methods that we use. In the hyper-interconnected world that we live in today, we are all in this planetary race together. There is still time. We can change the current trajectory. Focusing on positive and actionable stories and coordinating our efforts across disciplines and geographics will accelerate climate solutions. Let's write the most epic story of our time.

References

Abbott, H. (2002). *The Cambridge introduction to narrative*. Cambridge University Press.

Albanna, B., Heeks, R., Pawelke, A., Boy, J., Handl, J., & Gluecker, A. (2022). Data-powered positive deviance: Combining traditional and non-traditional data to identify and characterise development-related outperformers. *Development Engineering, 7*. https://doi.org/10.1016/j.deveng.2021.100090

Baden, D., & Brown, J. (2024). Climate fiction to inspire green actions: Tales from two authors. In E. Coren & H. Wang (Eds.), *Storytelling to accelerate climate solutions*. Springer Nature.

Bálint, K. E., & Bilandzic, H. (2017). Health communication through media narratives: Factors, processes, and effects. Introduction to the special issue. *International Journal of Communication, 11*, 4858–4864.

Bean, J. (2024). Instructional strategies for climate education in the classroom: Storytelling about our place in the earth system. In E. Coren & H. Wang (Eds.), *Storytelling to accelerate climate solutions*. Springer Nature.

Berry, T. (1988). *The dream of the Earth*. Sierra Club.

Berry, T. (2003). The new story. In A. Fabel & D. St. John (Eds.), *Teihard in the 21st century: The emerging spirit of Earth* (pp. 77–88). Orbis Books. First published in *Teihard Studies* no. 1 (winter 1978).

Bietti, L. M., Tilston, O., & Bangerter, A. (2019). Storytelling as adaptive collective sensemaking. *Topics in Cognitive Science, 11*(4), 710–732. https://doi.org/10.1111/tops.12358

Bilandzic, H. (2023). Stories about the environment for diverse audiences: Insights from environmental communication. In M. Schneider-Mayerson, A. W. von Mossner, W. P. Malecki, & F. Hakemulder. (Eds.). Empirical ecocriticism: Environmental narratives for social change (pp. 347–359). : University of Minnesota Press.

Bilandzic, H., & Kinnebrock, S. (2009). Narrative experiences and effects of media stories: An introduction to the special issue. *Communications, 34*(4), 355–360. https://doi.org/10.1515/COMM.2009.022

Bish, J. (2024). Positively life-changing stories today – Intergenerational benefits tomorrow. In E. Coren & H. Wang (Eds.), *Storytelling to accelerate climate solutions*. Springer Nature.

Bloomfield, E. F., & Manktelow, C. (2021). Climate communication and storytelling. *Climate Change, 167*, 34. https://doi.org/10.1007/s10584-021-03199-6

Brown, A. (2015). *Rising strong*. Harvard University Press.

Brown, N. (2024). The power of locally-driven narratives to support and sustain climate action. In E. Coren & H. Wang (Eds.), *Storytelling to accelerate climate solutions*. Springer Nature.

Bruner, J. S. (1990). *Acts of meaning*. Harvard University Press.

Charani, E., Shariq, S., Pinto, A. M. C., Farooqi, R., Nambatya, W., Mbamalu, O., et al. (2023). The use of imagery in global health: An analysis of infectious disease documents and a framework to guide practice. *The Lancet Global Health, 11*(1), E155–E164. https://doi.org/10.1016/S2214-109X(22)00465-X

Chichester, Z. A., Jewell, M. A., LePrevost, C. E., & Lee, J. G. L. (2023). The cost of diversity: An analysis of representation and cost barriers in stock photo libraries for health education materials, 2021. *Health Promotion Practice*, online first. https://doi.org/10.1177/15248399221150788

Climate Beacon Newsrooms. (2022, December). *Climate Beacon Newsrooms initiative homepage.* https://www.solutionsjournalism.org/topics/climate-beacons

Climate Central. (2021, June 18). *Climate Central launches realtime climate alerts to help journalists localize climate change stories.* https://www.climatecentral.org/press-release-realtime-climate

Climate Solutions Cohort. (2022, August). *Climate Solutions Cohort homepage.* https://www.solutionsjournalism.org/topics/climate-solutions-cohort

Collins, F. S., & Stoffels, P. (2020). Accelerating COVID-19 therapeutic interventions and vaccines (ACTIV): An unprecedented partnership for unprecedented times. JAMA, *323(24), 2455–2457.* https://doi.org/10.1001/jama.2020.8920

Coren, E. (2024). Rhythm and glue: An entertainment-education prototype for climate communication. In E. Coren & H. Wang (Eds.), *Storytelling to accelerate climate solutions.* Springer Nature.

Coren, E., & Tiwathia, A. (2023). The role of entertainment in changing social climate norms. ESS Open Archive. https://doi.org/10.22541/essoar.170224443.32061814/v1.

Corey, L., Mascola, J. R., Fauci, A. S., & Collins, F. S. (2020). A strategic approach to COVID-19 vaccine R&D. *Science, 368*(6494), 948–950. https://doi.org/10.1126/science.abc5312

Cosentino, M., Gal-Oz, R., & Safer, D. L. (2024). Community-based resilience: The influence of collective efficacy and positive deviance on climate change-related mental health. In E. Coren & H. Wang (Eds.), *Storytelling to accelerate climate solutions.* Springer Nature.

Dahlstrom, M. F. (2014). Using narratives and storytelling to communicate science with nonexpert audiences. *The Proceedings of the National Academy of Sciences, 111*(4), 13614–13620. https://doi.org/10.1073/pnas.1320645111

De Meyer, K., Coren, E., McCaffrey, M., & Slean, C. (2021). Transforming the stories we tell about climate change: From 'issue' to 'action'. *Environmental Research Letters, 16,* 015002. https://doi.org/10.1088/1748-9326/abcd5a

Dixon, C. E., Goldman, L. S., Crawford, S., & Lease, P. C. (2024). Music as a vehicle for climate change communication: The ClimateMusic Project. In E. Coren & H. Wang (Eds.), *Storytelling to accelerate climate solutions.* Springer Nature.

Eiseman, D., & Hoffman, M. (2024). Telling the story of climate change through food. In E. Coren & H. Wang (Eds.), *Storytelling to accelerate climate solutions.* Springer Nature.

El Amiri, N., Abernethy, P., Spence, N., Zakus, D., Kara, T.-A., & Schuster-Wallace, C. (2020). Community of practice: An effective mechanism to strengthen capacity in climate change and health. *Canadian Journal of Public Health, 111,* 862–868. https://doi.org/10.17269/s41997-020-00400-8

Falzone, P., Kiano, J., & Lukomska, G. (2024). Let's Go! Let's Know! N*Gen as an EE tool for climate education and agency. In E. Coren & H. Wang (Eds.), *Storytelling to accelerate climate solutions.* Springer Nature.

Figueres, C., & Rivett-Carnac, T. (2021). *The future we choose: The stubborn optimist's guide to the climate crisis.* Vintage.

Fisher, W. R. (1987). *Human communication as narration: Toward a philosophy of reason, value, and action.* The University of South Carolina Press.

Frank, L. B., & Falzone, P. (Eds.). (2021). *Entertainment-education behind the scenes: Case studies for theory and practice.* Palgrave Macmillan.

Friedman, P. (Producer & Director). (2013). *Poor Consuelo conquers the world* [Documentary film]. Yleisradio (YLE).

Garg, A., Godfrey, A., & Eko, R. (2024). *Kembali Ke Hutan* (Return to the forest): Using storytelling for youth engagement and climate action in Indonesia. In E. Coren & H. Wang (Eds.), *Storytelling to accelerate climate solutions.* Springer Nature.

Giaccardi, S., Rogers, A., & Rosenthal, E. L. (2022, October). *A glaring absence: The climate crisis is virtually nonexistent in scripted entertainment.* A report by USC Norman Lear Center Media Impact Project. https://sustainability.usc.edu/2022/10/17/a-glaring-absence-the-climate-crisis-is-virtually-nonexistent-in-scripted-entertainment/

Gottschall, J. (2012). *The storytelling animal: How stories make us human.* Houghton Mifflin Harcourt.

Green, M. C., Strange, J. J., & Brock, T. C. (Eds.). (2002). *Narrative impact: Social and cognitive foundations*. Psychology Press.

Green, L. W., Ottoson, J. M., Carcia, C., & Hiatt, R. A. (2009). Diffusion theory and knowledge dissemination, utilization, and integration in public health. *Annual Review of Public Health, 30*, 151–174. https://doi.org/10.1146/annurev.publhealth.031308.100049

Green, M. C., Bilandzic, H., Fitzgerald, K., & Paravati, E. (2019). Narrative effectives. In M. B. Oliver, A. A. Raney, & J. Bryant (Eds.), *Media effects: Advances in theory and research* (pp. 130–145). Routledge.

Gurney, C., & N'Diaye, M. (2024). LOLs: Secret weapon against CFCs and CO2? In E. Coren & H. Wang (Eds.), *Storytelling to accelerate climate solutions*. Springer Nature.

Harris, D. M. (2020). Telling stories about climate change. *The Professional Geographer, 72*(3), 309–316. https://doi.org/10.1080/00330124.2019.1686996

Hawken, P. (2021). *Regeneration: Ending the climate crisis in one generation*. Penguin Books.

Hendersson, H., & Wamsler, C. (2020). New stories for a more conscious, sustainable society: Claiming authorship of the climate story. *Climate Change, 158*, 345–359. https://doi.org/10.1007/s10584-019-02599-z

Hinderfeld, D., Slean, C., & Jacobs, K. (2024). Rewrite the future: Helping Hollywood accelerate climate solutions through storytelling. In E. Coren & H. Wang (Eds.), *Storytelling to accelerate climate solutions*. Springer Nature.

Hutchinson, P. L., Mirzoyants, A., & Leyton, A. (2022, December). *A multidimensional human capital approach to investing in youth: Estimating the return of investment (ROI) for the Shujaaz behavior change multimedia platform*. Presented at the SBCC Global Summit, https://www.shujaazinc.com/wp-content/uploads/2022/12/Shujaaz-final-poster-SBCC-summit-021222.pdf

IPCC. (2022a). *Climate change 2022: Impacts, adaption, and vulnerability*. https://www.ipcc.ch/report/ar6/wg2/downloads/report/IPCC_AR6_WGII_FinalDraft_FullReport.pdf

IPCC. (2022b). *Press release*. Retrieved on April 4, 2022, https://www.ipcc.ch/report/ar6/wg3/resources/press/press-release/

IPCC. (2023a). *AR6 synthesis report: Climate Change 2023*. https://www.ipcc.ch/ar6-syr/

IPCC. (2023b). *Press conference presentation*. Retrieved on March 20, 2023, https://report.ipcc.ch/ar6syr/pdf/IPCC_AR6_SYR_SlideDeck.pdf

Jones, M. D., & Peterson, H. (2018). *Narrative persuasion and storytelling as climate communication strategies. The Oxford Encyclopedia of Climate Change Communication*. Oxford University Press.

Keen, S. (2006). A theory of narrative empathy. *Narrative, 14*(3), 207–236.

Kreuter, M. W., Green, M. C., Cappella, J. N., Slater, M. D., Wise, M. E., Storey, D., Clark, E. M., O'Keefe, D. J., Erwin, D. O., Holmes, K., Hinyard, L. J., Houston, T., & Woolley, S. (2007). Narrative communication in cancer prevention and control: A framework to guide research and application. *Annals of Behavioral Medicine, 33*(3), 221–235. https://doi.org/10.1007/BF02879904

Lutkenhaus, R., Wang, H., Singhal, A., Jansz, J., & Bouman, M. P. A. (2022). Using markers for digital engagement and social change: Tracking meaningful narrative exchange in transmedia edutainment with text analytics techniques. *Digital Health, 8*, 1–18. https://doi.org/10.1177/20552076221107892

Maibach, E., Miller, J., Armstrong, F., El Omrani, O., Zhang, Y., Philpott, N., Atkinson, S., Rudoph, L., Karliner, J., Wang, J., Petrin-Desrosiers, C., Stauffer, A., & Jensen, G. K. (2021). Health professionals, the Paris agreement, and the fierce urgency of now. *The Journal of Climate Change and Health, 1*, 100002. https://doi.org/10.1016/j.joclim.2020.100002

Maibach, E., Uppalapati, S. S., Orr, M., & Thaker, J. (2023). Harnessing the power of communication and behavior science to enhance society's response to climate change. *Annual Review of Earth and Planetary Sciences, 51*, 53–77.

Martinez-Conde, S., & Macknik, S. L. (2017). Opinion: Finding the plot in science storytelling in hopes of enhancing science communication. *The Proceedings of the National Academy of Sciences, 114*(31), 8127–8129. https://doi.org/10.1073/pnas.1711790114

McComas, K., & Shanahan, J. (1999). Telling stories about global climate change: Measuring the impact of narratives on issue cycles. *Communication Research, 26*(1), 30–57. https://doi.org/10.1177/009365099026001003

Monoyios, K., Carlson, K., Litwak, T., Marien, T., & Martin, F. (2024). Visual storytelling as a catalyst for climate science communication. In E. Coren & H. Wang (Eds.), *Storytelling to accelerate climate solutions*. Springer Nature.

Morris, Z. S., Wooding, S., & Grant, J. (2011). The answer is 17 years, what is the questions: Understanding time lags in translational research. *Journal of the Royal Society of Medicine, 104*(12). https://doi.org/10.1258/jrsm.2011.110180

Morris, B. S., Chrysochou, P., Christensen, J. D., Orquin, J. L., Barraza, J., Zak, P. J., & Mitkidis, P. (2019). Stories vs. facts: Triggering emotion and action-taking on climate change. *Climate Change, 154*(1), 19–36. https://doi.org/10.1007/s10584-019-02425-6

Murphy, S. T., Frank, L. B., Chatterjee, J. S., Moran, M. B., Zhao, N., de Herrera, P. A., & Baezconde-Garbanati, L. A. (2015). Comparing the relative efficacy of narrative vs nonnarrative health messages in reducing health disparities using a randomized trial. *American Journal of Public Health, 105*(10), 2117–2123. https://doi.org/10.2105/AJPH.2014.302332

Osnes, B., Hackett, C., McDermott, M. T., & Safran, R. (2024). *Bird's eye view*: Engaging youth in storying a survivability future through performance and interspieces friendship. In E. Coren & H. Wang (Eds.), *Storytelling to accelerate climate solutions*. Springer Nature.

Polkinghorne, D. E. (1988). *Narrative knowing and the human sciences*. State University of New York Press.

Population Media Center. (n.d.). *How do we know if we succeeded?* https://www.populationmedia.org/our-approach/evaluation/

Project Drawdown. (n.d.) *Drawdown foundations*. Retrieved on April 23, 2022, https://drawdown.org/drawdown-framework

Rippberger, R. (2022). The power of storytelling: Social impact entertainment. Regent Press.

Riley, A. H., Sood, S., & Wang, H. (2022). Entertainment-education (effects). In J. van Weert, E. Ho, C. Bylund, N. Bol, M. D. Kruzel, & I. Basnyat (Eds.), *The international encyclopedia of health communication*. Wiley. https://doi.org/10.1002/9781119678816.iehc0625

Sachs, J. D., Karim, S. S. A., Aknin, L., Allen, J., Brosbol, K., Colombo, F., et al. (2022). The Lancet Comission on lessons for the future from the COVID-19 pandemic. The Lancet, *400(10359)*, 1224–1280. https://doi.org/10.1016/S0140-6736(22)01585-9.

Sarfaty, M., Duritz, N., Gould, R., Mitchell, M., Patel, L., Paulson, J., Rudolph, L., Nackerman, C., Harp, B., & Maibach, E. B. (2022). Organizing to advance equitable climate and health solutions: The medical society consortium on climate and health, *The Journal of Climate Change and Health, 7(100174)*, 2667–2782, https://doi.org/10.1016/j.joclim.2022.100174

Schneider-Mayerson, M., von Mossner, A. W., Malecki, W. P., & Hakemulder, F. (Eds.). (2023). *Empirical ecocriticism: Environmental narratives for social change*. University of Minnesota Press.

Scott, M. (2023, July). Hidden voices: Why inclusive storytelling is critical to accelerating climate solutions. *Project Drawdown webinar series*. Retrieved July 25, 2023, https://www.youtube.com/watch?v=qWZDDH8fGKo&list=PLwYnpej4pQF7IIubTTDBh3k8YXYhCS3Jl&index=5&t=1s

SIE Society. (n.d.). *Welcome to the SIE society*. https://siesociety.org

Singhal, A., Cody, M. J., Rogers, E. M., & Sabido, M. (2004). *Entertainment-education and social change: History, research, and practice*. Erlbaum.

Singhal, A., Wang, H., & Rogers, E. M. (2013). The rising tide of entertainment-education in communication campaigns. In R. Rice & C. Atkin (Eds.), *Public communication campaigns* (pp. 321–333). Sage.

Skoll Center for Social Impact Entertainment. (2019). The state of SIE: Mapping the landscape of social impact entertainment. https://www.thestateofsie.com

Smith, D., Schlaepfer, P., Major, K., Dyble, M., Page, A. E., Thompson, J., Chaudhary, N., Salali, G. D., Mace, R., Astete, L., Ngales, M., Vinicius, L., & Migliano, A. B. (2017). Cooperation and the evolution of hunter-gatherer storytelling. *Nature Communications, 8,* article 1853. https://www.nature.com/articles/s41467-017-02036-8

Solnit, R., & Lutunatabua, T. Y. (2023). *Not too late: Changing the climate story from despair to possibility.* Haymarket.

Sood, S., Riley, A., & Birkenstock, L. (2024). Entertainment-education and climate change: Program examples, evidence, and best practices from around the World. In E. Coren & H. Wang (Eds.), *Storytelling to accelerate climate solutions.* Springer Nature.

Spiegel, S., & Wang, H. (2024). Exploring climate science in the metaverse: Interactive storytelling in immersive environments for deep learning and public engagement. In E. Coren & H. Wang (Eds.), *Storytelling to accelerate climate solutions.* Springer Nature.

Stoknes, P. E. (2017). *How to transform apocalypse fatigue into action on global warming.* Retrieved June 12, 2022, https://www.ted.com/talks/per_espen_stoknes_how_to_transform_apocalypse_fatigue_into_action_on_global_warming

Storey, D., & Sood, S. (2013). Increasing equity, affirming the power of narrative and expanding dialogue: The evolution of entertainment education over two decades. *Critical Arts, 27,* 9–35.

Suran, M. (2022, September 9). Dr. Fauci and the art of science communication. *JAMA, 328*(13), 1286–1287. https://doi.org/10.1001/jama.2022.16280

Wang, H., & Coren, E. (2024). Storytelling as a catalyst for climate change communication and empowerment. In E. Coren & H. Wang (Eds.), *Storytelling to accelerate climate solutions.* Springer Nature.

Wang, H., & Singhal, A. (2016). East Los High: Transmedia edutainment to promote the sexual and reproductive health of young Latina/o Americans. *American Journal of Public Health, 106*(6), 1002–1010. https://doi.org/10.2105/AJPH.2016.303072

Wang, H., & Singhal, A. (2021). Mind the gap! Confronting the challenges of translational communication research in entertainment-education. In L. B. Frank & P. Falzone (Eds.), *Entertainment-education behind the scenes: Case studies for theory and practice* (pp. 223–242). Palgrave Macmillan.

Wang, H., Singhal, A., Quist, C., Sachdev, A., & Liu, S. (2019). Aligning the stars in *East Los High*: How authentic characters and storylines can translate into real-life changes through transmedia edutainment. *SEARCH Journal of Media and Communication Research, 11*(3), 1–22.

Wang, H., Safer, D. L., Cosentino, M., Cooper, R., Van Susteren, L., Coren, E., Nosek, G., Lertzman, R., & Sutton, S. (2023). Coping with eco-anxiety: An interdisciplinary perspective for collective learning and strategic communication. *The Journal of Climate Change and Health, 9*(9), 100211. https://doi.org/10.1016/j.joclim.2023.100211

Whitwell, J. (2024). Three ways to introduce more stories of climate action into climate change news reporting. In E. Coren & H. Wang (Eds.), *Storytelling to accelerate climate solutions.* Springer Nature.

Wolf-Jacobs, A., Glock-Grueneich, N., & Uchtmann, N. (2024). Mapping out our future: Using geospatial tools and visual aids to achieve climate empowerment in the U.S. In E. Coren & H. Wang (Eds.), *Storytelling to accelerate climate solutions.* Springer Nature.

Emily Coren is a science communicator and an affiliate in the Department of Psychiatry and Behavioral Sciences at Stanford University where she has been working to adapt entertainment-education strategies for health promotion and social change to create more effective climate communication. She has a BS in Ecology and Evolutionary Biology and is a certified professional Science Illustrator. She has worked in science communication for almost 20 years, contributing to collections at the Smithsonian Institution's Museum of Natural History, consulting on a World Health Organization clean air campaign, and developing educational content for children's films. In recent years, her work has led to new methods in developing frameworks at a national level, connecting community-led experiences to federal, local, and nonprofit sector programs for climate change communication. She is a member of the National Association of Science Writers and the Society of Environmental Journalists.

Hua Wang is a communication scientist who is passionate about using innovative strategies for health promotion, behavior change, and social justice. She specializes in the design, implementation, and evaluation of initiatives that leverage powerful storytelling, emerging technologies, and communication networks to facilitate positive change, particularly in the field of entertainment-education. She holds a Ph.D. from the University of Southern California's Annenberg School for Communication & Journalism and is currently a Professor in the Department of Communication at the University at Buffalo, the State University of New York. Her interdisciplinary research has been funded by federal agencies, private foundations, and nonprofit organizations, appeared in high-impact journals, and received prestigious awards from the International Communication Association and the American Public Health Association.

Index